WOOL HANDBOOK

VOLUME TWO

THIRD ENLARGED EDITION

A Text and Reference Book
For the Entire Wool Industry
Prepared by a Staff of Specialists
Under the Editorship of

WERNER VON BERGEN

J. P. Stevens & Co., Inc.,
Consultant

Part 2: Dyeing, Finishing, Carpets, Felts, Testing, Marketing

INTERSCIENCE PUBLISHERS

a division of John Wiley & Sons

New York • London • Sydney • Toronto

The paper used in this book has a pH of 6.5 or higher.
It has been used because the best information now available indicates that this will contribute to its longevity.

1 2 3 4 5 6 7 8 9 10

Library of Congress Catalog Card Number 63-11600
SBN 471 91016 3

Printed in the United States of America

Volume II of this book
is dedicated to the
employees of the J. P.
Stevens & Co., Inc.,
Woolen and Worsted Division

Foreword

Werner von Bergen and his collaborators have successfully carried out a most ambitious project in presenting this Third, enlarged, Edition of the *Wool Handbook*. Volume I of the work, which was completed and published in 1963, outlined the historical and economic background of sheep raising and the wool industry, and described the raw materials of the industry. In Volume II Part 1 and Part 2 the cooperating authors present a detailed, up-to-date technical account of wool manufacture and marketing.

The editor and authors of these volumes have presented the voluminous information in such a way as to be directly useful to the man in the mill. However, the books also provide to people outside the industry, including students, scientists, and business executives, a comprehensive picture of the handling and processing of wool.

Improved wool processing operations are being evolved rapidly through the accumulation of practical experience and the continuing development of refinements, and through breakthroughs into new methods. Notable contributions to wool technology are being made through applications of physics, chemistry, mathematics, and engineering. The enlarged Third Edition brings these contributions into sharp focus. It shows vividly the rapid pace at which knowledge is increasing and practices are changing in the wool industry.

Werner von Bergen's distinguished career in wool science and technology qualifies him eminently to define, select, and judge the significant aspects of modern wool processing. The contributors to these volumes are authorities on the subjects they discuss. Together with Werner von Bergen, they have prepared a handbook for which the entire wool industry may well be grateful.

On a personal note, I take this opportunity to express my high regard for Werner, not only in recognition of his lifelong dedication to advancement of the wool industry, but also because he has been a faithful friend and willing advisor. He is respected and beloved by all who know him.

HAROLD P. LUNDGREN, CHIEF

Wool and Mohair Laboratory
Western Utilization Research
and Development Division
Agricultural Research Service
U. S. Department of Agriculture

1969
Albany, California

Preface

Half a century has gone by since I started my first job as a chemist in a 100-loom vertical woolen and worsted mill in Switzerland. Within the last fifty years, the Wool Industry, in both the United States and abroad, has realized substantial growth and expansion. During this period the challenges and problems presented by an overwhelming technological growth rate, especially in the last ten years, have radically affected the Industry and have precipitated many important advances in the field. A reestablished recognition of the uniqueness of the wool fiber has resulted from this demand for new areas of development and application of wool products.

One of the major changes has been in the number of fibers the Industry is processing. In 1919, the fibers processed were limited to wool, specialty hair fibers, silk and cotton yarns, and a small amount of rayon. Today, the numbers normally encountered are around three times that with all the accompanying complications which these involve in any factory. Fiber blends now are a necessary part of the materials existence. Therefore, in many of the chapters the processing of man-made fibers and blends is woven into the text.

Considering the machines and processes over a span of fifty years, it is evident that we have advanced in rate of production per man-hour, and often use less operations between the raw wool and the finished cloth. For instance, in wool scouring 500 lb of scoured wool per hour was a good production in the 1920's whereas today we are getting 1000–2000 lb/hr. In carding, 20–28 lb/hr was regarded as high but today the same width card is carding 60 lb/hr and the use of wider cards up to 100 in. is allowing us to card 150 or more lb/hr. Similar increases could be quoted for gilling. In the 1920's, the classical gill box was operated at speeds of about 400 drops/min. Today, screw driven pindrafters operate at 1500–2000 drops/min while chain driven units are running 6000 drops/min. The most remarkable reduction in processing took place in worsted drawing where the operation, from top to yarn, has been cut from 8 to 10 steps to 4, utilizing autoleveling devices.

In spinning, superdrafting devices allow drafts of 40 to 100 against 10. Traveller speeds were considered high at 30–40 mph, but today they have reached over 100 mph. In woolen spinning, reduced balloon spinning has resulted in nearly doubling the spindle speed without excessive tension. In combination with various other factors the average yarn package today weighs more than 1 lb against 1–2 oz.

In twisting, it is possible to insert two turns of twist for the one turn of a spindle. Today, winding is automatic and self-knotting. Weaving is many times more productive and through the introduction of various auxiliary devices on looms, the efficiency in terms of picks per minute has increased over 50% and at the same time increasing the number of looms per weaver from 1 to 10 or more. The most revolutionary advance in weaving was in the introduction of the pirnless loom.

In finishing, the notable strides made in scouring, raising, shearing and pressing have been brought about through the application of better engineering rather than through a modification of the fiber. The surprising thing is that the natural fibers which formerly were processed under the old slow techniques are so adaptable that they can be processed under modern conditions.

Another change over the years is the trend in design. Piece-dyes are in the minority and colored blends, twists and cross-dyes are the modern trends. This was brought about through the tremendous advances in the production of dyes. Looking back to the complex dyeing methods of the old indigo trade we can appreciate the progress made in this field.

Great strides have been made in automation but it is only in the last 25 years that automation has been taken seriously in the Woolen Industry due to the advances in mechanical and electrical engineering. For example, the linkup of carding and gilling, comb-control devices, autoleveler devices, end-detectors, automatic winding, and linkup of finishing operations are some of the advances which have been made.

In the knitting section of the Industry the progress has been phenomenal.

Six years have passed since the first volume of the Third Edition was printed. During this interim period I have had the privilege of studying first hand the research and industrial developments in the Wool Industry around the world. As many of these developments were in their infancy it became difficult for me to decide upon the cutoff point for Volume II. The material gathered by my contributing authors and myself is so voluminous that it became advisable to incorporate the text into two additional volumes in order to provide detailed coverage of all the essential information. I am specially indebted to Professor P. P. Townend of the University of Leeds for his assistance in preparing the text for the two chapters covering Woolen Carding and Spinning and to Dr. C. Sustmann of the Hoehere Fachschule f. Textilindustry, Aachen, for supplying many of the line drawings incorporated in the chapters on Finishing.

Volume I covers the raw material prior to manufacturing. Its ten chapters include historical and economic background of the Industry, sheep raising, the physical and chemical properties of wool fiber, specialty hair fibers (including dehairing and bleaching), reclaiming wools and secondary raw materials, a chapter on wool blends with man-made fibers, the grading, production, and marketing of wool, and closes with a chapter on fiber testing.

Volume II, Part 1, covers the manufacturing processes of wool from the raw material to weaving and knitting. Its ten chapters include Sorting, Scouring and Carbonizing; Woolen or Carded Yarns; Woolen Spinning; Worsted Topmaking; Worsted Drawing; Worsted Yarn Spinning; Processes Preparatory to Weaving; The Principle of Weaving; Woolen and Worsted Weaves; and Knitting.

Volume II, Part 2, covers the basic processes involved in dyeing and finishing. Its seven chapters include Dyeing, Bleaching and Printing; Wet Finishing; Dry Finishing Operations, Special Wool Finishes, Shrinkproofing of Wool; Manufacture of Carpets and Rugs; The Manufacture of Felts, Nonwoven Felts—Woven Felts; Physical and Chemical Testing; and Marketing of Woolen and Worsted Fabrics.

Following the same layout as in Volume I, special value was laid on good illustrations and timely supplementary charts, current statistics and extended lists of references. I have done so in the belief that the information contained in Volume II, Part 1, and II, Part 2, is as current as it is critical. Naturally there are always new developments taking place but these volumes should serve their readers well for some time to come as a comprehensive text and reference source for people involved with the Wool Industry. I also hope it will set an example of sufficient value so that when developments in the field next dictate, another handbook will come into being.

My sincere thanks go to all contributing authors who assisted in the creation of this work and to my colleagues in the Woolen and Worsted Division and the staff of the Research and Development Division of J. P. Stevens & Co., Inc. for their valuable assistance. I am especially indebted to Mary Lee Wilkins and Katherine F. Earl for their assistance in proofreading and typing, and George Daniel Moro for preparing many of the photographs and line drawings.

WERNER VON BERGEN

March 1970
New York, N. Y.

Authors of Volume Two

Tibor Balazs

(Formerly Head Stylist, Director of Fabrication, Forstmann Menswear Division, New York and New Jersey)

Thomas N. Crowley

Research Director, Textile Chemical Division, Amchem. Products, Inc., Ambler, Pennsylvania

William Ebersold

(Formerly Chief Product Engineer, James Lees and Sons Company, Bridgeport, Pennsylvania)

Thomas Edman

Professor Philadelphia College of Textile Science, Philadelphia, Pa.

Thomas J. Gillick, Jr.

Manager Marketing Services, GAF Corp., Industrial Product Divisions, Glenville, Connecticut (formerly Manager, Engineering and Research Division, American Felt Company)

Carlton C. Gordon

Former Chemical Research Director, Albany Felt Co., Albany, New York

Fritz F. Kobayashi

Professor, Lowell Technological Institute, Lowell, Massachusetts

Hans W. Krause

Professor of Textile Engineering and Textile Industries, Swiss Federal Institute of Technology, Zurich, Switzerland

Leland S. Liang

Textile Marketing Consultant and Lecturer on Textile Marketing, City College of New York, New York, N. Y.

Henry E. Millson

Director of Research, Jorden Chemical Co., Inc., Bala Cynwyd, Pennsylvania (formerly Manager, Dyes Application Laboratories, American Cyanamid Co., Bound Brook, New Jersey)

Werner von Bergen

Consultant, J. P. Stevens & Co., Inc. (formerly Associate Director of Research, J. P. Stevens & Co., Inc., formerly Director of Research, Forstmann Woolen Company)

William W. Wetzold

Manager, J. P. Stevens & Co., Inc., Woolen and Worsted Division, Allendale Plant, South Carolina

Authors of Volume One

James R. Bercaw

Supervisor, Dacron Technical Division, Textile Fibers Department, E. I. du Pont de Nemours & Co., Inc., Kinston, North Carolina

Reginald J. Bray

Chief Scientific Liaison Officer, International Wool Secretariat, London, England

Glen F. Brown

Secretary, National Association of Wool Manufacturers, New York, New York

Joseph H. Dusenbury

Section Leader, Wool Research, Deering Milliken Research Corporation, Spartanburg, South Carolina (formerly Associate Research Director, Textile Research Institute, Princeton, New Jersey)

S. J. Golub

Assistant Director Fabric Research Laboratories, Inc. (Formerly Associate Director, ACH Fiber Service, Inc., Boston, Massachusetts) Dedham, Massachusetts

J. Menkart

President, Harris Research Laboratories, Inc., Washington, D. C.

Elroy M. Pohle
Head, Wool Standardization Laboratory, Agricultural Marketing Service, U. S. Department of Agriculture, Denver, Colorado

W. Brewster Southworth
Nichols & Co., Inc., Boston, Massachusetts

Clair E. Terrill
Sheep and Fur Animal Research Branch, Animal Husbandry Research Division, Agricultural Research Service, U. S. Department of Agriculture, Beltsville, Maryland

Werner von Bergen
Consultant J. P. Stevens & Co., Inc. (formerly Associate Director of Research, J. P. Stevens & Co., Inc., formerly Director of Research, Forstmann Woolen Company)

Contents

VOLUME TWO: Part 2

VOLUME TWO: Part 1

VOLUME ONE

CHAPTER 11

Dyeing, Bleaching, and Printing

Henry E. Millson and W. von Bergen

Dyeing has been practiced for thousands of years, and among the earliest peoples who dyed their garments were the Chinese and the American Indians. The dyes that were available to the ancients were produced naturally by plants and animals. There were two principal colors: blue indigo, which originated from a plant, and red kermas, obtained from the dried bodies of an insect.

Probably the most interesting documents on dyeing have been recovered in Egypt, the "Papyrus Graecus Homkensis," preserved at Upsala, Sweden, and the "Leyden Papyrus," preserved at Leyden, Holland. The former contains seventy recipes dealing with the cleaning, mordanting, and dyeing of wool, in which the following dyes are mentioned: alkanora (red), safflower (yellow and red), kermas (red), madder (red), and woad (blue). The art of dyeing was well developed, following the same principles that underlie modern dyeing.

The discovery of America added different dyewoods such as logwood, redwood, and fustic to the available dyes. The most important of the dyewoods, and the only one still used on a large scale, is logwood. The dye is extracted from the blood-red wood of the campeachy, a large tree which grows abundantly in the West Indies and Central American countries. The dye is sold today in the form of logwood extract.

In addition to dyewoods, the Spaniards found in Mexico an insect which produces cochineal, a beautiful scarlet with a tin mordant which has replaced the less attractive kermas red. There is every indication that the more cultured inhabitants among the Indians of Middle America and South America used these colors to a great extent. Garments and blankets found in the Inca graves in Peru and Chile, dating from before the Spanish Conquest, are examples of the various dyes used such as purple and indigo. The Incas were able to apply these dyes on wool as well as on cotton.

Modern Dyestuffs

The whole art and practice of dyeing was completely revolutionized in 1856 by the discovery of the artificial dye mauve (from the French name of the violet-colored mallow flower). The discovery was made accidentally by a young English chemistry student, William Henry Perkin.

When his discoveries were published, chemists all over the world began to manufacture and experiment with the new dye. Factories were started all over Europe. From this beginning, the manufacture of coal-tar dyes, and more recently their allied compounds, has become one of the most important and most profitable of all chemical industries.

Since that time not a year has passed without several new dyes being put on the market by some of the great dye concerns. In more recent years whole new classes of dyes such as fiber reactive, disperse, cationic basic, neutral dyeing premetalized have been discovered and produced for the dyeing of the natural and new synthetic, hydrophobic fibers.

The large dye concerns furnish the trade with excellently madeup sample shade cards showing actual dyed wool samples in the form of loose wool, yarn, or pieces and the shades and strengths offered by the company. With the trend toward pastel shades, several dyestuff manufacturers have improved their shade cards by including swatches which show the shade produced by each dye ranging from a pastel through deep shades.

In addition to shade cards, a number of the larger manufacturers also supply manuals, pamphlets and technical bulletins describing improved methods of application and the fastness of each dye when subjected to a large variety of tests.

Designation of Dyes

Trade names

In these sample cards and manuals, the dyes are grouped according to the class of application. In each class of dyes they are arranged according to the color and their relative shades beginning with yellow, orange, red, violet, blue, green, brown, and black. For proper identification of a dye the manufacturer gives each dye a trade name. The trade name usually bears a reference to the class, property, and color of the dye, as "Acid Light Red G," or to its chemical composition as "Anthraquinone Blue B" or "Alizarine Yellow GG." However, in many cases it is simply an arbitrary and nondescriptive name assigned by the manufacturer or the jobber.

Letter designations

The letter or letters following the name generally refer to the shade, for instance, B for blue, R for red, Y or G for yellow (German *gelb*). Methyl Violet is sold in brands running from 6B to 6R—that is, from a shade very close to blue with types becoming increasingly redder to a bright reddish-violet shade. Sometimes the letter refers to a fastness property such as Alkali Green 2G to indicate fastness to Alkali or Red L where the L is used to indicate good fastness to light; milling fast where the word milling is used to indicate good fastness to fulling or milling.

In other instances the letter refers to its class such as Wool Green S or Acid Blue both of which means German *Sauer* or applied from an acid bath. Quite frequently letters have no significance and are used merely to identify a type supplied by a specific manufacturer.

Also, it should be remembered that letters used for a specific dye can be misleading since one manufacturer's Yellow 3G may be redder in shade than a competitor's Yellow 2G or G.

Abbreviations and percentages

In addition to the letter designations there are such terms as *conc.* and *extra conc.*, which are abbreviations for the concentration of the dyes. These terms were further broadened by the addition of a percentage figure such as 100 percent or 125 percent, meaning in this case that the 125 percent dye is 25 percent stronger in its color value than the 100 per cent. During World War II most manufacturers increased their dye concentration in order to conserve packing material. In buying dyes the concentration factor is one of the most important things to consider, because the same dye may be sold in various concentrations and the prices are based accordingly.

In more recent times concentrations and terminology have changed very greatly so that a blue called extra conc. may be weaker than one called conc. or one without any letters to indicate its strength. The true money value of a dye is dependent upon its strength and performance and not necessarily by price.

Index numbers

The Colour Index originally prepared by Rowe and published in 1924 by the British Society of Dyers and Colourists has been revised jointly by the Society of Dyers and Colourists and the American Association of Textile Chemists and Colorists. This new revision lists a very large number of dyes, many discovered and introduced within the past thirty years.

One of the main functions of the Colour Index is to correlate the different names for the same dye. For example, color index number 31 refers to the dye "Amido Naphthol Red G" which is manufactured by most dye companies. This same product is sold under many different trade names. By this numbering system, a dyer is able to find a specific manufacturer's trade name for dyes which are on the market.

The Colour Index is today the standard reference on dyes used throughout the world and contains a vast amount of information of value to dyers and textile chemists.

Theory of Dyeing

The wool fiber has an affinity for almost all natural and synthetic coloring matters. This becomes apparent when wool is dipped into an aqueous or solvent dye solution. The affinity varies with differences in the chemical constitution of the dye and consequently many methods of application are used.

The mechanism of the process of dye adsorption and diffusion into the wool fiber is not entirely understood by the chemists. A number of different theories have been advanced to explain the process as already discussed in Volume 1 of this book, pp. 261–264, but none of the theories proposed to date completely satisfy all of the known facts about wool dyeing. A particularly exhaustive study by Delmenico and Peters (15) suggests that the Donnan theory is the best approach. Others claim the Gilbert-Rideal treatment is the more attractive.

Fig. 1. Morton system, raw stock dyeing installation. (Courtesy of J. P. Stevens & Co., Inc.)

Wool Dyes

Wool dyeing may be carried out on loose material, slubbing, yarn, and piece. Selection of dyestuffs and of the dyeing procedure to be employed is largely decided by the end-use of the finished goods and the stage of manufacture at which dyeing is carried out. The extensive ranges of dyestuffs available from manufacturers today make it possible to produce most shades with the fastness properties required, provided their cost is acceptable to the customer.

A new classification of dyes which are applicable to wool was given in the *Ciba Review* by Kehrer, each class representing a virtually complete range of colors. To the five groups listed by Kehrer (29) a sixth has been added.

1. Acid dyes (applicable from a strongly, moderately, or weakly acid dyebath)

2. Chrome dyes (applied by the chrome mordant, afterchroming, and one-bath methods)

3. Chrome complex dyes applied from a strongly acid dyebath

4. Neutral or weakly acid dyeing chrome or other metal-complex dyes

5. Vat dyes

6. Reactive dyes

Acid dyes

The "acid dyes," mostly sodium salts of sulfonic acids, are dissociated to the free color acids in an acid dyebath. In view of the amphoteric nature of wool, it is a safe assumption that the color acids combine with the basic groups of the wool protein to form saltlike compounds. Besides this relatively simple reciprocal effect, other, considerably more involved reactions take place owing to the complex character of the wool fiber.

It is likely, for instance, that following on the process of salt formation other reactive groups in the dyestuffs combine with certain further groups of atoms in the wool. The actual dyeing mechanism is no doubt seen in its true light if one regards the wool surface as a semipermeable membrane placed in a liquor. Only certain ions pass through the membrane, i.e., dyestuff ions diffuse from the dyebath into the fiber. In all probability, the size and the shape of the dyestuff molecules exert some influence on the manner in which diffusion takes place.

According to Elöd (17), the reactions involved in dyeing wool with acid dyes may be summarized as follows:

1. Adsorption of the dyestuff anion onto the fiber surface

2. Diffusion of the anion through the cortical layer

3. Chemical reactions within the wool fiber (salt formation and other types of combination)

The affinity of water-soluble acid dyes is a function of the temperature and the pH of the dye liquor. Dyes which exhaust onto the fiber slowly are applied with strong acids such as formic or sulfuric acid; for dyes which have a high rate of exhaustion the use of acetic acid (which is weaker) is preferred. Alternatively, an ammonium salt, such as ammonium acetate or ammonium sulfate, which dissociates at the boil and yields the corresponding acid may be used. Normally Glauber's salt is added which acts as a control for the dye adsorption.

Individual acid dyes have different affinity for wool. Some exhaust completely from an acid dyebath and are fast while other dyes exhaust only in part and can be partly removed from the fiber by the Glauber's salt contained in the dyebath. The latter dyes level particularly well as the result of this two-way process and are sometimes known as "leveling dyestuffs" on that account.

The acid dyes which exhaust rapidly and completely onto the fiber and are stable to Glauber's salt generally give unlevel dyeings when applied with strong acids. The general rule holds that dyestuffs with marked affinity for wool exhibit poor leveling properties but good fastness to water, while dyes with poor affinity for wool level well but are not fast to water.

Acid dyes applied from a strongly acid dyebath in part have good fastness to light and are used for dyeing dress goods, carpet yarns, certain upholstery materials, and the like, which do not require a high degree of wet fastness.

The use of dyes applicable from a moderately acid dyebath is indicated when a substantial degree of fastness to water and washing is demanded. Shades produced with such dyes even withstand very light milling.

The weakly acid dyeing dyes, also known as milling colors, which have correspondingly poor leveling property possess good fastness to water, seawater, and washing, and are fairly stable to moderately severe milling. The field of application of these dyes includes weaving and machine and hand knitting yarns, swimsuits, and bunting.

Up to the present the normal dye procedure was to wet out the material in bath with the liquor ratio to the weight of the wool between 20 and 50 to 1. The starting temperature is set between 70–100°F (20–40°C). The dissolved dye is added through a filter cloth, the process started by running the piece goods or circulating the liquor for 10 min then adding the salt and acids—then opening the steam bringing the bath to boil within an hour and boil for 1 hour. When the proper shade is reached the bath is slowly cooled down with fresh water to room temperature, rinsing off any excess dyes and cooling the material at the same time. To correct the shade the necessary amount of dye which cools well can be added to the boiling bath and needs an additional 20 min of boiling.

Modern methods of application call for periodic pH measurements as a means of assuring better reproducibility from batch to batch. Dyeings should actually be run by pH control rather than by percentages of acid on the weight of materials. Such procedures make it possible for the dyer to know before the lot is started that the acid has been added and the amount is correct. Thus, differences in alkaline or acid content of the wool, longer liquor ratios, variations in hardness and other variables are controlled by adjustment of dye baths to a definite pH before the dyeing is started.

A second pH reading may be taken when the temperature of the dye bath reaches the boil. Such controls reduce redyes, speed up dyeing cycles, saving time and money.

Correcting acid color dyeings. For various reasons, the dyed material may be off-shade, meaning that the dyer has not properly matched the standard or desired shade. Under these conditions the lot must be worked longer in the machine, redyed or stripped and redyed without spoiling the appearance of a fabric or damaging the wool by prolonged processing.

As a general rule acid dyes will level readily at the boil under the correct dye bath conditions. If the lot is uneven due to nonuniform distribution of the dye, an addition of Glauber's salt and/or acid and longer time at the boil will produce the desired levelness. Occasionally a feed of ammonia may be required to cause desorption of the dye. The desorbed dye can then be exhausted uniformly by careful additions of well-diluted acid.

When the shade is too dark and off-shade, it is often good practice to correct the shade by boiling dyed and undyed materials together in a fresh bath containing Glauber's salt and acid. During the boiling period dye will be removed from the dyed pieces and absorbed by the undyed pieces. Dyes can be added to this bath to adjust the lot to the desired shade.

If the desired shade is bright and the shade of the lot is quite dull, there is little that can be done to produce a brighter shade. The longer such a lot is boiled the duller the shade becomes. Pastel blues and baby pinks can be brightened by the addition of an optical bleach in a fresh bath.

If none of these corrective measures produce the desired result the lot should be dyed into a darker shade or black.

Correcting milling color dyeings. Milling dyes have much greaer affinity for the wool than they have for the dye bath and therefore once they are absorbed by the wool they are difficult to remove. Prolonged boiling in the dye bath rarely ever levels an uneven lot sufficiently to make it satisfactory and therefore whenever possible, the lot should be dyed black or a dark shade immediately. When this is not possible, a fresh bath containing 1–2% ammonia and 20% Glauber's salt may be used. An addition of a good nonionic surface active agent sometimes causes sufficient desorption and redistribution of the dye to produce a level result.

Stripping with hydrosulfite or sulfoxylate may destroy sufficient dye and produce a level bottom so that it is possible to cover by redyeing.

Chrome dyes

As the acid dyes satisfy medium fastness requirements only, they may have to be fixed more durably on the fiber. This is accomplished by bringing about "color lake formation," that is, the dyes are converted into insoluble metal complexes.

Metal atoms can be incorporated in an azo dye if the latter exhibits a certain molecular structure. The metal introduced into the molecule can be made to combine, through primary or secondary valencies, to form an inner complex salt, five- and six-membered ring systems resulting in the process.

In practice, the goods are dyed and afterchromed, that is, aftertreated with potassium dichromate or sodium dichromate and acid, or the dyebath is set with so-called Synchromate (Metachrome) mordant, which decomposes during dyeing and liberates the chromic acid which has the lake-forming property. This one-bath method calls for dyestuffs which exhaust evenly from a neutral or weakly acid liquor and do not form an insoluble color lake in the dyebath itself. The earliest method by which the wool was premordanted with potassium dichromate and sulfuric acid is no longer in favor owing to the prolonged boiling period required by prechroming and dyeing.

Dyeings produced by the afterchrome and one-bath methods exhibit fastness properties of a high order. They are distinguished by fastness to water, washing, and milling, even in white styles with colored effects, and possess good to very good fastness to pressing and decatizing. This makes chrome dyes specially suited for goods that have to be subjected to milling such as uniform and coating materials, and hats.

Correcting mordant colors. Because of the good fastness properties of these mordant colors, it is extremely difficult to correct shades that are too dark or streaky. For this reason it is important to start the dyeing operation at the proper pH so that the dye will exhaust slowly and be absorbed uniformly. Microscopical studies have shown that the dye is absorbed first as the acid or unchromed form in metachrome, and bottom-chrome dyeings. This allows the dye to redistribute itself and level as an acid color during the time the temperature is being raised to the boil. When the boil is reached, chromation of the dye takes place more rapidly and there is less chance of leveling thereafter. Once the insoluble chromium complex of the dye is formed, little if any leveling can take place.

As a general precaution it is wise to keep the shades on the light side and to use small amounts of level dyeing acid colors for matching. When the shade is too light, the dyeing procedure has to be started over again with chrome colors as with undyed goods.

A rather large number of chromable dyes can be demetalized, (returned to their original state) and will then level as though they were acid dyes. When such fabrics are rechromed the spots do not reappear and the fabrics are generally level and satisfactory. A method of leveling by the simultaneous demetalization of the metal dye complex and the removal and sequestration of metallic impurities has been developed. The method found to give the best results for the greatest number of chromable dyes tested was a 2-hour boil in a 40:1 liquor ratio bath containing 8% oxalic acid, 20% anhydrous sodium sulfate, and 2% of a sequestering agent (tetrasodium ethylenediamine tetracetic acid) based on the weight of the material.

Experiments have shown that desorption and redistribution of dye occurs during this treatment and thus leveling is promoted in raw stock, slubbing, yarns and piecegoods.

1:1 Metal-complex dyes

Dyestuffs chemists succeeded in devising a simpler method of dyeing wool by incorporating the metal required for lake formation in the dyestuff molecule: they created premetallized azo dyes such as the Neolan and Palatine Fast dyes. In these chrome-complex dyes the chrome is combined with a sulfo group to form an internal salt. The chromium atom is not completely saturated and combines with the amino groups of the wool to produce permanent fixation of the dye on the fiber. The Neolan dyes are derived from sulfonated azo dyes, one azo dye molecule containing one chrome atom. For example, see the chemical formula for C. I. Acid Blue 158.

The introduction of the chrome-complex colors simplified dyeing procedures and eliminated the need for chrome salt, which attacks the wool molecule. A further advantage is seen in the superior fastness properties of the chrome-complex dyes as compared with metal-free acid wool dyes.

These dyes require a strong acid dyebath, 6–8% sulfuric acid, which has an adverse effect on the physical and chemical properties of the wools. In Table 1 are given the breaking strength, the flex abrasion and the alkali solubility data of four worsted pieces dyed with 1% Palatin Fast Tan IWP and 8% sulfuric acid, dyeing time 1 1/2 hr to boil, 2 hr boiling. These data were obtained during the Princeton Wool Research Project and reported by Wakelin and von Bergen (64).

There is no significant differences in the properties of the four pieces. The breaking strength loss is around 20%, whereas there is a decrease of over 50% in the abrasion resistance. A clear indication of the chemical damage is the threefold increase in the alkali solubility. This is probably the result of a relatively small amount of peptide hydrolysis produced under these conditions.

Correcting chrome-complex color dyeings. This class of dyes can be leveled rather easily and whenever possible the lot should be leveled in the original dye bath. If the pieces appear uneven or skittery it may be due to the presence of insufficient sulfuric acid. Take a pH reading and if the bath has a higher pH than 2.0, a further addition of sulfuric acid should be made. The lot should be boiled 30 min and examined again. If the lot is still uneven, add 1–3% of a good leveling agent and boil 30 min longer.

TABLE 1

Changes in Physical and Chemical Properties of Four Worsted Fabrics, 2/2 Twill Dyed Together with 1% 1:1 Metalized Dye and 8% Sulfuric Acid—pH of Dyebath 2.0

Finishing state	Weight per Linear yd (oz)	Strength (lb)[a]		Flex abrasion (cycles)[b]		Alkali solubility[c] %
		Warp	Filling	Warp	Filling	
			Australian 70s (20.4 μ)			
Ready-to-dye	12.9	100	110	5226	5333	11.7
Finished	12.9	76	90	2241	2270	29.1
			U.S. 64s (22 μ)			
Ready-to-dye	12.3	86	104	5584	6212	10.4
Finished	12.4	75	83	2500	2538	30.9
			U.S. 58s (26.3 μ)			
Ready-to-dye	12.5	93	102	3921	4681	10.0
Finished	13.1	81	80	1841	2247	36.6
			New Zealand 58s (26.6 μ)			
Ready-to-dye	12.1	93	101	5704	6490	9.6
Finished	12.6	78	83	2103	2283	35.6

[a] ASTM D-1682, Grab Test, 4 × 6 in. samples.
[b] ASTM D-1175.
[c] ASTM D-1283.

Redyeing, however, will reduce the tensile strength and harshen the hand.

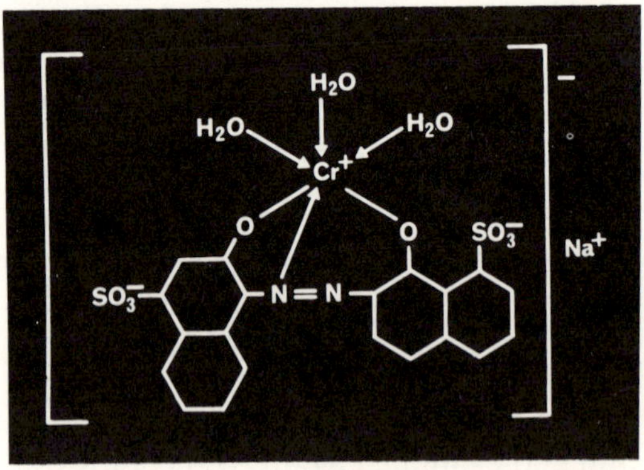

C. I. Acid Blue 158

1:2 Metal-complex dyes

The development of neutral or weakly acid dyeing metal-complex colors made it possible to dispense with large additions of acid and to shorten the boiling period. As regards chemical constitution, these new types of dye differ from those just described by the proportion of chromium atoms to dye molecules. They are azo dyes free from sulfo groups and contain two dye molecules to one metal atom. Not all members of the group are premetallized with chromium; other metals, cobalt for instance, can take its place. Dyes of this type, e.g. (Cibalan, Irgalan, and Isolan colors, are finding wide application in bulk. An example of a 1:2 chromium dye complex is C. I. Acid Violet 78. Wool dyers, too, have become accustomed to working with small amounts of acid or of ammonium salts. As a result, the wool fibers felt less and retain their tensile strength and fastness to abrasion. The fastness properties obtained closely approach those of chrome dyeings; with the exception of fastness to potting and acid milling, the fastness figures range between 4 and 5. The very high light fastness deserves special mention.

Correcting neutral metalized color dyeings. The neutral dyeing premetalized dyes do not migrate readily during the normal dyeing cycle. The leveling agents mentioned previously will cause some desorptions and redistribution when added to the dyebath or to a fresh bath, but rather high percentages are required for good leveling.

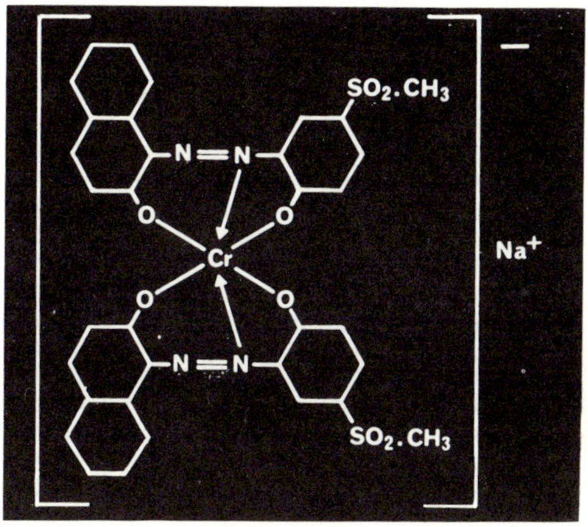

C. I. Acid Violet 78

Stripping may yield a level bottom for redyeing but wherever possible the lot should be redyed into a darker shade or black in order to save the wool from excessive damage.

Vat dyes

In the dyeing methods heretofore discussed the dye baths are prepared with dyes that are soluble in water. Vat dyeing, on the other hand, employs dyes that are generally insoluble in water. The consequence is that the methods of dyeing are different in both their chemical aspect and actual procedure. The dyes suitable for vat dyeing are coloring substances that are capable of being reduced to alkali-soluble leuco compounds and are fixed in this form upon the fiber. They are then retransformed by spontaneous oxidation in the open air into insoluble color pgiments and thus produce the color on the fiber.

Chemically, the vat dyes may be divided into two groups; the anthraquinone vat dyes and the indigoid vat dyes. The anthraquinone vat dyes, owing to the strong alkaline nature of the dye bath, are generally not suitable for wool, although it is possible through the use of protective colloids to minimize damage to the wool fibers.

Dyeings produced with vat dyes possess the best all-around fastness and are recommended whenever maximum resistance to the various manufacturing processes and wearing is required as in Army or Navy uniforms. The oldest representative of this type of dyestuff is indigo, the blue color which sets the standard in light fastness. In wool dyeing, two types of vats are employed (*1*) the hydrosulfite vat; (*2*) the warm fermentation vat.

The hydrosulfite vat. The hydrosulfite vat is generally used today because it is easily prepared and operated, preserves the quality of the wool, requires little room and steam, and gives a large production.

Since the vat dyes are insoluble in water, it becomes necessary to transform them with the help of caustic soda and hydrosulfite into a reduced leuco compound. Up to 1920 this process had to be done by the dyer himself, but since then dye concerns have introduced vat colors in powder form, which are soluble in hot water. In addition, there are various auxiliaries required for proper operation of the hydrosulfite vat, as follows:

Glue, which serves as a protective colloid to keep the dye in solution in the vat. Alkyl naphthalene sodium sulfonates may be used in place of glue.

Ammonia (25%) is used to keep the vat on the alkaline side.

Hydrosulfite conc. (powder) is a powerful reducing agent which regulates the condition of the vat and prevents its oxidation.

Sulfonated oil soaps or similar products serve in the dissolving of the vat dyes, when preparing stock vats. The proper alkalinity of the bath is checked with phenolphthalein solution (5:1000).

The vat dyes in soluble form are dissolved with boiling water to which 5% glue and 5% hydrosulfite conc. (powder) on the weight of the dye have been added. The solution obtained is allowed to stand for a short time before it is added to the dyebath. The dyebath proper is brought to a temperature of 120–150°F. For 65 lb of loose wool 550–660 gal of vat liquor is needed. To this liquor are added, in the order given:

3% freshly dissolved glue
3% ammonia (25%)
2% hydrosulfite conc. (powder)

These percentages are based on the weight of the goods. At this point the bath is ready to receive the stock vat solution. The liquor is then well stirred and should have the following characteristics: (*1*) a clear transparent appearance with a light foamy film of oxidized dye on the surface; (*2*) a slightly reddish reaction when tested with phenolphthalein. Into this healthy vat (as called by the dyer) the wool is entered dry or preferably wetted out and hydroextracted, and worked for 20–30 min. After it is taken out of the liquor, it is passed between squeeze rollers and dropped into boxes, where it is allowed to stand for 10–15 min to oxidize.

When oxidation is completed, the wool is either rinsed or given a second dip for deep shades. Thorough squeezing is essential after each dip. The fastness to rubbing may be improved by dropping the wool into cold running water after the squeeze; if this is done the oxidation proceeds more slowly.

To run the vat economically it should be kept going continuously, averaging about 10 dips per day; the same vat liquor can be kept for about a week when thoroughly scoured wool or goods are used. Before making an additional dip the vat has to be brought back into condition by adding the necessary quantities of ammonia and hydrosulfite to make the liquor once again transparent and slightly alkaline to phenolphthalein. Additions of dye are made only after the vat is corrected. Additions of glue are not necessary when working with a standing bath. The vat liquor is not exhausted, and, therefore, the quantity of dye required to bring it to normal strength may be reduced in the following proportions:

	Second bath	Third bath	Fourth bath
For most indigo brands	30%	10%	10%
For all other vat dyes	15%	10%	—

Commencing with the third batch the dye quantity for replenishing remains practically stationary. The shading of the color in vat dyeing is quite a simple matter. The stock vat necessary for additions is diluted with vat liquor in a pail. This diluted solution is then added to the vat liquor while working the goods. After 15–20 min the shade is matched again.

To increase production vat colors can be dyed also in circulating machines in which the stock is stationary and the vat liquor is forced through the wool by means of a pumper propeller. In such machines about four times the quantity of wool may be dyed in the same volume of liquor as can be dyed in the open vat. The general makeup of the bath is 1% glue, 0.5–1.5% ammonia (25%), and 1.5% hydrosulfite conc. (powder). Otherwise, the working procedure is the same as in the open vat. Special precaution has to be taken against entry of air into the circulating liquor, which causes oxidation and uneven dyeing. The vat liquor is renewed more often because of the greater amount of material treated.

For blacks and navies the best procedure (when using circulating machines) is to replace the ammonia right from the start with 1–4% of ammonium sulfate. Two dips of half an hour each are normally required to get the depth of the shade and good exhaustion of the liquor. Cheap blacks may be obtained by using natural brown wool as a raw material and overdyeing it with indigo.

Several dye concerns have developed a limited number of vat colors covering all the shades from yellow to blacks. The best known are the Helindon dyes from the General Dyestuff Corporation. Helindon dyes are separated into two groups: the HN group and the HW group.

Two continuous processes for dyeing with vat dyes, specifically indigo, have been successfully done during the second world war. One described by von Bergen, Crowley, and Brommelsick (63) for loose wool and one described by Linberg (33) for piece goods. They can be considered the forerunners of most of the continuous dye methods for wool used today.

That indigo dyeing preserves the quality of the wool better than chrome dyes was proved by the results obtained at the Forstmann Woolen Company given in Table 2.

The interest in the United States in wool vat dyes has for years been mainly limited to indigo, the oldest of all vat dye colors. The

TABLE 2

Differences in Breaking Strength of Yarns and Pieces for 30-ounce Kersey Using Different Blends and Dyed by Different Methods

	1		2		3		
Wool blends 64's grade	25% Australian 25% Cape 50% 12-mo. Texas		100% Cape		100% 12-mo. Tex		
Blend No.		1			2	3	
Dye process	Natural	100% chrome	50% Algosol, 50% Chrome	Chrome	Indigo	Indigo	
Yarn 2.5 run							
Breaking strength, lb	185	147	134	161	186	173	
Percent	100	79.5	72.5	87	100.5	93.5	
Elongation	29.7	20.8	19.2	19.8	27.1	32.2	
Percent	100	70	64.5	66.5	91	108.5	
Breaking strength in piece, lb							Navy Specifications
Warp	112	81	102	98	121	113	90
Percent	100	72	91	87	108	101	
Filling	98	73	89	98	113	98	80
Percent	100	74	91	100	115	100	

only military specifications calling for the use of indigo vat dyes are the ones issued by the U.S. Navy Department for the manufacture of the navy blue flannel, melton and kersey. In Europe the military authorities of several countries, including Germany, for years have specified the use of vat dyes in the production of military cloth as a result of tests for their fastness properties, their influence on the quality of the wool, and the wearing properties of vat-dyed woolens.

Solubilized Vat Dyes

The solubilized vat dyes are a group of dyes marketed as Indigosol (Durand & Huguenin) and Algosol (General Dyestuff Corporation). They are readily soluble in water and are applied to animal fibers without vatting. Chemically, they are sulfuric esters of leuco vat dyes. Being derived from indigoid and anthraquinone vat dyes, they possess high fastness properties.

The solubilized vat dyes are applied to wool from a weak acid bath and under these conditions generally have such a high affinity for the wool fiber that the dyebath can be completely exhausted. The dyes are then developed or oxidized with sodium nitrate or bichromate and sulfuric acid similarly to the ordinary afterchrome colors. They have not found much application to wool.

Reactive dyes

This new class of dyes was introduced in 1956. They are available in bright colors, such as a turquoise blue, a shade which formerly could not be obtained fast to light. On wool the reactive dyes are very fiber selective by dyeing "tippy," which means when applied by conventional method, they dye the sunlight damaged tips very much darker than the rest of the staple. They show little or no migratory power. The tippy dyeing behavior can be largely overcome when the reactive dyes are applied from a weak acid bath in the presence of a cationic compound which forms a complex with the anionic dye, which is less fiber selective.

On wool, the wet fastness is not quite as high as for chrome and vat colors and it is often difficult to obtain completely level dyeing. So far, their use has been limited to medium to full shades in self-colors on loose wool and tops for knitwear and carpets. The reactive dyes are excellent for producing two-tone or tone-in-tone effects by making use of their tippy dyeing behavior (see page 700). However, Lewis and Seltzer (73) have now developed a procedure for pad/batch (cold) dyeing of wool piece goods with all classes of reactive dyes and procedures of this type may well allow much wider use of reactive dyes on wool than has been possible in the past.

The wet fastness of wool dyed with reactive dyes depends very greatly on efficient washing off of any unfixed dye. Such procedures are well established for the cotton trade for example, but are virtually unknown for wool dyeing, so that it is important for suitable procedures to be emphasized to wool dyers who may contemplate the use of reactive dyes to obtain very bright and very fast shades.

There are two groups of reactive dyes manufactured specifically for wool, the Procilans (ICI, introduced in 1964) and the Lanasols (CIBA, introduced in 1966), which together offer a very wide range of shades with wet fastness properties, at least equal to those obtained with chrome dyes.

Mosimann (75) presented a paper "Reactive Dyes on Wool" at the 1968 AATCC International Technical Conference at Montreal. Over the years a multitude of reactive dyes with different reactive groups have been developed and introduced on the market. Table 3 summarizes the reactive groups of some of the dyestuffs on the

TABLE 3
Reactive Systems Used in Wool Dyeing

	Predominantly used in		Since
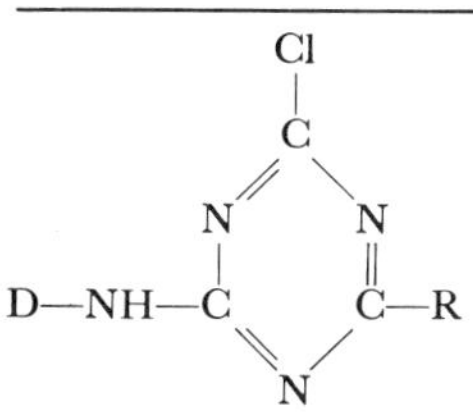	Procion H Cibacron	ICI CIBA	1957 1957
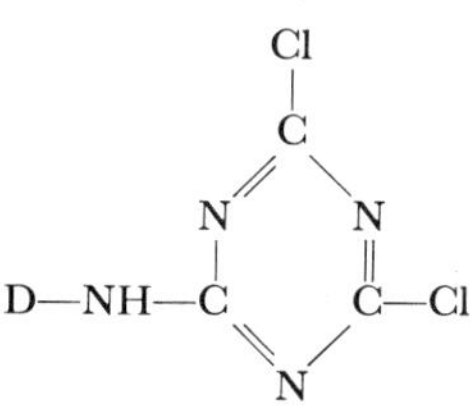	Procion (M)	ICI	1956

TABLE 3—Continued
Reactive Systems Used in Wool Dyeing

	Predominantly used in		Since
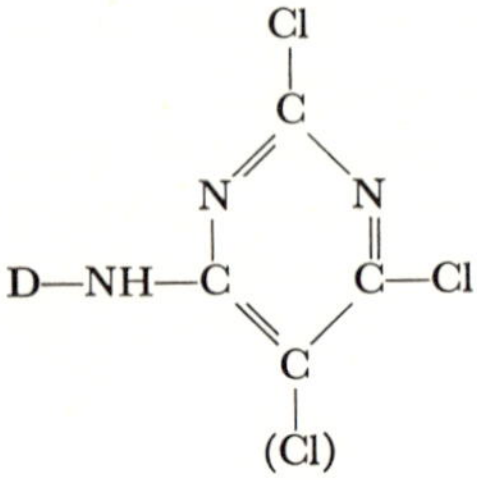	Drimaren Reactone	Sandoz Geigy	1960 1960
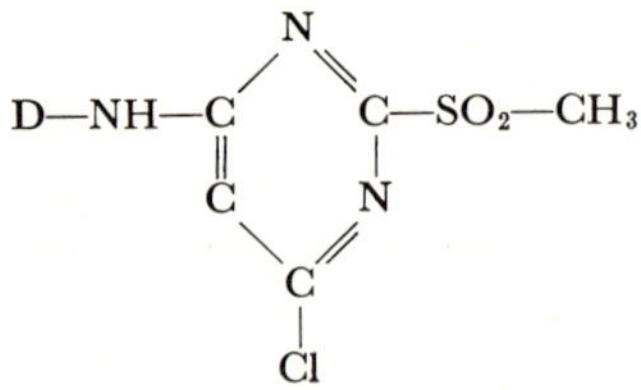	Levafix P	Bayer	1967
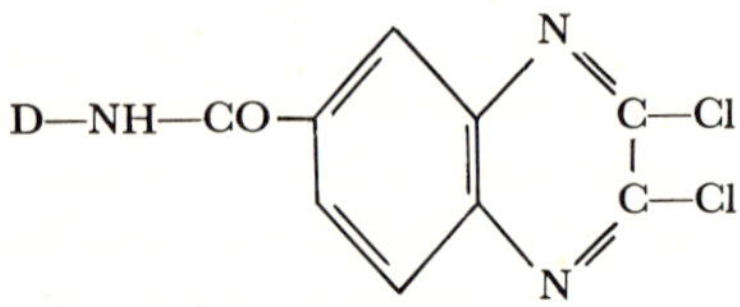	Levafix E	Bayer	1961
$D-NH-CO-CH_2-Cl$	Cibalan Brilliant Drimalan	CIBA Sandoz	1954 1962
$D-NH-CO-CH/\!/CH_2$	Procilan	ICI	1964
$D-NH-CO-\underset{\substack{\vert \\ Br}}{C}{=}CH_2$	Lanasol	CIBA	1966
$D-NH-CO-CH_2-CH_2-O-SO_3H$	Primazin	BASF	1961
$D-SO_2-CH_2-CH_2-O-SO_3H$ $(D-SO_2-CH{=}CH_2)$	Remalan Remazolan	Höchst Höchst	1953 1962
$D-SO_2-NH-CH_2-CH_2-O-SO_3H$	Levafix	Bayer	1960
$D-CH_2-NH-CO-CCl-CH_2$	Lanasyn Brilliant	Sandoz	1962
$D-CH_2-NH-CO-CH_2-CH_2$ $\vert$ $HO_3S-O-CH_2-CH_2-SO_2$	Lanafix	Sumitomo	1967

market today. It is limited to those groups of known structure which are applicable for wool dyeing or at least discussed with reference to wool dyeing. Mosimann covered in detail the properties and uses of the Lanasols.

Influence of Fiber Fineness on Quantity of Dye Required

As stated in *Wool Science Review* No. 32, naming the concentration of the dye, C, the absorption coefficient at a specific wavelength, α, and the diameter of the fiber, L; then the optical density, which is a measure of the absorption of light of that wavelength, is αCL. This generalization combines Beer's and Lambert's laws. Assuming the absorption is the same, one can deduce that for the same depth of shade, fine fibers will require a higher concentration of dye than the thicker fibers for the same optical density:

$$\alpha C_f L_f = \alpha C_t L_t$$

where the subscripts f and t indicate fine and thick fibers, respectively. By cancelling and transposing, it follows from this equation that

$$C_f/C_t = L_t/L_f$$

That is, the dye concentration required is inversely proportional to the thickness of the fibers. Thus a 70s wool of 19μ diameter would need 40% more dye than a 56s wool of 27μ diameter to obtain the same shade. Similar reasoning explains why light marks, appear if wool fibers are fibrillated in wear.

The data on practical dyeings obtained during the Princeton Wool Research project as reported by von Bergen and Wakelin (63a) and later by Brommelsick, von Bergen, and Millson (71) are in excellent agreement as seen from Table 4.

The Princeton project involved the manufacturing of worsted pieces from four wool types representing three fineness grades of American, Australian, and New Zealand origins.

In addition to the differences in the depth of a shade there occur differences in the hue of the shades (produced with the same formula), between fleece wools of the same grade but of different origins. The degree of tippiness is the main cause of these differences as clearly seen from Fig. 2.

Lambs wools and pulled or slip wools can be another source of dye differences. Therefore, it is essential to keep the dyer informed about the type of wools and grades present in the loose stock, top, yarn, or pieces he is expected to dye and match perfectly.

TABLE 4

Depth of Shade Relationship of Four Wool Types
Dye formula: 1% Palatin Fast Tan IWP[a]; 8% sulfuric acid

Wool type	Australian merino	Idaho merino	Idaho crossbred	New Zealand crossbred
U.S. grade	70s	64s	58s	58s
Average fineness increases	19.8	21.1	26.2	26.3
Calculated depth of shade $L_t/L_f \times 100$	100	107	132	133
Actual depth of shade	100	110	135	139

[a] Palatin Fast Tan IWP	%
P.F. Orange AENA Pr 315	14.9
P.F. Claret RNT No Pr	11.7
P.F. Blue RRNA Pr 319	18.5
P.F. Yellow ELNA Pr 330	35.8
Glaube's salt	19.1

Some Sources of Faulty Dyeing

In addition to the inherited differences existing between wools of various grades, origin, and types, each different operation to which wool is subjected during its long journey from fleece to finished fabric is a potential source of trouble. When materials from the dye house are unsatisfactory it is the dyer's responsibility not only to correct the faulty lots but also to take steps to prevent a recurrence. The following sections have been prepared to assist the dyer to trace the causes of trouble.

Tippy dyeing

An ideal dyeing is one in which all dyes are distributed uniformly along each fiber from tip to body end. However, this is rarely ever attained in practice and, instead, distinct differences in shade and depth of color are obtained on the same fiber or fleece. These variations in affinity for dye are inherent in each fiber and one of the main causes of these differences is unequal exposure to sunlight, weather and other agencies which affect the scale struc-

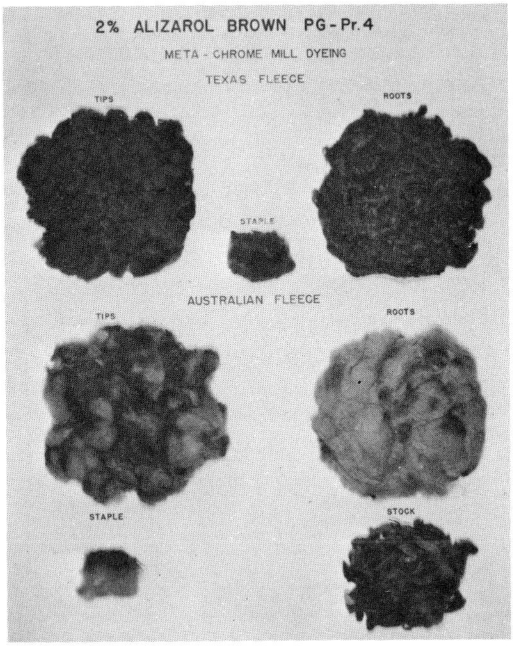

Fig. 2. Differences in tippy dyeing between Texas and Australian merino wools with the same dye formula and procedure.

ture while the wool is growing on the sheep. Such differences in affinity for dyes result in what is called "tippy dyeing."

Werner von Bergen (62) was the first to conduct an exhaustive scientific study into the causes of "tippy dyeing." Von Bergen found that all wools regardless of source or quality exhibit tippy dyeing effects to a greater or lesser extent. Tippy dyeing occurs in several forms, the first and most serious being those dyeings in which the tips of the fibers are deeply dyed, while the body sections remained undyed or are only lightly dyed (Fig. 2, and Fig. 3). A second variation is obtained where two or more contrasting colors are used to produce a shade that results in pronounced variations in hue and color value between the tips and body sections. A third and less troublesome though undesirable effect is caused by the tips remaining undyed or lightly dyed, while the body portions are heavily dyed (e.g., indigo).

Some of these extremes in color contrast are minimized when the stock is carded or combed; but such operations cannot entirely eliminate or mask these defects and therefore skittery, frosty, or heathery effects are obtained which result in weaker shades. Tippy dyeing has long been recognized as a major problem and many attempts have been made to eliminate this defect.

In an extensive study of this problem by the New York Section AATCC it was found that tippiness can be minimized and often eliminated by the careful selection of the dyes and by the use of surface active leveling agents (62).

Tippy effects are greatly minimized by the use of neutral dyeing premetalized dyes such as Irgalans, Cibalans, Calcofast, and other groups of 1:2 complex dyes. However, these dyes are not suitable for the bright, high-fashioned shades now in demand.

(a) (b) (c) (d)

Fig. 3. Effect of sunlight tippy dyeing. Tip and root ends of the same fiber dyed metachrome with 2% Calcochrome Fast Red ECB. The presence of the dye-resistant cuticle over most of the fiber and the bulbous root (*a*) is revealed by the absence of uniform dye penetration. Only the extreme root end and the weathered tip (*b*) of the fiber have absorbed dye deeply and uniformly. Tip (*d*) and sheared body end (*c*) of the same fiber dyed metachrome with 2% Calcochrome Fast Green SGV. The tip end is completely dyed and appears black in photomicrograph. The sheared body end is undyed, except where dye has penetrated through the cut end.

Scouring

Scouring plays a very important part in the production of satisfactory work. Soaps generally cause more trouble than sulfatep alcohols and nonionic scouring agents. Calcium, magnesium, aluminum, and iron salts react with ordinary soaps, forming insoluble compounds. These metal soaps attach themselves to the wool and are difficult to remove. When pieces containing such insoluble materials are dyed they may cause the finished pieces to appear cloudy and uneven. The addition to the scouring and rinsing baths of sequesterants such as the sodium salt of ethylene diamine tetracetic acid or various phosphates helps to prevent the reaction between the soap and metal ions when hard or contaminated waters must be used.

Heavily dyed blotches and spots may be caused by improper addition of soap or alkali solutions. When pieces or skeins are splashed with such solutions the wool is damaged and its affinity for dye is increased.

Wetting out

Yarns and pieces that have been scoured, carbonized, bleached, etc., and then dried must be wet out carefully and thoroughly prior to dyeing. The material should be wet out at 160°F or higher in a bath containing a surface active agent. Twenty to thirty minutes is usually long enough to completely saturate all parts of the material. This bath may be used for dyeing since many nonionic or anionic surface active agents also promote leveling during dyeing.

Incomplete wetting can cause light areas in dyed pieces or skeins. Microscopical studies have shown that the dye is deposited first on the outer parts of the wool fiber and then penetrates when the critical diffusion temperature is reached. When parts of the wool are not completely saturated deposition and subsequent absorption of the dye may not occur and these areas may remain lighter in shade throughout the entire dyeing cycle. This is especially true of those dyes which do not desorb and level during the boiling period.

Mixed stocks

Mixed stocks can also be a source of Barré effects, warp streaks, and other faults not caused by dyeing. When fabrics are made from yarns of different stocks, different twists or numbers these mistakes are generally not discovered until after the pieces are dyed. Such defects are different in appearance from normal unevenness caused by dyeing because the lines occur with regularity and follow the direction of the warp or filling. It is practicallly impossible to select dyes that will completely eliminate these faults.

Stripping with soluble sulfoxylate either before or after dyeing will tend to equalize the affinity of different wools for dyes and assist in producing level shades.

Dyeing into a dark shade or black will also minimize differences in stocks and yarns.

Differences in stocks present in a blend or in a dyeing may be accentuated for examination by dyeing the material in a bath containing 0.02% Calcofluor White RW, 2% acetic acid for 15 min at 100–110°F. When examined in a dark room with an ultraviolet lamp wide variations in fluorescence will be observed. This test is also valuable for determining the uniformity of blending of various stocks.

Carbonizing and neutralizing

Carbonizing and neutralizing processes can be responsible for many of the difficulties encountered by a dyer when using acid-sensitive dyes. The drying and backing operations are an important part of the carbonizing process and can cause considerable trouble if not properly controlled. Differences in temperature and the acid content during baking can so change the affinity of wool for dyes that even though the pieces are dyed with the most level dyeing types of acid dyes the pieces show side to center shading. Unevenly baked pieces are best dyed with acid-insensitive metalized dyes or with selected acid colors which minimize these differences in affinity for dye. Experiments have shown that it is difficult to neutralize wool completely and uniformly after it has been carbonized. Unless the neutralization is uniform and complete, pieces may be quite uneven after dyeing. This is because many neutralizing baths cause pieces to be completely neutralized on the outer part but incompletely neutralized on the inside of the fabric. When pieces are incompletely neutralized the acid migrates to the surface as the temperature of the dye bath is raised to the boil and acid-sensitive dyes exhaust and fix on the outer parts causing unevenness and two-sidedness. Blotches and side to center shading have also been traced to careless or incomplete neutralization prior to dyeing.

Carbonized unneutralized wool may be dyed level by using the acid-insensitive Calcofast dyes, Neolan and Palatine colors which require high concentrations of sulfuric acid for best results.

Sun-bleached yarns and fabrics

When rolls of undyed pieces are stored near windows or artificial lights without protection the edges are often light-bleached. When such pieces are dyed with certain dyes these areas may appear as a light streak running the full length of the piece. Other dyes may cause the light-streaked area to dye darker. Stripping or redyeing to a darker shade is necessary to overcome this defect and best results will be obtained when dyes which color exposed and unexposed wool about the same shade are used.

Lime in pulled wools

Pulled wools are often contaminated with lime which will cause trouble in dyeing. Heavily limed wools may remain undyed or lightly dyed because the outer part of the fiber is coated with the insoluble compound and the dye bath cannot come in contact with the fiber to permit adsorption of the dye to occur. A pretreatment with hydrochloric acid will often correct this condition by dissolving the lime forming soluble calcium chloride. Rinsing will remove the calcium compound and make it possible for the wool to absorb dye.

Effects of faulty steaming

Single wool yarns are sometimes steamed to set the twist. When the steam is applied unevenly those areas subject to excessive steaming will have greater affinity for dyes than will the partially steamed areas. The steaming operation increases the affinity of the wool for dye and therefore colors such as the Sulfon Cyanine Blues will dye lighter or darker resulting in a mottled effect when dyed in yarn or fabric. Unfortunately, redyeing to the same shade does not overcome these variations. Therefore, when the fault can be traced to uneven steaming it is advisable to redye into black.

Matching shades

Certain shades of unknown origin may exhibit a peculiar changeable cast that is extremely difficult to match in all lights. If the sample is large enough a spectrophotometric reflectance curve will often tell the experienced operator the types of dye that were used to produce the shade.

If such an instrument is not available the sample should be examined in daylight and artificial light. Shades that change radically in different light often contain orange, green, or violet. When an accurate match is required the standard should be matched in the laboratory first before proceeding with the lot.

Examination of the shade in ultraviolet light often gives a clue to the dyes used to produce it. If the shade fluoresces, one or more of the dyes used in the combination may have been an acid or basic Rhodamine, Acid Eosine, Brilliant Sulfo Flavine, an optical bleach or other dyes with lesser brilliance of fluorescence.

Additional information as to the nature of the dyes may be obtained by removing the dye by boiling with ammonia. This solution is concentrated by evaporation. The end of a strip of filter paper is placed in the solution and the dyes will separate by capillarity due to the differences in their water solubility. Examination under ultraviolet rays will show which of the dyes are fluorescent.

A more accurate method is to pass the unknown dye solution through a chromatographic column that separates the dyes by their differences in solubility forming definite rings of different colors. Examination of the column in ultraviolet light will make it possible to detect fluorescent dyes including optical bleaches which may be present. If desirable the column can be removed from the tube and cut into its different colors. These colors can be dyed on wool and compared to colors in shade cards or can be measured spectrophotometrically to obtain curves for comparison with the master of known dyes.

Conditioning samples before matching

One of the important requirements of a good colorist is an ability to match shades quickly and accurately. Poor matches often result in redyeing, delays and loss of production. Incomplete curing before matching to a standard shade may be responsible for off color batches. Colors applied to wool often exhibit shade changes depending upon the amount of moisture contained in the sample. For example, Wool Green S is almost a reddish-blue when bone dry and a yellowish green when the wool is damp. Most dyes including acid, milling, chrome and metalized, exhibit a shade change when the wool is bone dry. Blowing on a dried sample can also cause excessive color changes.

A simple method of drying is to attach hooks to the front grid of an air fan and then blow air through the sample until it is dry. Little if any further color change takes place when samples are dried in this manner.

Feeding dyes

The addition of dyes to a boiling dye bath for the purpose of matching a lot to a standard sample requires an intimate knowledge of the working properties of the dyes to be used. Level dyeing acid dyes may be added to a boiling dye bath without great risk because these dyes desorb and adsorb readily so that redistribution leveling can take place. They also exhaust readily so that each addition moves the shade in the desired direction. If excessive quantities of correcting dyes are added it is not too serious because the excess dye can be made to desorb and lighten the shade by an addition to the dye bath of Glauber's salt or by reducing the volume of the bath and refilling with hot water.

The problem is more complicated if the lot is being dyed with milling, neutral dyeing metalized types or direct or other acid sensitive dyes. These dyes have great affinity for the wool fiber and once they have been adsorbed they tend to remain and not desorb. Thus these dyes must be applied levelly by exhaustion techniques where dyeing is controlled by careful attention to the pH of the dye bath. Additions for the correction of shade must be made by first cooling the bath below the diffusion temperature and by raising the pH if it is considered too low. The dyes must be well diluted, fed on slowly in the temperature range of 140–160°F, run 15 min, and then the bath should be gradually raised to the boil and boiled for 20–30 min.

Microscopical studies have shown that acid-sensitive dyes exhaust rapidly at the boil and this is especially true at low pH. These dyes are adsorbed on the outer part of the fiber and then with time diffuse into the fiber where fixation takes place. Therefore, sufficient time must be allowed after each addition of dye for these steps to take place.

At times the working properties of the dye may necessitate its addition in a fresh bath at low temperature. The bath should be raised slowly to the boil and boiling continued for at least 30 min to ensure levelness and the full development of shade.

Ring dyeing

The first step in the dyeing of any fiber with any class of dyes is the movement of the dye from the bath onto the fiber. Under most conditions of conventional dyeing the color builds upon the outer areas of the fiber prior to its diffusion into the fiber. This is called ring dyeing. With time and temperature the dye molecules diffuse away from the outer part of the fiber toward the center or core of the fiber. The time required for uniform diffusion of the dye within the fiber varies but is influenced by the structure of the dye, the type of fiber, the pH of the dye bath, the dyeing temperature and the time cycle. Some acid dyes complete the diffusing within a few minutes, while certain milling dyes may require 10–20 hr for diffusion to be completed.

The diffusion of dye is a very important feature of the dyeing operation because in the case of milling and other slow diffusing dyes the depth of shade and at times the fastness properties are determined by the position of the dye, i.e., whether it is on or in the fiber. With the same amount of dye a ring dyeing may be only 50% as strong as a completely penetrated dyeing. Time, temperature, and pH are important factors in controlling the rate of diffusion of all dyes into wool fibers.

Excessive crocking

All of the causes of crocking are not known. Millson, Watkins, and Royer (41a) found that materials that crock excessively do so because some of the dye has remained on the outer parts of the fiber and material. This lack of penetration may be due to too rapid exhaustion of the dye on the outer periphery of the fiber, insufficient time at the boil, using an acid which causes the dye to precipitate or incompatibility of materials which were added to the dye bath. Crocking can generally be reduced by controlling the rate of exhaustion and dye penetration. If a fast-to-washing color crocks excessively a light soap or detergent scour will often remove the loose color with an improvement in fastness. In vat-dyed prices, especially navy blue and indigo shades, the crocking can be minimized by dry cleaning in a Derby unit.

Chlorinated wool

Wool is generally chlorinated to decrease its felting properties, but also to increase the affinity for dyes in piece printing. Microscopical studies have shown that the chlorine first attacks the distal edges of the scales and then with increase in time the attack spreads toward the provincial end of the scale to completely chlorinate the entire fiber.

Chlorinated wool is quite different in its dyeing properties than the normal wool. Since the scales have been affected and in many cases obliterated, chlorinated wool dyes much more rapidly, at a lower temperature and requires less acid for dyeing than untreated wool. See page 696.

Metal contaminants

The effect on dyeing of calcium and magnesium salts in hard water is well known. The presence of these salts in the dye bath reduces the solubility of dyes and may react with dyes to form the insoluble lake or salt.

Iron is probably the most undesirable metal salt contaminant found in dyehouse waters. Iron is generally present in natural waters; dyes, Glauber's and other materials may contain it as an impurity, iron steam and water pipes rust and yield undesirable ions, iron bacteria may thrive in water storage tanks and other ions of iron may come from dyeing machinery. Iron is objectionable because its reaction products with chromable and other dyes are generally brownish and dull in shade. In addition, compounds of iron have affinity for wool and exhaust from a dye bath and dull the shade of the wool itself.

The iron complexes of chrome dyes are generally less fast to light and washing than the desired chromium complex and therefore in addition to causing undesirable shade changes, the fastness of the dyeing may be reduced in proportion to the parts per million of iron present.

Copper is not so prevalent as iron but may be present as an impurity in natural waters or may be derived from copper, brass, or bronze propellers, fittings, etc. In general, copper changes the shades of chromable dyes and reduces their fastness properties.

Aluminum is generally present in natural and treated waters. In the treatment of water aluminum sulfate is often added to assist in the purification process. It is not always consumed or removed and when present can form insoluble aluminum lakes with acid milling chrome and direct dyes.

Managanese is also an undesirable metal to have in the dye bath especially when dyed materials are to be rubberized. Natural waters often contain traces or more of manganese. Manganese bronze parts can also be a source of such metal ions.

If it is known that metal ion contaminants are present they can be inactivated by an addition to the dye of a sequesterant such as Cyquest 40, Trilon B, Sequestrene, Versene, and similar compounds. In order to obtain the maximum benefit from the use of sequesterants they should be added to the dye bath prior to the addition of anything else. The material should be run in the bath for 15–20 min to allow time for the deactivation reaction to occur.

Stripping dyed wool

Wherever possible unsatisfactory work should be redyed into black. This leaves the wool in the best possible condition for the amount of processing it has had. If stripping is necessary, treatment at the boil with ammonia may remove sufficient dye for leveling and rematching of shade. When more drastic treatments are required the color may be destroyed chemically with formaldehyde sulfoxylates. All of these treatments degrade the wool and make it more difficult to dye level with acid-sensitive colors. When dyeing with difficult-to-level wool dyes, special precautions should be taken because stripped wool absorbs dyes more rapidly and fixation takes place at lower temperatures than it does with normal wool. Acid-sensitive dyes must be exhausted slowly by pH and temperature controls.

Tightly woven fabrics are especially difficult to penetrate after stripping and therefore additional precautions should be taken when redyeing such materials.

Abrasion marks

One of the most common causes of faulty dyeings is abraded wool. Hadfield and Lewin (20) made a study of the effect of abrasion of wool on the characteristic properties of various acid dyes on wool fibers. When abrasion takes place before dyeing, the abraded areas dye initially more heavily than the unabraded areas. When abrasion occurs after dyeing, the abraded areas appear less heavily dyed, owing to an optical effect caused by fibrillation of these fibers.

The time required to produce solid dyeings on abraded wool varies widely from dye to dye, and the degree of sulfonation of the dye has an important effect. Certain cationic auxiliary agents reduce the time required to produce solid dyeings.

Many microscopic studies such as shown in Volume 1, pages 471–472 and by Hadfield and Lewin and many others have shown that abrasion or mechanical wear causes marked changes in the surface characteristics of the wool fibers. Scale removal and fibrillation of the cortex are accompanied by increased wetability resulting in an increased adsorption of dye. The effect to a certain degree is similar to that for tippy dyeing of wool. The influence of the scales in the dyeing of the wool fiber is well illustrated by Figs. 4a,b mechanical breaking of epicuticle and 4c,d descaling of sunlight damaged tips reported by Millson and Turl (41b). To trace abrasion marks in piece-dyed material is often difficult because of the many processes the pieces have been exposed to before the dyeing operation such as wearing, scouring, carbonizing, fulling, and raising.

The following are practical examples dyers were confronted with:

In fine clear worsteds such as gabardines abrasion marks appeared in the form of short, dark, narrow warp streaks. Occasionally a single streak was observed running the whole length of pieces from the same warp beam. The short inches long streaks were traced to the mending department. After the mender had repaired the faulty place, she straighted the yarns with a burling iron or needle thereby damaging the surface of the fiber. By stopping this practice the

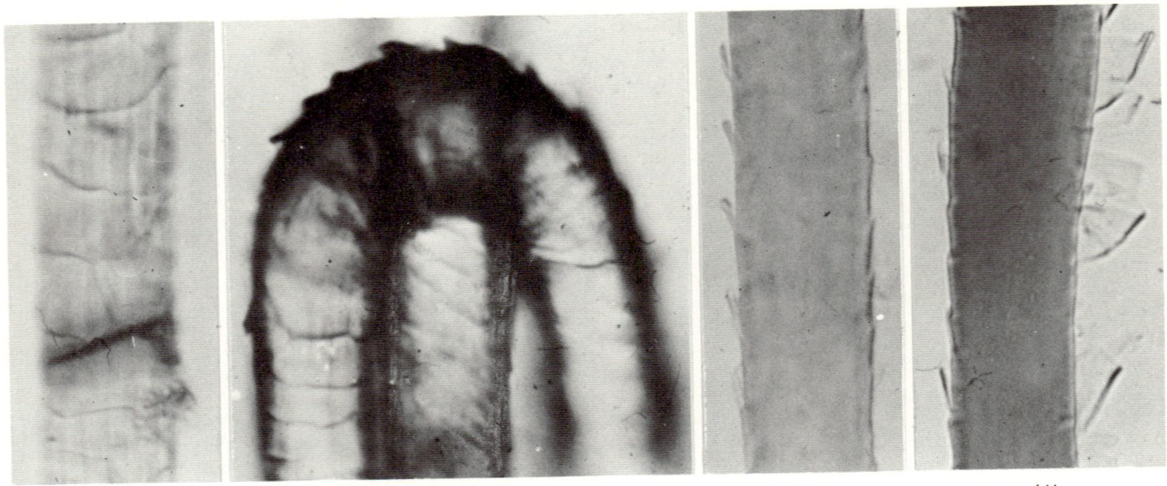

(a) (b) (c) (d)

Fig. 4. Effect of scales on dye penetration. Mechanical breaking of epicuticle. (*a*) Incised wool fiber showing penetration and diffusion of a milling dye after boiling 30 min. Note diffusion of dye away from point of penetration. (*b*) Part of knot tied in wool fiber showing ratchet-like separation of scales. Note local penetration of dye in upper and lower areas. (*c*) Protruding scales on tip of wool fiber after dyeing with 2% Calcochrome Alizarine Red SC; 8% anhydrous Glauber's salt; 5% acetic acid, 28%; boiled 45 min. Not chromed. (*d*) Wool fiber showing scales sloughing off the tip, after dyeing with 2% Calcochrome Alizarine Red SC, 8% anhydrous Glauber's salt, 5% acetic acid, 28%; boiled 45 min, top-chromed with 1% sodium bichromate, boiled 40 min.

marks were eliminated. The full piece length streaks had their origin in the weaving itself but could never be traced to a specific part. In heavy woolens fulled and carbonized the abrasion marks usually consists of ill-defined warp streaks caused by slippage of the rollers in the dolly washer or in the fulling machine. These defects are referred to in Chapter 12, page 831. If the pieces are tacked (bagged) before being wet processed, the protected inside face of the cloth may dye evenly, whereas the exposed outside of the material shows streaks. Even when the exposed outside is uniformly dyed very often extreme shade differences occur between the two sides with the exposed outside being much darker.

Microscopical studies have also shown that abrasion marks can be caused by the rubbing of the fabric over pitted stainless steel. The fibers in the resist or lighter area are cut or shredded by the needle-sharp edges of the pitted areas. This increase in fiber surface area causes an increase in light reflection and hence the shade appears lighter.

Stainless steel 18–8 is more liable to pit during use than is the molybdenum fortified stainless steel. When this condition occurs the pitted areas should be buffed or ground down and then polished. The stainless steel can then be passivated by treating in a bath containing nitric acid and bichromate. The bath is left in contact with the metal overnight. If this treatment is repeated about every six months the rate of pitting is considerably reduced.

Boiler compounds

Most mills use boiler compounds to control scale formation. Little trouble is experienced unless these materials contain free caustic soda or soda ash. Caustic soda is especially troublesome because it is liable to be entrained in the steam and carried into the dye bath. When this occurs, the amount of alkali in the bath increases with time and gradually neutralizes the acid, retards the rate of exhaustion and progressively damages the wool fiber. When dye baths become difficult to exhaust by standard procedures, the steam condensate should be tested for the presence of alkali. If alkali is present the boiler may be blown down, or an addition of phosphoric acid made to neutralize the alkali and form phosphates which are helpful in controlling scale. An alkali-free boiler compound should then be adopted.

Machine and spinning oils

Greases and oils are a constant source of trouble to the dyer and must be removed before dyeing if satisfactory results are to be achieved. Oils and greases usually cause lightly dyed or resisted areas in pieces. When the dye bath is prevented from coming in contact with the fiber, deposition of dye is prevented and light areas result. Dry cleaning pieces before dyeing has greatly reduced the imperfections caused by the presence of oils and greases.

It is sometimes difficult to detect oil stains and spots in greige goods by examination on the rough perch. Examination of suspected fabrics with ultraviolet light has proved of value because many oils and greases have a distinct fluorescence of their own.

The Matching of Shades

Dyeing is an art and not a science. It requires years of study and actual practice at the art to become proficient in color matching. The art of color matching has been made increasingly difficult in recent years because of the widespread practice of blending many other fibers with wool on which the dyer is expected to produce solid or contrasting shades. Thus, reproducibility from lot to lot has become a serious problem. There are certain basic principles that govern the reproduction of various shades and a brief description of these may be of assistance to the dyer.

In theory, all colors can be matched with the three primaries, yellow, red, and blue. However, in practice it will be found that the shade of violet produced by a mixture of red and blue dyes will not produce the brilliance of shade that can be obtained with a single violet dye. However, even though the three primaries and mixtures thereof do not give all the shades a dyer is required to match, the concept is of value in the art of color matching. The greater the number of shades that can be obtained with combinations of the three primaries the less complicated the formula and the more easily it can be reproduced. Carpet dyers follow this concept and have proved it to be beneficial.

There is a general terminology in use by dyers to describe a shade and its deviations from the standard. In describing hue, the terms used are "redder," "bluer," "greener," and "yellower." Stearns (56) prepared the table of deviations in hue shown below.

Deviations in Hue

Red	Bluer	Yellower
Orange	Redder	Yellower
Yellow	Redder	Greener
Green	Yellower	Bluer
Blue	Redder	Greener
Purple	Bluer	Redder

In addition, "duller," "brighter," "grayer," "browner," and other terms are in common use.

There is another equally important factor in color matching in addition to hue. This relates to the total amount of dye present and may be called the "strength," "weight," "body," or "depth" of the dyeing. The comparative strength of two shades can be determined visually by examining the samples by looking down on them or by looking across them at eye level.

Brightness is also an important aspect of color matching and often causes the wool dyer considerable trouble when dyeing pastel shades. Wool tends to boil yellower and duller the longer it is run at elevated temperatures. The brightness of a shade can generally be controlled by reducing the amount of blue, gray, or brown dye component in a formula. Prebleaching is a help, but peroxide bleached wool may produce duller pastel shades than unbleached wool if it contains residual peroxide.

The brightest shades on wool are obtained by running at elevated temperatures for the shortest possible time cycle. The presence of iron, copper, and manganese metal ions in the dyebath also dulls the shade and makes it difficult to obtain brightness equal to the standard.

Other impurities such as tannins, suspended organic and inorganic matters, mineral salts, soaps, dung extracts, chlorophyll and other agents may cause dulling or alternation of shade. This problem has been especially troublesome in the dyeing of carpet wools. Extensive studies have been made by Stewart (58).

The moisture content of the wool at the time of matching plays an important part in the ultimate match obtained. Many wool dyes are quite sensitive to the presence or absence of moisture and if a shade is passed before the dyehouse sample has regained its normal amount of hygroscopic moisture, the shade may be quite different after it has aired and is in equilibrium with the surrounding air. A note on the standard formula card stating the difference in shade between bone dry and air samples will alert the dyer and provide information of value for subsequent matching.

Matching and shade control by instruments

The evaluation of colored samples and the formulation of instrumental match predictions requires instruments to make the measurements and computers to perform the calculations. St. John (59) of Geigy dyestuffs gives the following descriptions.

There are two types of instruments available to the industry to provide basic information about the quality of light reaching the eye of the observer.

Spectrophotometers measure the amount of light reflected or transmitted by a sample at specified wavelengths. These are the measurements of standards which would be used in the formulation of color matches. A fairly large number of wavelengths are required for match prediction. Generally, 16 or more are used to insure that the match will resemble the standard as closely as possible.

Colorimeters measure the amount of light transmitted or reflected by a sample under a standard light source. They simulate the response of the human eye under specified conditions. These instruments in their various forms are used to evaluate small differences among various samples. They find their biggest use in the evaluation of production batches, faded samples, or other cases where an aid is desired for measuring small differences andmaking numerical ratings. These measurements are generally made by using three or four filters in combination with appropriate electronic circuitry. The values obtained are called tristimulus values. They are usually labeled X, Y, and Z. These values approximate the amounts of red, green and blue light needed to match the standard by the average observer. Some colorimeters have special circuitry to provide numbers which aid in the computation of color differences.

Following is a list of instruments with general distribution in the textile industry.

Colorimeters: Colormaster Differential Colorimeter; Hunter D-25, Color Difference Meter; Gardiner Color Difference Meter; Photovolt Reflectometer.

Abridged Spectrophotometers Colorimeters: Color Eye, I.D.L. Elrepho; Zeiss.

Spectrophotometers: Bausch & Lomb Spectronic 505 Beckman Models DK, DU, DB; Cary 14, 15; General Electric.

The use of spectrophotometers in formulation relies upon a series of equations which relate the amount of light reaching the observer's eye to the amount of dyestuff present in the sample.

In the case of transparent samples the equations used, derived from Beer's Law, state that if everything is constant the absorbance of light is proportional to the concentration of dyestuff present. They also state that the total absorbance of the solution is equal to the sum of the absorbances from all the dyestuffs present. The equation is generally used in the form

$$A = C \times K$$

where

$$A = K\ (\log 1/T)$$

where

A = absorbance
K = a constant describing the conditions of the system
C = the concentration of dyestuff
T = the amount of light transmitted by the sample

When dyestuffs are applied to a fiber the scattering of light by the substrate must be considered and the equations become more complicated. These equations are derived from the work of Kubelka and Munk. They state, when simplified, that the absorbance of light by a dyestuff fiber combination is proportional to the concentration of the dyestuff, modified by subtracting the scattering effect of the substrate. These equations are generally used in the following forms:

$$\theta = \frac{C \times A \text{ dye}}{S \text{ subs}} - \frac{A \text{ subs}}{S \text{ subs}} \qquad \text{or} \qquad \frac{A}{S} = \frac{(1 - R)^2}{2R}$$

where R is the light reflected from the samples, S is the amount of light scattered, and θ is the total amount of light absorbed or scattered.

The use of the above equations and spectrophotometers in instrumental match prediction has been greatly simplified by the development of special purpose analog computers and by the development of Color Matching Programs suitable for use with digital computers.

Several analog and special purpose digital computers have been designed to provide formulations using these equations or variations thereof. These are currently in use in many locations.

Name	Company
COMIC I, COMIC II	Davidson & Hemmendinger, Easton, Pennsylvania
Pretema color computer	Pretema AG, Zurich, Switzerland (Fig. 5)
Color matching computer	Redifon Limited, London, England

Many large-size digital computers have also been programed to deal with the equations necessary to relate colorant concentration to amount of reflected or transmitted light. A number of dye houses and several dyestuff suppliers are currently using these

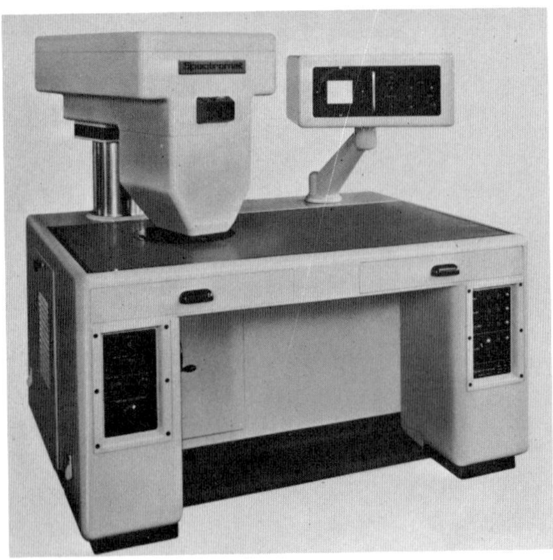

Fig. 5. Pretema color computer. (Courtesy of Pretema, A. G., Switzerland.)

large scale digital computers to aid in selecting dyestuffs and to provide formulations for matching shades. In the field of production control there are several instruments which have been specially prepared to provide information from an on line location. These instruments provide measurements which indicate the variations which occur during production runs and which enable the pin-pointing of problem areas. They are marketed under the following names:

Colorede: I.D.L., Attleboro, Massachusetts

D-44 Color Monitor: Hunter Associates Laboratory, McLean, Va.

Chromotronic Color Monitor: Teledyne Inc., Charlottesville, Va.

In the postproduction area there are specially prepared instruments which use basic colorimetric information to sort various production lots into color groups. This enables shipment of orders containing samples which differ as little as possible in color.

Chromosorter: Teledyne Inc., Charlottesville, Va.

Color Eye Shade Sorting System: I.D.L., Attleboro, Mass.

All these instruments and systems have been devised to aid in the job of evaluating colored samples. They are tools which allow the dyer to do more work much faster than he was ever able to do before and to have available numerical descriptions of samples which can be kept without fear of future deterioration.

Rizzo and coworkers (50,51) of the U.S. Army Natick laboratories developed in cooperation with the Hunter Associates laboratory a tristimulus colorimeter seen in Fig. 6 for the control of the color and colorfastness of military textiles. The performance achieved with this instrument in the Natick Laboratories makes possible the use of instrumental methods of shade and color fastness evaluation in lieu of visual methods with assurance that the results will be acceptable in terms of visual ratings.

Low temperature dyeing

Realizing that the normal dyeing of wool in a boiling liquor even in a weak acid bath impairs the breaking strength, considerable research efforts over the years have been made to reduce or even eliminate this damage. A break through in this field came with the introduction of the Irga solvent process by Geigy.

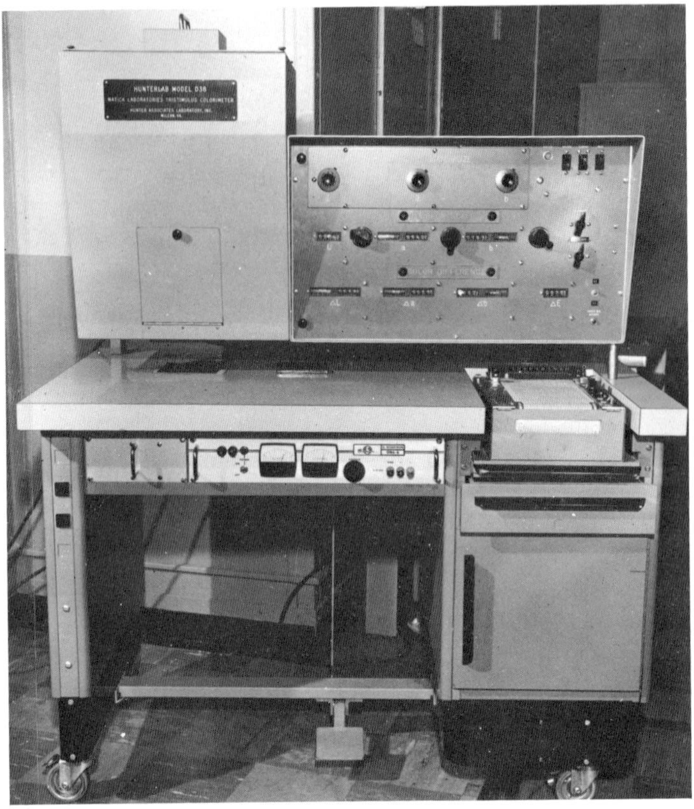

Fig. 6. Tristimulus colorimeter hunter. (Courtesy of U.S. Army Natick Laboratories.)

Irga solvent process

It has been known for several years that the rate of absorption of dyes by wool is affected by the presence of an organic solvent. Peters and Stevens (48) showed that certain alcohols, phenols, and amines, when added to an aqueous dyebath, markedly increased the rate of absorption. They also showed that very good exhaustion of the dyebath could be obtained in a relatively short time and that this occurred at temperatures much lower than those normally used for the dyeing of wool, i.e., appreciably below the boil.

Similar results were obtained by Karrholm and Lindberg (28) who examined the effect of the wool cuticle on the rate of sportion of dyes from baths containing alcoholic potassium hydroxide, butanol and amyl alcohol.

In Europe, particularly in England, the technician of T. R. Geigy, A. G. Hirsbruner (26), Beal, Dickinson and Bellhouse (6,7) developed this method into a commercial scale process. After trying many other solvents they decided that benzyl alcohol gives the best possible results. A study of the cost of the solvent required for dyeing by this process showed that it varied inversely with temperature and directly with the liquor: wool ratio. The most favorable contents were found to be 175°F and a liquor ratio between 10 and 20:1. Lately a considerable amount of dyeing has been done at 195°F. The relationship between the solvent and the dye temperature is as follows:

Temperature		Bensyl alcohol, vol %
60°C	140°F	3.5
70°C	158°F	3.3
80°C	175°F	1.8
90°C	195°F	0.65

A certain amount of alcohol is absorbed by the wool. The absorption decreases with the increase of temperature as shown in the following table:

Temperature		Absorption, %
60°C	140°F	11.5
70°C	158°F	8.5
80°C	175°F	6.0
90°C	195°F	4.0

As stated, for reason of economy it is necessary to reuse the exhaust dye liquor after suitable replenishment of solvent and dyeing assistants. There is one dyehouse who has reused the same liquor for 25 dyeings of 500 lb. Of the many wool dyes which have been found to be satisfactory for this method, the neutral premetallized dyes of the Irgalan type are excellently suitable and most acid milling dyes of the Polar type are also satisfactory. In the case of chrome colors, chroming has to be done separately, which adds ½ to ¾ hr to the overall dyeing time.

Several requirements have to be met in order to obtain level results:

1. It is essential to ensure that the rate of strike of the dye under the prevailing conditions is such that it is possible to obtain a level result with the circulation available. The rate of strike (which can be unusually high in solvent-assisted dyeing) must be adjusted to the efficiency of circulation, which varies widely in machines of different types. For example, very high rates of strike have been found to give level results with slubbing dyed in top form on a spindle machine, but the same method resulted in failure when dyeing similar tops in a can machine, unless the rate of strike was greatly reduced. Similarly, yarn may be dyed very quickly in cheese form with level results, but a much slower dyeing process is required on yarn in hank form.

2. It is preferable to establish a uniform pH throughout the material to be dyed *before* commencing dyeing by the solvent method, because the rate of strike is usually high, and nonuniform pH causes unlevel dyeings. It is not always easy to establish a uniform pH in bulk practice; it is especially difficult where circulation is poor and a pH near 7 is required on material which was previously markedly alkaline. It is easier to establish a lower pH uniformly but when, as is often the case, dyeing has to be carried out near pH 7, e.g., on a top initially at pH 9–10, the use of a buffer is advantageous, unless the circulation is really efficient.

Collins process

In 1963 Collins was granted U.S. Pat. 3,112,983 for a process of low temperature dyeing of wool. This process overcomes some of the limitations of the solvent dyeing, such as high cost of the solvent or the necessity to include a retrieval step to enable the solvent liquor to be reused. Collins (13) low cost, low temperature dye method is also based on the use of solvents such as tridecyl alcohol and ethylene glycol mixed with detergents.

The dye assistant K-12 as given in the patent and used in the various recipes stated by Collins is composed of compounds shown in Table 5.

By using the dyeing system for stock, top and yarn dyeings, it was found that the following conditions were of significant influence: (*1*) rate of circulation; (*2*) pH buffer system; (*3*) liquor:

TABLE 5

Composition of Dye Assistant K-12

Compound	Percent
Tridecyl alcohol	27.5
Ethylene glycol	35.0
Dicthylene glycol	12.5
Hexylene glycol	10.0
Igepal CO-730[a]	7.5
Tergitol[b]	7.5
	100.0

[a] Nonylphenol/polyoxyethylene/ethanol.
[b] Trimethyl/nonanol/polyoxyethylene/ethanol.

stock ratio. The procedure in general is quite similar to conventional dyeings at the boil. However, dye exhaust characteristics are different. The temperature for initial exhaust is lower, the rate of exhaust is accelerated. This dyeing system minimizes the difference in exhaust rates of individual dyestuffs. The rate of exhaust is proportional to temperature. For example, the specific procedure for stock, top and yarn, indicates a time–temperature cycle of 20–30 min at 180–185°F. One plant, however, is successfully operating on a 15 min at 190–195°F cycle for all depths of shades.

As dyers develop certain variations in the application of chrome, milling and neutral premetallized dyestuffs for their specific equipment and requirements, the procedures outlined by Collins are generalized. Two percent of dye assistant K12 or 1.5% dye assistant N (a newer product) on the weight of the material is required.

Neutral Premetallized and Milling Dyeings
Set bath at 120°F
Add 1.5% Dye Assistant N
  or 2.0% Dye Assistant K-12
Add 1.0–2.0% acetic acid (56%)
Circulate 10 min
Add Dyestuff
Circulate 5 min
Raise to 180–185°F in 40 min
Hold 25 min and sample

Procedural variations:

1. For neutral premetallized colors, start with
2.0% diammonium phosphate
2.0% acetic acid (56%)
or
1.0% ammonia
2.0% acetic acid (56%)

This variation permits a slower rate of exhaustion and insures better levelness, top to bottom, for kettles with slow rate of circulation.

2. The rate of temperature rise is dependent upon the rate of flow through the stock. This varies from 25 to 50 min.

3. Acid may be added before or after dyestuff. Adding before dyestuff allows a longer period of circulation and neutralization of any alkaline stock in a blend of mixed wools. It has been observed that a rapid rate of exhaust at lower temperatures of specific neutral premetallized colors is undesirable. The condition appears to mitigate against maximum fastness.

4. On heavy shades it is recommended that the temperature be maintained for 30 min at 185°F. Heavy shades of Acid Red 114, dyed on oily or greasy stock, will require 20 min at 200°F for maximum fastness.

Chromate dyeings

Set bath at 120°F
Add 1.5% Dye Assistant N
or 2.0% Dye Assistant K-12
Sodium bichromate
1.0–2.0% acetic acid (56%)
Run 10 min
Add Dyestuff
Run 5 min
Raise in 40 min to 180–185°F
Hold 20 min and sample

Top chrome dyeings

Same as chromate dyeings except add bichromate after 20 min at 180°F and hold additional 20 min. For heavy shades, up to 6.0% acetic acid (56%) can be added at the start of the dyeing, preferably added in two increments before and after the addition of dyestuff.

Results of Production Dyeings

Significant mill results are summarized by Collins as follows:

1. The reduction in fiber damage resulted in improved efficiency in carding, spinning and weaving noted in every plant. The mills have found not only the advantage of greater efficiency, but also the added advantage of increasing spinning rates. One mill reported a 15% increase on front roll speed on a woolen spinning frame. On the worsted system, a mill reported increasing the ring traveller speed and at the same time reported ends down as low as 5/1000 spindle hours.

On low grade woolen stocks, the use of this method has been the difference in processing a 100% wool blend and the necessity to process it as an 85% wool/15% nylon blend.

2. Use of cheaper wools. As a direct result of improved fiber quality after dyeing, mills are now able to process lower cost blends.

3. Lower cost dye formulations.

4. Reduced steam consumption.

5. Higher rate of dyehouse production, 25–30%.

6. Less packing of stock in dye kettle. The loftier stock is in better condition for carding. Second, the lack of packing has permitted some dyehouses to increase the poundage per dip on fine wools.

Piecegoods Low Temperature Dyeing

This is possible with level dyeing acid dyes in presence of products which reduce the critical dyeing temperature and promote a high rate of dye migration. Collins claims that with Product 220 it is practical to apply the full range of level dyeing acid colors at temperatures as low as 175°F (80°C) with good levelness, penetration and equal fastness to the conventional dye method. The following is the procedure given:

Cloth neutral
Set temperature at 80–90°F
Add 10.0% Glaubers salt
Acid
1.5% Product 220

Run 10 min
Add dyestuff
Run 10 min
Raise in 30–45 min to 175°F
Hold 45–60 min
For adds: Temperature does not need to be lowered for the addition of dyestuff add. Hold 30 min after add.

The preferable acid systems for dyeing may be any of the following:

a. 4.0–5.0% acetic acid (56%)
b. 3.0% acetic acid (56%)
1.0% sulfuric acid
c. 3.0–4.0% acetic acid (56%) followed by exhaust with 1.0–2.0% sulfuric acid after 25 min at temperature.

For the dyeing of strong acid 1:1 metal dye complex it is possible to reduce the boiling time to 30 min.

C.S.T.R.O. Process

Hine and McPhee (24) developed another rapid dyeing method for wool below the boiling point.

Damage to wool in boiling aqueous solutions is at a minimum when the pH is in the range 3.5–4. Damage is always less below the boil and they have found that losses in strength and abrasion resistance are usually very low at pH 3.6–4.2 at 90°C for times up to 30 min. The influence of pH and time is not so critical at lower temperatures so that, if the maximum temperature is about 85°C damage may be negligible at pH 3–5 for times up to 1 hr.

For proper penetration and to give level and fast dyeings under the above condition, a certain type of nonionic detergent has to be used as an assistant.

Some members of the series of polyoxyethylated nonylphenol surfactants, having 6–10 moles ethylene oxide per mole nonylphenol (abbreviated to NP6–NP10), give excellent results. Using 0.5–2% on the weight of wool, good dyeings are obtained with the 1:2 metal-complex dyes, at dyebath pH values of 3.6–4.2 and a maximum temperature of 90°C for 30 min. Levelness of dyeing depends on the rate of temperature rise. The dyeings are usually started at about 40°C and the temperature raised to 90°C at a rate of about 1°C/min.

Polyoxyethylated nonylphenols containing about 10 moles ethylene oxide per mole nonylphenol (NP10) are cheap and readily available under a variety of trade names. Among common products in Australia which have been tested and found to be satisfactory are Antarox CO 630 (General Aniline and Film), Triton N-100 (Rohm and Haas) and Lissapol N (I.C.I.), but there are many others which could undoubtedly be used.

Chrome Dyes

Chrome dyes can be applied, using 1% NP10 on the weight of wool, in a similar way to the 1:2 metal-complex dyes. The dyeings are started at about 40°C at pH 3.6–4.2; the temperature is raised at 1°C/min to 85°C and held at this temperature for 20–30 min. The normal amount of chrome is then added and chroming carried out at 85°C for a further 30 min. Dyeings with the chrome added at the same time as dye have given adequate levelness but the shades are sometimes not as bright as those obtained with the after-chrome method and wash-fastness may not be as good. High concentrations of dye (e.g. 8% C. I. Mordant Black 11) can be applied using the same 1% concentration of NP10 in the bath as is used with lower dye concentrations.

Acid Milling and 1:1 Metal-Complex Dyes

The acid milling dyes can be applied in exactly the same way as the 1:2 metal-complex dyes but, with the 1:1 metal complex dyes, it is generally better to take the temperature of the bath to at least 95°C to obtain a shade identical with that found in normal dyeing.

The wool is wet out in the dyebath containing 1% NP10 on the weight of wool and sufficient acetic acid to give a pH between 3.6 and 4.2. (The addition of a small amount of "anti-foam" is sometimes desirable.) After running for 5–10 min at 40°C, the dissolved dye is added and the temperature raised at 1–2°C/min, the actual rate depending on the efficiency of liquor circulation in the particular machine. (For example, the rate can be 2°C/min in a pressure machine, but in some can-dyeing machines, it is better to use 1°C/min.) The levelness of dyeing is controlled by the rate of temperature rise so that this is a critical feature of the dyeings.

The temperatures, and times at these temperatures required for good fiber penetration and wash-fastness are:

1:2 metal-complex dyes: 90°C, 30 min
acid milling dyes: 90°C, 30 min
acid leveling dyes: 85°C, 30 min
1:1 metal-complex dyes: 95°C, 30 min
chrome dyes: 85°C, 30 min

In Table 6 are listed some of the strength changes of yarns and fabrics when dyed below boil with 1:2 metal-complex dyes.

TABLE 6

Results of Industrial Dyeings with 1:2 Metal-Complex Dyes

Procedure		Yarn breaking load, g wt	Extension at break, %	Abrasion resistance of fabric, no. of rubs to hole on Martindale tester
Top-dyed, made into 2/32s worsted count hosiery yarn	Undyed	410	23.4	13,400
	Dyed, pH 4, 90°C, 30 min.	415	20.8	14,000
	Dyed, pH 6, 100°C, 1 hr	352	18.0	10,500
2/28s worsted count weaving yarn dyed on packages	Undyed	509	30.1	
	Dyed pH 4, 90°C, 30 min	480	22.2	
	Dyed by normal mill procedure	401	18.8	
2/32s worsted count hoisery yarn, dyed in hanks	Undyed	331	20.2	
	Dyed pH 4, 90°C, 30 min	340	18.9	
		Warp tensile strength (1 in. grab test), kg wt		
Plain-weave woolen fabric, 32s Yorkshire skeins yarn with 16 tpi, 5.7 oz/yd	Undyed	11.8		8,500
	Dyed pH 4, 90°C, 30 min	11.3		8,000
	Dyed pH 6, 100°C, 1 hr	9.5		5,500

Reactive Dyes

Hine and McPhee (26) found that reactive dyes can be similarly applied to wool and, furthermore, that some classes give very bright dyeings of high washing fastness on tops and worsted fabrics, even when the maximum temperature of dyeing is as low as 60–70°C. Previously such dyes have been applied either at the boil or at 70°C in strongly acid solutions.

Dyeing was carried out by allowing wool samples to equilibrate at 30°C in dyebaths brought to pH 4.0 with acetic acid and containing 1% (on wt of wool) of a polyoxyethylated nonylphenol detergent (with 10 moles ethylene oxide per molenonylphenol). The dissolved dye was added and the temperature raised at 1°C/min to the required maximum. Generally the baths were then kept for 30 min at this temperature, except for the dyeings at 60°C, where they were held for 1 hr.

Good washing fastness is obtained with reactive dyes by dyeing below the boil at pH 4 and not necessarily washing-off with alkali or soap. Under these conditions, all the dyeings have washing fastness comparing favorably with that of dyeings of the Carbolan and Irgalan dyes. Outstanding results were obtained with dyes containing sulfate ester side-chains, which may be converted during dyeing into reactive vinyl sulfone, vinyl sulfonamide, or acrylamide groups.

Lewis and Seltzer (73) have recently demonstrated that a pad dyeing system using concentrated urea solutions can be adapted to allow a very high fixation of reactive dyes to be obtained on wool piece goods by impregnation followed by batching at room temperature for periods from 6 to 48 hr, dpending on the reactivity of the dye. The most reactive dyes, which, therefore, require the shortest batching times are Procion M (ICI), Reactofil (Geigy), Cibacron Pront (CIBA) and Remazol (Farbwerke Hoechst). This process combines a very high dye fixation with all the advantages to wool of dyeing in the cold.

U.S. Pat. 3,363,972 (1968) refers to a similar process of treating wool at a temperature from 15 to 40°C with an aqueous preparation containing a reactive dyestuff and at least 5% calculated on the whole preparation of urea and then subjecting the material in the wet state to a storage of at least 1 hr at 15–40°C.

The Dyeing of Wool at High Temperatures

The high temperatures dye processes were first introduced to overcome the difficulties of dyeing man-made fibers such as polyesters and acrylics. The interest of dyeing wool above the boiling temperature of water came with the increased popularity of the wool–polyester blends. The dyer was forced to find a shorter dye cycle. The dyeing time for such blends ranged between 8–14 hr in a dyebeck; by going over to high temperature the dye cycle could be reduced to 2–3 hr. However, the wool fiber suffers a considerable higher damage than even at normal boiling temperature as shown by experiments reported by Bauer (5), Millson (40b),and von Bergen (61). Millson in his microscopical studies with the high temperature (pressurized) microdyeoscope found that wool swells and decomposes in aqueous media at elevated temperatures. When wool fibers were dyed blue with a premetallized dye and then immersed in distilled water, at elevated temperatures it was observed that the shade changed from blue to a brownish color as the treating time was increased. After a period of time at 290°F colorless sacs or blisters formed which appeared to be independent of the scale structure. As the time of the treatment increased the blisters increased in number and also in size and the fiber became more plastic. Figure 7 shows (*a*) the fiber at the start, (*b*) after an hour and a half, (*c*) after 2 hr, and (*d*) after 2 hr and 20 min at 290°F, respectively.

To reduce the damage to the wool fiber, the addition of certain cross linking chemicals such as formaldehyde and other protective agents such as glucose and albumen hydrolysate have found good use. Bauer (5) found that there occurs a substantial loss of wool due to protein hydrolysis especially in strong acid and high temperature dyeings. The use of formaldehyde in appropriate strengths (about 0.5–5%) offers a practical way of minimizing weight loss while maintaining improved fiber properties. High temperature dyeings on wool in the pH region 4–6 in the presence of formaldehyde makes it possible to apply 1:1 premetallized dyes, which otherwise would require strong sulfuric acid.

Even though high temperature dyeing of wool has certain advantages such as more rapid rate of penetration of dye into the fiber and more uniform distribution of color from fiber to fiber and a

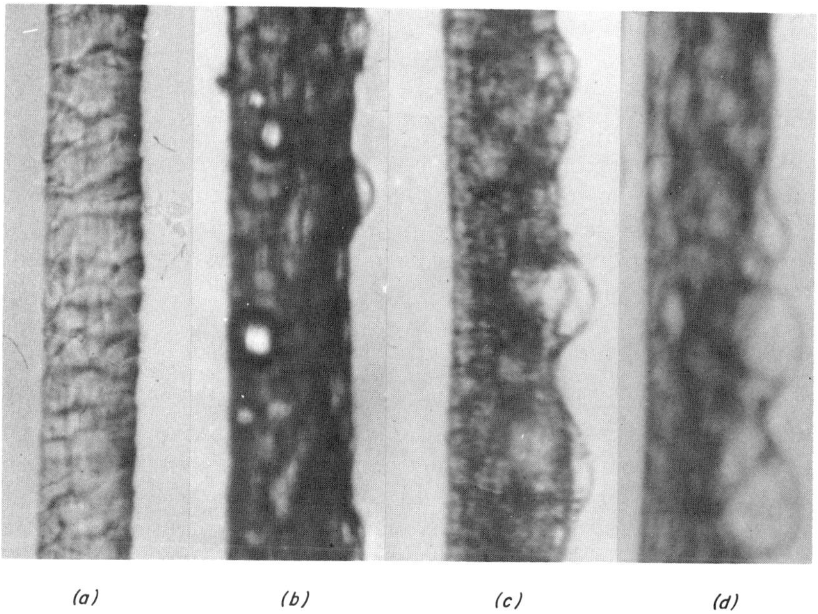

Fig. 7. Changes in wool fiber structure during high temperature dyeing at 290°F (150°C). (*a*) Fiber at start; (*b*) after 1½ hr; (*c*) after 2 hrs; (*d*) after 2 hr 20 min. (Courtesy of American Cyanamid Company.)

reduction in dyeing time, the high damage factor outweighs the advantages. With the successful application of wool dyes below the boiling point, the dyeing of wool above the boiling temperature should be limited to blends with man-made fibers.

Ultrasonic Dyeing

Many experiments have been run in an effort to use ultrasonics for dyeing wool and other textiles. To date none have proved practical.

Wool dyed with acid dyes using an apparatus consisting of an ultrasonic generator and transducer showed local areas of deeper color than was obtained in a control dyeing using heat instead of ultrasonics. Microscopical studies on longitudinal sections where local areas were deeply dyed showed that there was extreme scale damage and therefore dye penetrated quite readily in the damaged areas.

Further studies using ultrasonics may make possible the development of improved methods of dyeing with this form of energy.

Pad Dyeing Methods

The forerunner of pad dyeing of wool is Vigoureux printing of wool slubbing or top. It was therefore logical that the two continuous pad dye methods developed recently were first applied to the dyeing of worsted top. The two methods are: The Cibaphasol technique and the Irga Pad process.

Cibaphasol technique

The Cibaphasol method as described by Casty (11) is based on a new two-phase system belonging to a special class known as the coacervation systems. The difference between an aqueous coacervation system and other aqueous/liquid two-phase systems, such as the water/organic solvent system, is that both phases consist mainly of water. The term coacervation is used in colloid chemistry to denote the separation of a colloidal system into two liquid phases. The concentrated "oil" phase is known as the coacervate and the more diluted "aqueous" phase as the equilibrium liquid. The two phases are solutions in the same solvent, i.e. water. In order to bring about coacervation a special surface-active agent (e.g. electrolyte) has to be used. The product of Ciba which has the capacity of coacervation is a fatty acid derivative named Cibaphasol AS (acid stable).

The outstanding property of the system is the fact that the coacervate has an extremely high dissolving power for dyestuff. Therefore it lends itself excellent for the preparation of padding dye liquors for continuous dyeing of wool, especially in loose or top form.

In order to stabilize the coacervate emulsion it is necessary to add a thickening agent. Estrified locust-bean gums have proved especially suitable for the purpose. In the preparation of the padding liquor for wool, it has been found that, as a rule 10 g/liter Cibaphasol AS is required to produce the desired "aqueous two-phase system" for recipes containing up to 1½% dyestuff in the presence of acid and thickening agent. A normal recipe calls for:

up to 1½% dyestuff
10 g/liter Cibaphasol AS

10 g/liter Acetic Acid (glacial)
Poly Gum 260

In the padding the wool picks up 80% of its weight of the dyepaste or in other words, per 100 lb of wool 80 lb of padding liquor.

Steaming, the second main step in the process, is carried out immediately after impregnation. The steaming time necessary to fix the color runs from 20 to 35 min depending on the depth of shade, using moist steam at a working pressure from ¾ to 1½ lb with a temperature of 102/103°C. The basis of the Cibaphasol process is the diffusion of the dye from the wet surface film during steaming, which brings about even dyeing of the fiber right into the cortex, also along the fiber which eliminates tippyness. Washing-off is the third and last step. In this operation the unfixed dye, gum, and the Cibaphasol AS have to be washed off. If the wool has been adequately steamed, a thorough rinse is all that is necessary, the Cibaphasol AS itself acting as a powerful detergent. An acid rinse is recommended as the last wet treatment.

According to Ciba's experience, nearly all water-soluble wool dyes are suitable for the technique. The most used today are the neutral 2:1 metallized and the milling dyes for carpet wools and slubbings. These dyes generally exhaust 98%. The amount of dye used is exactly the same as in the traditional dyeing procedures; all that is necessary is to convert the percentages into the quantities required for padding.

After chrome dyes, with exception of black, still produce some obstacles for a one-step heat fixation method.

In the dyeing of wool-cellulosic union no difficulites are encountered, particularly when cotton dyes are used which reserve the wool. Also wool polyamid and wool polyester blends can be dyed successfully with the Cibaphasol process.

Irga pad process

This process developed by Geigy is a true water solution system especially suitable for continuous dyeing of loose stock and top. The auxiliary products forming the chemical basis of the method are:

Irgapadol P—for wool, polyamides and acid dyeable acrylics
Irgapadol A—for basic dyeable acrylics

The Irgapadols serve as solvents for the dye, an excellent wetting agent, have good stability in acid media and assist dyestuff fixation within a short steaming time, between 20–30 min.

C.S.I.R.O. methods

Two pad dyeing methods for wool top have been developed in Australia; namely, the Formic Acid Method and the Urea Process. Milligan (391) reported on the Formic Acid Method, which is covered by Australian Patent No. 23540, (1962) granted to Hallack (21). With this process it is possible to dye wool at room temperature in a matter of seconds. This is achieved by using a padding liquor in which the dye is dissolved in formic acid instead of water.

Although this process is technically successful, formic acid is toxic and corrosive, and must be recovered because of its cost; therefore, it is not a commercially suitable process.

Further research at the C.S.I.R.O., Division of Textile Industry, led to the discovery that concentrated solutions of urea (35%) used as a vehicle for the dye are as successful as formic acid solution in accelerating wool dyeing. The details of the urea process for continuous dyeing of wool shoes was given by Angliss, Brady, Delmenico, and Hine (3).

Essentially, the sequence of operation is to pad the wool in sliver form with the dye liquor, convey it into the steaming chamber at approximately 100°C and fold it onto a stainless steel conveyor which takes up to 10 min to travel the length of the steamer. After leaving the steamer, the dyed sliver is washed in a 3-bowl backwasher and then dried.

So far milling and 1:2 metal-complex dyes appear to offer the greatest promise. A wetting agent such as Aerosol OT is essential and in some cases 2% of acetic acid (60%) is required.

Altering the Affinity of the Wool Fiber for Dyes

Multicolored woolen and worsted goods as a rule are produced either from dyed stocks, tops, or yarns. The manufacturer generally weaves them in hundreds of different designs, which compels him to keep a large proportion of his capital tied up in stocks of yarns for long periods. Making pieces from undyed wool and dyeing them into plain shades is cheaper and involves a much lower risk because of certain changes in fashion. Nevertheless, plain piece-dyed cloth satisfies the public only to a limited extent in spite of innumerable changes and variations made in the design of the cloth. Multi-colored effects in piece dyeing might be introduced by the use of colored or white silk, cotton, or man-made fibers that are not dyed by the piece-dyeing colors, but the material is no longer all wool. To introduce color and novelty effects in piece-dyed woolens, manufacturers have put in their cloth woolen knops, kemps, natural brown and black wools, fur fibers, goose and ostrich feathers, cellophane, and other substances. An excellent and popular method of increasing the variety of woolens is to weave in colored yarns with undyed yarns and then to cross-dye them in the piece. The dyes used for dyeing these colored effects must withstand the cross-dyeing without appreciable bleeding into the other fibers. The colored yarns are generally dyed with suitable chrome browns, blues, or blacks. To avoid bleeding the dyeing is done below the boiling point and in the presence of organic acids. This cross-dyeing is usually limited to pale shades.

Decreasing the affinity of wool for dyes

For many years, wool manufacturers have searched for a process by which the wool fiber could be treated to such an extent that it would resist the absorption of the normal wool dyes as used in piece dyeing. Such a process would make it possible to obtain in a cheap and simple way a large variety of multicolored effects on all woolen cloths. Considerable research has been devoted to this problem, but so far no process has been developed that would decrease the affinity of the wool fibers toward dyes to such an extent that the treated wool would stay pure white when dyed together with normal wool. Of the numerous methods, it is worth while to mention the following:

1. Acetylation. The process of acetylation of wool for the purpose of immunization was developed in 1920 by Cassella (DRP 380, 994 and 384, 103). The loose wool or yarn is treated with a mixture of acetic anhydride and glacial acetic acid in the presence of sulfuric acid as a catalyst. The resist effect can be improved through the addition of a small amount of a tertiary base such as pyridin or dimethylanilin. Niederhauser (43) gives the following recipe:

1 kg wool is treated for 4–5 hr in 1 liter acetic anhydride, 70 ml concentrated sulfuric acid and 80 ml dimethylanilin at 58°C

This process is up to now the best resist method known but necessitates special machinery and is quite expensive. It is successfully used by a French concern for immunization of rabbit's hair, kemps and wool neps, not only for white but also bright colors cross dye effects.

2. *Sulfamin acid process.* This process developed by Sandos (DBP 933,022 BP 723,690, F.P. 1,059,340) is claimed by Frieser (19) to be the most widely used process in Europe. The wool is treated in water solutions of sulfamin acid or sulfamin derivatives as amino disulfonacid, respectively, the salts of them to reduce reaction temperature. It was found that the addition of wool swelling agents such as formic acid, urea, thiourea, phenoles, foramid or acetamid helps. Also the addition of isooktylphenylpolyglycol ether has a beneficial effect. The following recipe is quoted:

20% sulfamin acid

12% urea

0.2% isooktylphenylpolyglycol ether treated at 100°C, extracted and dried, followed by a 5 min exposure at 150°C

Wools and hairs in forms of loose wool, wool neps, tops and yarns immunized by this process are marketed by Sandos under the name of "Resilan." As strong acids impair the treatment piece dyeing should be done with acetic or formic acid. It is recommended that carbonization should be carried out in the piece after dyeing. Alkali treatment should be done below pH 9.

3. *Glyoxylation.* Speakman and Elliott (55) stated that the use of glyoxal produces a resist that is only slightly inferior to acetylation. They gave the following recipe:

Treat 1 lb wool for one hour at 185°F in 6½ ounces glyoxal, 12½ oz sodium bisulfite, in 4¾ gal water, with ¼ gal acetic acid.

4. *Wool resist with condensation products of naphthalene sulfonic acid with formaldehyde.* Various processes have been developed using formaldehyde combined with different salts or synthetic tanning materials made from naphthalene sulfonic acids with formaldehyde. Even better results are obtained by the formation of the tanning material in the presence of the wool fiber. The best results are produced by the formaldehyde and R salt process. The procedure is as follows:

100 lb wool are boiled for 30 min in 100 gal of water containing 50–100 lb formaldehyde (40%—the larger amount gives a better

resist, i.e., 43:1 compared with 30:1), 2–3 lb R salt, and ½ gal sulfuric acid.

The wool is then centrifuged, dried, and baked for 15 min at 225°F. The effectiveness of the resist is estimated by comparing the depth of shade on the normal wool to that on the resist-treated wool, e.g., 3–0.1% equals a resist ratio of 30:1. The wool is discolored but as Frieser (19) found shows excellent immunization against acid dyes.

Other accounts have been given in articles in *Wool Science Review* (68) to obtain on an all-wool material tone in tone or multicolored effects by combining "resisted" wool with untreated material.

A more comprehensive list of resist treatments for the various textile fibers is given by Frieser (19).

Increasing the affinity of wool for dyes

It is obvious that, tone-in-tone and multicolour effects can also be produced by modifying the fiber so that it will absorb more dye than untreated wool when both are dyed together in the same dye-bath, and in practice chlorination has been found to afford one of the most convenient ways of achieving this end.

Treatments intended to increase the amount of dye absorbed by wool when treated and untreated wool mixtures are dyed depend upon the fulfilment of either one or both of the following conditions: First, the total dye combining capacity of the material may be increased by introducing into the wool additional dye-combining groups. Second, the rate of absorption of the dye may be enhanced by increasing the accessibility of the fiber to the dye. This presumably involves modifying appropriately the outer surface structure of the fiber, which presents the most important physical barrier to dye penetration.

1. Chlorination. Of the various methods in practical use chlorination has been found to afford one of the most convenient ways of achieving the above end. Barritt and Elsworth (4). The following is a summary account from Wool Science Review (68):

"Chlorination affects both the total dye combining capacity of the fiber and its accessibility to dyes. Which effect is the more significant depends upon the conditions of chlorination and the type of dye used. Thus, commercial chlorination, which usually is based either on treatment of the wool in dilute acid solutions of hypochlorites or with gaseous chlorine, has been shown to affect primarily the accessibility of fiber to dyes, certainly as far as anionic acid, chrome and premetallized dyes are concerned. Even though drastic treatments with chlorine may reduce the total acid dye combining capacity of the fiber, the chlorinated constituent will nevertheless, under commercial cross dyeing conditions, still dye deeper than the untreated, simply because the amount of dye normally present in the bath is always much less than would be required to saturate both the chlorinated and the unchlorinated components of the mixture.

On the other hand, if chlorination is carried out using hypochlorite solutions under neutral or alkaline conditions, wool so treated is dyed less deeply than untreated wool when a mixture of the two is dyed with many acid premetallized and chrome dyes.

The molecular structure of dyes is important because the difference in the rate and extent of absorption of dye by the two types of material is a function of the dyestuff as well as the dyeing conditions, such as time, temperature, etc. This effect is perhaps best illustrated by the three acid dyes, Naphthalene Red J, Red EA and Fast Scarlet 4R. All are simple mono-azo dyes and differ only in the number of sulfonic groups in the molecule thus:

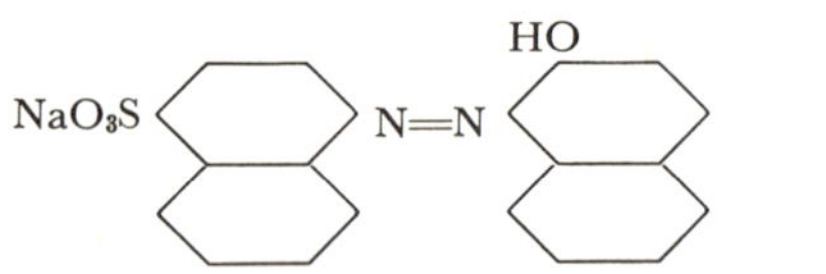

Naphthalene Red J (ICI)
C. I. Acid Red 88

This dyestuff is absorbed to virtually the same extent by both chlorinated and unchlorinated wool.

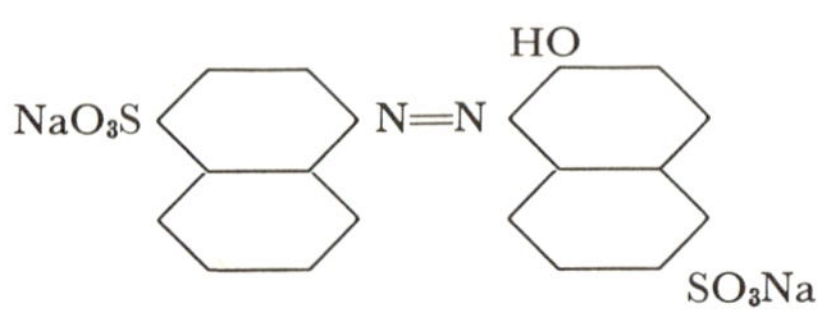

Naphthalene Red EA (ICI)
C. I. Acid Red 13

This one is absorbed to a greater extent by chlorinated wool.

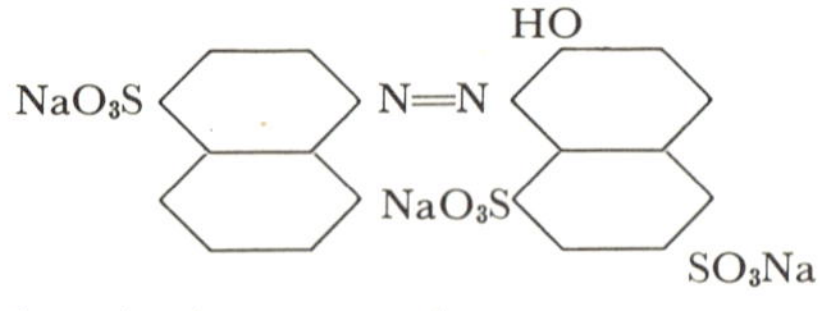

Naphthalene Fast Scarlet 4R
C. I. Acid Red 18

This is absorbed to a much greater extent by chlorinated wool.

The same effect is noted with dyes from other dyeing classes. Thus polysulfonated chrome dyes give excellent contrast between chlorinated and ordinary wool but dyes which contain neither sulfonic nor carboxylic acid groups in their molecule, dye both types of material to the same depth. Among the direct dyes, it is found that those with the minimum number of acidic solubilizing groups give solid dyeings on wool/chlorinated wool unions. Polysulfonated direct dyes, on the other hand, which are unsuitable for dyeing ordinary wool, will dye chlorinated wool and give good contrast on wool/chlorinated wool mixtures.

Secondly, the rate and extent of absorption of dye by the two types of material is markedly affected by the dyeing conditions employed. At low temperatures, all dyes are absorbed more rapidly by chlorinated wool than by normal wool. This difference is particularly noticeable with the colloidally aggregated types of dye, the particles of which find difficulty in penetrating the fiber structure at low temperatures and require to be broken down into single molecules or something approaching this state before they can enter the fiber.

At high temperatures, the contrast initially produced may be reduced owing to some dye initially absorbed by the chlorinated wool diffusing out again into the solution and being reabsorbed by the untreated wool from which it subsequently diffuses less readily. This effect is most likely to occur with the level dyeing acid dyes which have high migrating power. It is much less likely to occur with the neutral dyeing acid dyes or with the chrome dyes, especially when the latter are applied by a single bath method. In this instance, a high degree of permanence of contrast is achieved since the dye-chromium complex formed inside the fiber is virtually insoluble and nondiffusing. Thus Solochrome Brilliant Blue B, dyed by a single bath method (which would not be considered suitable for normal wool) gives very good contrast on wool/chlorinated wool mixtures.

With the 1:1 dye-metal complex type of dye (Neolans, Palatins and Ultralans) applied by the normal method, namely, from a strongly acid dye bath, little contrast is obtained. If, however, these dyes are applied from very weakly acid dyebaths, e.g. at pH 5 using 1% of acetic acid and 10% of sodium acetate, a considerable contrast is obtained.

The neutral dyeing 2:1 dye-metal complex premetallized dyes give solid shades on wool/chlorinated wool mixtures.

Since the main effect of chlorination is so to change the fiber structure that dyes can diffuse more rapidly into the fiber, it is to be expected that rediffusion into the surrounding solution will also be facilitated. Consequently dyeings on chlorinated wool may be expected to exhibit lower fastness to wet treatments than those of the same dyes on untreated wool. This effect is particularly noticeable with the level dyeing acid dyes. With the fast-to-milling and chrome types the reduction in washing fastness is negligible."

Townend (60) has shown that the degree of contrast may be accentuated by the addition to the dye bath of retarding compounds, for example, the complex phenol-sulfur condensates. The retarding compound is absorbed by both the normal and chlorinated wools, but it slows down the absorption rate of certain dyes more with normal wool. In this manner, material may be piece-dyed, not only into tone-in-tone effects but also into very attractive two-color effects.

By selecting suitable combinations of contrast-producing and level-dyeing dyes, Townend produced a series of multicolored dyeings as summarized in Table 7.

Emphasis has been laid on the importance of strict control of dyeing conditions in order that maximum contrast may be initially obtained and then subsequently retained by reducing to a minimum the migration of dye already taken up by the chlorinated wool. Wet treatments carried out prior to dyeing should also be carefully controlled. Thus, the dye absorption of unchlorinated wool is much more sensitive to variations in scouring than is that of chlorinated wool. Treatment of ordinary wool in alkaline solutions promotes absorption of dye and the contrast obtainable on mixture fabrics is thereby reduced. Scouring should, therefore, be as mild and as reproducible as possible. Even very dilute alkaline solutions have the same effect, if used hot, and thus such processes as crabbing may lead to reduced contrast. Carbonizing before dyeing, on the other hand, is stated to have only a slight effect on the degree of contrast subsequently obtained.

2. Mordanting. By premordanting the wool fibers with various chrome salts, such as sodium bichromate U.S. Pat. 3,310,361 (1967) and chromium fluoride Ger. Pat. 530,394 (1931), and then

TABLE 7
Typical Contrast Effect Recipes Using 5%. Taninol WR Added to All the Dyebaths

Shades		Components	
Chlorinated wool	Normal wool	Contrast producing	Level dyeing
Scarlet	Gold	1.0% Naphthalene Scarlet 4RS	0.25% Metanil Yellow YKS
Ruby	Steel	1.0% Naphthalene Scarlet 4RS	0.25% Solway Ultra Blue BS
Grey	Gold	0.04% Lissamine Blue BFS	0.05% Metanil Yellow YKS 0.08% Solway Rubinol RS
Blue-green	Grey	0.1% Lissamine Blue[a] BFS 0.2% Lissamine Green SFS	0.01 % Metanil Yellow YKS 0.10% Solway Rubinol RS 0.05% Solway Blue 2GS
Nigger	Green	0.75% Naphthalene Scarlet 4RS 0.15% Naphthalene Blue Black CS	1.0% Quinoline Yellow AS 1.0% Solway Blue 2GS

mixing such fibers in yarn forms, tone-in-tone effects can be produced by selecting the proper chrome and acid colors. This method is limited by the dyes that can be used. The affinity can also be increased by the use of potassium sulfocyanide (mentioned in Ger. Pat. 231,885). For completeness, it may be noted that salts such as thiosulfate, sodium sulfite, borax, or caustic soda will lead to increased affinity.

Bicolored Tippy Dyeing

A third way of producing tone-in-tone or two-tone effect in 100 percent wool materials is making use of the tippy dyeing properties of the natural wool itself. As shown by von Bergen (62), due to the action of sunlight the wool on the sheep's back during its growth undergoes a photochemical decomposition in the tip region of the wool staples which is responsible for the change in the affinity of the tip portion for wool dyes (Vol. 1, pages 257–261). By proper selection of dyestuff from the list given in the paper "Tippy Dyeing of Wool and Its Control," (Fig. 5) heathers and other bi colored shades may be produced with comparatively good reproducibility. For example by combining Palatin fast pink BNA and Alizarin Astrol B, the tips are dyed red and the body section blue. Millson (40) found the effect can be increased by the presence of metals.

With the introduction of the reactive dyes, a class of dyes which have a generally very high affinity for the tip portion, any attractive tone-in-tone and two-tone effects can be produced by the one bath method. Harding was granted a patent (U.S. Pat. 2,999,731 (1967)) for his observation that contrast between the tip and root portion can be increased by addition of compounds consisting of condensation products of aldehydes and aromatic sulfonic acids, a water-soluble organic alphatic acid to dyestuffs which contain at least two sulfuric acid groups. Products available on the market are for example Dispergator CC of Ciba and Tinegol BAN of Geigy.

Ciba, in their color card, gives the following instruction:

Colored and white or off-white effects are produced with Cibracon dyes alone, and it is possible to imitate Vigoureux printing. Two-color effects are produced by applying level-dyeing acid dyes, which are fast to light and washing, along with Cibacrons.

Method 1 (*with Univadine W*). The dyebath is set at 60–80°C (140–175°F) with the Cibacron dye

1–2% Disperser CC*
5–8% acetic acid 40%.

The bath is brought to the boil within 20 minutes and maintained at the boil until exhausted of Cibacron dye, formic acid being added, if necessary. The steam is then shut off and a dilute solution of

3% Univadine W added, followed by the acid dyes.

(Please note that the Univadine W should not be added until *after* the bath has been exhausted of Cibacrons.) Boiling is then continued for 30 min. Topping with acid dyes can be carried out without difficulty when Univadine W is present in the liquor, which naturally facilitates dyeing to shade. The goods are rinsed when dyeing is completed.

Method 2 (without Univadine W). The dyebath is set with the acid dyes

1–2% Disperser CC*
10% ammonium sulfate or
2% acetic acid 40%.

The bath is brought to the boil within 45 minutes and maintained at the boil until exhausted of dye. The bath is then cooled down to 60–70°C (140–160°F) by dropping some of the liquor and topping up with fresh water, whereupon the Cibacron dyes and

5–8% acetic acid 40%

are added. The bath is then brought to the boil within 20 min and maintained at the boil until exhausted of Cibacron dye, formic acid being added, if necessary. Any topping which may be necessary should be carried out with Cibacrons only. The goods are then rinsed.

Method 3 (Cibracron dyes only). The dyebath is set at 60–80°C (140–175°F) with the Cibacron dyes,

1–2% Disperser CC and
5–8% acetic acid 40% **or**
1–2% formic acid 85%.

The bath is brought to the boil within 20 min and maintained at the boil until practically exhausted of Cibacron dye. The goods are then rinsed.

Dyeing Wool Mixtures

Of the various fibers that the dyer is called upon to bring to greater perfection, sheep's wool plays by far the largest part. A less extensive use is made of the fleeces of other animals, the most important of which are (*1*) mohair, obtained from the Angora goat; (*2*) cashmere, derived from a species of goat living in central Asia; (*3*) camel's hair, (*4*) alpaca, vicuña, and other hairs of the llama family; (*5*) cow, calf, and horse hair, used occasionally for certain goods, and (*6*) rabbit hair, employed in the manufacture of hats, and also in women's wear (Volume 1, page 445).

Although the chemical character of the various kinds of wool and hair is pretty much alike, there are nevertheless slight differences in their behavior in dyeing, so that the affinity of the dyes for the various fibers differs more or less. As a general rule, the finer the fiber, the more dye it will absorb in producing a given shade, since under equal conditions *the finer fiber, in presenting a larger surface in proportion to the bulk, takes up more dye than a coarser fiber*. Hence, it follows that it is advisable to use materials of as uniform a fiber as possible, especially for goods to be piece-dyed, and also that difficulties will arise in employing wools from different kinds of animals simultaneously, unless it be intended to produce a contrast between the various fibers.

Generally speaking, the process of dyeing sheep's wool applies to all other animal fibers, apart from slight variations which their structure necessitates for obtaining greater or smaller affinity for the dyes. Therefore, it is not necessary to discuss the mode of dyeing of these special fibers, since their treatment can be easily derived from what is stated about general dyeing of wool.

Wool and silk mixtures

Silk is ordinarily used in wool and worsted yarns and fabrics to obtain stripes, as in men's wear fabrics, or to create light-weight fabrics such as silk cashmere. Silk noils and tussah silks are used occasionally in woolen fabrics. Such mixtures occur most frequently in woven or knitted piece goods. Silk is an animal fiber and therefor closely related to wool. Hence, most of the wool colors such as acid and mordant colors are also utilized in the dyeing of silk.

The dyeing procedure varies according to whether solid or two-color effects are wanted, in other words, either to dye the silk the same shade as the wool, or to have the silk undyed or cross-dyed, if the silk is already colored.

1. Dyeing silk and wool the same shade. The best results are obtained by dyeing with acid colors in one bath. The general procedure is as follows: The dye bath is set with 5% Glauber's salt (calcined) and 4–6% sulfuric acid (66° Bé). The goods are entered at 100°F and raised to the boil in about half an hour and dyed to shade just below the boil. By severe boiling with reduced quantities of acid, the wool dyes deeper than the silk. At temperatures between 180–200°F and with a higher quantity of acid, the silk absorbs more dye. For deep shades, like dark browns and marine blues, where silk is difficult to match with wool, it becomes advantageous to use a strong acetic acid bath (10–15%), the bath is started with half of the necessary quantity of dye and after exhaustion the remaining dye is added during cooling to bring the silk to the same shade as the wool.

2. *Dyeing silk and wool, leaving silk white.* To accomplish this, dyes have to be selected that do not or only slightly stain the silk. The normal procedure is to dye the wool first with acid or chrome colors as described before. The dyeing must be done at a boil. Should the silk be slightly stained, it is cleared up in a fresh bath consisting of 1–2 lb hydrosulfite per 100 gal liquor at about 90°F. Even better results are obtained with the addition of Setamol W.S. of General Dyestuff Corporation, after the boiling has proceeded for $^1/_2$ to 1 hr. Depending on the amount of silk staining, 2–10% of Setamol is recommended. An additional boil of at least $^1/_2$ hr is necessary. The best effect is obtained by the use of acetic or formic acid at a high temperature. It is practical to dye wool fabrics with white silk effects intended for dress goods with acid colors, whereas for men's wear fabrics chrome colors are chiefly employed.

Wool and vegetable fiber mixtures

Most of the fabrics in this group constitute the "unions," containing wool and cotton in various proportions already mixed in the raw stock, twisted-in yarns, cotton warp and wool filling, or vice versa. The preparatory processes of the various types of union before dyeing vary according to their makeup. They are carried out in a way similar to all wool goods.

Cotton being a vegetable fiber, in most instances requires an entirely different class of dye, as well as a different dyeing procedure. Similar to wool and silk mixtures, the "unions" may be dyed in *solid* shades, dyeing the wool and cotton simultaneously. The second condition is to dye the wool and leave the cotton white and finally to dye both wool and cotton to different shades.

Dyeing wool and cotton the same shade. The most common method is based on the fact that certain groups of dyes have an affinity for wool and cotton from a boiling Glauber's salt bath. However, the affinity of the individual dye varies widely, some dyeing wool and cotton uniformly, others dyeing either the wool or the cotton more deeply. Consequently, the dye manufacturers divide their colors into these various groups. As a rule the wool dyes deeper at the boil, whereas the cotton dyes better at 180–200°F. By regulating the temperature, it becomes possible to force the dye either onto the wool or onto the cotton at will.

Wool and man made fiber mixtures

The dyeing of such blends has become a major operation. The numerous methods are well covered in the literatures supplied by the Dyestuff companies. One of the outstanding books, covering this field is by R. C. Cheetham, *Dyeing Fiber Blends*, which covers the processing of blends, unions, and mixtures containing natural or man-made fibers (D. van Nostrand Co. Ltd., London, 1966).

Wool Dyeing Machinery

In order that goods may be dyed evenly, the dye should be deposited uniformly at a controlled rate throughout the material being dyed. This necessitates a movement of either the dye liquor or the material to be dyed. All dyeing equipment therefore, must provide for the circulation of the dye liquor through a mass of wool either in the form of loose stock, top yarn, or piece; or for the transporting of the woolen material through the dye liquor. The dyeing machine must also be equipped with a means of heating the dye liquor because temperature plays an important part in wool dyeing. In some equipment the bath is heated directly by means of a steam jacket, enclosed steam coil, hot water system or by passing live steam into it. In the latter case the dye bath tends to increase in volume due to the condensation of steam to water. In other cases the liquor is heated externally and circulated through the material by means of a pump or propeller.

The water used in wool dyeing must be of the same high purity as that used in wool scouring. For the chrome-dyeing process water of zero hardness is preferred because some of the chrome dyes are quite sensitive to calcium and magnesium salts. Many chromable dyes are also affected by the presence of metal salts, and especially those of iron since such impurity metal ions can cause radical changes in shade and have an adverse effect upon the fastness properties.

Hardness is less detrimental when applying acid dyes from a strongly acid dye bath. The presence of iron salts may, however, dull the shades of most dyes.

Wool is dyed in all stages of manufacture. After scouring as loose stock; after combing as top, or after spinning as yarn in form of skeins, or packages; and after weaving or knitting, in form of pieces or cuts. Dyeing machines have been designed to produce the best results for each form in which wool is dyed. In years past, wool was dyed in raw stock form in open kettles where men poled the wool to produce even dyeings. Wool skeins were dyed on sticks in hand kettles and turned by hand thrusting a stick through the yarn and then lifting it over a stick.

Present-day wool dyeing machines have been designed to protect the wool fiber, prevent matting or felting and to preserve its valuable properties and soft hand. The felting of loose wool is prevented, which likewise guarantees less fiber breakage during the subsequent carding, gilling and spinning operations.

No one type of apparatus can be used for wool in all the different forms in which it is dyed. Package dyeing machines are generally designed to handle raw stock. Husson-type machines can be used to dye wool in raw stock, skeins on stick or for pack-dyeing in which skeins are placed and dyed in the kettle without the use of sticks.

More recently wool raw stock has been dyed continuously using modified scouring trains for the purpose.

Compressing wool yarns is also injurious and should be avoided; for this reason the dyeing of hand knitting yarns where softness and bulkiness is of paramount importance is carried out in a Hussong-type of Klauder-Weldon machine. Wound packages, of tops and yarns, may be compressed in the direction of their axis without impairing the wool fiber.

Construction of dyeing machines

The metal of which dyeing machines are constructed plays a very important part in the results obtained. Certain metals such as iron, copper, brass, certain bronzes and plain aluminum should be avoided when building machines for dyeing wool. All of these metals release undesirable metal ions into the bath at a rate often determined by the pH and temperature of the dye bath. Phosphor-bronze pumps and propellers have proven satisfactory because of their resistance to acid dye baths.

The 18-8 type of stainless steel also releases sufficient nickel and iron ions to change the shade of certain chromable dyes. The molybdenum fortified stainless steels 316 SS contains:

16–18 parts % of chromium
10–14 parts % of nickel
2–3 parts % of molybdenum

are the best for wool dyeing apparatus and the slightly higher initial cost of the fortified stainless steels is more offset by the improvement in clarity of shade and better reproducibility from batch to batch. The 18-8 type of stainless steel linings may also pit badly in time and microscopical studies have shown that streaks and marks that look like resist shots can be caused by fabric rubbing over such pitted surfaces.

Large sums of money are spent in the designing and equipping of modern dyehouses with the finest stainless steel equipment available only to offset some of the advantages of such steels by installing iron pipes which carry water and steam to the machines. Many stains, streaks, dull and off-shade batches have been traced to rust and metal ions originating in iron pipes. Stainless steel piping would more than offset the additional cost by freedom from troublesome redyes.

Manually controlled steam valves have now been replaced by automatic timing and temperature control devices in modern dyeing machines, thus reducing labor costs, saving in steam and assuring better reproducibility. These controls also assist in preserving the quality of the dyed wool.

Loose-stock dyeing machinery

Loose stock is now dyed in two different types of machines: (*1*) the open type, and (*2*) the closed or pressure type. In both systems the stock remains stationary and the dye liquor is forced through the mass by means of a pump or by a propeller.

In the past, most of the dye vats and kettles were of the open type which meant that the dyehouse was often filled with a dense fog on rainy or cold days. Most of the modern wool stock and yarn dyeing machines are enclosed and this has meant the elimination of fogging with a resultant saving in steam and a reduction of collision-type accidents.

Open stock-dyeing machine. The wool to be dyed according to the open system is packed in one or several layers and is prevented from boiling over by a perforated cover placed (without exerting too much pressure) on top. The circulation of the liquor is brought about by pumps or propellers. Figures 8 and 9 illustrate the two systems. The use of a pump or a propeller permits an alternation circulation of the dye liquor. This reverse action is necessary in the open-type kettle to free the wool from air. Air bubbles remaining in the material cause undyed spots and, in most instances, by proper alternating of the circulation the white dots may be avoided. The alternating circulation is beneficial only for the wetting out. Once actual dyeing has started it is much better for the preservation of the strength of the wool to run the machine only one way, from the inside out.

When working with apparatus with the steam coil fixed inside, it is advisable to shut off steam while the liquor is penetrating the material from below upward, since otherwise there would be danger of the lower layers becoming too deep a shade because of excessive heat and of the material suffering in consequence. The installation of a separate heating chamber for heating the liquor is

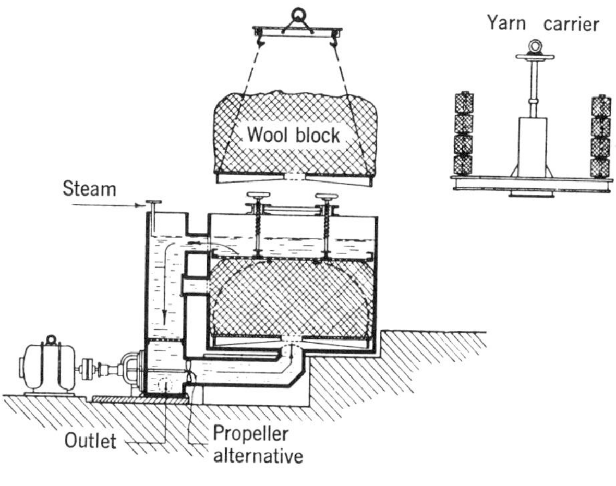

Fig. 8. Open stock and yarn dyeing machine.

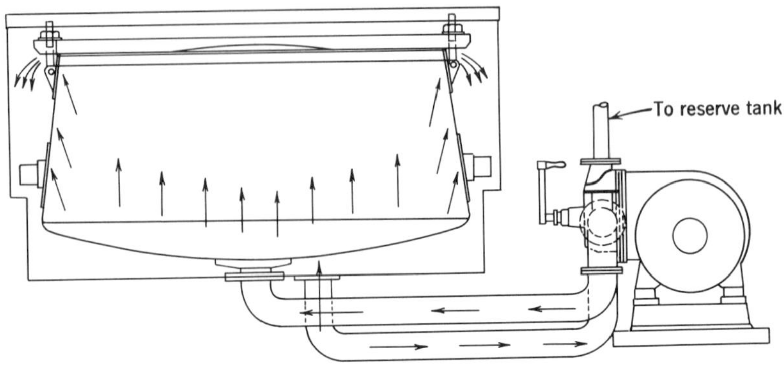

Fig. 9. Conial pan loose-stock dyeing machine. (Courtesy of The Longclose Engineering Co., Ltd., Leeds, England.)

for this reason more advantageous. Dyeing in one layer is generally preferred when dyeing loose wool, since larger quantities may thus be dyed in one lot.

A well-known representative of the open machine in the United States is the Hussong machine (Fig. 10). This machine consists of a rectangular vat with two sections; the smaller section on one end of the box carries the circulation equipment, consisting of a motor that drives two propellers situated in the lower part of the box. The wool is packed in a removable metal basket that rests on a protruding ledge inside the larger section in the machine.

The procedure is as follows: The cage is loaded away from the machine by manual labor. The loaded cage is then closed and, with the help of a hoist, transported into the dyeing chamber in which the dyeing liquor has been previously prepared. The motor is started and the propellers force the dye liquor through the material in the cage. The circulation is easily controlled by a switch for starting, stopping, and reversing. The liquor is heated by perforated steam pipe running along the bottom of the dye chamber. When an addition or a change of the bath is necessary, the cage can be elevated in two minutes. The unloading of the cage after completion of the dyeing is done very rapidly by lifting the cage out of the machine and running it on the overhead track away from the machine. The bottom of the cage is so constructed that it can be easily opened and the stock emptied either in trucks or into a pit directly on to a conveyor belt, which delivers it automatically to squeeze rollers and later onto a continuous drying machine (Fig. 10). The unloading time is less than 5 min after which the cage is available for a new lot. The machines are built in seven different sizes, from 20 lb up to 500 lb wool capacity. When used for rags, noils, or similar material the capacity increases.

Another machine of the open type is seen in (Fig. 8). The material is packed into a circular vat and rests on a perforated bottom. This bottom acts as a carrier and can be lifted out of the machine, after the dyeing process, with the help of the chains fastened to it. By replacing this loose wool carrier with a yarn carrier, the machine can be converted into a yarn-dyeing apparatus.

Pressure stock-dyeing machine. They are built by dye machine manufacturers throughout the world. Figure 11 shows diagrammatically the general construction and the liquor flow of the pressure machines. Most machines have the expansion tank separated from the main kettle, whereas in one instance the main kettle and expansion tank are one unit, eliminating thereby all outside pipes, valves, and fittings except steam, water, and drainage connections.

The expansion tank or reserve tank is used for feeding and dissolving the dyes and chemicals. In loading and unloading the same principles are used as in open machine. Depending on the type of carrier, the machines can be used for stock as well as yarn dyeing. The pressure range is from 10 to 50 lb/sq in. The machines are equipped either with pumps, or a propeller to circulate the liquor. The one-way circulation produces the best results in stock dyeing. The pressure machine preserves the material better than the open machine. Yarns spun from stock-dyed wools are generally from 15 to 30% weaker than the same yarns spun from white wool, depending on the depth of shade and the dyeing procedure.

The drawing in Fig. 9 shows a conical pan loose material dyeing machine. The special conical shape of the stock container ensures against "channeling" and prevents movement of the wool during dyeing. Probably the world's largest dyeing machine with a capacity of 9,000 lb of carpet yarn has been built by Longclose.

Vat-dyeing apparatus. For dyeing vat colors, the so-called circulation vats are used which are built on the same principle as the two systems just described. Contrary to other wool-dyeing machines,

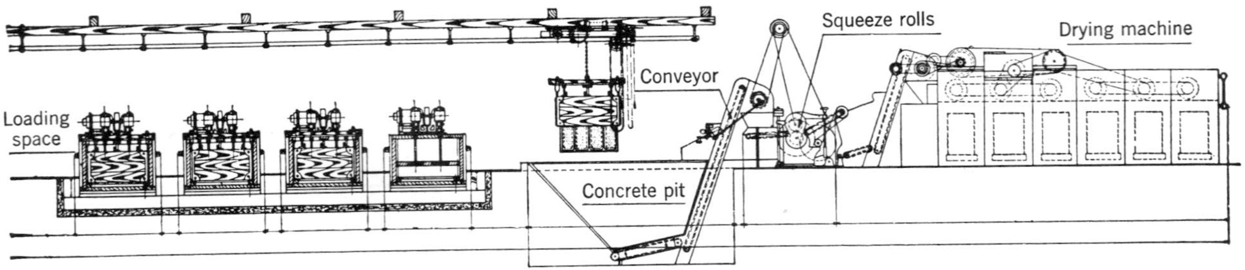

Fig. 10. Hussong stock dyeing installation including drying. (Courtesy of Hussong-Walker-Davis Co.)

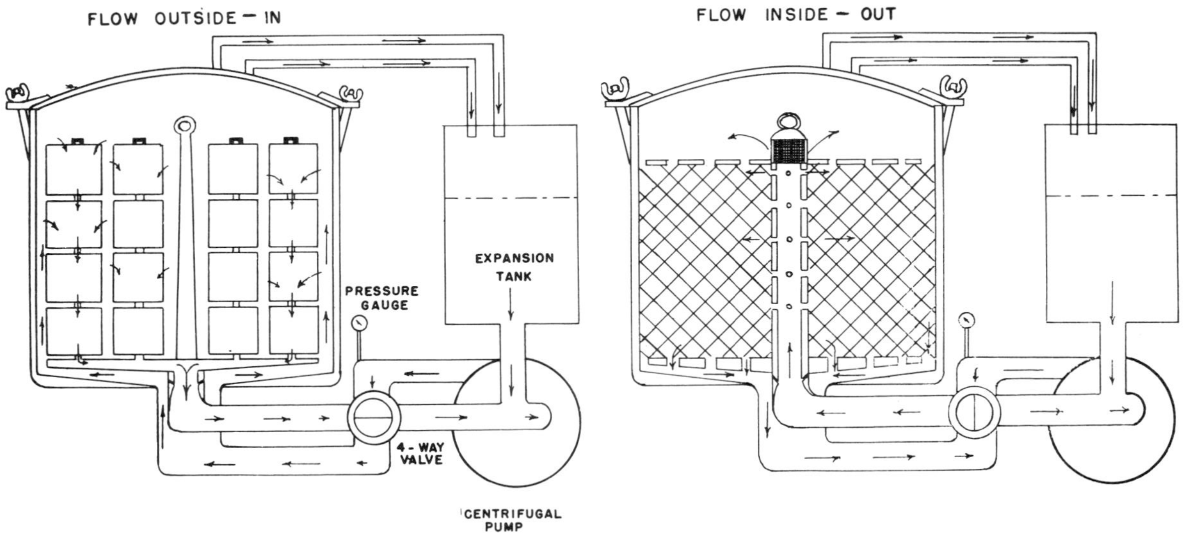

Fig. 11. Liquor circulation in a pressure-type stock-yarn dyeing machine indicating flow of dye liquor through loose-stock as well as yarn packages.

they are generally built of iron, as iron resists the weak alkaline vat liquors best and does not adversely affect the vat colors themselves.

These special vat-dyeing apparatus, like the open dye vats, are provided with an inner receptable to hold the wool. It may be closed with a perforated lid and lifted by means of a winch. Connected with the apparatus is a pair of strong squeeze rollers, which provide for very thorough extracting of the dyed wool. The circulation of the liquor is effected exclusively by pumps with reversible action. The bottom part of the inner receptacle fits tightly against the outer dye vessel; its sides are solid and only the bottom and the removable lid are perforated, thus forcing the liquor to circulate through the material either from above down or in the opposite direction.

Continuous processes. In the early 1920's the trend toward mechanization of the small batch machine was the light of progress. Today mechanization has reached the point where dye houses are entering the era of continuous processing in dyeing, a natural development in this age of streamlining. Probably the first fully continuous dyeing process for wool stock was developed during World War II. Its application was limited to indigo, which would mean low-temperature dyeing. The process was based on the passing of the loose wool through a dye train approximately 200 ft long. The machine (Fig. 12) was more or less a duplicate of a raw-wool scouring train and consisted of eight bowls, interconnected by squeeze rollers and mobile aprons in which the wool is transported through the bowl liquor at a fixed rate of speed by the propelling motion of harrow forks. It took the wool 45 min to travel from the first feeder of the machine to the outlet of the dryer. The delivery is regulated so as to assure even feeding and to maintain a production of 550 lb of dyed wool per hour.

There is an increasing interest in the continuous dyeing of wool in all its various forms (Dauble and Theiling (14)). The progress which has been made was due to the use of dye solvents, dyebath accelerants and surface active agents.

Carpet wool raw stock is dyed continuously in modified scouring trains with acid and with neutral dyeing premetalized dyes with varying success. The main problem in this kind of dyeing is to allow the wool to remain in the dye bath sufficiently long for

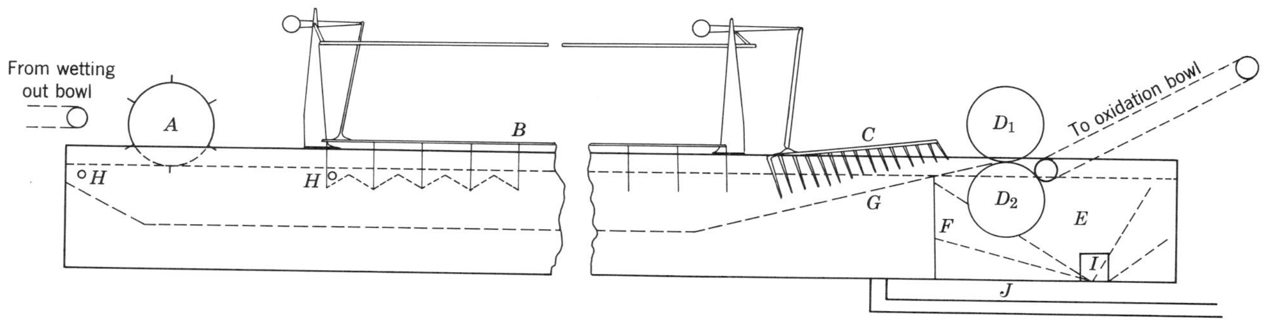

Fig. 12. Construction detail of the continuous indigo dye bowl.

Fig. 13. General view of Smith continuous loose-stock dyeing unit. (Courtesy of Jacques M. Schloss Associates, Agent for F. Smith & Co., Ltd., Rochdale, England.)

diffusion of the absorbed dye to take place. Merely staining the wool does not produce the maximum potential fastness properties of these dyes. One manufacturer scours, dyes and moth proofs carpet wools in a single five bowl train.

The most recent continuous loose stock dyeing unit is the Smith range as shown in Figs. 13 and 14. Loose stock is delivered to the unit by means of a conventional hopper feeder and is fed onto a rubber conveyer belt, which passes through the mangle nip. The dye liquor is applied before the wool enters the nip formed between the belt and the rubber covered padder bowl. After padding the material is conveyed into a direct coupled piston type tubular

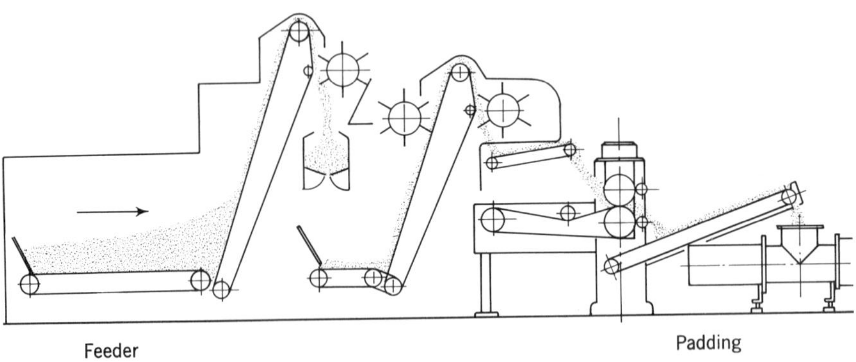

Fig. 14. Diagram of Smith loose-stock dyeing range.

compression steamer, where the stock is automatically compressed and passed through preheating, steaming and fixation zones. This machinery has been developed in collaboration with Ciba Clayton Ltd., using the Ciba phasol process for dyeing wool and blends containing wool by a simple pad-steam sequence.

The large unit, 25 ft tube length, has a production capacity of 1000–1200 lb/hr. The production is regulated by the capacity of the dryer. Typical recipe for pad liquor:

Up to 1½% dyestuff
10 g/c glacial acetic or 12 g 80% acetic acid
10 g/c Ciba phasol A A
Polygum 260

Pickup 80% equal to 0.8 lb dyepaste per 1 lb of wool. The piston works best between 2 and 3 strokes per minute pushing forward approximately 10 lb of wool per stroke. The packing density in the tube is 18–22 lb/ft^2.

The most suitable dyes are the neutral 2:1 metallized and the milling colors, which exhaust 98%. The steaming time runs between 20–35 min using moist steam 6 lb pressure resulting in a working pressure in the tube between ¾ and 1½ lb with the temperature around 102–103°C. It is important to keep the pH of the wool around 7 or slightly on the acid side.

The dyed wool leaving the tube is best delivered into a storage bin with a capacity between 1000–1500 lb from which it is then fed into a short scouring train—two bowls for wool and 3 bowls for acrilic fibers where the third bowl is used to add a spinning lubricant. The scouring trains are directly connected with a dryer, the most used is a Fleissner 4 drum unit, drying temperature may vary between (55–75°C). The change over to a new shade takes approximately 15 min involving mainly the cleaning of the padder

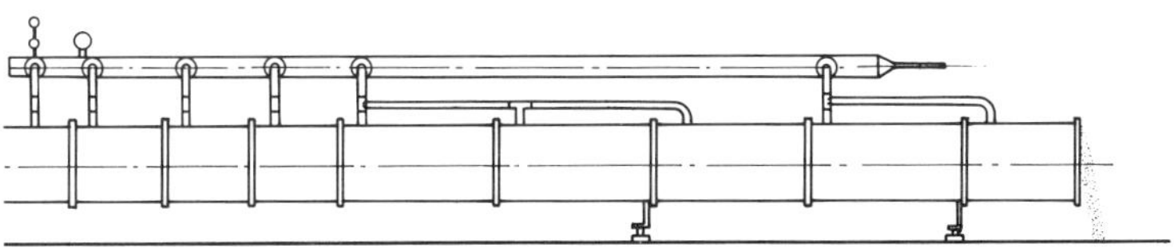

Steaming

not the tube. To separate the two lots, a circular pillow is inserted into the tube.

Matching is made much easier due to nontippyness of the wool. Most carpet mills in the United States which are using this type of a machine are equipped with spectrophotometers and digital computers to do the shade control.

Top or slubbing dyeing machinery

In the days following World War I probably the largest percentage of slubbing was still dyed (in the Klauder-Weldon machine) in the form of hanks mounted on revolving carriers that were fixed

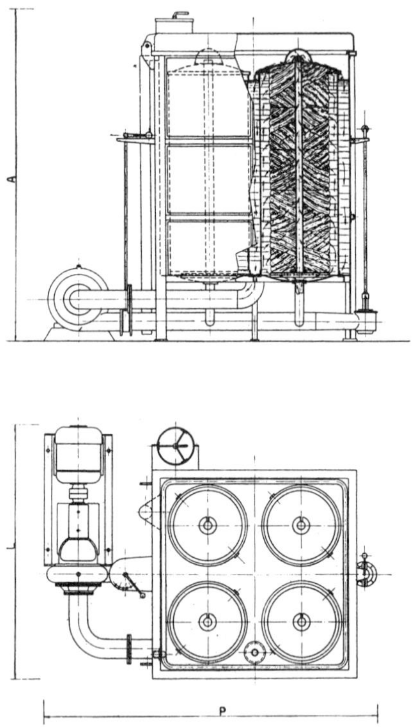

Fig. 15. Ilma Dyeing Machine for TOP. (Courtesy of Industria Lavorazioni Metalli Antiacidi Schio, Italy.)

on dye wheels. With half of the wheel in the dye liquor as it turned, the slubbing passed continually through the stationary dye liquor.

In this type of machine a certain amount of felting was unavoidable and this, coupled with the trouble of unwinding the tops and balling them again after dyeing, brought into existence numerous machines in which top was dyed in the form of balls. Various systems are in use. In one, several balls are placed over perforated pipes, called spindles, which are enclosed within a cylindrical tank, the lid of which is securely tightened. The lower ends of the spindles are inserted into a perforated bottom. The dye bath, made up in an auxiliary tank, is circulated by pumping it through the balls with enough pressure or suction to overcome the resistance of the wound top and to get a desired rate of flow. The

Fig. 16. Gru hydraulic device for packing loose-stock. (Courtesy of Edda Intern. Corp., New York.)

difficulties with this machine are in the loading, which, if not properly done, leads to uneven penetration (Fig. 15).

During the dyeing operation the tops are compressed and held in place by the conventional cap and bar arrangement. Unloading may be accomplished by means of a chain hoist fastened to the clamping bar, which in turn lifts all the tops in one operation with the aid of chains that are fastened to lifting plates under each column of tops. In order to increase the loading capacity various hydraulic and pneumatic devices are in use such as shown in Figs. 16 and 17.

A new way to top dyeing was introduced by the Abbott Machine Company in 1929 under the name of the Abbott dyeing process.

Fig. 17. Gru pneumatic device for packing TOPS into perforated cylinders on carrier. (Courtesy of Vald. Henriksen A/S, Denmark, represented by Edda International Corp., New York.)

The top-dyeing equipment comprises a specially designed gill box through which the top sliver runs (in an untwisted form) and from which it is then wound under tension on a perforated-metal dye spool. The dyeing is carried out on this perforated metal spool which is locked in place over an orifice in a specially designed dye kettle. The Abbott dyeing machine was the only one that was constructed of an aluminum alloy, which had proven quite suitable for regular chrome dyeings.

Pot or Can Dyeing Machinery

The pot or can dyeing system, which is considerably simpler, originated in Europe in the early 1900's but has been used in the United States on a large scale only since World War I. Figures 18 and 19 illustrate the Longclose versions of this type. The "English Type" takes two Bradford tops of 6½–8½ lb each per can, the "Continental Type" one larger top of about 15 lb (7 kg) per can. Each can has a suitable perforated lid and bottom through which the liquor enters and penetrates the material. The machines are built with one to forty cans and have a capacity of 10–600 lb. Enclosing covers are available to cut down steam consumption. In this type of apparatus the presence of metal is so considerable that only the best alloy is suitable.

The advantages of these machines are many: (*1*) quick loading and unloading: two men may load and unload a thirty-two can machine in 20 min; (*2*) the evenness of the dyeing and the penetra-

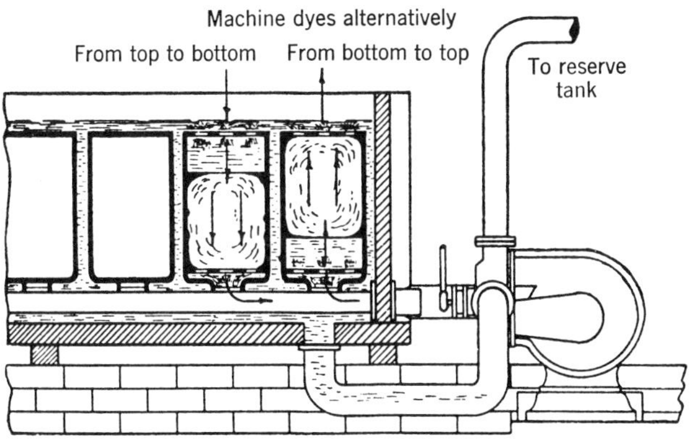

Fig. 18. Can top-dyeing machine. (Courtesy of Callebant and DeBlicguy.)

tion is of a high uniformity, and the top is preserved to such an extent that recombing is generally not necessary; (*3*) flexibility; in order to dye a smaller weight than the one for which the machine was intended, half of the cans are fitted with a device that stops them off by simply turning the cans half way around on the conical base.

Continuous top dyeing

The combined efforts of dyestuff manufacturers and machine builders over the last few years have been culminated in solving most of the physical-chemical and mechanical problems associated with continuous dyeing of worsted slubbing.

Khachoyan (30) lists four types of production machines built today for continuous dyeing of worsted slubbing:

The Machine Built by Fleissner (Fig. 20)

The continue-dyeing method used is based on the "Cibaphasol" dyeing process (CIBA). With it, the dyestuff solution is applied by the two-phase system avoiding all the disadvantages of padding aqueous dyestuff solutions on wool. The contribution of the textile industry in this team work was made by the worsted spinner van Schuppen, Veenendaal, Holland, who developed the so-called WAF-continue steamer for worsted tops, which is now built under license by Messrs. Fleissner.

The method of procedure of this process is, as indicated in the schematic drawing as follows: up to 48 worsted slivers, each 30 g/m coming from an unrolling frame or preferable from a turn table to insert a slight twist for increased strength, enter a raised horizontal padder. Here the koaservat-dyestuff solution is applied by padding: this is followed by a squeezing operation. The dyeing speed corresponds with that of the backwasher which varies between 0 and 10 m/min (approx. 11 yd/min). The dyestuff consumption is the same as for conventional dyeing processes. Two dissolving vessels for the dyestuff Cibaphasol preparation are required per padder in continuous operation; each with a capacity of 500:1. With this method the dyeing operation actually becomes part of the backwashing process.

Fig. 19. Loading of a Longclose 24 cans-top dyeing machine.

Dyeing is followed by treatment in the WAF steamer, which has the form of a J box and swings in a pendulum motion about its center gravity. The slivers enter from the top by two carrier rolls, which serve as a steam tight sealing joint, and are packed in even folds. During the passage from the top to the bottom, steam is introduced from the bottom, and the dye slubbing is steamed at 98–102°C along the vertical part of the J box. Directly following is the traditional backwasher for washing the worsted tops, to remove unfixed dyestuff. After passing through the last squeezing rolls, the slivers enter the rotary screen drier, and when dry are deposited in cans.

If the slubbing is to be evenly plaited down, the steamer should be filled almost to capacity. It contains 70–75 kg slubbing in this condition.

The production capacity of a steamer is qualified by the necessary steaming time and by the performance of the backwasher.

The average figures have been established as shown in Table 8.

TABLE 8

Fleissner Top Dyeing Machine Production Capacity

Backwasher speed, m/min	Steaming time, min (approx.)	Performance I, kg/hr	Performance II, kg/hr
3	60	72	144
4	45	96	192
5	36	120	240
6	30	144	288
7	26	168	336
8	23	192	384
9	20	216	432

Performance I in a steamer with a single working width of 500 mm, furnished with 0.4 kg slubbing/running meter (e.g. 20 bands at 20 g/m).

Performance II in a dual steamer with two adjacent working widths of 500 mm, furnished with 0.8 kg/running meter (e.g. 2 × 18 bands at 22 g/m).

These performance figures are calculated on continuous operation. It takes only a few minutes to make a change of color on the same slubbing.

The following steaming times are required in the Cibaphasol process to ensure perfect dyestuff fixation:

Pale shades: 10 min
Medium-depth shades: 15–20 min
Heavy shades: 20–30 min
Chrome dyes: 30–45 min

The performance of the dryer determines the speed of the backwashing operation. The normal running speed of a Fleissner backwasher attached to a dryer with two sieve drying drums is about 4 m/min. This speed provides for a sufficiently long steaming time in the continuous steamer, even for Chrome Black dyeings. Even if a backwasher with three drying drums is installed the fixation time is still sufficient for very heavy dyeings at a running speed of 5–6m/min. It is only necessary to select a slightly lower running speed in the case of heavy chrome dyeings.

Fig. 20. Fleissner Continuous Top Dyeing Range. (Courtesy of Fleissner GMBH & Co., Maschinenfabrik, Germany.)

Segard—Serracant Tunnel Equipment

The equipment developed by the Textile des Francs at Tourcoing (France) is built by S. A. Serracant in Sabadell (Spain). The machine consists of a padder connected with an approximately 50 ft long square casing built within a horizontal cylinder or tunnel. As seen from Fig. 21 the slivers of top (*4*) are fed into the tunnel by means of feed-rollers and tightly packed in folds. A stainless steel no-end chain conveys the sliver through the various sections—presteaming, steaming, washing, chroming, washing, and rinsing. The casing is perforated at the points of treatments. The steaming chamber is made vapor-tight by the material which serves as a steam sealer on both ends of the section. The temperature of the steam can be set between 105 and 140°C (220–280°F).

The production depends on the necessary steaming time and the numbers of slivers fed into the tunnel. In Table 9 are listed the possible production figures at a set sliver speed and corresponding steaming time based on 4 slivers 20 g/m or approximately 3 oz/5 yd or 23 grains/yd.

TABLE 9

Production and Steaming Time of Segard-Serracant Top Dyeing Machine[a]
Sliver Weight 4 × 20 g/m

Sliver speed, ft/min	50	100	150	200
Production				
lb/hr	154	306	462	616
kilo/hr	70	140	210	280
Steaming time, sec	22	11	8.5	5.5

[a] Good results have been obtained with 1:2 metalized dyes on wool top, with 8 sec steaming at 220°F and polyester 8 sec steaming at 260°F.

The running speed of Wooltop was 109 ft/min at a steaming temperature between 220 and 225°F and of Polyester 105 ft/min at a steaming temperature of 266°F—a liquor ratio of 1:1.

In bulk practice, the method of preparing the dye solution is to make a stock solution of the thickener, the solvent, and an antisceptic. As thickener, Polygum 270 has proved to be very good. The following are recipes for the gum solutions for wool and polyester tops:

	Wool	*Polyester*
Polygum 270	0.31%	0.62%
Isopropylalcohol	0.25%	0.50%
Orthophenylphenol	0.25%	0.50%

These are dissolved in half the amount of water. The balance of the water, together with the dyes, the acid and a carrier (in case of polyester), is then mixed into the gum solution shortly before dyeing.

During a 5-day, 3-shift week representing a 90% efficiency the following production results were obtained:

Wooltop	46,629 lb	389 lb/hr
Polyestertop	37,414 lb	312 lb/hr

Ilma Range

The machine built by Ilma developed in cooperation with Geigy as a continuous dyeing, backwashing and drying unit for worsted tops is based on the "Irga-Pad-Method" of Geigy; see Fig. 22 as described by Leddy (31).

In the standard model, 24–48 combed slivers are padded in the wedge of a horizontal padder with the dye solution to which a special wetting agent has been added. The roller pressure is adjustable by means of a control valve by way of pneumatically adjustable cylinders, which act on the roller journals. The pressure applied is shown on a pressure guage. The feed speed of the slivers to be treated can be adjusted between 0 and 2 m/min.

The horizontal padder is followed by a special steaming chamber made of stainless steel. The slivers enter the steamer through slots at the top of the chamber. The width of these slots can be adjusted according to the volume of the slivers.

In the steamer the slivers first move vertically, then they are deposited in regular folds onto a slide to be moved along on a lattice. The drive of the lattice is controlled by a variater which makes it possible to regulate the time of the slivers in the steamer from 5 to 40 min at temperatures up to 110°C.

After leaving the steamer, the slivers pass through a water lock which not only prevents an undesirable loss of steam but which also provides the first washing effect. After passing through a

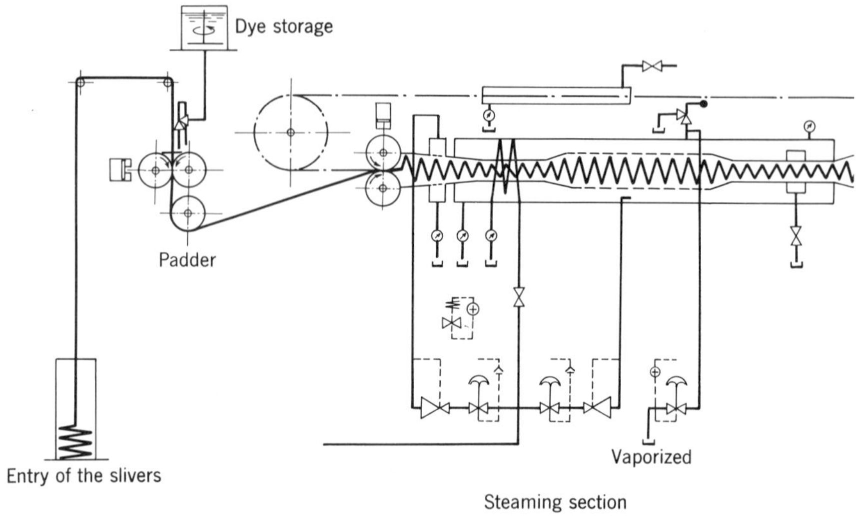

Fig. 21. Segard-Serracant continuous tunnel top dyeing equipment. (Courtesy

squeezing unit, the pressure of which is again pneumatically adjustable, the slivers go through 3 washing vats which, however, can be increased to 4 or 5 if required. In each vat is a large diameter perforated drum which is rotated by means of a speed variater. A pump sucks the wash liquor through the perforated drum and returns it by way of rinse kettles on the side of each vat. Due to the vacuum produced, the combed slivers firmly adhere to the perforated drums and in this way are carefully carried forward and penetrated by the wash liquor at the same time. At the exit end of each vat are squeezing rollers the pressure of which is adjustable up to 5 tons. The drive of the squeezing unit is by means of infinitely variable speed motors.

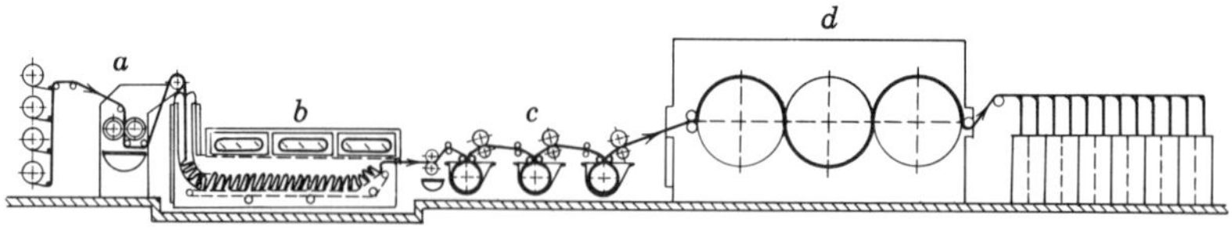

Fig. 22. Ilma continuous top dyeing range. (Courtesy of Industria Lavorazioni Metalli Antiacidi Schio, Italy.)

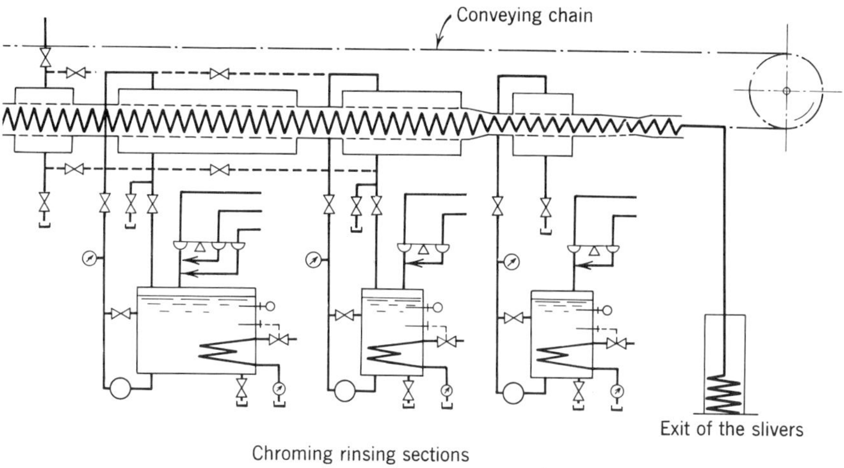

of S. A. Serracant Sabadell, Spain.)

The process is concluded on a perforated drum drier where transport and drying are again provided by the vacuum created. If required, the drying unit can be equipped with a winding frame or the slivers may be deposited in cans.

The dye solution at the padder is kept at a temperature around 58°C and replenished from two stock bath tanks holding 500 liters each. A normal production of the 24 sliver machine and 30 min steaming time is 220 kg/hr or approximately 1500 kg per 8-hr shift.

Most suitable dyes are the neutral or weakly acid dyeing metal-complex colors. The starting up time for the machine is 45–60 min and the change over from one light shade to the next requires about 10 min.

Konrad Peter Range

The dyeing range of Konrad Peter which consists of a padding mangle, a high temperature steaming unit, and two different washing units. The sealproof devices at the entrance and exit of the steamer enable working on the continuous principle under pressure at a temperature of the saturated steam of 120°C.

The equipment contains a continuous apron which allows for forward motion of the sliver without tension or pull-breaks.

General Experiences

So far the experience has shown that with most of the steamers as built at present for continuous dyeing of slubbing that with chrome dyes, with the exception of chrome black T, no satisfactory results have been achieved. The use of dyes of the fluorchrome type, even though it would be an elegant solution has its limitations due to the corrosive action on stainless steel and the limited number of dyestuffs. The use of chromates with ammonium sulfate would make a higher number of colors available, but in the steaming alkali damage occurs, possibly through the liberation of ammonia gas.

A study by Mazingue (36) on the viscosity and pH of the padding solution showed that for some chrome dyes of the crinolane (fluorchrome) type an increase in the viscosity improves the penetration and reduces tippyness. There is a large drop in the viscosity of the padding solution when the slubbing reaches the steamer and the temperature increases to 95°C by turning watery (acid hydrolysis). In cases where this did not occur, it was established that a change in the pH was responsible. The main parameters which influence the pH value are: the alkalinity of the top which can range from pH 7.5–9.5 and the dye concentration. With increasing dye concentration the pH is reduced. The explanation is that as the dye combines with the wool, acid is set free.

The experiments covering the spinning of the continuous dyed top in comparison with normal and pressure dyed as well as undyed top gave the following results:

The data in Table 10 show that in regard to the mechanical properties of the ringspun yarns such as breaking strength, elongation and number of end breaks the continuous dyeing comes closest to the undyed yarn. The continuous dyeings were made on a Ilma laboratory machine.

A similar study on the effect of the top dyeing process upon the formation of neps was reported by Primentas (49). The result showed only an approximately 3% increase in the nep count when dyed in an open type machine.

In the conclusion Mazingue points out certain difficulties such as poor evenness and tippyness or spots which in most cases can be traced to a poor fixation of the dye due to a too high pH value during steaming.

TABLE 10

Comparison of Spinning Results of Different Dyeings of Top

	Undyed	Normal dyeing	Pressure dyeing, 105°C	Continuous dyeing
Neps in top	39	167	72	65
Yarn, Nm	39.9	39.9	40.0	39.6
Turns, pm	522	513	513	506
Breaking strength, g	134	124	122	134
C.V., %	17	19	18	20
Elongation, %	12	8.5	7.5	8.8
C.V., %	36	38	36	39
End breaks per 1000 spindle hours	173	363	377	334

From a cleaning standpoint, the dye lots should weigh a minimum of 1,000 lb. As advantages for the dyer are counted, the elimination of one gilling, the reduction of the dyeing time and saving on water and power.

The question of cost is difficult to assess, the necessity of certain agents and expensive dye may even increase the costs—so far the spinner does not admit that the improvement in the yarn quality, even the elimination of recombing, is worth the price.

One of the main factors influencing the cost is the reduction in the dyeing time. For slow developing dyes, the steaming time ranges from 40 min at 100°C, or 20 min at 108°C, or 10 min at 115°C, or finally 5–10 min at 120°C. There are mechanical problems involved, to guide a top under pressure into a closed chamber and to guide it out again without impairing its evenness properties. Machines are already available but still need plenty of refinements.

Yarn-Dyeing Machinery

The dyeing of yarn in machines is not done to the extent of raw-stock and top dyeing, because it is more difficult and needs special attention. Apart from the fullest preservation of the material, and the avoidance of felting and crushing, the demand for greatest uniformity of the shades has to be met, as no blending is possible to correct differences in shades, as with stock- or top-dyed material. Special consideration in yarn dyeing should be given to the selection of suitable apparatus, because the different systems vary considerably.

Dyeing of Yarn According to the Packing System

The yarn is laid into the vat in layers and placed in alternate directions. The hanks of the second layer are at right angles to those of the first, thus ensuring a uniform packing of the yarn, and at the same time, a more uniform penetration of the yarn by avoiding the formation of open channels.

Dyeing of Yarn According to the Hanging System

The yarn is either suspended on narrow rods, sticks, or cords, and then is grouped into blocks which are placed in the machine.

The machines which, with the exception of a few circular machines, are of the same fundamental design and are generally referred to as Hussong machines. Instead of using a cage, the top cover or carrier is provided with self-unloading yarn racks. The yarn is suspended from sticks fitted at the top of the carrier only "single-stick system," (Fig. 23) or is suspended from the top of the rack and attached to a second set of sticks fitted at the bottom of the carrier—"double stick system." The distance between the sticks is a little less than half the reel length of the yarn. These carriers are easily lifted out of the machine with the help of a pneumatic lifting outfit and transferred to special trucks, where the yarn can hang until completely drained.

Liquor circulation has been measured on single stick and double stick hank-dyeing machines under various loading conditions by Derbyshire and Lennin (16).

The work has shown the dependence of flow on the weight of yarn, and also on the fineness of the wool and the number of the yarn.

In all Hussong type of machines the liquor is circulated by propeller and may flow either upward or downward through the yarn, depending on the direction of rotation of the propeller. Measurements of flow of liquor, both upward and downward through the yarn discloses marked differences in rate of flow between upward and downward circulation obtain in single-stick hank-dyeing machines. In double-stick machines no such marked differences in flow occur.

Derbyshire and Lennin investigated further the effect on the flow of liquor in a single-stick machine of variations in both count and quality (Fig. 24). The yarns used are detailed in Table 11. Samples 1–3 show the effect of wool quality only, the yarn count (and twist) being similar, while samples 4–6 show changes only in yarn count.

Tests were carried out using upward flow only in a single-stick machine at 40°C, the system being buffered to pH 6.5 with sodium dihydrogen phosphate.

The flow of liquor is markedly affected by the yarn loading of the machine and there is an optimum loading for each machine. Variation in yarn count for a given wool quality does not significantly affect the rate of flow, but changes in wool quality may do so. The finer the wool the less yarn can be loaded in a machine to maintain a given flow rate.

This confirms the dyer's observation that machines must be more lightly loaded with fine quality wool than with coarse. Many hank

TABLE 11

Count and Quality of Yarn Samples

Sample no.	Count, *Nw*	Wool quality
1	2/24	56s
2	2/24	60s
3	2/24	70s
4	4/8	56s
5	3/13	56s
6	2/20	56s

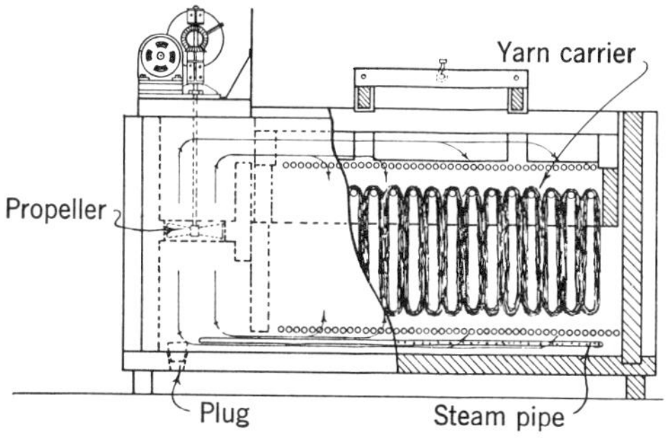

Fig. 23. Hussong skein dyeing machine.

dyers load up to 20% more yarn of 54–56s quality than of 64s. Moreover, in the publications of machinery manufacturers, capacity ranges are quoted, e.g. 700–850 lb or 400–600 lb, depending on the machine. The problem of hank packing when dyeing small weights of yarn is also dealt with empirically by many dyers, either by doubling the hanks to obtain a more uniform package or by placing the hanks over two sticks instead of the normal one. The reasons for these variations in technique are now clearly related to the flow of liquor through the mass of yarn.

There are marked differences in flow rate from one region of a machine to another and this can result in significant differences in dye uptake.

An approximate relation has been derived relating the degree of unlevelness of a dyeing in a hank-dyeing machine with the rate of dyeing of the dyes used and the time taken to produce one complete change of liquor in the machine.

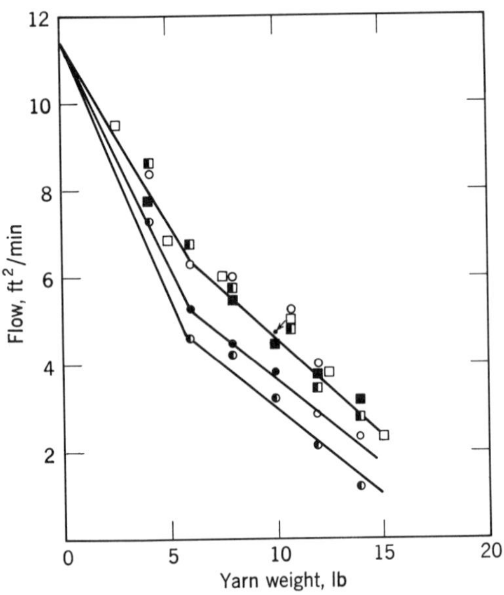

Fig. 24. Effect of wool quality and yarn count on upward flow in single-stick machine at 40°C. (Courtesy of Derbyshire and Lemin.) (○) Sample 1. (●) Sample 2. (○) Sample 3. (□) Sample 4. (■) Sample 5. (□) Sample 6.

Figure 25 illustrates that the time required for one complete change is a function of yarn loading.

Many machine builders have worked out systems of coupling pairs or groups of four identical machines together based on scientific assessment of rates of flow required and employ positive mixing of liquor from the coupled machines. One such system is the Longclose coupling-mixing system. The largest dyeing range built by Longclose has a batch capacity of 9,000 lb of carpet yarn, consisting of four 2400 lb units.

Another innovation is the Brown-Longclose double stick system. Being of an open construction they minimize contact with the yarn and restriction of liquor flow. The grid of the bottom sticks has a degree of free vertical movement at any of its adjustable positions while both sets of sticks can oscillate through a wide angle. The combination of these free movements prevents stick markings.

A deviation from the normal Hussong construction is the Ilma T/Fn hank dyeing machine (Fig. 26). The machine consists of two yarn compartments. The turbo-propeller set with horizontal shaft fitted in the middle of the tank bottom for a symmetrical liquor flow. The liquor ratio is about 25:1.

The F.T.M. "Plusflow" hank dyeing machine is a very unique new design (Fig. 27). By positioning a separate hank cage box cen-

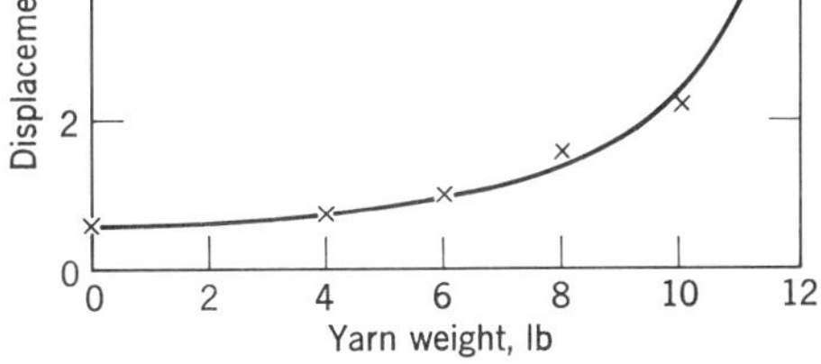

Fig. 25. Time required for one complete change of liquor in single-stick machine at 40°C as a function of yarn loading, 10 lb nominal capacity. (Courtesy of Derbyshire-Lemin.)

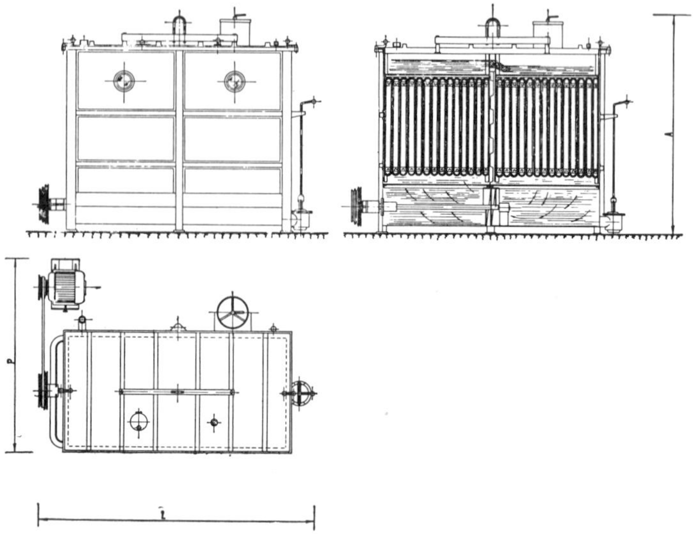

Fig. 26. Ilma T/FN hank dyeing machine.

trally within the machine so that a clear annular type space is formed between it and the walls of the tank, it has been possible to obtain a four sided weir having a total length of 15¼ ft for the 200/300 lb machine, which is at least four times the normal weir length. This increased length of weir ensures a uniform, low velocity, large volume flow. The flow is further balanced by being passed through grille plates of a special V slotted design situated at the top and bottom of this box. These two features provide a flow in excess of 4 gal per lb of material per minute and is so smooth and even that felting is completely eliminated and the dye take remarkable even without the necessity of turning the hanks. Under normal conditions this machine will accommodate hanks varying from 47 in to. 72 in. reel length.

The liquor is circulated by means of right and left handed propellers situated in the bottom of the tank, by means of a four speed drive using two squirrel cage motors intercoupled and driving the propellor shaft through vee belts. Control of speed is by push button and speed preselector by which any forward speed will be combined with any reverse speed. This is carried out automatically by the F.T.M. "Autorev" Time Cycle Reversing Controller.

To enable dyeings to be carried out at temperatures in excess of 212°F (100°C) the machine can be lightly pressurized to 2 lb/sq in. (218°F/103°C). To accelerate this pressurization a steam operated air injector has been fitted.

Specially high rates of liquor flows are obtained in the Smith Drum pulsating and the Pegg "Pulsator" hank dyeing machines up to 30 gallons of water per pound of yarn per minute at a liquor ratio of 18:1.

The general design of the Pulsator is shown in Fig. 28. The circulation of the dye liquor is induced by two propellers arranged horizontally. The basic principle is the intermittent flow of liquor, which results in a pulsating action. The circulation is in one direction only—best for woolen and worsted knitting yarns. The hank carrier consists of a top and bottom duct joined by telescope tubes. The top duct supports fixed hank poles, which are slotted and through which the liquor is circulated, and is fitted with adjustable flow posts which allow a portion of the liquor to circulate over the top of the hanks when required. The bottom duct supports a row of poles which control the length of the hank. Whereas in general the hank dyeing machines are square, some manufacturers offer circular apparatus such as built by Henriksen, (Fig. 29) and Scholl.

The circular form of the kettles requires a special circular carrier for yarns in hank form to make loading and unloading simple and economical. Such a carrier is illustrated with Fig. 30. The units through the use of specific carriers are suitable for dyeing loose materials, tops, or packages.

Dyeing of yarn according to the spindle system

In this system the yarn may be dyed in the form of cops, packages, cheeses, or spools. The dyeing of cops is not very practical. The dyeing of packages or cheeses is generally done in the same type of machines as described under stock dyeing. The builders of raw-stock-dyeing machines furnish special yarn carriers (Fig. 8,) so that the machine can be used efficiently for stock as well as yarn dyeing. These dual machines are of particular value in mills that manufacture stock and top-dyed goods with multicolored effect stripes. Only small amounts of these effect yarns are needed and, therefore the most economical way is to dye them in the yarn.

Fig. 27. F. T. M. Plusflow hank dyeing machinc. (Courtesy of Freeman, Taylor Machines, Ltd., Leicestee, England.)

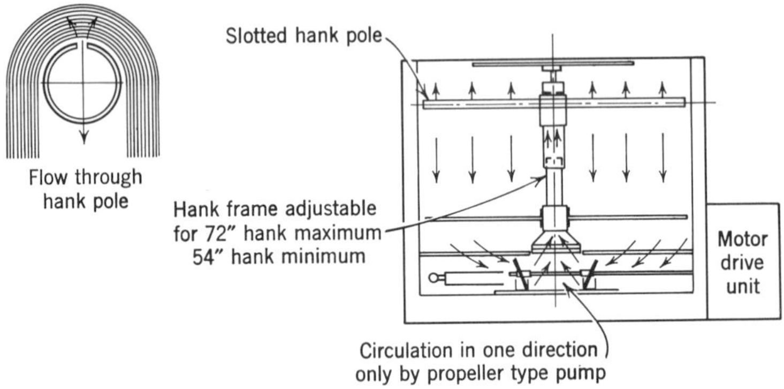

Fig. 28. Pegg pulsator hank dyeing machine GSH Mark II. (Courtesy of Samuel Pegg & Son, Ltd., Leicestee, England, represented by U.S.A. Bentley Machinery, Inc., Providence, R.I.)

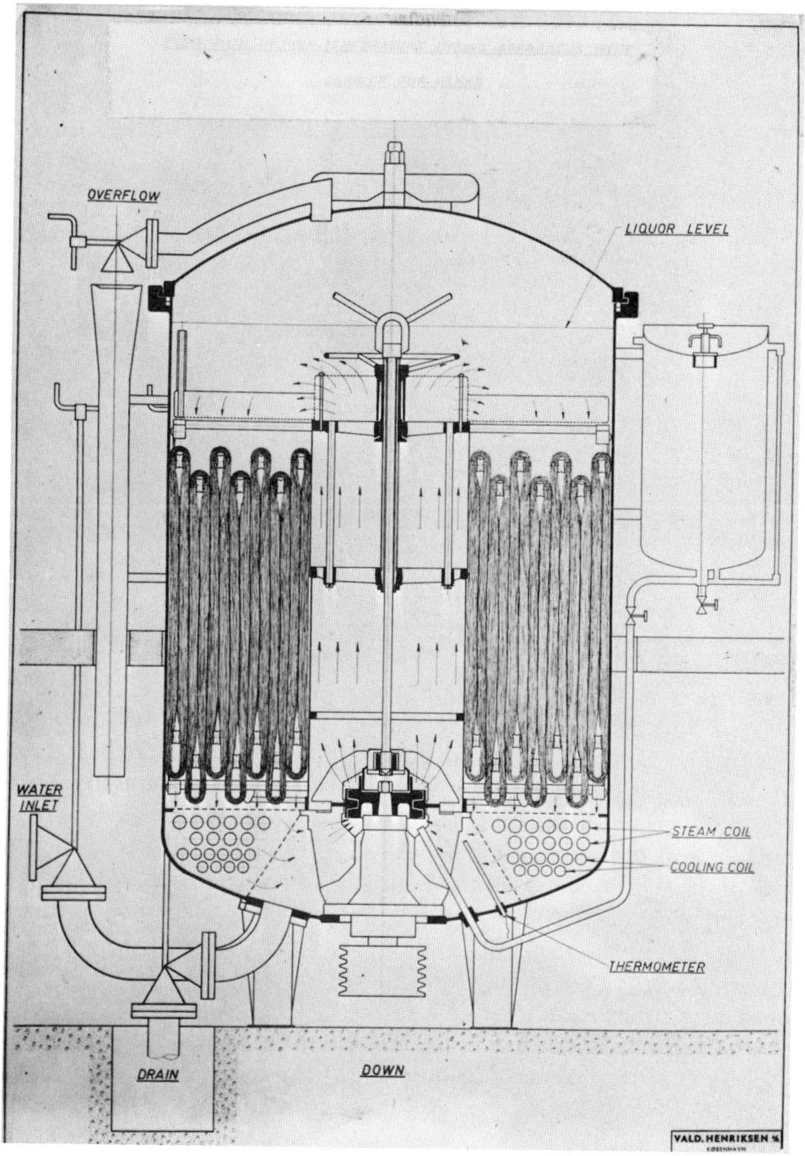

Fig. 29. Henriksen high-temperature dyeing apparatus with carrier for hanks. (Courtesy of Edda International Corp.)

Fig. 30. Yarn carrier for Henriksen dyeing machine.

The carriers are built to accommodate from 24 to 500 packages, normally four per spindle. These spindles are either perforated pipes, as described under top dyeing, or fluted metal rods with three or four ribs as illustrated in Fig. 31. In making up the packages, the yarn is wound upon perforated rigid cardboard or metal tubes or spools or upon spiral springs covered with a knitted jacket or special paper. Impregnated cardboard tubes have proven very suitable and most economical. Figure 31c illustrates one method of loading in the carrier system. Four packages and three spacers are dropped over each of the fluted spindles rising vertically from the bottom of the carrier.

The spacers between each package are stainless-steel collars that connect the perforated tubes and avoid deforming of the packages by pressing or shrinkage, as well as possible leakage. A clamp nut is screwed on top of each spindle to fasten and seal the whole column, thus preventing the liquor from going anywhere except through the yarn.

When fully loaded, the carrier is lifted from the floor and lowered into the dye tub with a hoist. When dyeing is completed the car-

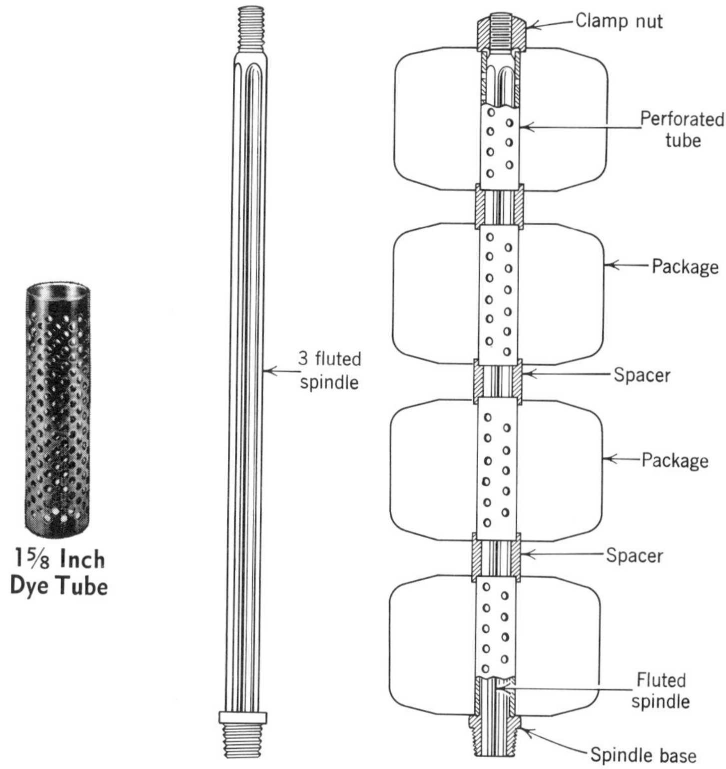

(*a*) (*b*) (*c*)

Fig. 31. (*a*) Perforated dye tube. (*b*) Fluted spindle. (*c*) Mounting of yarn packages with tubes. (Courtesy of Venango Engineering Co., Inc.)

rier is lifted out of the tub and the yarn vacuum-extracted and dried, while still on the frame. The packages are not removed until ready for use. The use of stainless steel dye springs is illustrated in Fig. 32 is today considered the safest way to eliminate the danger of liquor leakages.

When dyeing packages consisting of wool–synthetic blends or only synthetic fibers such as Polyester or Polyamid the main difficulties encountered are high shrinkage and uneven shrinkage within a package. The center of the spools become much harder than the outside layer leading to uneven dyeings and other errors. To pre-

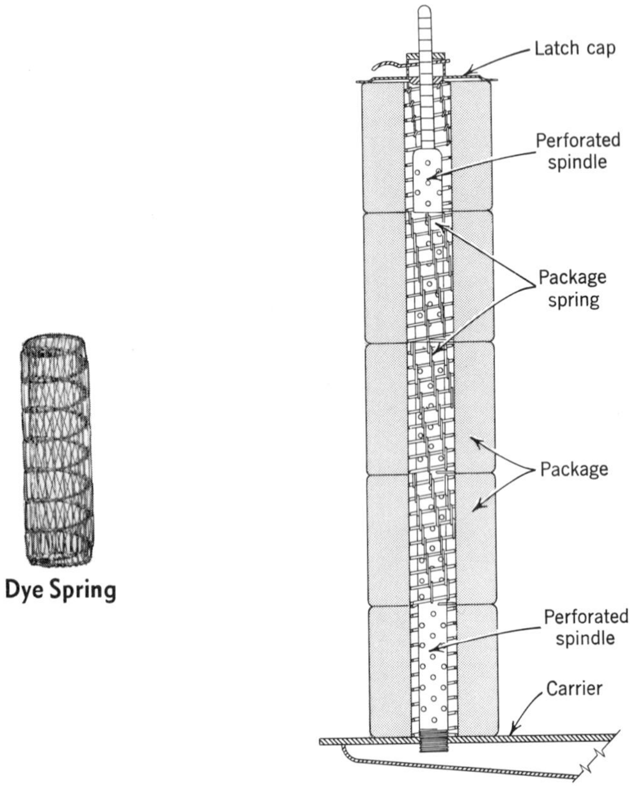

(*a*) (*b*)

Fig. 32. (*a*) Stainless steel dye spring. (*b*) Mounting of yarn packages with springs. (Courtesy of Gaston County Dyeing Machine Co., Stanley, N. C.)

vent this the yarn had to be preshrunk before winding it on solid perforated dye tubes.

To avoid these extra costs special spring tubes as seen in Fig. 33 were developed and their advantages are described by Lutzel (34). The newest construction of the spring tube RASTA 1/F has a circular 15% elasticity range which is sufficient for most commercial yarns.

High temperature pressure dyeing machines have become a must in dyehouses where Polyester and acrilic yarns or blends with wool

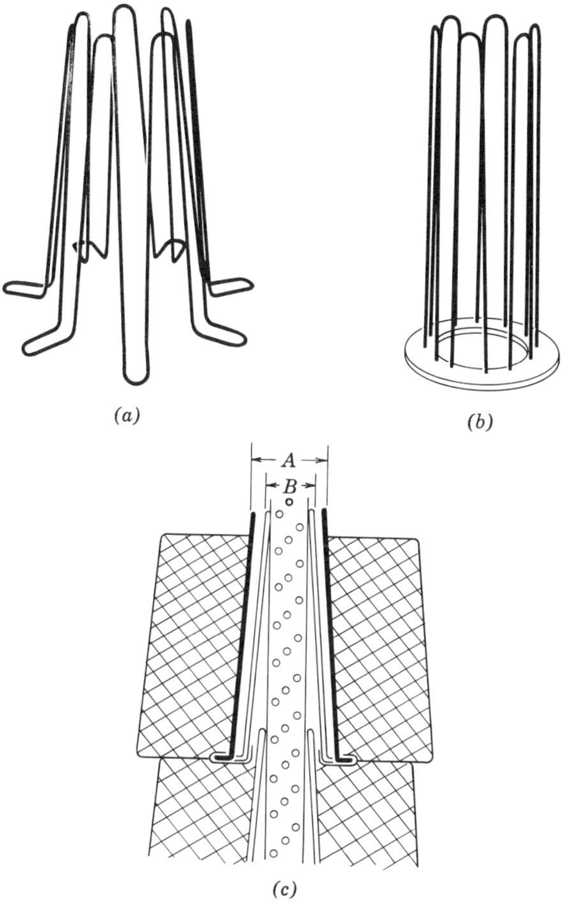

Fig. 33. (*a*) Cross-spool spring tube (Radia III K). (*b*) Cross-spool spring tube with base ring (Rasta 1/F). (*c*) Spring tube Rasta 1/F Conical, shrinkage possibility A to B.

have to be dyed. The Bi-Ferma type dyeing machine as seen in the diagram Fig. 34 is a typical example.

Dyeing in jackspools or large dye spools is favored in the United States because of its great time-saving factor. The yarns to be dyed are wound from regular jackspools (see Chapter 2), or from bobbins, cones, or parallel tubes onto a jackspool with a perforated barrel. These spools carry from 10 to 35 lb of yarn, allowing great latitude

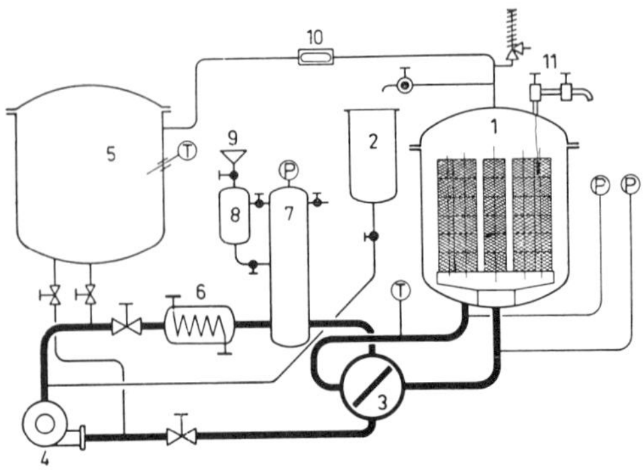

Fig. 34. Diagram of the BiFerma type. (Courtesy of Scholl A. G., Switzerland.)

in the weight of the lots to be dyed. The spools are placed vertically in the dyeing machine.

One of the important recent studies on the dyeing of yarn packages was made by Johnels (27). Some of the basic facts established by Johnels were as follows:

Types of Pumps

In comparing the characteristic properties of different types of pumps Johnels found that reversible pumps are convenient but do not have the same efficiency as an ordinary centrifugal pump. The most reliable pumps are the centrifugal pumps and some vortex pumps.

Types of Packages and Centers

The choice of package and the center on which the yarn is wound depends on several factors. The solid centers are very handy and give a minimum of waste yarn. Flexible centers require more labor per kilogram of yarn dyed and will presumably give more waste. Yarn on solid centers can be used directly after the dyeing, if the drying of the dyed yarn is done properly, with consequent saving of labor. It has been shown that yarn on flexible tubes will give a little more even dyeing within each package.

In dyed packages the yarn next to the outside and inside of a package will differ significantly in color from the rest of the package. These differences seem to be biggest for packages on solid centers and especially in the yarn next to the tube. This fact suggests that the color difference next to the center is related to density of the yarn. This idea is supported by experience with coaxial dyeing.

Which type of center a manufacturer of dyed yarn chooses is, however, not only based on the dyeing conditions, but also how the yarn is used in the further manufacture of dyed fabrics. If the fabric is made of twisted yarn, differences in shade in the dyed, single yarn will be evened out through the twisting of single yarn to a twofold or multifold yarn. If the yarn is woven on looms using two or more shuttles another mixing effect might also occur in the weft. Many modern shuttleless weaving machines will in this way mix yarn from several packages. The tolerance with which a manufacturer shall dye in the dyehouse is consequently a problem which is effected by the twisting and weaving and finishing procedure following the dyeing.

Johnels expresses some idea for the reconstruction of yarn dyeing machines in order to make automatic loading and unloading of carriers possible. He suggests a horizontal carrier of the porcupine type.

Machines for Drying of the Dyed Materials

For proper drying, in whatever form—of the dyed raw stock, the dyed top, or the dyed yarn—drying machinery has to be provided. In drying of dyed stock, the same machines are used as in drying of scoured wool. The procedure to be followed after dyeing to restore the top or slubbing to the proper condition for drawing and spinning depends on the dye process.

A sufficient rinsing cannot be given the top-dyed ball in the dye machine itself, therefore the rinsing is combined with the drying, gilling, and mixing by running the ball through a backwasher (Chapter 4) especially built for dyed top. Instead of only two scouring bowls as built for backwashing white top, the machine is provided with five compartments. The first bowl is filled with an alkali-containing liquor (½ lb soda per 100 gal) to neutralize any excess of acid; this is especially necessary for deep blacks dyed with sulfuric acid. The next two bowls are filled with a very dilute soap solution of 4–8 oz flake soap per 100 gal. Synthetic detergents are preferred when soft water is not available. The last two bowls are supplied with soft water and act as rinsers. The temperature of each of the five baths is kept around 100°F. The drying of the top is then accomplished by running it over perforated cylinders or it is carried on a wire apron. In both instances it is dried with hot air. A gilling operation finally follows for opening the fibers and to facilitate drafting in the succeeding operations. When carefully handled by the dyer, the recombing of can-dyed top is not necessary.

The drying procedures for jackspools are very economical, because after dyeing to shade and proper rinsing, the spools with their dyed contents undisturbed are taken from the kettle and, while still wet, are placed in the dryer (Fig. 35).

The spool is placed over a horizontal arm or stud, in principle the same as the spool is placed in the dye kettle, and in like manner it is locked over a special orifice at the rear of the drying compartment and held in rigid position by a plate clamp and hand screw at the outer end of the spool. All air, by reason of this arrangement, passes through the spool perforations.

The dryer accommodates eighteen spools at one time (500–540 lb) or, as with the dye kettle, the spool openings may be blocked off. In this case an unperforated "dummy" spool is used, so that

Fig. 35. Top and yarn drying apparatus with blower. (Courtesy of Abbott Machine Co.)

any number of spools may be dried. In operation, the air is blown through the spools and expels the water. This is accomplished in a very few moments, then warm air is introduced and passed through the material until it is dry. Fine top will be dried to a definite moisture content in about 3 hr, coarse top and yarns in from 1 to 1½ hr. Back-washing is dispensed with entirely. After drying, the spool is placed in a six-spool creel and condensed either 5 or 6:1 through a plain balling gill, and thence to the drawing proper.

Hank dryers

There are numerous constructions built such as the Fleissner dryer with revolving rods in the shape of two discs at each end joined together by three skeleton rods, which move slowly through the dryer without the help of chains and without the use of grease by revolving left and right hand alternating on smooth guide rails. The hot air circularing system is illustrated in Fig. 36. The machines are built in four sizes ranging in capacities from 2200 to 8800 lb of wool hank yarn in 8 hr.

For the yarn dyed on jack spools the drying procedure is the same as described for the Abbott-process dyed top. After drying,

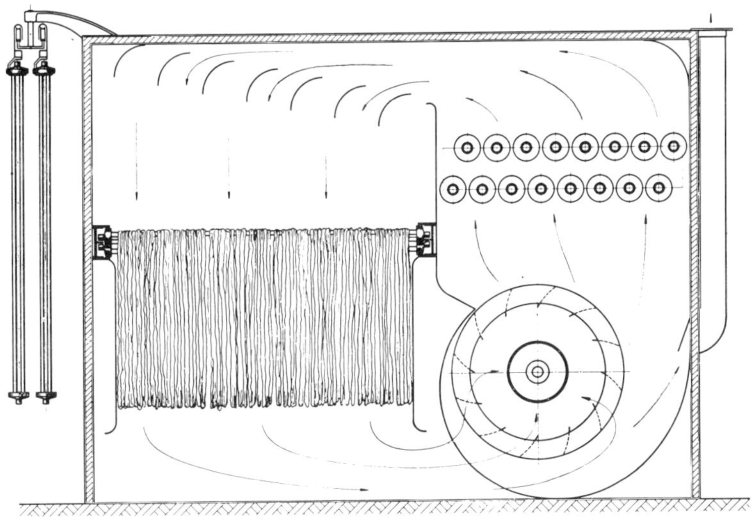

Fig. 36. Hank yarn dryer with revolving rods. (Courtesy of Fleissner & Son.)

the yarn may be run from the spools on a rewinder to a dresser spool or it may be placed in a jackwinder and run onto bobbins, or it may be run on the cheese or cone rewinder directly into the desired package form.

Similar drying systems are followed for the package-dyed yarn. The carrier, loaded with 18–72 cheeses, is lifted out of the dye kettle and first placed on top of a vacuum tank. In a few minutes the excess water is extracted down to approximately 50% moisture in the yarn. The carrier is then connected with a large blower unit and warm air blown through the spools until they are dry.

One of the most elaborate package yarn dryer is the Gaston County AVESTA Static Pressure Rapid Dryer as seen in diagram Fig. 37.

The yarn carrier after it is placed in the kier and the quick-acting lid is secured and all instruments are pre-set to desired time and temperature settings, the blower is started, the "Instrument Power" switch is turned on, and the "process start" button is pressed. From this point the operation is completely automatic.

A "static pressure rapid-dryer" requires an average of 100 blower hp when operated at 5 atm (75 psia) and has an inlet capacity of 4000 cfm. If operated at 280°F the blower will carry five times the weight of air or 1250 lb containing five times the Btu, or a total of 40,000 Btu to the yarn each minute of operation. The average drying time of this unit for cotton packages is 45–60 min, depending on package size and batch size only. It is not subject to other limitations since it has a closed system.

Wool and worsted yarns are dried by the same methods used for cotton, but the drying cycles are shorter. With the exception of rayon, yarns made from all of the man-made fibers dry faster than cotton.

Piece-Dyeing Machinery

Wool pieces whether all wool or all worsted, on blends with other fibers are generally dyed in open or closed becks in which the fabric moves through the stationary bath. The dye beck consists of a rectangular box with a curved back sloping inward toward the bottom. A slat reel or winch that is mounted on the top, lengthwise of the machine, is usually driven by individual motors on each machine.

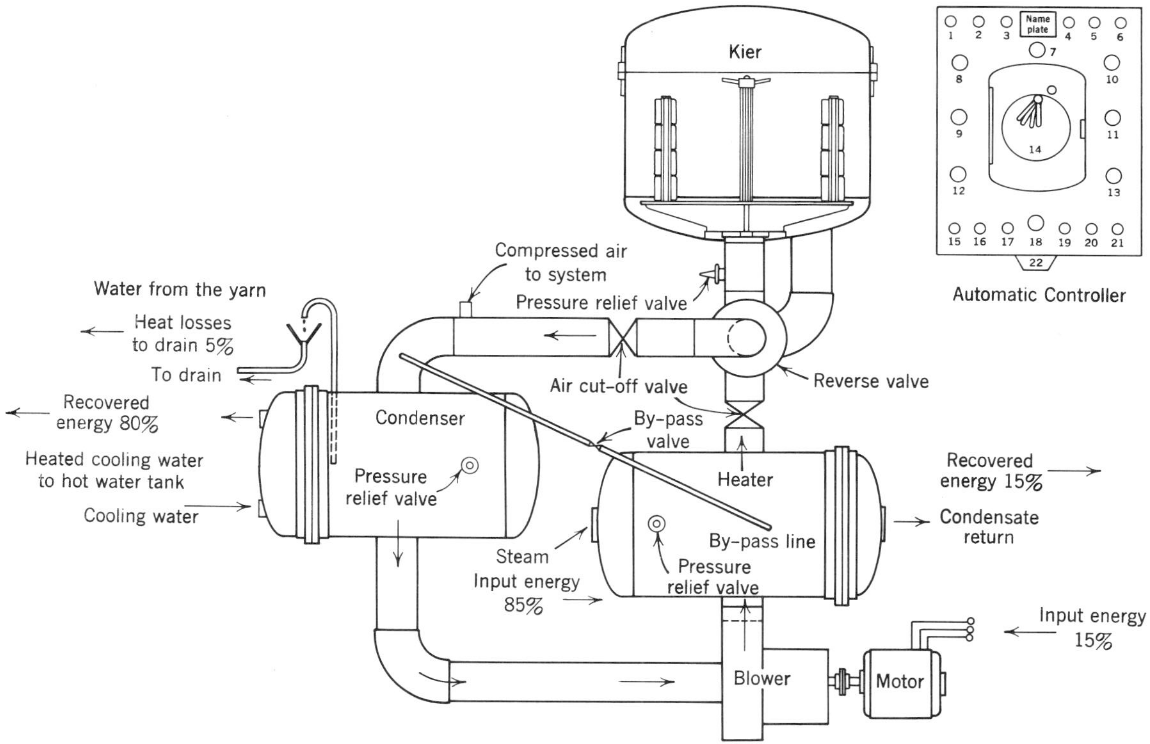

Fig. 37. Avesta static pressure rapid yarn dryer. (Courtesy of Gaston County Dyeing Machine Co.)

The piece is sewed end to end over the reel to form a continuous rope which moves through the dye bath. Figures 38 and 39 illustrate such a machine. Kettles are built to carry two, four, eight, twelve or more pieces depending upon the weight, and each measuring 50–100 yd in length. Larger becks are now in use which carry up to 50 pieces of Dacron-wool fabrics. The kettle is heated by an open steam pipe located in the front of the machine which is separated from the actual dyeing chamber by a perforated partition which prevents the cloth from coming in direct contact with the steam and allows the bath to become heated uniformly. These machines are now constructed of the 18–8 molybdenum fortified type of stainless steel. For economical and practical reasons the machines are totally enclosed. This reduces the consumption of steam (a saving of 30% has been obtained) and has an important advantage in that the dye bath temperature can be made to approach 212°F at which temperature many dyes diffuse more rapidly into the fiber and level more readily.

Opinion is divided among piece dyers as to the best type of reel to use, some prefer the plain slatted reel and others a hump reell In skilled hands both reels give good results. A more compact ree. has been developed, consisting of hexagonal stainless-steel rollers mounted on an arch in the kettle itself. The number of rollers

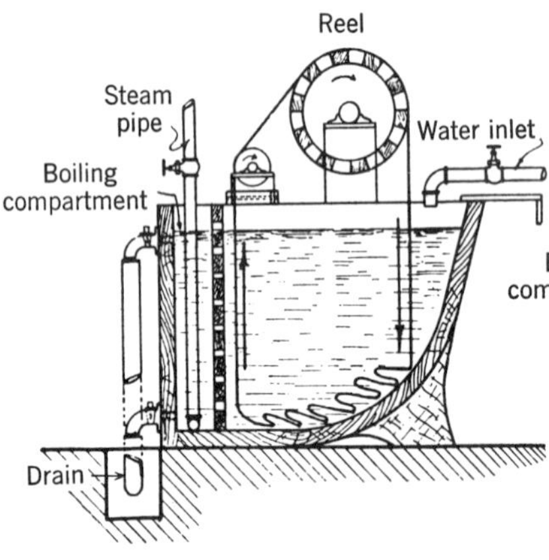

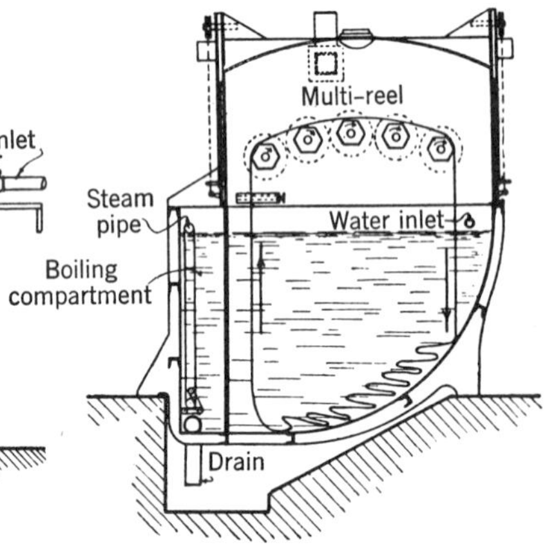

Fig. 38. (*left*) Wooden piece dye beck.
Fig. 39. (*right*) Multi reel stainless steel dye beck.

varies from three or more according to the size of the machine. This reel has the advantage of keeping the cloth in the dye liquor most of the time and avoids an unnecessary stretching of the fabrics such as fine crepes.

For proper observation of the pieces during dyeing, the top of the machine is fitted with an acid-proof light. Sliding doors in the back and the front have small windows. The front door is normally set back, making it possible to add dyes or chemicals without opening the machine. The drain valves are outside of the machine thus facilitating the closing or opening of the valves. In some machines the dye bath is heated through a system of hot-water pipes instead of by direct steam with its disadvantage of condensation. The closed system makes it possible to maintain the same liquor ratio through the dyeing cycle and prevents dilution of the acid with rise in dye bath pH.

To avoid unnecessary felting of delicate goods, machines are fitted with low- and high-speed drives. For greater security and accuracy, automatic time-, temperature-, and steam-control systems are considered a necessity.

Another new machine "Modesta" for processing fabrics in rope form, which is equally suitable for scouring, desizing, bleaching, dyeing, aftertreatment and rinsing, without the need for the goods to be removed between operations, has been designed by Hausemann and Hablutzel (23). The new design differs markedly from the design of conventional winch vats. The diagram illustrates its mode of operation (Fig. 40).

The goods in rope form, leaving the winch are deposited on the perforated double bottom and are then pulled upward in countercurrent with the liquor cascading freely down from the uppertank. A thorough all-round rinsing action is thereby created that also penetrates deeply into the goods. Since the goods run through the cascade and the upper tank, the distance over which the goods and liquor are in contact is sufficient for uniform treatment even with a goods-to-liquor ratio of 1:10. The quality and duration of the treatment are influenced to an exceptional degree by the liquor being suctioned through the goods deposited on the perforated base, the intense rinsing in the cascade and the constant circulation of the liquor.

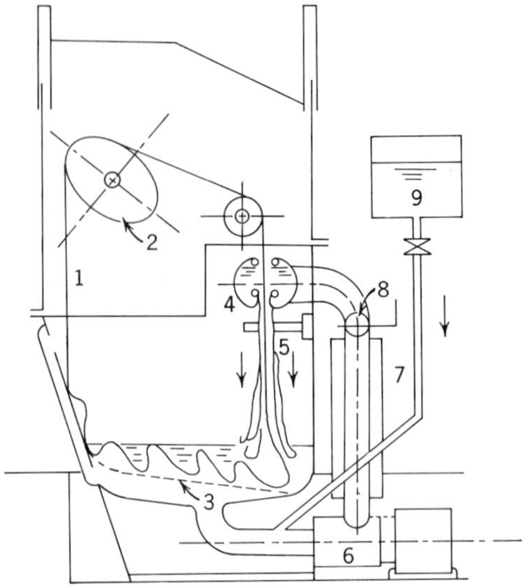

Fig. 40. Modesta piece dye kettle.

Practical trials carried out over a long period by an industrial firm have proved the eminent suitability of the described machine for the acid wool dyeing of delicate worsted fabrics and Indigosol dyeings on woolen fabrics.

Control of the dye process by instrumentation has probably reached the nearest point to perfection in large piece dye houses such as is the case in the one shown in Fig. 41. The schematic diagram (Fig. 42) illustrates how all the becks are controlled from a central point. A close circuit television unit also enables the head dyer to observe each beck separately.

A further step forward to improve dyeing in rope form was made in 1961 when Ordway (47) described the twin pressure becks built by Burlington Engineering. Figure 43 shows a front view of the low pressure beck, which is quite similar in appearance to a standard-type closed beck except for the loading doors. The inside consists of a main, slightly elliptical reel, idler reel, and pin rail. These low pressure becks operate on a range of standard atmospheric temperatures up to 218°F with 29 oz of pressure when

Fig. 41. Piece dyehouse, all dye becks automatically controlled. (Courtesy of J. P. Stevens & Co.)

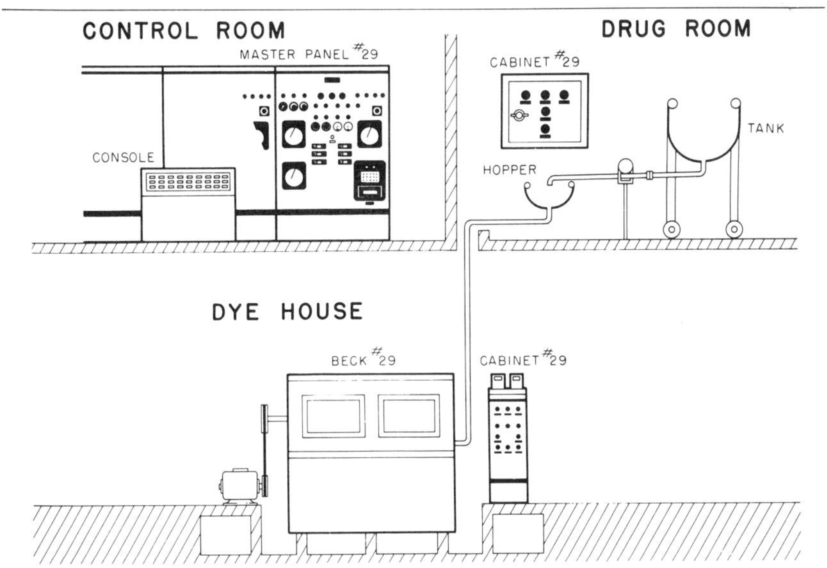

Fig. 42. Schematic control layout of J. P. Stevens & Co., Inc. piece dyehouse.

Fig. 43. Low-pressure beck. (Courtesy of Burlington Engineering Sales Co., Inc., Graham, North Carolina, U.S.A.)

closed. The Hy-pressure beck operates under atmospheric temperatures up to 250°F and 20 lb pressure. In the 12 ft wide Hy-pressure beck it is possible to dye 36 pieces, 7–8 oz goods, polyester fabrics and blends.

That the high pressure winch becks have already found wide acceptance is proven by the fact that at the start of 1968 sixteen firms manufactured such machines. The International Textile Bulletin (67) published the construction details of 13 such high temperature winch becks.

Jet dyeing machines

The basic idea of dyeing fabrics by a jet action principle is a recent development of Burlington Industries. The machines are built by Gaston County Dyeing Machine Company (66) (Fig. 44). The machine can be operated above or below the boiling point as required. The fabrics move entirely by jet action and dye liquor flow without reels or drives. Friction is eliminated since the material is constantly engulfed in the dye liquor, making a very low liquor ratio of 10:1 possible which also means considerable saving on chemicals and steam.

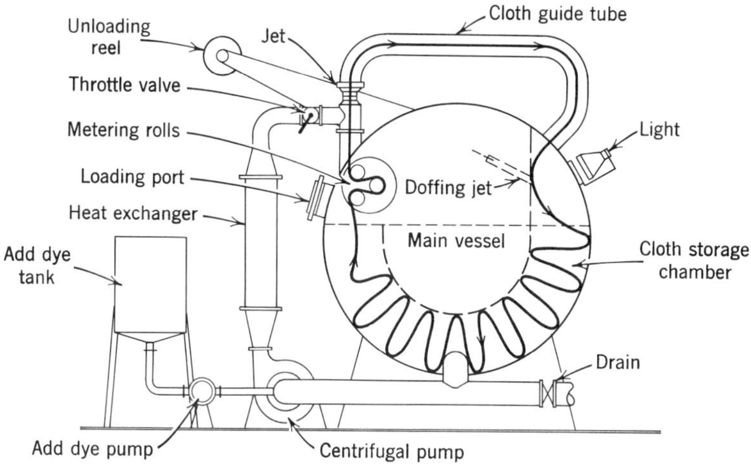

Fig. 44. Schematic view of jet dyeing machine. (Courtesy of Gaston County Dyeing Machine Co., Stanley, North Carolina, U.S.A.)

The jet is the principal element which produces the distinct differences and advantages in this new method of dyeing. The cloth being passed in rope form vertically upward through a circumferential converging jet of liquid which simultaneously effects a liquid exchange with the fabric and propels it through the cloth guide tube.

The cloth is guided to the cloth storage chamber by a stainless steel tube with special flush joint connections. During passage through the tube, the rope of cloth is surrounded by a continuous ring of fluid which is moving faster than the cloth. This results in the cloth being floated through the tube with no cloth to metal contact. Considerable turbulence is generated by the elbows and the straight pipe friction providing an extended penetration and exchange period after leaving the jet.

After leaving the cloth guide tube, the impingement upon a bar rack separates the cloth from most of the conveying fluid and removes any entrapped air. A doffing jet of the liquid assists in the orderly piling of the cloth in the storage chamber. During passage through the storage chamber, the cloth is completely submerged in liquor. Special dimpled perforations in the bottom sheet of the storage chamber allow circulation and drainage of fluids without impeding cloth movement or snagging delicate fabrics.

The metering rolls consist of three polished stainless steel rolls which assist the jet in moving the cloth and closely controlling the cloth speed. The metering rolls will not move the cloth if the jet does not hold the cloth tight against them and the jet cannot pull the cloth through the metering rolls faster than the rolls are running. Variable metering roll speed range of 30–110 yd/min is available.

A centrifugal pump recirculates the entire bath every 20–30 sec.

Basic dyeing cycles are employed in jet dyeing which are similar to those used in conventional equipment but are considerably shortened due to the high penetration, extreme leveling action, and its ability to operate at temperatures up to 250°F. The gentle handling of fabric and lack of tension makes it possible to piece dye and process many delicate fabrics formerly impossible to dye in this fashion.

Woven and knitted fabrics of all synthetic blends and blends of synthetic and natural fibers have been processed with complete levelness and brilliance of shade.

Many different weights and constructions of 100% wool, worsted and cotton fabrics have been successfully dyed and processed in the jet dyeing machine. These range from 4 oz/yd women's wear of quite loose construction to 22 oz/yd overcoat material. Woolen and blended knit goods have been dyed with no lengthwise stretch.

Beam dyeing

Until the development of the Burlington machine no machine builder had solved the problem of producing a satisfactory machine for dyeing wool piece goods in which the dye liquor is circulated and the fabric kept stationary.

Since the granting of the United States Patent 2,707,383 on May 3, 1955 to Newcomb and Williamson and Burlington Engineering Co. beam dyeing has become universal the world over. Beam dyeing equipment according to Ordway (46) may be divided into two types and uses:

A. processing of tricot knit fabric or similar open weave fabrics, mostly made from man-made fibers at standard temperatures of 170–212°F and in some cases up to 230°F.

B. processing of polyester and polyester blend fabrics in knit and woven construction at advanced temperatures, up to 250°F.

Type A is basically built similar to the Gessner Wet-decating unit used by the wool finisher, consisting of a rectangular tank with rounded bottom. For higher temperature dyeings the machine is fitted with a cover, which is bolted to the kier (Fig. 45).

Type B is a cylindrical unit, known as the Hy-Press machine and can be compared with the pressure blowing or vacuum steamer used in dry finishing of wool fabrics, only in this case the machine is filled with water instead of steam (Figs. 46 and 47).

The practical aspects of dyeing fabrics on a beam were first discussed by Meunier (38) and the more recent advances by Newcomb (42) and Ward (42), Ordway (46), and by Limbert (32) and Maurer (35), the latter two discussing European built machines.

All papers stress the importance of preparation of the cloth for dyeing.

Preparations for Dyeing

Success in the final result may depend a great deal on the thought and care devoted to the initial preparation of the cloth. Whatever preliminary treatments are undertaken, the batch, when wound on the beam, should be left with the tension uniform along the whole length of the cloth and normally with the ends smooth and in a plane at right-angles to the beam axis. Fabrics with a high potential for shrinkage should first be boiled off. This is necessary to take out the initial shrinkage before dyeing, to give a beam which will

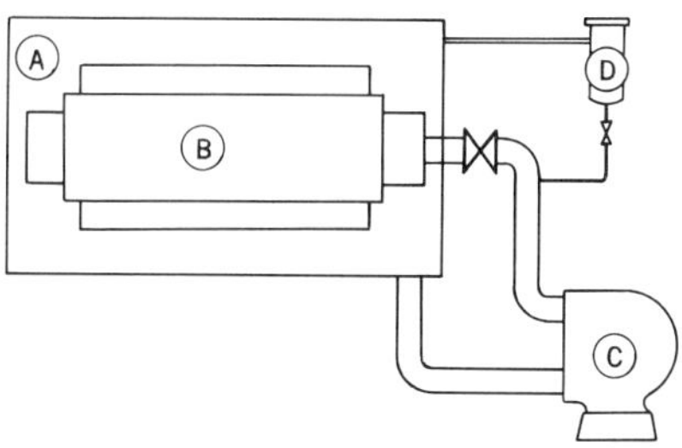

Fig. 45. Longclose beam dyeing machine, atmospheric pressure. (*A*) Main tank; (*B*) beam; (*C*) main pump; (*D*) sample tank.

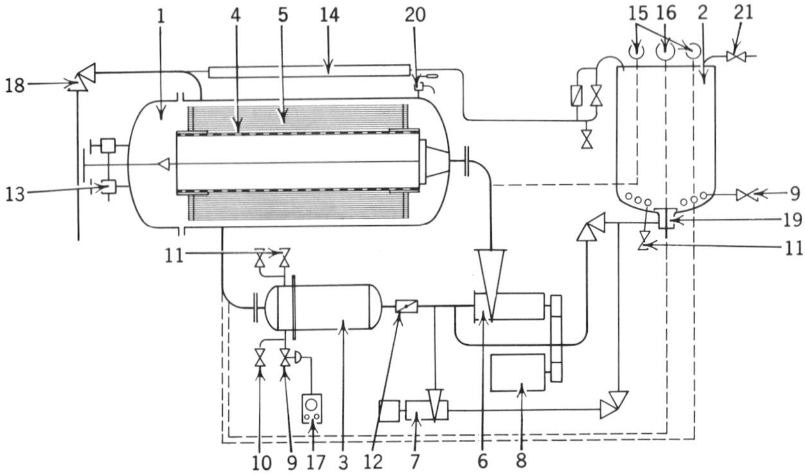

1. Autoclave: working pressure 4.5 atü, temperature up to 130°C (266°F)
2. Overflow vessel with indirect or direct heating
3. Heat exchanger for the heating-up or re-cooling
4. Piece beam for various material widths
5. Dye goods
6. Centrifugal pump resp. axial pump for reversing possibility
7. Additions pump for static pressure
8. Motor
9. Inlet for steam or hot water
10. Inlet for cooling water
11. Outlet for condensate or hot water resp. cooling water
12. Throttle valve for the pressure regulation
13. Patented sampling device
14. Recooling
15. Manometer
16. Thermometer
17. Automatic temperature program-regulator
18. Quick-rinsing device
19. Bottom valve
20. Safety valve
21. Water inlet

Fig. 46. Scholl high-pressure beam dyeing machine.

maintain a constant density throughout the dyeing process, and to give a final fabric of good dimensional stability.

The fabric should, if possible, be batched wet or damp, since the fabric seems to lay better in this state. Each selvedge must be guided at positions as near as possible to the point at which the cloth is taken on to the batch, either by selvedge straighteners alone or, with knitted fabrics, with uncurlers as well.

In dealing with a cloth whose selvedges are appreciably thicker than the cloth itself, it may be necessary to wind-on with a small

Fig. 47. Burlington high-pressure beam dyeing machine. (Courtesy of J. P. Stevens & Co., Inc.)

traversing movement. Otherwise, the wound batch will have a greater diameter at the sides than in the middle, and this can lead to poor dye penetration at the edges and an undesirable slackness of the rest of the cloth. Before winding-on the cloth to be dyed, the cylinder is first lapped with 8–10 layers of a porous bleached cotton cloth. This lapping spreads out the liquor as it comes through the perforations and prevents irregular dyeing of the batch proper, especially as a replica of the pattern of the perforations; it also filters out undissolved dye and any other solid matter which may be present in the liquor. The outside of the batch, after winding-on, is similarly lapped with cotton cloth to give an outer layer which is firm but not too hard. This procedure helps to minimize the retention of pockets of air which may be displaced during the first part of the circulation and then trapped in the outer layers of the batch, thus leading to irregularities of flow and hence to uneven dyeing.

The perforated zone must be less in width than the width of the cloth itself. The overlap of the wound cloth on to the unperforated parts of the cylinder should be equal to the thickness of the layer of wrapped cloth, with a maximum overlap of 2 in. at either end. This will normally give level dyeing at the edges of the cloth, avoiding excessive leakage but ensuring proper penetration.

Sets of beams of the same overall length, but with the perforated zones of different lengths, will in many cases provide for the satisfactory dyeing of cloths in a variety of widths. However, the use of a stainless-steel sleeve, in lieu of a beam of the proper width, has been used with success. In the latest Longclose designs, the whole outer surface of the barrel is free from projections, the effective width being adjusted by means of the telescopic end-shields.

Removal of Water

This stage consists in the removal of entrained water from the batch after the dyeing sequence, and not complete drying. There are four main methods, viz.

1. Centrifugal extraction, by rotation of the batch about its axis by means of a suitable spindle and housing. There are drawbacks attached to this method. If the cloth is not wrapped-on to give an even distribution of weight round its axis, excessive whip can occur, especially with long beams.

2. Vacuum extraction through the beam. Although frequently employed, this method suffers from the drawback that the outer layers may filter off very fine dirt suspended in the dyehouse atmosphere.

3. Partial drying with hot air. The air is fed under pressure to the inside of the beam and passes radially through it. This air can be both filtered and heated, and if the beam is in a closed stove, various systems of recirculation are possible.

4. The suction slot. The cloth passing from the batch to the stenter for final drying and setting passes over a slit suction extractor.

Limbert (32) describes the present position of beam dyeing, acknowledging its limitations as follows:

Warp-knitted fabrics in filament polyamide and polyester yarns, of weights and constructions suitable for shirtings, lingeries, etc., are eminently suitable for beam dyeing. Batches of up to 5,000 yd of fine denier, lightweight materials can be dyed on the larger machines.

Heavier shirtings in batches of 3700 yd can be dyed at full width and split to give 7400 yd of making-up width.

Raised-loop warp-knitted nylon fabrics can be dyed successfully.

Materials which are regularly processed on the beam include rainwear fabrics, dress goods, furnishings, outerwear fabrics from texturized yarns, jersey fabrics, etc. In Scandinavian and Western European countries, great strides have been made in beam-dyeing worsted suitings in wool–polyester blends.

The dyeing of acrylic jersey fabrics has already been mentioned as an example of materials for which beam dyeing is specially suitable. Double-jersey fabrics in the same textile material, but knitted from spun yarns, can also be beam-dyed successfully.

With beam dyeing, it is often possible to get more uniform and better penetrated dyeings in a shorter time than when the jigger or the winch machine is used, and much valuable detailed working information on the subject is available from the manufacturers of fibers and dyes, as well as from machinery makers. Beam dyeing is however, not the universal answer to all piece-dyeing problems. Two instances of the kind of difficulties that may arise with certain kinds of fabric are as follows.

1. The fabric may not be sufficiently porous in the wet state for the dye-liquor to pass through it readily. This may often occur when it is necessary to dye fabrics made from blends which contain appreciable proportions of normal cellulosic fibers. These can swell, particularly in the early stages before high temperatures are reached, thus closing up the fabric. Very tightly woven materials composed of man-made fibers, e.g. sailcloths, are also unsuitable for beam dyeing, owing to the inadequate rate of flow through the material.

2. Pile or raised fabrics, ribbed fabrics, and those with other weaves in which the effect is dependent on peculiarities of the surface structure, may suffer from treatment in firmly wrapped batch form. Pile fabrics may show unwanted crushing, which is serious when the dyeing conditions can cause this crushing to be set more or less permanently, for instance when the pile contains wool. Repps, cords, twills, etc., may, from the fact that one layer of cloth is pressed firmly on its neighbors, develop moire effects. A pronounced surface pattern may also lead to small undyed patches, owing to pockets of air being trapped between the layers of fabric.

These drawbacks are not necessarily absolute, and much may be done to avoid or minimize them by careful control of winding-on tensions, dyeing temperatures, and pump pressures.

The Pad-Roll Piece Dyeing Machine

The pad roll machine invented in Sweden includes a padding unit consisting of two 220-gal storage tanks containing an open or closed steam coil. The padding bath is fed from the storage tanks to a deep padder pan to increase the time of immersion. The padded fabric then passes through infrared tubes which raise the cloth temperature to that of the batching chamber. A second unit is called a residence chamber and can hold from 4,000 to 12,000 yd of fabric. The padded cloth is wound on a batch roll located inside the residence chamber and when the chamber is full the roll will be approximately 6 ft in diameter. During residence in the chamber the roll of cloth is turned slowly for several hours in a moist atmosphere at a temperature of 200–205°F. During residence in the chamber the padded dye diffuses into the fiber and the dyeing is completed when the fibers are uniformly penetrated.

The pad-roll machine appears to be particularly suitable for the Cibaphasol technique. Wool blends may present some problems because one-bath pad techniques must be developed for economical use of the machine (Erikson, Landquist, and Mellbin (18).

Continuous woolen piece dyeing

No other dyeing method lends itself so readily to continuous operation as does piece dyeing. Whereas continuous dyeing of cotton, rayon, and mixed piece goods has been successfully solved, the only really successful *continuous woolen piece goods dyeing* in the U.S.A., on a large scale was done during World War II in the dyeing of the 11-oz Navy flannel with indigo. The difficulties encountered in developing this process at the Hampshire Company were reported by George Linberg (33).

The essential parts of the machine were a rectangular, 3000-gal tank, 70 in. wide, through which the pieces pass in full width, threaded over and under a series of guide rollers, fully immersed in the reduced indigo liquor. Above the tank was a sky frame which is needed for the proper oxidation between dips (see Fig. 18). To produce the full shade, the flannel was given two dips of three minutes each, with a 5-min skying or oxidation after each dip. Two sets of squeeze rollers were necessary to recover the excess dye liquor and to assist in the proper penetration of the dye through the cloth. The speed of the machine was 10 yd per minute and the goods coming off the end of the machine were plaited on trucks. After dyeing, the cloth was scoured in regular dolly washers.

An improved continuous method has been described recently for dyeing wool and wool blends. The process involves the following operations—padding, intermediate drying by dry heat, "acid-shock" fixing, and finally washing to remove the thickening agent. The "acid-shock" fixing may be replaced partly or completely by steaming under pressure or by a wet heat treatment, e.g. by the pad-roll method. Pale, medium, and heavy shades with high fastness properties have been produced by this method.

The main reason why continuous piece dyeing of woolen and worsted fabric has not progressed further is the relative small number of large dye lots. A minimum of 20 pieces, approximately 2000 yd per shade is essential for economical dyeing. Even in large companies the number of dye lots of that size per week are limited, which means that a continuous machine would be idle 50–80% of the time. As stated by Noble (44), automation and continuous processing are clearly linked and it is worthwhile to rethink the whole question of continuous versus batch-wise processing by studying the articles by Stevenson (57) and Berndt (9), as a result of which it may be found that correctly planned discontinuous processing is the better.

Bleaching Wool

The wool fiber in its natural state always contains some pigment matter. Even the best grades of so-called white wool have a slightly yellow tint. In order to produce a clear white, it is necessary either to remove these coloring matters by bleaching or mask them by tinting, or by the use of optical bleaches.

In naturally colored brown and black wools there is a fairly high concentration of pigment matter and, although the amount of these colored wools is small compared with the normal white wools, a considerable portion of the pigmented wool is bleached, especially cashmere, camel hair and alpaca fibers, by the ferrous sulfate process developed by the Harris Research Laboratories (see page 360, Vol. 1).

To bleach wool by the tinting process means that the yellow cast of the natural wool is masked or neutralized by dyeing the fibers with a delicate tint of reddish-blue, or violet dye. It is, therefore, not a removal or a destruction of the natural pigment, but simply a change of the yellowish tint to a bluish tint, thus causing the wool to appear whiter to the eye.

The dyes suitable for the tinting of wool are reddish-blues or bluish-violets. Methyl Violet- and Formyl Violet 4BXN-types produce the best results. The actual amount of color required is very small and care must be exercised to avoid over-tinting, or over-bluing the wool to a strong bluish tone. Wool whitened in this manner does not possess the same clear white color as when the natural pigment is actually decolorized and destroyed. Destruction of pigments can be accomplished either by reduction or oxidation. Depending on the type of the bleaching agent, differentiation is made between oxidation bleach and reduction bleach. When the two are combined the process is known as "double bleach" or full bleach.

The main oxidation process is based on the use of hydrogen peroxide which is the widely accepted bleaching method in the wool industry.

Of the reduction bleach methods, the oldest using sulfur dioxide in gas form which is seldom met with anymore, has been replaced by the much simpler hydrosulfite bleach.

Hydrogen Peroxide Bleach

A cold bath containing per 100 gal, 10–12 gal hydrogen peroxide, 3% by weight (12 vol); or 1 gal hydrogen peroxide, 30% by weight (100 vol), is prepared and made slightly alkaline (pH 8–9) with sodium silicate, or ammonia. In place of hydrogen peroxide, sodium peroxide may be used in which case a cold bath is made up with 12–13 lb sulfuric acid per 100 gal and then 10 lb sodium peroxide is slowly stirred in. At this stage the bath, now slightly acid, is made alkaline by the addition of a small quantity of sodium silicate. The well-scoured material is saturated and then steeped overnight for some 12 hr.

Yarns and tops are often bleached in machines. The latter, like the open-bleaching vessels, must be made from wood, stone, earthenware, or stainless steel. Connecting pipes are made from clay or earthenware, the pumps for circulating the liquor may be earthenware or porcelain, and the heating pipes should be lead, and detachable. Stainless steel is also used for the heating coils and pumps or propellers.

It is advantageous to use an apparatus for bleaching that is provided with a special liquor tank which is fitted with a stirring arrangement, and which serves for setting the bleaching liquor and also for storing the bleaching solution. For 100 lb of wool and 100 gal of bleaching liquor, 27 lb of sulfuric acid and 20 lb of sodium peroxide [or 2½ gal of hydrogen peroxide (30 vol)] are used.

The bleaching liquor is then pumped from the storage into the bleaching tank in such a manner that it is first forced upwards through the wool, after which the pump is reversed. While the bleaching liquor is being circulated it is gradually heated to 115°F and kept at this temperature for 3–6 hr, according to the degree of bleach desired. When the bleaching is finished, the liquor is pumped back into the storage tank. The wool is removed and thoroughly rinsed first in clean, warm water, then in cold water slightly soured with sulfuric acid, and again rinsed. The pumps should not make more than 500 or 600 rpm.

For piece goods it is customary to use large earthenware tanks fitted with a winch. The goods, in string form, are run in the bleach bath from 4 to 6 hr at a temperature of 125°F. The bleach bath is prepared as follows: 250 gal of water are heated to 125°F and the necessary chemicals are added:

20	qt hydrogen peroxide (100 vol)
2	qt ammonia (free of iron), or
2½	qt silicate of soda (28°Bé)

To prevent too rapid liberation of the oxygen, a so-called "buffer" may be added, which acts as a stabilizer—for example, 4 lb of sodium pyrophosphate.

The exact strength of the bleach bath is determined by titration. The above bath will show:

3.1 g of oxygen per liter, or

2.15 vol with a pH of 9.4, and a total alkalinity of 8.

This is an ideal bleach bath for piece goods. After the desired bleach effect is reached the goods are taken out, brought to a string-washing machine, rinsed well with fresh water, and acidified with sulfuric, or formic acid. In some cases a small amount of dye, either methyl violet or ultramarine blue, is added during rinsing to brighten the white.

Peroxide "Dry-In" Bleach

The carpet industry has adopted a method of treatment for wool raw-stock called a "dry-in" peroxide bleach. The bleaching is carried out in the last bowl of a scouring train. This bowl contains a 1–2 vol peroxide bleach bath. The wool is scoured and rinsed in the first four bowls and then passed into the bleaching bowl where it remains for a short time. The wool is then dried and stored in bins where the peroxide continues its bleaching action for several days gradually improving the whiteness of the bottom.

In some cases a small amount of blue or violet dye is also added to assist in masking the yellowness of the wool. The method has been worked out by Shanley, Kaufmann, and Kibbel (54).

Hydrosulfite Bleach

Hydrosulfite bleaches are used in two ways, (*1*) alone, and (*2*) as an aftertreatment for peroxide bleach. A hydrosulfite bleach used alone is generally cheaper and less damaging to the wool fiber than peroxide. The compounds used contain sodium hydrosulfite which in addition to stabilizers sometimes are combined with optical bleach.

The amounts of sodium hydrosulfite required vary to some extent but 1.0–4.0% of hydrosulfite (on the weight of wool) is generally sufficient. The bath is run at 80–120°F for 1–8 hr., and the lot is then rinsed.

Double Bleach

In order to achieve the best possible white with the least amount of fiber damage, it was established by the results of Ziegler (64) that the prebleaching with hydrogen peroxide followed by air reduction afterbleach with sodium hydrosulfite works best.

Using the degree of white and the alkali solubility as a critical measure, Ziegler reported the following data:

Prebleach with H_2O_2 (30%), 10 ml/liter, 5 hr 50°C, pH 9
After bleach with 5 g/liter hydrosulfite, 5 hr 50°C

	Degree of white, %	Alkali solubility, %
Without after bleach	66.5	23.0
With after bleach	72	23.9

Alkali solubilities around 25% are considered normal for a bleached wool with a slightly impaired harsher hand. Any increase in the concentration of hydrogen peroxide or temperature and bleaching time will be reflected in an increased alkali solubility.

The dependence of the alkali solubility from the bleach temperature for a 4 hr treatment at 50, 55, and 65°C with increasing hydrogen peroxide concentration and the treatment time at a fixed hydrogen peroxide concentration with changes in the bleaching temperature from 40, 50, 55, and 65°C are seen from the curves in Figs. 48 and 49.

Under the assumption that the bleach degree is the same, it is possible to reduce the bleaching time down to 1 hr at 65°C with the same favorable alkali solubility.

Continuous Process

This approach opens the way to continuous piece bleaching processes as successfully used today by the cotton bleachers with the Becco and DuPont systems.

In 1962 Brit. Pat. 974,689 was granted to the *FMC Corp.* for a continuous process where loose wool after scouring is saturated with hydrogen peroxide solution (pH 7.5–9.5) containing an alkali–metal phosphate, steamed at 190–212°F for 1–3 min and then dried. The same technique is now in the pilot plant stage for bleaching wool piece goods.

Potassium Permanganate Bleach

Wool may also be whitened by the use of potassium permanganate. This method is sometimes used to clear the wool when used in

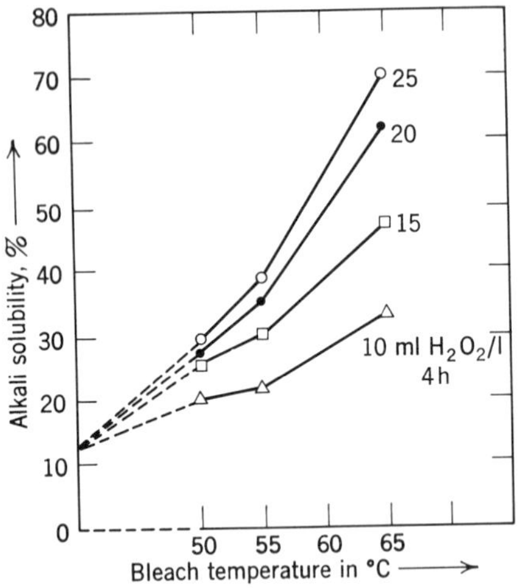

Fig. 48. Dependency of the alkali solubility on hydrogen peroxide concentration and temperature. (Courtesy of Ziegler.)

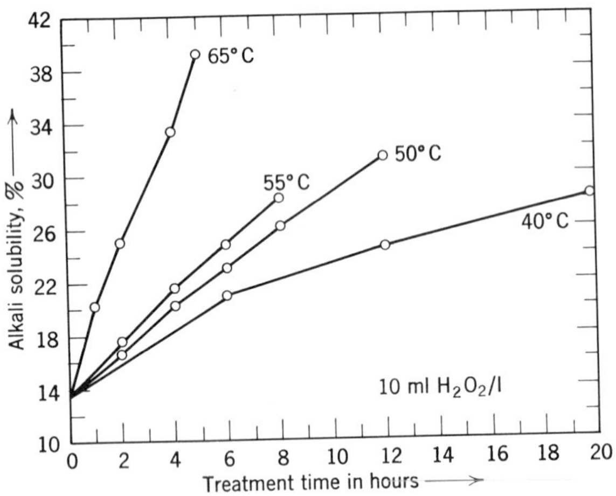

Fig. 49. Dependency of the alkali solubility on treatment time and temperature. (Courtesy of Ziegler.)

acrylic or polyester fiber blends. The method if comparatively simple as outlined below:

0.3 g/liter of potassium permanganate
0.5 g/liter of sulfuric acid
Apply cold for 20 min. Rinse thoroughly
Treat in a fresh bath with sodium *meta*-bisulfite to dissolve and remove the manganese dioxide which is formed in the first treatment
3–5% sodium *meta*-bisulfite (o.w.f.)
100–120°F for 20 min
Rinse thoroughly.

Bisulfite bleach

When maximum whiteness is not required but a less yellow cast is desired, wool may be partially bleached with sodium *meta*-bisulfite. This method has serveral advantages in that it is cheap, requires a short period for treatment and does not cause excessive damage to the wool fiber. The procedure follows:

3–5% sodium *meta*-bisulfite (o.w.f.)
1–3% sulfuric acid
Run at 100–120°F for 30 min
Rinse thoroughly

The white may be further improved by the use of a violet or blue tint in a fresh bath.

Sodium chlorite bleach

Serafirnoff, Valtocheva, and Mahljanska (53) have shown that urine-stained white wool can be bleached with sodium chlorite at a concentration of 3–10 g/liter preferably in a solution of acetic acid (4.8 ml/liter of 60%) and sodium acetate (5.44 g/liter) as activating agents (pH 3.5), treatment being carried out at 20–40°C for 5 hr.

Similar results can be achieved with hydrogen peroxide (30%) in concentration of 30 ml/liter, pH 8 temperature 50°C, time 4 hr, alkali solubility 22.4%, strength loss around 6%, elongation loss about 25%.

Optical bleaching wool

In recent years the use of optical bleaches for whitening wool has increased tremendously. There is a reason for this popularity because now, for the first time, it is possible to produce a true white on wool. Regardless of how white a chemical bleach may make wool it can always be greatly improved by an addition of an optical bleach having the required chroma and fluorescence.

Optical bleaches do not tender wool, or change them chemically, because they are actually colorless dyes, which in some cases have structures somewhat similar to those of direct and disperse dyes. These compounds exhaust from the dye bath quite readily and are generally quite fast to washing. Optical bleaches cause wool and other fibers to appear whiter because they fluoresce and emit blue to reddish-blue visible light which masks the natural yellowness of the wool fiber. The efficiency of an optical bleach for wool is dependent upon the color of the fluorescent visible light emitted by the compound. If an optical bleach has a white or greenish fluorescence it will not mask the yellowness of the wool and, therefore, will not whiten to the same extent as an optical bleach which has a reddish-blue fluorescence and chroma. Of the various chemical classes of fluoro-brighteners, stilbene derivatives

are the best.

Wool which has been bleached with peroxide, sulfur dioxide or other chemical bleaches can be greatly improved by treatment with an optical bleach, such as Calcofluor White RW. A general formula is as follows:

0.02–0.3% Calcofluor White RW
2.0% acetic acid, 28%
Start treatment at 100°F and gradually raise the temperature to 120–140°F, and run at the desired temperature for 30 min. Rinsing is at the discretion of the dyer.

The optically bleached wool should not be dried at elevated temperatures. The lower the drying temperature the clearer and more desirable is the white.

In the past, baby blues and pinks were dyed and then stoved (exposed to sulfur dioxide fumes) overnight. Optical bleaches have eliminated such treatments because they can be added to the dye bath, or they may be applied as above from a fresh bath.

When a partial white is desired the wool can be given a half bleach followed by treatment with an optical bleach.

Wool–acrylic fiber blends cannot be bleached by the usual "Textone" (sodium chlorite) nitric acid bleach because of its damaging effect on the wool. In such cases a peroxide bleach followed by treatment with certain optical bleaches will produce an excellent white on such blends.

Wool–Dacron (polyester fiber) blends can be made to appear whiter by treatment with a suitable optical bleach.

Bleaching and Dyeing in One Bath

To produce clear and brilliant pastel shades on wool, prebleaching generally preceded the dye process. This double process often produced tender pieces especially in woolen flannels which have a low strength level (around 25 lb) before dyeing; a real trouble for the dyer. Products to help the dyer to overcome this are mild reductive bleaching compound such as Erioclarit B which make it possible to combine the bleaching and dyeing. The results of such as procedure are reported by Schaffner (52). Another advantage of this product is its compatibility with moth proofing agents such as Mitin FF conc.

Effect of Stabilizers

The influence of Stabilizer C (a mixture of sodium oxalate and sodium pyrophosphate [Laporte]), sodium pyrophosphate, trisodium phosphate, and sodium silicate on the decomposition of solutions of hydrogen peroxide, as well as on the bleaching effect produced on a wool fabric, has been studied by J. Cegarra, J. Ribe, and J. Gacen (12). The pH of the peroxide solutions remains constant, except for the solutions containing trisodium phosphate, whose pH increases. The release of oxygen is regulated by the nature of the stabilizing anion and the pH of the solution. In the pH range 8.5–9.5, the order of stabilization of oxygen release by the four compounds was stabilizer C, sodium pyrophosphate, sodium silicate, and trisodium phosphate.

The presence of wool in bleaching solutions decreases the pH of the solution, the oxygen release being regulated by the nature of the stabilizing anion. No relation was found between the degree of whiteness obtained and the loss of oxygen from the bleaching solution. The cystine content of the wool decreases during bleaching and its alkali solubility increases, both values being related to the whiteness obtained; the tryptophan content of the wool also decreases considerably. Stabilizer C gave the highest reflectance for a given oxygen loss, the order then being sodium pyrophosphate, sodium silicate, and trisodium phosphate.

The properties used to evaluate the efficiency of each stabilizer in bleaching are given in Table 12.

Merkle (37) studied the relationship between the chemical and physical properties of wool bleached with hydrogen peroxide. As in many other finishing operations, the chemical changes are easier proven than the physical. In general the mechanical properties

TABLE 12

Efficiency of Bleaching Stabilizers

			Quality of bleaching				
Wool bleached in presence of	Stabilizer concn. (g/l.)	pH	Reflectance, %	Alkali solubility, %	Tryptophan, %	Cystine, %	Cysteine, %
Unbleached	0		35.2	9.48	0.382	9.53	0.28
Stabilizer C	0.6	8.5	48.2	21.40	0.079	8.36	0.22
	6.0	9.0	50.9	26.64	0.085	7.85	0.22
Sodium pyrophosphate	0.3	8.5	49.1	19.59	0.070	8.55	0.27
	2.0	9.0	49.4	27.14	0.070	8.40	0.22
Trisodium phosphate	0.25	8.5	48.7	21.65	0.093	7.90	0.23
	0.80	9.0	47.7	21.76	0.082	8.51	0.24
	2.85	9.5	47.6	21.63	0.075	8.43	0.23
	0.2	8.5	45.9	22.62	0.075	8.37	0.22
Sodium silicate	1.0	9.0	47.9	24.89	0.082	7.56	0.37
	2.8	9.5	52.3	42.88	0.110	6.22	0.22

alter very little when the bleaching is carefully done. With treatments ranging from 10, 20, and 30 ml of hydrogen peroxide per liter at 50°C and pH 9 for 8 hr some of the results are given in Table 13.

These results correspond very closely to the findings of Ziegler (90).

The permanence of wool bleaching was the basis of a study by Norton and Nicholls (45). There is a difference in the efficiency and permanence of the bleaching between wools discolored by heat or alkali. In general alkaline yellowing was more easily removed than heat yellowing. The bleaching action of peroxide was found to be more efficient than the reducing agents but less permanent especially when followed by further alkali treatment.

Anti-Yellowing Treatment

When exposed to sunlight, wool turns yellow, particularly when the wool is wet and alkaline, even as stated above when bleached with peroxide. The same is true for wool treated with fluorobrightening agents which actually accelerate the yellowing. In the last few years observation made at the C.S.I.R.O. Wool Research Laboratories in Melbourne and Geelong has helped them to develop an antiyellowing treatment. They found that Thiourea and formaldehyde applications retard the yellowing of wool and

TABLE 13

Chemical and Physical Properties Changes in Bleaching

Material	Worsted top				Fabric			
ml H_2O_2/liter	0	10	20	30	0	10	20	30
Relative abrasion numbers					100	80	76	74
Alkali solubility %	16	26	33	48	12	20	25	36
Urea bisulfite solubility %	32	37	49	59	10	16	20	55
Cystine content % 10^{-5} mol/g	11.6	10.7	9.9	9.1	11.2	10.6	10.2	9.4
Methylene blue absorption, 10^{-5} mol/g	6	17	28	41	11	19	25	31

also of wool treated with a fluoro-brightening agent. Reductive bleaches are better than oxidative bleaches, since the latter accelerate the yellowing.

Combining the advantages of the three findings a reductive bleach followed by a treatment with a stilbenetype fluoro-brightener, Leuchophor PAF, plus thiourea and formaldehyde leads to the production of a much better white together with a marked retardation of the yellowing on exposure to sunlight. A similar treatment can be used to improve the brilliance of bright shades of wool. Here the improvement is obtained by removing the yellowness of the untreated wool fibers. Fuller details can be obtained from C.S.I.R.O. Technical Report G.16.

Previous to these findings Lundgren (74) reported that suitable yellowing inhibiters are certain phosphatesters and butylitanate but the compounds are uneconomical. Pardo (76) found that yellowing and heating up of moist grease wool can be inhibited by formaldehyde, applied by spraying as an aqueous solution or dusting in the form of paraformaldehyde. For maximum protection the grease wool should be treated as soon as possible because significant yellowing, associated with growth of microorganisms, can occur within 2 days in a mass of grease wool containing over 20% moisture. An extensive review of the chemical aspect of the yellowing of wool was made by Hoare (72).

Vigoureux or Melange Printing

In general, oxfords and other blends are produced by dyeing wool top to a desired depth of shade and blending it with various proportions of undyed tops to obtain a wide variety of oxford shades. This method has a rather serious drawback in that it is impossible to obtain uniform distribution of the dyed and undyed fibers. The resulting fabrics have a somewhat streaky appearance owing to the fact that bunches of colored or white wool fibers tend to show on the surface of the cloth. This defect can be overcome by Vigoureux printing of the wool top. In 1863 patents were granted to Stanislas Vigoureux for a machine for printing slubbing.

Vigoureux printing consists of printing a web of wool slubbing with wide or narrow bars of color. The printing machine is operated on the same principle as that used for printing other materials, except that instead of using an engraved roller to carry the color paste to the slubbing, a roller with raised bars is employed. The extent of color in a mixture is dependent upon the size and depth of the colored areas as compared to the portion which remained uncolored.

Machines now in use often have two rollers engraved with alternating spiral ribs. This technique makes it possible to obtain wide variations in the degree of coverage obtained. Buxtorp and Wiazmitinow (10) give the following description:

In the printing of woolen cloth the wool is normally chlorinated before being printed, but this treatment is totally unsuitable for slubbing to be printed by the Vigoureux process, although it has often been suggested. The reason is, of course, that slubbing printed by this process must possess the same spinning and weaving properties as normal slubbing, and such clorinated wool would not be suitable for production of the highest grade fabrics.

It is essential during the Vigoureux process that good penetration of the slubbing web should be obtained, and for this reason the printing pastes should not be too thick. Before transferring printed slubbing from the Vigoureux machine to the steaming cabinet, it is advisable to allow it to lie for a short while as it is delivered from the machine, so that the printing paste may fully penetrate the wool fibers.

The first requirements of dyes for Vigoureux printing are good working properties, involving good solubility and rapid fixation by the steaming process allowed, and also a high standard of fastness to washing, milling, light, perspiration, and rubbing. The classes of dyes which meet these requirements are the fast-to-milling acid dyes and the more recently introduced metal complex dyes of the Cibalan (Ciba), Irgalan (Gy), Lanasyn (S), and Neutracyl (DuP) types, and also the afterchrome dyes. Of the latter class, it is necessary to select those members with which development of the chrome lake is most easily effected. Recommendations have also been made for the use of the Reactive dyes such as the Procions and Cibachrones attaining good wet fastness within a short steaming time.

The following recipe (in parts by weight) may be regarded as typical for the acid and the metal-complex dyes.

20	Dye
40	Thiodiglycol or pyridine
320	Water
50	Glycerol
500	British gum (1:1)
60	Ammonium tartrate
10	Sodium chlorate
1000	

Thiodiglycol, pyridine, and water act as solvents; glycerol keeps the slubbing moist, and prevents it from drying out before steaming; while British-gum thickener can be replaced in the case of acid dyes by Sennar gum, dextrin, tragacanth, etc. Ammonium tartrate or oxalate, and oxalic acid where suitable, are used with acid dyes for fixation. Sodium chlorate prevents any reducing action either by the fiber or by the thickener during steaming. The addition of an agent to counteract foaming has proved of value for dyes having this tendency.

The following is a typical recipe for chrome dyes:

30	Dye
40	Ammonia (25%), thiodiglycol, or pyridine
190	Water
50	Glycerol
500	British gum (1:1)
20	Ammonium oxalate
30	Chromium fluoride, $CrF_3 4H_2O$ – Heat to dissolve
130	Water
10	Sodium chlorate
1000	

Ammonium oxalate and oxalic acid act as fixing agents; British gum and certain lichenous preparations, marketed under the name *Sennar gum*, as thickeners; and chromic fluoride, formate, and acetate as chrome mordants. Similar recipes and details of the most suitable solvents can be found in the pattern cards issued by dye manufacturers.

In bulk practice there are two methods for preparing the printing paste. Sometimes a stock solution is prepared from the dye, the solvents, glycerol, and British gum, and the other chemicals are mixed into this shortly before printing. This procedure has the advantage that larger quantities of stock color can be stored at one and the same time, but the drawback is that the tinctorial power does not remain constant when the solution has been standing for some time, even after thorough stirring. Other printers prefer to make up the paste separately for each run, to make sure of getting a consistent shade.

Normally, for printing, 16 tops of slubbing are mounted on the creel, and from each of these tops a sliver of about ½ oz/yd average weight is run through the gillbox attached to the machine, which usually has a draft of 1:4. A web 12–20 in. in width, weighing approx. 3 oz/yd, is delivered from this gillbox and passed between the printing and the color rollers.

The color roller, made nowadays of metal covered with wool felt, is impregnated with print paste from a rubber roller rotating in the trough. The printing roller is of bronze and is engraved with raised spiral or lateral ribs.

The web is thereby impregnated with the print paste as the raised surfaces of the printing roller or rollers exert pressure on it. The printed slubbing is then conveyed by a mechanical folding device on to metal trays or cages standing on a lattice which has a lateral traverse. See Fig. 50.

The color is fixed on the wool by steaming, but, as previously mentioned, it is always advisable to allow the slubbing to lie for a short time before the batch is actually conveyed to the steaming cabinet. It is important, however, that the color should not dry into the slubbing, otherwise difficulty will be experienced in obtaining full development of the shade. Steaming is usually carried out for two separate periods of 1–1½ hr each with damp non-pressurized steam, and the slubbing should be allowed time to cool between the two steaming operations. Fast-to-milling acid dyes and dyes of the metal-complex types previously referred to require only one steaming to attain their full fastness properties.

The British Patent 930,429 covers the use of the Cibaphasol system for Vigoureux printing of wool slubbing. In an Australian mill, using this technique in applying Cibalan and chrome-dyes the steaming, as done in an horizontal autoclave at 5 lb pressure, ranged from twenty five minutes for light greys to a maximum of 55

Fig. 50. Vigoreux printing. (Courtesy of A. B. Skanska Yllefabriken, Sweden, Dr. B. Melin.)

min for dark shades up to 90% coverage. What helped the steam penetration through the printed top was its loading into open stainless steel carriages; 32 lb per carriage are loaded in four layers of 8 lb each, divided by wide open screens easily removable, resting on the solid frame.

In Fig. 51 are shown the main construction details of a cabinet shaped autoclave normally used for the steaming process of Vigoureux printed top. They have capacities from 50 kg (110 lb) to 200 kg (240 lb), ranging in size from approximately 4½ ft width, 6½ ft height and 6½ ft length to 4½ ft width, 7½ ft height and 12½ ft length. Maximum working pressure, 0.5 atm.

Steaming has been a patch process since the inception of Vigoureux printing a hundred years ago. Unquestionably, many experiments have been made to make it continuous by using backwasher type of dryers; but it is only since the perfection of steamers such as the Ilma and Fleissner used for continuous dyeing of top that it can be done successfully. The final goal is to combine printing, steaming, backwashing and drying into one continuous range. (See Figures 18 and 20.)

After being steamed, the wool is passed through the backwashing machine, and for thorough and complete removal of surplus dye, and thickener, it is desirable to use four- or five-bowl machines for this purpose. The backwashing machine provides for subsequent drying and also for the first gilling operation in the gillbox which is attached to the drying section of the backwashing machine.

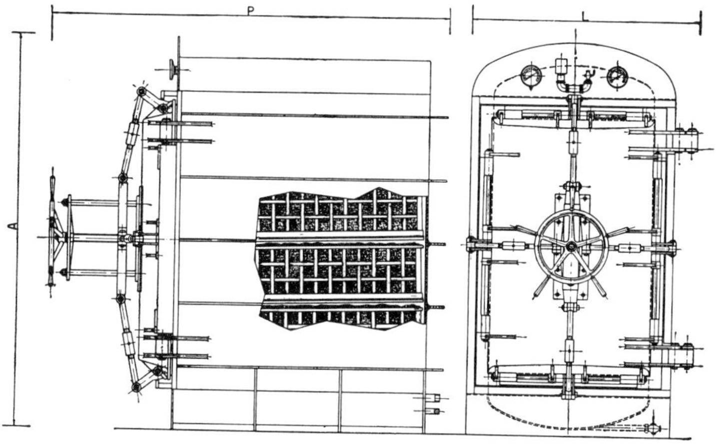

standard models capacity in Kg.	allround dimensions (in mms.) L	A	P	number of baskets
50	1250	1940	2000	3
100	1320	2240	2330	4
150	1320	2240	3100	8
200	1320	2240	3700	8

Fig. 51. Ilma autoclave for steaming Vigoreux top.

Before being passed to the spinners, the wool is gilled several times and matched to shade in the form of a small felt pattern.

It is desirable that slubbing for Vigoureux printing should not be of too high a pH value, and generally speaking 8.5 should be considered the maximum. The reason for this is that higher pH values would tend to produce yellowing of the white during steaming. For similar reasons, it is not desirable to use an alkaline soap in the backwashing machine, a synthetic detergent being preferable.

The oil content of the slubbing as delivered to the Vigoureux printing machine is also important: too high an oil content tends to give prints with poor rubbing fastness. On the Continent, dry combed tops are generally used where Vigoureux printing is to be undertaken.

The Printing of Wool Piece Goods

Pretreatment

Special steps are required in the preparation of wool fabrics for printing. In its normally scoured condition wool is not receptive to print colors and, therefore, wool is generally chlorinated to increase its natural affinity for printing colors.

The problem in chlorination is to properly control the reaction so as to get even application of the chlorine to all the fibers across the piece. When this is not achieved the result will be plotchy prints. To overcome this hazard finishers prefer to use retarding compounds such as Melafix (Ciba), Chloregal (Geigy) or Nikrulan CLF (Cassella).

In the Melafix process wool pieces can be treated in a string washer by immersing them in a bath containing diluted hydrochloric acid, a wetting agent and Melafix a melamine formaldehyde resin. Dilute sodium hypochlorite is then run into the bath and chlorination proceeds over a period of ½ to 1 hr, followed by dechlorination with sodium bisulfite. The melanine resin acts as a reservoir for the chlorine, releasing it gradually. For control from bath to batch the increased affinity for dyes should be tested with Benzo purpurin 4B (CI23500).

Bleaching with peroxide will also make the wool more receptive to the print color. However, unless the effect of the bleaching operation is desired for whiter grounds, chlorination is usually preferred.

Most of the woolen dress fabrics in the United States are screen printed. The process of screen printing is a modern adaption of an old Chinese method of applying designs to cloth.

Screen printing is done on 60–80 yd long tables in order to permit the printing of one whole piece of cloth. They are covered with ¾ in. soft felt and this covering protected by greige cotton which represents the permanent covering of the table. Another temporary protective covering is put on before the print cloth is applied which is changed with patterns of color combinations. The print tables have a guide rail to which stoppers are attached. The distance between the stoppers is governed by the repeat of the pattern. The printing screen comprises a wooden or aluminum frame to which a suitable fine woven cloth screen is attached, either made of silk, nylon, or vinyon. The material is stretched tightly over the frame, then a pattern is painted in, each color requires its own screen. The unfilled, unpainted areas permit the print paste to be squeezed through the cloth with a scraper. For proper fitting between the various colors, the screen is provided with a metal projection fastened to the frame. In putting down the screen, the metal projection fits flush with the stopper on the guide rail. After the cloth is printed it is hung above the table for drying and a new layer of cloth is put on the table. After drying, the cloth is steamed and aged by hanging it on star frames which are put in a steam chamber and exposed to moist steam, fixing the color. The pieces are then washed to remove the thickener.

Acid Dyes

A line of acid dyes having adequate fastness and shade requirements may be selected from the large number of these dyes which are available. The formula for printing varies according to the available equipment and steaming conditions. Acid dyes are printed from an acid print paste with hygroscopic materials present to aid in fixation.

Where a long steaming process is feasible the following simple formula is satisfactory:

Long Steaming Procedure

4-Ounce Color
3 lb acid dye
½ gal glycerine
3 gal water
7 gal gum tragacanth, 6 oz/gal
½ gal acetic acid, 56%
2 lb oxalic acid
1 gal water

The above 4-ounce color is reduced for lighter shades with thickening containing glycerine and acid as follows:

Reduction thickening
7 gal gum tragacanth, 6 oz/gal
½ gal glycerine
1 lb oxalic acid
4½ gal water

The prints are dried as lightly as possible and steamed 45–60 min without pressure, supplying as much moisture as possible by maintaining a strong flow of steam.

The steamed prints are washed well in water to remove the thickener and may be soaped lightly at 100°F followed by rinsing and finishing off in an acetic acid rinse.

Where long steaming is not feasible, satisfactory results can be obtained with a 10-min aging by increasing the amounts of hygroscopic and swelling agents. The following formula is satisfactory:

Short Steaming Procedure

4-Ounce Color
3 lb acid dye
10 lb urea
1½ gal thiodiethylene glycol
2 gal water
7 gal gum tragacanth, 6 oz/gal
1¼ gal ammonium sulfate, 4 lb/gal

Reduction Thickness
10 lb urea
1½ gal thiodiethylene glycol
2¼ gal water
7 gal gum tragacanth
1¼ gal ammonium sulfate

The prints are dried moderately, aged 10 minutes in a vat ager, rinsed well in cold water, soaped lightly, rinsed and finished in weak acetic acid solution.

Basic Dyes

Shades of greater brilliance but of lesser fastness may be printed on wool with basic dyes. These are also particularly useful in discharge patterns as a full range of hydrosulfite-resistant shades can be obtained by blending suitable dyes.

Discharge Formula

Hydro Paste
12 oz basic dye
10 lb urea
1½ gal thiodiethylene glycol
2½ gal water
7 gal gum tragacanth, 6 oz/gal
1 gal hydro paste, 4 lb/gal

The amount of hydro paste is regulated to suit the depth of the dyed ground.

For application printing the hydro paste is omitted altogether and replaced by plain gum tragacanth paste. The prints are aged 10 min and washed in the usual manner.

Direct Dyes

Very few direct dyes exhibit any advantages over acid dyes of similar shade for the printing of wool. The principal interest lies in their use for printing fabrics of mixed fibers such as wool–cotton or wool–viscose unions. More level coverage with moderate fastness to washing and good fastness to dry cleaning is thus possible. Many of the fast-to-light direct dyes show good fastness to light when printed on wool.

Printing Formula for Direct Dyes for Wool

Concentrated Paste
3 lb direct dye
10 lb urea
1½ gal thiodiethylene glycol
2 gal water
7 gal gum tragacanth, 6 oz/gal
1¼ gal ammonium sulfate, 4 lb/gal

Reduction Thickener
10 lb urea
1½ gal thiodiethylene glyclo
2¼ gal water
7 gal gum tragacanth
1¼ gal ammonium sulfate

The prints are dried moderately, aged 10 min in the vat ager, washed well in cold water, soaped lightly, rinsed and dried.

Where pressure steaming is available, the amounts of urea and thiodiethylene glycol can be reduced according to the time and steam conditions.

Vat Dyes

Vat dyes are of minor importance in the printing of wool. The use of caustic soda or sodium carbonate results in degradation of the fiber which limits the use of these dyes. The use of ammonia or triethanolamine in place of the carbonates has been suggested but no considerable amount of commercial production has been run by this method.

For small peg prints where the detrimental effect of the alkali on the wool can be tolerated, a typical rayon formulation may be used. No caustic alkali should be included and for this reason many of the anthraquinone vats are unsatisfactory. However, the indigoid class of vat dyes gives good results as to shade and color yield.

Soluble Vat Dyes

Very little wool is printed with soluble vat dyes although satisfactory results may be obtained by printing with the chlorate neutral-aging method. A steaming of one-half to three-quarters of an hour may be required for good fixation and shade development.

Oxidation Colors

Aniline black may be printed on wool if a deep black shade at low cost is desired. The ordinary cotton formulae do not develop on wool. However, a more rapid developing, vanadium-black type, which is less stable in the print paste, will produce a satisfactory black shade.

Print Formula for Aniline Black on Wool

12 lb aniline salt
2½ gal water
½ pint aniline oil
5 gal gum tragacanth, 6 oz/gal
12 oz *para*-phenylenediamine
1 gal acetic acid, 56%

$6\frac{1}{2}$ lb sodium chlorate
3 gal water
1 qt ammonium vanadate, 1:1,000 solution

The aniline salt is dissolved in 2.5 gal of water and the aniline oil added. After stirring until the aniline oil is taken up, the solution is added to the gum tragacanth. The *para*-phenylenediamine is dissolved by warming in the acetic acid, and is then added. The sodium chlorate is dissolved and added next. At this point the color is fairly stable. The ammonium vanadate is added to the color immediately before using.

The aniline black is printed on chlorinated wool and dried to an olive green shade. The prints are then aged 3 min and washed well in water, soaped lightly, and scoured in weak acetic acid solution.

Other types of oxidation colors have not found any use on wool.

Additions of Solvents

Belle, Smith, and Stevens (8) made a study of the effect on dye transfer in wool printing by the addition of solvents to the printing paste. Addition of up to 12% benzyl alcohol to a printing paste containing a relatively hydrophobic dye, C.I. Acid Blue 140, greatly increased the rate of dye fixation, even at room temperature. A similar increase in rate of dye fixation was observed with a 2:1 dye–metal complex. The effect was much less marked with the highly sulfonated dye C.I. Acid Blue 92, but when a cationic agent was incorporated in the print paste the rate of fixation of this dye was also increased by addition of benzyl alcohol.

The Australian formic acid dye method by Harrap and Rosman (22) lends itself also to wool printing without the need of after steaming. This printing method has been successfully applied by a wool blanket mill to stamp each blanket with their trademark. Polyethylene glycol has been found a suitable thickener for the printing paste.

Spray Printing

A new process of applying dye to wool fabrics is being developed by the C.S.I.R.O. Wool Laboratories Division of Textile Industry. The product gives unique marl and mottle effects. In this process, dye solution is sprayed onto a fabric in a series of overlapping weft-ways bands. After each traverse of the spray head the fabric is moved forward to present a fresh area to the next traverse of the spray head.

Dye solution is fed to a narrow wire mesh band by a pair of flexible tubes. Excess dye is removed by pairs of mangle rollers on each side of the air jet. The air pressure at the jet defines the size of the droplets of dye solution sprayed from the gauze. The lower the pressure, the larger the droplets. This is the main way of controlling the pattern of dye on the fabric.

The dye solution is similar to that described for the continuous dyeing of tops with the Urea Process. To complete the dyeing of the wool, the sprayed fabric must be steamed and then scoured to remove urea and loose dye.

References

1. Angliss, I. B., and Delmenico, J., *J. Soc. Dyers Colourists*, **80**, 543 (Oct. 1964).
2. Angliss, I. B., and Delmenico, J., *C. S. I. R. O. Wool Textile News*, No. 12 (1966).
3. Angliss, I. B., Brady, P. R., Delmenico, J., and Hine, R. J., *Textile J. Australia*, **43** (No. 4), 17 (April 1968).
4. Barritt, J., and Elsworth, F. F., *J. Soc. Dyers Colourists*, **64**, 19 (1948).
5. Bauer, J., *Am. Dyestuff Rept.*, **45** (April 1956).
6. Beal, W., *J. Soc. Dyers Colourists*, **83**, 3 (1967).
7. Beal, W., Dickinson, K., and Bellhouse, E., *J. Soc. Dyers Colourists*, **76**, 6, 333 (June 1960) 341.
8. Bell, J. W., Smith, P. J., and Stevens, C. B., *J. Soc. Dyers Colourists*, **79**, 305 (1963).
9. Berndt, F., *Z. Ges. Textil Ind.*, **66**, 669 (1964).
10. Buxtorf, F., and Wiazmitinow, A., *Proc. Symp. Textile Printing*, **1953**, 26–32.
11. Casty, R., *Melliand Textilber.*, **41**, 1365 (1960); *Can. Textile J.*, **77**, 47 (1960). *Ciba Rev.*, **113** (1956). *Textil.-Rundschau* (1963).
12. Cegarra, J., Ribe, J., and Gacen, J., *J. Soc. Dyers Colourists*, **83**, 189 (1967).
13. Collins, D. P., *Am. Dyestuff Rept.*, **53**, 219 (1964).
14. Dauble, M., and Theilig, D., *Melliand Textilber.*, **47**, 279 (1966).
15. Delmenico, J., and Peters, R. H., *Textile Res. J.*, **34**, 207 (1964).
16. Derbyshire, A. N., and Lemin, D. R., *J. Soc. Dyers Colourists*, **80** (7), 363 (1964).
17. Elod, E., *Trans. Faraday Soc.*, **29**, 327 (1933).
18. Erikson, C. O., Landquist, N., and Mellbin, B., *Am. Dyestuff Rept.*, **44**, 877–881 (1955); *Am. Dyestuff Rept.*, **46**, 770–776 (1957).
19. Frieser, E., *Textil-Rundschau*, **17**, 1–12 (1962).
20. Hadfield, H. R. and Lemin, DR, *J. Soc. Dyers Colouristt* **79**, 715 (1961).
21. Hallack, B.S., Australion Pat. 23540 (1962).
22. Harrap, B. S., and Rosman, R. E. J.,C.S.I.R.O. Report No. M4 (1964).

23. Hausermann, J., and Hablintzel, E., *Melliand Textilber.*, **46**, 519 (1965).
24. Hine, R. J., and McPhee, J. R., *Dyer*, **132**, 523 (1964); C.S.I.R.O. Report G12, Dec. (1964).
25. Hine, R. J., and McPhee, J. R., *J. Soc. Dyers Colourists*, **81**, 268 (1965).
26. Hirsbrunner, H. R., *Textil-Praxis*, **1**, 52 (1962).
27. Johnels, A., Discs, Chalmers Technical University, Goteburg, 1967.
28. Karrholm, E. M., and Lindberg, J., Theses, Chalmers Tekinska Hogskola, 1960; *J. Textile Inst.*, **51**, T1323 (1960).
29. Kehrer, *Ciba Rundschau.*
30. Khachoyan, J., *Cirtel*, **3**, 233 (1965).
31. Leddy, J. A., *Am. Dyestuff Reptr.*, **56**, 52 (1967).
32. Limbert, K., *J. Soc. Dyers Colourists*, **82**, 97 (1966).
33. Linberg, G. O., *Am. Dyestuff Reptr.*, **32**, 33 (1943).
34. Lutzel, L., *Melliand Textilber*, **46**, 504 (1965).
35. Maurer, H., *S.R.F. Fachorgan*, **17**, 593 (1962).
36. Mazingue, A., *Melliand Textilber*, **47**, 166–172 (1966); *Textil Veredlung*, **3**, No. 5 (1968).
37. Merkle, R., *Textil-Rundschau* (1964).
38. Meunier, P. L., *Am. Dyestuff Reptr.*, **45** (1956).
39. Milligan, B., *J. Soc. Dyers Colourists*, **77**, 106 (1961).
40a. Millson, H. E., Calco Tech. Bulletin No. 809, 1949.
40b. Millson, H. E., *Am. Dyestuff Reptr.*, **47**, 339 (1958).
41a. Millson, H. E., Watkins, W. H., and Royer, G. L., *Am. Dyestuff Reptr.*, **36**, 45 (1947).
41b. Millson, H. E., and Turl, L. H., *Textile Reptr.*, **21**, 685 (1951); *Am. Dyestuff Reptr.*, **39**, 647 (1950).
42. Newcomb, W. J., and Ward, G. C., *Am. Dyestuff Reptr.*, **49**, 55 (1960).
43. Niederhauser, J. P., *Tiba*, **1**, 21 (1923).
44. Noble, P. G., *Rev. Textile Prog.*, **1964**, 288.
45. Norton, G. P., and Nicholls, C. H., *Cirtel.*
46. Ordway, Ch. B., *Am. Dyestuff Reptr.*, **51**, No. 43 (1962).
47. Ordway, Ch. B., *Am. Dyestuff Reptr.*, **53**, 110–114 (1964).
48. Peters, L., and Stevens, C. B., *Dyes*, **115**, 327 (1956); *J. Soc. Dyers Colourists*, **72**, 100 (1956).
49. Primentas, N., *Textile Inst. Ind.*, **2**, 80 (1964).
50. Rizzo, F. J., and Ramsley, A. O., *Am. Dyestuff Reptr.*, **50**, 374 (1961).
51. Rizzo, F. J., Merola, A., and Evans, R. J., Progress Report U.S. Army Natick Laboratory, 1966.
52. Schaffner, K., *Textil Veredlung*, **1**, 334 (1966).
53. Serapinoff, S., Valtschewa, R., and Mahljanska, T., *Melliand Textilber*, **44**, 53 (1963).
54. Shanley, E. S., Kauffman, H. O., and Kippel, W. H., *Am. Dyestuff Reptr.*, **40**, 1–4 (1951).
55. Speakman, T. B., and Elliott, G. H., *J. Soc. Dyers Colourists*, **59**, 185 (1943).
56. Stearns, E. I., *Am. Dyestuff Reptr.*, **53**, 668 (1964).
57. Stevenson, F. M., and Stevenson, M. C., *J. Textile Inst.*, **53**, 649 (1962).

58. Stewart, R. S., *Am. Dyestuff Reptr.*, **55**, 361 (1966).
59. St. John, personnel communication.
60. Townend, F., *J. Soc. Dyers Colourists*, **61**, 144 (1945).
61. Von Bergen, W., *Textil-Rundschau*, **14**, Heft 2 (1959).
62. Von Bergen, W., *Melliand Textilber.*, **4**, 23, 77, 123 (1923); **7**, 451 (1926).
62a. Von Bergen, W., et al., *Am. Dyestuff Reptr.*, **27**, 485 (1947).
63a. Von Bergen, W., and Wakelin, J. H., Text. Res. J., 22, 123 (1952).
63b. Von Bergen, W., Crowley, T., and Brommelsick, W., *Am. Dysetuff Reptr.*
64. Wakelin, J. H., and von Bergen, W., Int. Wool Project Summary Report, *Textile Res. J.* (1953).
65. Ziegler, K. E., *Textil-Praxis*, **17**, 376 (1962).
66. F. M. C. Becco Bulletin #35, 1952.
67. International Textile Soc., *Dyeing/Finishing*, **1** (Item 358), Page 23 (1968).
68. *Wool Science Review*, **15** (1956); **16** (1956).
69. *Wool Science Review*, **30**, 1 (1966); **31**, 32–39 (1967); **32**, 36 (1967).
70. *Wool Science Review*, **30**, 16 (1966); **31**, 1–11 (1967).
71. Brommelsick, W. F., von Bergen, W., and Millson, H. E.
72. Hoare, J. L., *Am. Dyestuff Reptr.*, **44**, 73 (1955); Communication No. 2, Wool Res. Org., New Zealand, 1968.
73. Lewis, D. M., and Seltzer, I., *J. Soc. Dyers Colourists*, **84**, 501 (1968).
74. Lundgren, H. P., *Proc. Intern. Wool Text. Tes. Conf., Australia*, **374** (1955).
75. Mosimann, W., *T. C. & C.*, **1**, 182–189 (1969).
76. Pardo, C. E., U.S.D.A., A.R.S.–74–29 Report 65, 1964.

CHAPTER 12

Wet Finishing

WERNER VON BERGEN

Introduction

Developments in the finishing sector of the Wool Textile Industry since 1950 have been twofold. First in the further development of continuous finishing and its automation, and second in the introduction and perfection of chemical or special finishes such as high luster, water and oil repellent, washable, stretch and presensitized for permanent creasing. Bell (1) and Elsworth (7) give a general account of the most recent developments. The goal of a 100 percent wool wash and wear fabric has been nearly reached. This is specially true in the case of woolen sport shirts. Carter (3) expressed the following thoughts on the basis of continuous wool finishing:

"Though some research work still remains to be done, it is our judgment that satisfactory versatile continuous finishing for *worsteds* may soon be technically achieved. On the other hand *woolen* finishing still seems to be far away over the horizon, even from the most purely technical standpoint.

Perhaps, however, the most important point to which we need turn our attention is the economic feasibility of such plants. Continuous finishing ranges have, of course, been profitable when used on cotton cloths, especially on long runs such as are common in the U.S.A. But the wool trade is very different and a few hundred meters is usually reckoned to be a reasonable batch, while finishers must often handle single 50 meter pieces or even half pieces!

But continuous finishing is only possible when a sufficient length of cloth can be assembled at one place and time, all requiring the same routine. Alternatively, we must think of finishing ranges controlled from one central console which can automatically select temperatures, concentrations, dwell times, etc., for each section of the plant.

As a final stage we can readily envisage a magnetic ink code at the end of each piece signalling the required finishing program to the central console.

But such installations must represent heavy capital investments and must run with little interruption for most of the working week if an adequate return on capital is to be earned. Their yearly output might be of the order of 1½ to 3 million meters per year. It is difficult to know exactly how many such plants could justify their existence even if they were versatile enough to readjust for single pieces. Our own studies suggest that in the wool industry even an output of 1½ million meters/annum represents an unusually large weaver and there may be only 80 such units in the world. But they produce only something over 20% of the cloth. Evidently, in some concentrations of industry there will be commission finishers who could also take up such machines—perhaps the equivalent of another 30 firms and a production increase to 30%.

These figures are for woolen and worsted combined. More research is needed before continuous finishing of worsteds will give consistently the same quality of cloth as the present batch methods. We are carrying out some of these researches. Continuous finishing of woolens seems not yet possible.

Even if as many continuous plants were installed as is economically justified, at least two-thirds of wool fabric production would continue to be finished on a batch basis."

Kittan (10) in his talk on the possibilities of continuous finishing of wool fabrics stated: "The further development of the wool industry and the maintenance of the ability to compete, especially the commission finisher, demand an ever so critical, but basically a positive attitude toward continuous finishing, in order to avoid error in judgment."

Due to the new developments in the last few years, today the second step can be taken and a thought given on the installations of ranges for specific fabrics as well as for part ranges for all wool and wool mixture fabrics. At present it seems best and least risk to combine processing groups rather than the full range of operations. In any case there must be still a separation between wet and dry finishing. Between the two are often processes such as specking and knotcutting which can only be done manually. The pause serves also for the quality improvement. It is also necessary to balance the possible differences in the speed of the machines between the wet and dry finishing. An excellent analysis of the principles, organization and economics involved in the continuous finishing of wool cloth is given in *Wool Science Review*, No. 34 (27).

Finishing relates to all those processes to which woolen and worsted goods are subjected after they leave the loom and before they are presented for sale. The primary object of finishing is to enhance the quality of the cloth, imparting to the woven fabric a specific appearance and handle in order to render it attractive to the purchaser, and it must be suitable for tailoring into garments by contemporary methods.

Involved Factors

Finishing processes are extremely varied. The number of operations and the degree of treatment depend on the following factors:

1. The type and quality of the raw wool used to make the fabric.
2. The type of yarns used in the fabric.
3. The design or structure of the fabric.
4. The specified surface character and handle to which the finished fabric must adhere.

Type and Quality of the Raw Wool

The type and quality of the raw wool influences the felting and shrinking properties of the fabric during the finishing operations.

The ability to felt is peculiar to wool and related animal fibers. The degree, however, to which animal fibers will felt depends on their nature, quality, and physical and chemical structure. The felting properties are more pronounced in the fine merino wools than in the crossbred or in the coarse, long wools. The fibers of the goat and the camel families, such as mohair and camel's hair, have a low felting ability due to their much smoother scale structure.

Cotton, rayon staple fibers, protein fibers, and synthetic fibers are nonfelting in character. Therefore, when a cloth is a mixture or combination of wool and nonfelting fibers, such as staple rayon, the finishing treatments have to be of such a nature that the natural properties of each are preserved and exploited to the best advantage. Not only does each fiber type and subtype differ in its felting and its shrinking abilities but also in its behavior toward the numerous chemicals employed in the wet-finishing and dyeing processes. These facts impose upon the wool and worsted finisher, if the best results are to be obtained, the necessity of considering the character of each material with which he is dealing.

Yarn Type

Wool is spun into woolen and worsted yarns which are dissimilar in formation. The handle and appearance they produce in the finished cloth are markedly different. A woolen fabric will felt more readily and to a greater degree than a similar worsted fabric, as the fibers of the woolen yarns, due to their crosswise relationship, tend easily to become further entangled with each other. In a worsted-yarn fabric the fibers lie almost parallel to one another throughout the whole length of the yarn and, therefore, have not the opportunity to felt so readily.

The amount and the direction of twist present in a yarn also have a definite influence on the felting property, as well as on the surface character and the handle of the finished cloth. The fibers of a soft twisted yarn react in an entirely different way from the fibers of a hard-twisted yarn. Consequently the appearance imparted by finishing to fabrics such as crepes, which are made from hard-twisted yarns, is different from that of flannel, composed of soft, twisted yarns of the same material. In twills the direction of the twist has a definite bearing on the clearness of the twill line in the finished state.

There is quite a difference in the felting properties of yarns in the natural or undyed and the dyed condition even when spun from identical material. The yarn composed of dyed fibers has already been subjected to some fiber shrinkage while being dyed in the stock, top, or yarn state; thus it is necessary, when making a piece-dyed cloth that is to be equal to a mixture, or one composed of solid colored yarns, to make suitable allowances.

Fabric Structure

The design or structure of the cloth controls, to a large extent, the influence of the finishing processes. The structure varies according to the weave, the yarn count of the warp and the filling, and the number of ends and picks per inch. These factors greatly influence the varying widths and lengths of fabrics as they pass from loom to finished cloth. There is great diversity in the contracting and shrinking properties of weaves.

A study reported a variation of contraction from reed to gray cloth width of from 4 to 12% for nine different weaves, using the same warp and filling yarns. The additional width shrinkage from the gray to the scoured cloth varied from 3 to 10%. Similar contractions and shrinkages are caused by using the same weave but changing the yarn count of the filling and the number of picks per inch.

Character and Handle

In general, all classes of woven fabrics change in character, handle, weight, thickness, strength, and elasticity during the finishing processes. The wool fabrics as they come from the loom in the raw or gray condition have a dull, lifeless appearance and a loose thready structure which may be easily unraveled. When a woolen fabric is subjected to the fulling and raising operation, the thready appearance totally disappears, being altered to resemble a felt with a nap of fibers drawn from the body of the cloth. Consequently, finished jerseys, doeskins, and fleece fabrics have very little resemblance to the respective fabrics as they left the looms.

The finishing of worsted fabrics generally does not result in the distinctive change produced by finishing most woolen fabrics, for the character of the worsted weave or effect is not only maintained but further enhanced, in serges and gabardines, for example. Fancy-colored worsted suitings in their finished state and in the gray are very similar but it is not difficult to distinguish the unfinished from the finished fabric. The latter has a very slick, soft handle and a smart appearance which is brought about by careful scouring and pressing and removal of the fiber fuzz from the surface by singeing and shearing.

Types of Finishes

The types of finishes applied to woolen and worsted fabrics are almost innumerable. For general classification the various types of finishes as far as the physical character of the final fabric is concerned, may be subdivided into three main groups: (*1*) clear finish, (*2*) face finish, and (*3*) modifications of the clear or the face finish.

Clear Finish

The main characteristic of a clear-finished fabric is its clear, even, and threadbare surface. The weave and the colors are prominently developed and distinct. The clear-finished goods receive either no fulling at all or only sufficient to produce the necessary firmness and handle. Most worsted fabrics fall into this class.

Face Finish

The main characteristic of the face-finished fabric is that it has a nap or pile of fibers on the surface. The weave does not show at all or is indistinct, and the colors are subdued and softened. These effects are brought about by fulling and raising. In some instances. for example, velours, the pile stands erect from the fabric, whereas in others, broadcloths and kerseys, the fibers are laid flat in one direction and treated to give a lustrous appearance. When full shades are employed, the character of the cloth is very rich. The face finish is generally applied to woolen fabrics.

Modifications of the Clear or the Face Finish

The finisher can produce a great variety of finishes in between the two main types, by simple modifications of them. Among the worsted fabrics, for example, there is the popular unfinished worsted. The weave and the color designs are more or less subdued because the face of the fabric is not shorn clear; it still carries a surface fuzz of loose and straggling fibers. The handle is soft and flannellike. The melton and the covert-type finish of woolen fabrics is produced through heavy fulling, ending with steaming, brushing, and pressing. The surface shows a fibrous appearance with the weave not visible, similar to a felt.

Fabrics may even carry both types of finishes: an example was the 22-oz serge, which was made especially for U.S. Army uniforms worn in Alaska during World War II. The face of the cloth was clear, whereas the back was finished with a velour like nap.

In addition to these groups of finishes produced by mechanical means, there are the finishes designed to overcome the traditional drawbacks of wool textiles, namely, susceptibility to insect damage by larvae of the cloth-moth and carpet beetle as well as the felting in laundering. To these are added a second group of finishes designed to give wool textiles the new properties made popular by synthetic fibers. As stated by McPhee (14), among these are the methods of imparting permanent creases and pleats to garments, and also methods of giving wool fabrics the ability to retain their smooth appearance during laundering, that is imparting a minimum-iron effect.

As all these finishes depend on the use of chemicals, they can be classified as chemical or special finishes. To this group belong also waterproofing, oil repellent, permanent luster, and some permanent stretch finishes. These finishes are discussed in Chapter 13, page 935.

Flow chart, Fig. 1, shows the normal processing steps for clear finish as well as face finish fabrics.

Numbering, Perching, Picking, Burling, Mending

Prior to the finishing and the dyeing processes, woven fabrics are subjected to numbering, perching, picking, burling, and mending. All cloths as they leave the loom contain imperfections to some degree. The defects generally found in wool cloths may be classified into three groups:

1. Those due to impurities such as burrs, straws, skin pieces, pitch stains, and kemps in the raw materials.

2. Those incurred in the spinning processes that show up in the woven fabrics as knots, slubs, and thick and thin yarn.

3. Weaving defects such as missing warp and filling threads, "ends out," "picks out," or mispicks, holes, harness skips, and wrong draws, and also oil and graphite stains.

To establish and to correct their imperfections, woven fabrics after leaving the loom are subjected to the following processes.

Numbering

In order to identify and to establish a record of each piece, it is necessary to stitch or mark it with reference numbers. The numbers

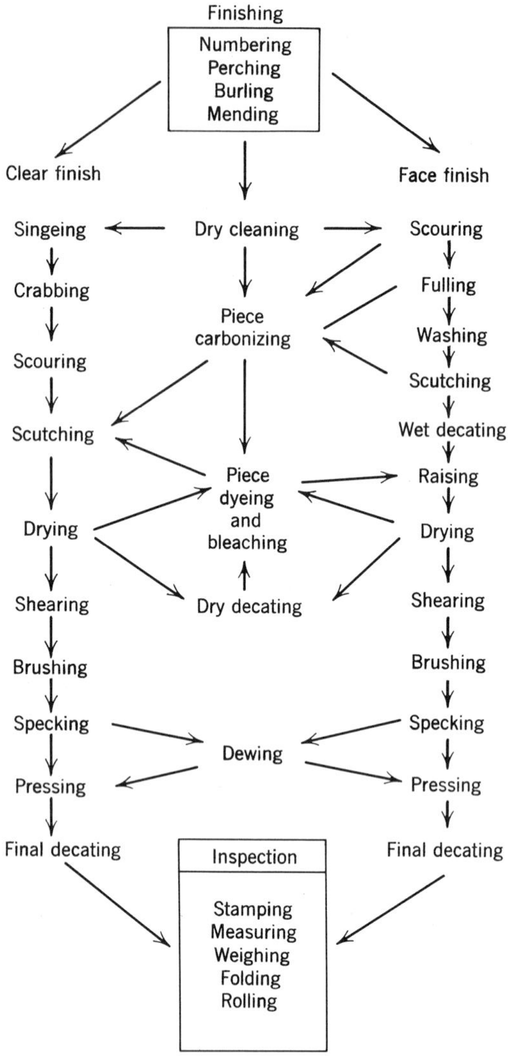

Fig. 1. Flow chart of finishing and piece dyeing.

generally indicate (*1*) the number of the "cut," and (*2*) the number of the style. The numbers are usually sewn with cotton thread on the face of the front end of each piece. For each sample piece of a new style, a numbered record is kept referring to the particular finishing processes and degree of treatment to which each sample has been subjected. By using this record as a reference, the finisher is able to subject the regular pieces to the same processes and degree of treatment undergone by the sample piece, in this way obtaining identical results.

Perching

Perching consists of examining the cloth and marking with colored chalk all imperfections which should be eliminated in the subsequent processes. The machine used for inspection is known as a perch. The modern perch is designed so that the operator may give the cloth careful and thorough inspection with a minimum of effort. The machine, which is power driven, has a forward, a reverse, and a variable-speed drive. The machine is controlled either by a small switch that the operator holds in his hand, or by a foot switch, either of which can start or stop the cloth instantly. The cloth passes before the operator at full width and free of wrinkles. Inspection is made in two ways: (*1*) the goods between the light and the inspector examining against the light; (*2*) inspecting the surface of the goods only, with the light on the goods. The importance of proper perch lighting has been stressed by Poole and Turner (17).

The measuring and the weighing of the cloth may be combined with perching. In such cases the perch is fitted with a yardage counter and the delivery of the cloth is so arranged that it drops from the folder onto the platform of a scale, the beam of which hangs between the frame of the perch.

Picking

This process consists of extracting from the face of the cloth all extraneous matter such as neps, slubs, burrs, hairs, and straws. In some mills this process is referred to as specking. The removal of these substances is generally done by hand, but in fabrics having a high nep count, it is economical to run the cloth through a sanding or a despecking machine first.

Burling

In this process the back of the cloth is first examined by the burler and all snarls, slubs, straws, etc., are removed. The thick warp threads and the filling threads are drawn out and replaced, and the knots opened. Afterward, the burler examines the face of the cloth, pushing such irregularities as remain (loose ends and curls) to the back, in order that the face of the cloth may be as free as possible of irregularities and imperfections.

Usually burling is carried out on a table with a sloping top and a smooth surface which allows any thick places or knots to be easily detected when the hands are passed over the cloth. Warp threads must not be cut below the knot. The knot should be opened to allow the two ends of yarn to hang together, thus avoiding a small hole which would be created if the two ends of yarn were apart. This work is done with the aid of burling irons and scissors.

Mending

This word is used to describe the insertion of yarn in the woven fabric where any warp or filling threads are missing. Every missing thread must be carefully hand woven into the piece in exactly the same manner as if it had been machine woven. Knowledge of weave construction, therefore, is very important. The faulty places are repaired by sewing in new lengths of yarn with a needle, beginning and finishing at the back of the cloth so that the loose ends of the thread can be cut off. This corrective work is absolutely necessary on clear finished fabrics. However, on heavy fabrics and on fabrics to be fulled, smaller imperfections may be left in and they will be invisible in the finished cloth. This work is very delicate and time consuming, which makes it expensive. It is performed entirely by women. In some cases it becomes necessary to return the goods after scouring for additional mending of places that escaped inspection in the gray. Missing threads are always mended in high-priced, quality goods.

Wet-Finishing Operations

The goods are now ready for the finishing processes and are subjected to a definite routine of operations selected to produce the desired finish or character of the goods. The following procedures for finishing woolen and worsted goods are given in the general order in which they occur but not necessarily in the definite sequence, since the procedure varies according to the desired result.

The operations in wet finishing are: dry cleaning, singeing, crabbing, piece scouring, fulling, carbonizing, scutching, beaming, wet decating, blowing and raising.

Dry Cleaning

The successful application of dry cleaning to woolen and worsted fabrics is a recent development. The process, also known as the Derby (6) process, was introduced in 1939. Because installation is quite expensive, however, only a few mills have made it. The diagram (Fig. 2) illustrates the method used. The cloth enters the totally enclosed machine through a seal into a soaking compartment containing about 100 gallons of trichlorethylene solvent. The cloth travels in its full width over and under rollers, with the lower roller immersed in the trichlorethylene. The cloth then passes between a pair of draw rollers, that keep it under tension as it is drawn over a vacuum slot and passed through a second pair of draw rollers, which are synchronized with the first. Through a seal it then enters a drying compartment and runs over five dry cans.

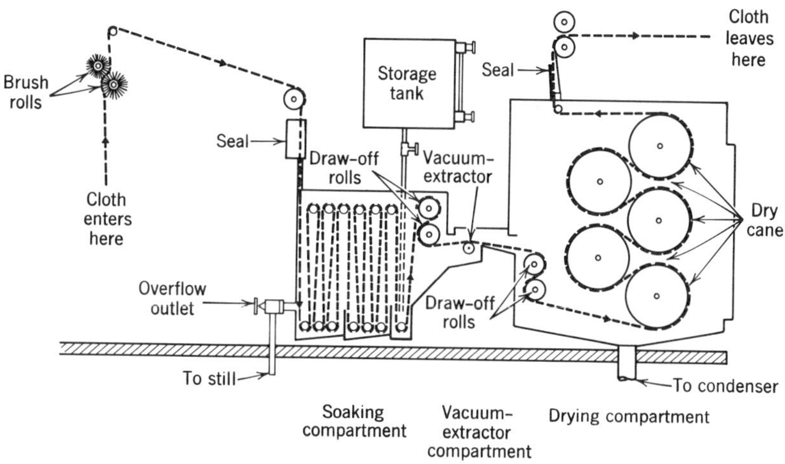

Fig. 2. Derby continuous dry cleaner.

The clean, solvent-free cloth leaves the drying section through a seal and between a pair of draw rollers at the top.

From a storage tank above the machine, clean solvent is fed automatically into the last section of the soaking compartment. The solvent then flows countercurrent to the direction of the cloth until it reaches the first section of the compartment. The solvent, now saturated with spinning oils and other impurities from the cloth, overflows into a still where it is recovered continuously by ordinary steam distillation. The solvent vapors, mainly from the drying compartment, pass into a condenser and then through a separating chamber for removal of any water that may have been collected. The solvents from the still pass from the condenser and are collected in a sump tank from which they are pumped back into the storage tank at the top of the machine, where they are ready for reuse.

The specific advantages of dry cleaning or trichlorethylene scouring that is more effective and more uniform than aqueous scouring, this is especially true for blends of wool and synthetic fibers which hold spinning and combing oils tenaciously. As Whewell (24) has stated, it appears that, in general, the more hydrophobic the fiber, the greater the binding force between oil and the fiber. Low cost mineral oils which are readily soluble in trichlorethylene, can be used in place of the costly animal or vegetable oils.

The use of depitching solution is eliminated. Certain tints applied in spinning or weaving as well as chalk marks can be removed.

The speed range is 10–30 yards per minute adjusted to the linear ounce weight as follows:

Worsted and crepes	6–9 oz/yd	25 yd/min
Suitings	9–12 oz/yd	15–20 yd/min
Coatings	13–20 oz/yd	11–15 yd/min

The ether extractable matter within the approximate 3 min in the solvent is reduced to less than 0.5%.

To maintain peak efficiency, which is measured by the pounds of cloth per one pound of trichlorethylene used a periodical check of several points must be maintained. Based over several years of experience by Scibior and von Bergen at the Forstmann Woolen Co., these check points are:

1. All doors, cloth entrance and exit posts, windows and joints should be checked weekly for leakages. Teflon gaskets are best.

2. The flock catcher, and dry cans should be cleaned once a week.

3. The condenser should be cleaned quarterly.

4. The vacuum pump gland should be checked weekly and repacked as needed.

5. The refrigeration unit should be checked weekly. The importance of such checkups to prevent heavy solvent losses is best illustrated by a three year record graphically shown in Fig. 3. The best monthly average solvent efficiency was 1 lb solvent per 32 lb of cloth equal to 3 lb/100 lb of cloth.

A new solvent scouring unit built by ICI (Imperial Chemical Industry, England) was exhibited in September, 1965 in Atlantic City. The process consists basically of paning fabrics continuously—open width, through boiling trichlorethylene and then either through a hot water vessel (see Fig. 4) or steam chest (see Fig. 5) to flash-off entrained solvent and finally through a single unit open width washer. A more detailed description is given by Crowder and White (5).

The main difference between the ICI and the Derby process is the use of boiling solvent and the flash-off of the solvent with hot

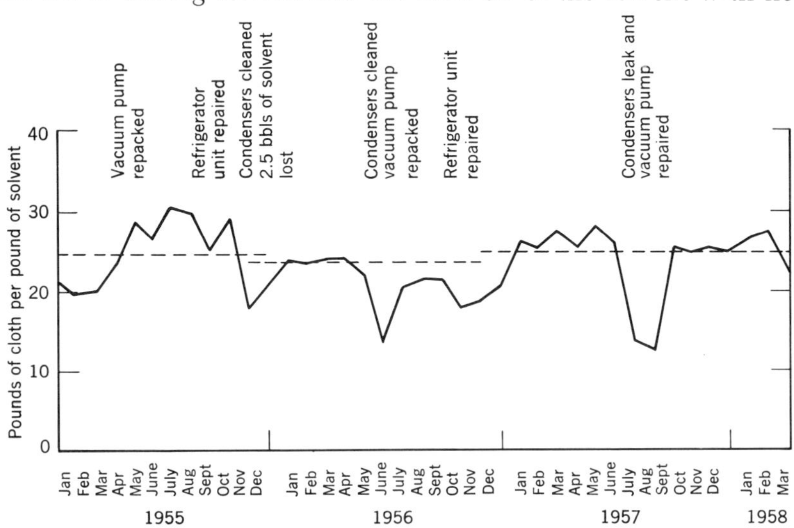

Fig. 3. Solvent consumption control of Derby Dry Cleaner.

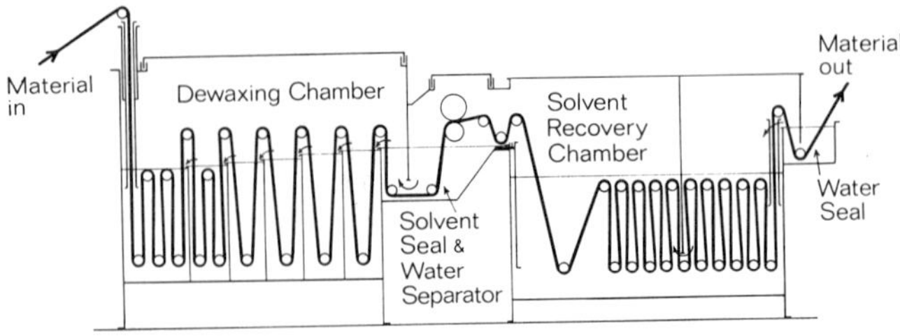

Fig. 4 (*top*). Imperial Chemical Industry Solvent Scouring Unit with Hot Water Flash-Off of solvent.

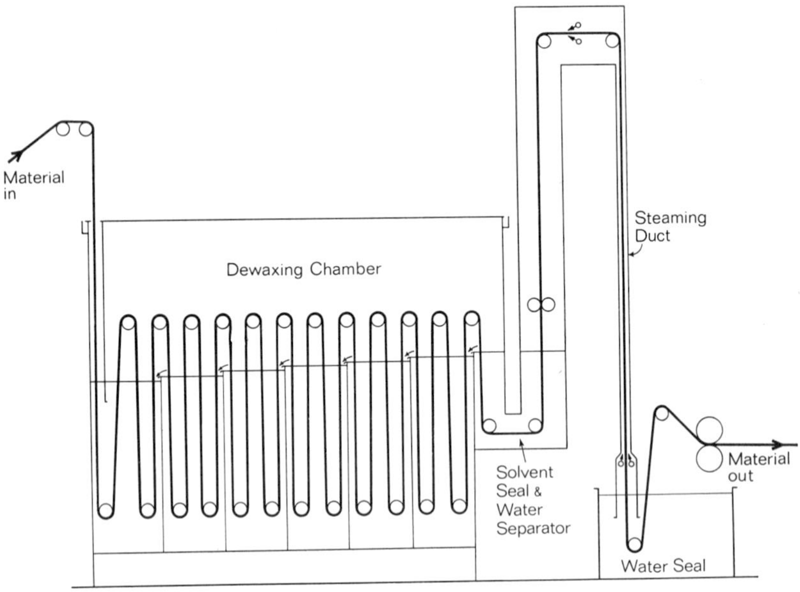

Fig. 5 (*bottom*). I.C.I. solvent Scouring Unit with steam flash-off of solvent.

water or steam against cold solvent and dry heat. The advantages claimed are lower solvent loss and better fabric consolidation, the latter being possible through the wet treatment.

Speeds up to 60 yd/min are claimed to be possible with the machine shown in Fig. 4 and with the stream flash-off, up to 120 yd/min.

The following are the requirement and consumption data given.

TABLE 1
I.C.I. Solvent Scouring Unit

	Requirement/Consumption
Labor	1 full-time operator
Solvent consumption[a]	3 lb/100 lb cloth
Steam (30 psig)	150 lb/100 lb cloth
Water loss to waste	50 U.S. gal/100 lb cloth
Cooling water requirement	200 U.S. gal/100 lb cloth
(can be reused)	(Assuming inlet temp. 60°F.; outlet temp. 115°F.)
Power requirement	15–35 horsepower

[a] Trichloroethylene, it is continuously purified and recycled. There are inevitable minor losses; water overflowing from the water separators contains up to 0.1% w/w of solvent and a similar proportion of solvent is held in the treated cloth. Small amounts of solvent vent to outside atmosphere via ducts from each section of the machine. Total solvent losses from all sources should be less than 3% on the weight of fabric treated.

Singeing

Singeing consists of burning off the free, projecting fibers from the surface of the cloth, leaving it perfectly smooth and bare. It is applied only to fabrics to be clear finished. There are two methods for removing this fuzz from fabrics, gas-flame singeing and plate singeing.

Gas-flame singeing is the most common method used. It is performed by passing the cloth in full width at a rapid speed over open gas flames. The machines are built to provide singeing on one or both sides of the cloth in one run.

Plate singeing is done by passing the flat goods over gas- or oil-heated oval plates and is usually used on clear finished goods. It produced a clear and highly lustrous surface on all kinds of linings, particularly those of mohair, or mohair and cotton used in men's and ladies' suitings, prior to their replacement by man-made fibers.

Crabbing

The object of crabbing or boiling is to fix the warp and the filling threads in a regular manner by the removal of the irregular hidden strains, thus preventing the formation of cockles, crimps, creases or other forms of uneven shrinkage in later stages of finishing.

The method of crabbing and the degree of treatment to which the worsted fabric is subjected, depend on the quality of the fibers, the twist of the yarns, and the cloth construction. Special precautions have to be taken with fancy-colored worsteds, because of the risk of bleeding of the colors. Patch crabbing and continuous crabbing are the two methods used.

Patch Crabbing

The goods are passed in full width through boiling water onto a roller with or without top-roller pressure. The machine may be used in one unit or in series of three or four units. Working in series has the advantage that both ends of the pieces receive equal treatment. Through the transferring of the piece from one machine to the next, the top end of the first machine will be the bottom end of the second machine, thereby preventing shadiness (end to end) in the succeeding dyeing process. In a three-series crab, the first two units operate with boiling water for five to ten minutes; the last unit employs cold water to complete the setting of the fabric, with either slow or rapid cooling as desired.

A more advanced double unit of German construction is shown in Fig. 6. In Fig. 7 are shown a series of units generally known as Yorkshire Crab, built by Sellers. In the double unit the pieces are wound onto a perforated tube and steam blown through for a few minutes which produces a more permanent setting than is possible in the boiling units.

Continuous Crabbing

In this method the cloth passes in full width through a series of tanks. Each tank is equipped with immersion rolls, steam piping, cross spraying, and overflow and at the end is a pair of squeeze rollers. The pressure is applied by means of air cylinders. Before each nip there is a pair of spirally fluted, opening rollers for the removal of wrinkles and double selvages. A spray pipe for rinsing is also set ahead of each nip. This type of machine is used not only for the crabbing of worsteds but it also has been found suitable for wool—man-made fiber blends, where the goods are to be set, scoured, and rinsed in preparation for dyeing.

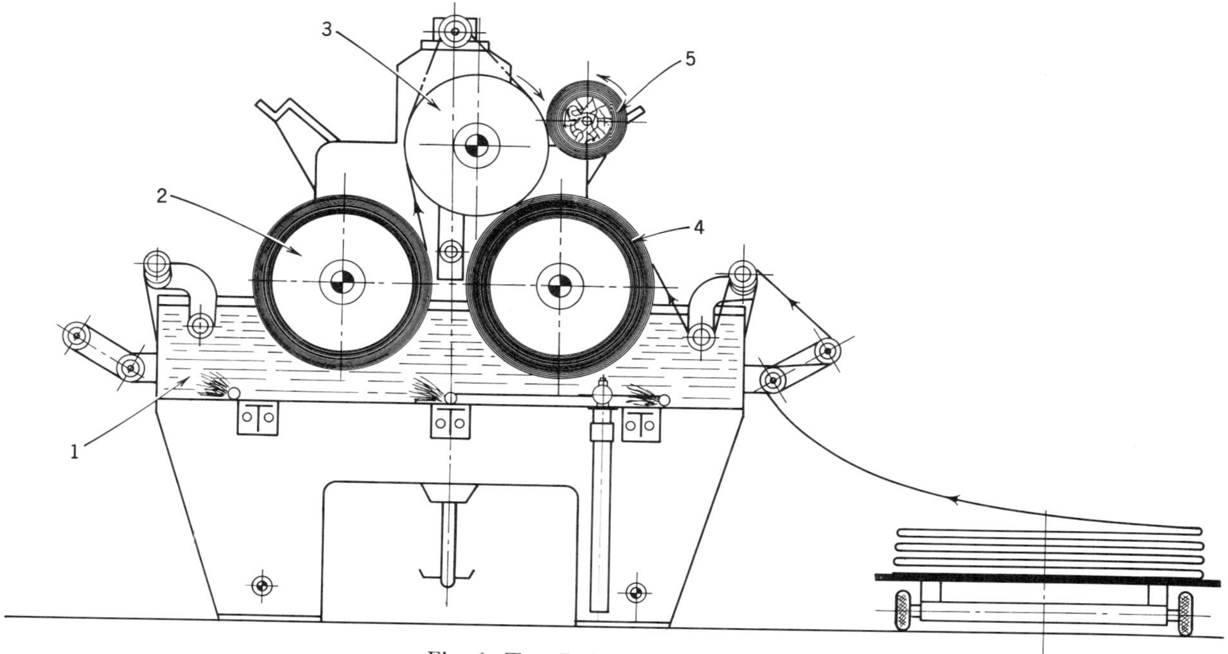

Fig. 6. Two Roll Crabbing Machine.

1. Warm Water Tank
2. Cloth Roll
3. Pressure Roll
4. Start of cloth rolling
5. Rolling of crabbed cloth on cloth beam

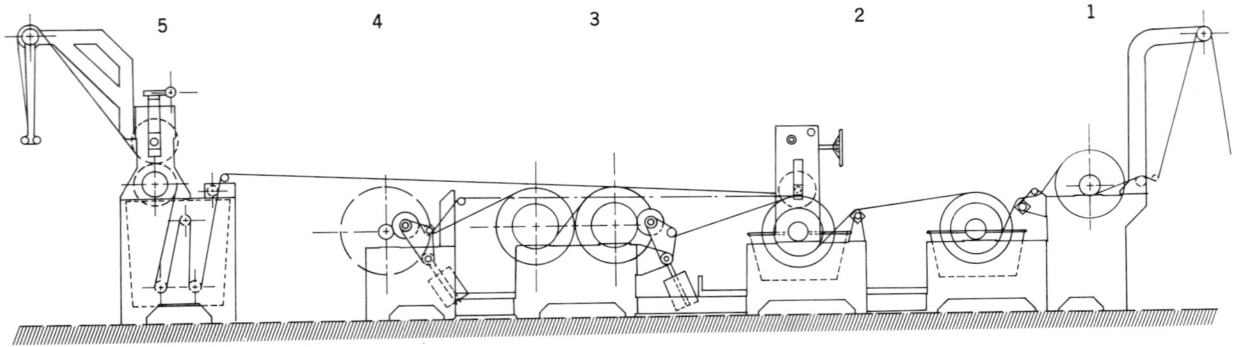

Fig. 7. Yorkshire Crabbing Range. (Courtesy of Sellers Huddersfield, England.)

1. Rolling Unit
2. Boiling Unit
3. Double Steamer
4. Rolling after crabbing
5. Cooling and squeezing unit

An entirely new design is the Konticrab. On the Konticrab the oncoming belt of cloth is passed first, via appropriate guiding devices, through a hot water trough, then nipped by a pair of rubber rolls with pneumatic pressure control. From there the fabric runs over a mechanical expander roller on to the stainless steel surface of a heated drum (260cm = 102½ in.). Simultaneously, the drum surface and thus also the fabric is covered up by a thick endless rubber blanket. By it the cloth is pressed firmly against the polished steel surface. While it is held firmly in position by the rubber blanket, it rotates with the drum and is exposed to the activity of heat and moisture. After completion of a whole revolution, the cloth leaves the drum and is guided through a cooling bath, as shown in diagram (Fig. 8).

In accordance with the finishing procedure, the cloth may then be batched on rolls, folded out on a cart or fed directly into a tenter frame. The standard width of the drum is 180 cm = 72 in. The maximum steam pressure 21 lb/sq in. 221°F. The machine can be operated at speeds from 5 to 24 yd/min. The Konticrab can also

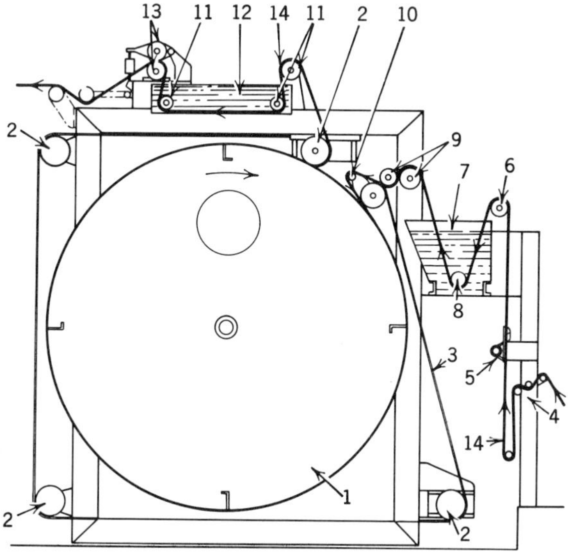

Fig. 8. Diagram of "KONTICRAB" continuous crabbing machine. (Courtesy of L. Ph. Hemmer, Aachen, Germany.)

be used for impregnation purposes or chemical setting such as Siroset, provided its hot water trough is adapted to such requirements.

Piece Scouring

Piece scouring is necessary to remove the oil or emulsion with which the wool was treated in spinning, the sizing applied to the warp before weaving, and any other foreign matter, such as oil spots and dirt, which the yarns and fabrics have acquired during manufacturing operations.

In woolen goods these impurities may be from 15 to 20% of the raw weight of the cloth. In qualities containing much reworked wool, the loss, due largely to the grade and the amount of oil applied, may amount to as much as 30%. In worsted fabrics, where less spinning oil is used than on woolens, the loss of weight is between 2 and 10%. Occasionally, before scouring goods containing pitch or tar stains, it is necessary to run the goods through a depitching solution to loosen up the stains. Although in the previously discussed piece dry cleaning operation most of the oily matters are removed, a wet treatment has to follow to get rid of the non-solvent soluble impurities, such as sizes and dirt, but also to break the stiffness of the cloth by swelling and relaxing the yarns and fibers, thereby bringing out the desired hand and surface character or finish of the fabric.

Water for Scouring

The type of water in scouring piece goods is just as important as in wool scouring. A soft water, with hardness of less than 5 grains per gallon, should be available to avoid the formation of insoluble calcium and magnesium soaps on the surface of the goods, which would cause serious trouble in piece dyeing. Even with such soft water, the use of products such as sodium hexametaphosphate and sodium tetrametaphosphate has been found advantageous, especially in the high-speed washers. These water softeners are generally used in the final rinse to remove the last traces of soap.

Scouring Agents

The scouring agents used are the same as for scouring loose wool, namely, soap and weak alkalies such as soda and ammonia and synthetic detergents. It is an emulsification and saponification process combined with mechanical action. The amounts of chemicals are adjusted in accordance with the impurities present and are generally used in dilute solution. For fine worsteds 2–4 gal of a 5% soap solution per piece will be sufficient, whereas for heavy woolen coatings 5–9 gal of soap per piece and a soda solution 1½–3° Be are required. In great numbers of finishing plants, the synthetic detergents of the nonionic type have replaced soap because they are unaffected by hard water and are of special value in scouring goods when felting is to be avoided and also in scouring yarn-dyed goods of delicate shades. A further advantage of synthetic detergents for scouring is that, because of their nonalkalinity, the harmful effects of alkali on both fibers and certain dyes can be avoided. These products have good cleansing action in a neutral or even acid bath. Furthermore, they can be rinsed more quickly and easily than any common soap. If desired, the detergent action of these compounds can be improved still further by the addition of sodium tripolyphosphate, tetrasodium pyrophosphate, or trisodium phosphate.

Scouring is done in three different types of machines; the string washer, the broad washer and the cylinder washer.

String washer

The most common is the string washer, also known as the rope or the dolly washer, in which the cloth is sewed into an endless string or rope which passes between two rollers (a combination of plastic bottom roller and rubber-covered cast iron or stainless steel top roller) as illustrated in Fig. 9. Depending on the size four to ten pieces are handled side by side in one machine. According to the weight and structure of the cloth (heavy, medium, or light), top rollers are used, respectively. The following top roller weight (based on a four-piece load) are suitable: medium woolens (16 oz.) 550 lb; worsteds (14–19 oz) 750 lb; heavy woolens (over 20 oz) 1450 lb. Each string is kept in its position by guide rings or sticks to avoid entanglement with adjacent pieces. The top roller serves as a squeeze roller and the bottom roller is positively driven. The strings are then allowed to float in the scouring liquor through the bottom

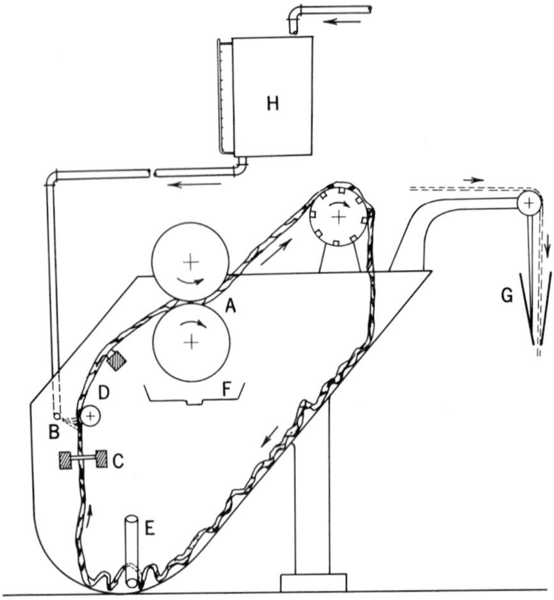

Fig. 9. Diagram of a conventional string washer.

A — Squeeze Rolls
B — Detergent and water inlet
C — Guide rings or roads
D — Guide Bar
E — Overflow pipe
F — Sud Box
G — Cloth Folder
H — Detergent supply tank

of the washer, till they are returned to the rollers in their continuous passage.

The running time for different goods varies according to the weight from 1 to 4 hr. The temperature of the scour-bath is kept at about 100°F (blood temperature) during this first phase of scouring to speed emulsification and saponification of the oils, which take from 10 min to 1 hr and are characterized by the formation of lofty soap suds.

In order to remove the emulsified oil, rinsing is necessary. This is done by dropping the scouring liquor and opening the warm water valves slowly until the greater part of the soap is flushed away. Rinsing may require ½ to 2 hr according to the weight of the goods. The last traces of soap and alkali are then rinsed away with large amounts of cold water.

Jet High Speed Washer

The jet high speed washer developed by Hemmer (8) represent a novel change in the construction of the string washer. The machine as shown in diagram Fig. 10 is fully enclosed resembling a dye kettle. The standard jet high speed string washer Model SLD 180 is equipped with either five or six jet box apertures of 10 cm (4 in.) interior diameter each, i.e., for 5 to 6 ropes each approximately 50–60 m long and a maximum weight of 20 oz/yd. With the exception of the wash rollers all parts are made from stainless steel. The three possible cloth speeds are 110, 150, and 220 yd/min. Either mechanical or pneumatic pressure control the top roller. The heating of the liquor is done through a steam pipe in the sud trough below the rollers and the temperature controlled with a thermostat. For an intrinsic reduction of the rinsing times the strings are guided through the jet spray rings which force a jet of warm water with 45 lb of pressure through the cloth (Fig. 11).

The standard equipment includes a control board with built in timer, on which the scouring and washing-off times are set, and a manometer dial to reproduce and maintain the roller pressure. Fundamentally the main wash and rinse action in the jet washer is transferred into the space between the jet spray rings and the squeeze rolls. Ladders with porcelain rings have been used for many years to achieve a better wash effect.

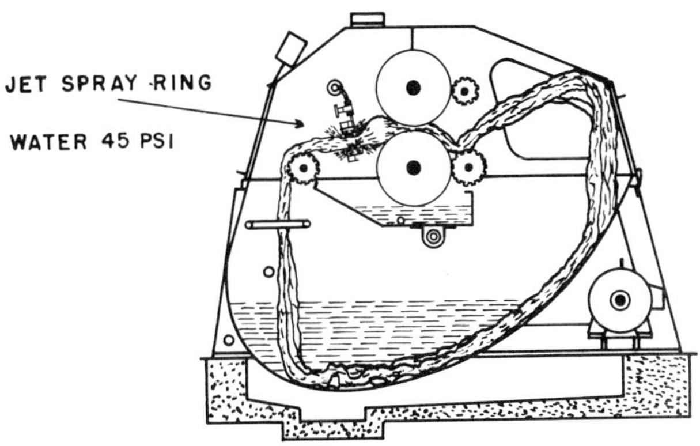

Fig. 10. Jet High Speed Washer. (Courtesy of L. Ph. Hemmer, Aachen, Germany.)

Fig. 11. Jet Box aggregate in operation. (Courtesy of L. Ph. Hemmer, Aachen, Germany.)

The following are the advantages gained.

1. Wetting out of the dry fabrics is achieved within one to two minutes due to the high pressure of the water stream preventing the danger of breaks.

2. Due to the high string speed and the turbulence in the bath, large amounts of the well homogenized liquor are carried up and through the jet rings to be squeezed right after by the rollers. This leads to the formation of bubbles through the bulging out and resulting in a high increase of the scouring action as the fabric passed between the squeeze rollers whereby the forced motion water and air are pressed out and therefore a considerable reduction in the washing and rinsing times. Another factor is the mechanical action produced as the fabric bounds on the knock off roll and rebounds from the back wall of the machine. This produces a felting action sufficient for Worsted fabrics which generally need a short fulling time to achieve the proper compactness. With light weight fabrics such as wool and wool–polyester blends up to 12 oz with a 9 string machine, production capacities up to 72 pieces per 8-hr shift have been reached. The working cycle for fine worsted up to 14 oz per example is as follows:

Wetting out time at a temperature between 75 and 85°F for 5 min. Full detergent, 15 min at a cloth speed of 220 yd/pm, warm rinsing 40 min at 160 yd/m and cold rinsing 5 min followed by acidification.

For Cheviot Worsteds 56s wool quality—twill and related weaves, the Jet Washer has been found especially suitable and beneficial.

With 20 min in full detergent and 40 min warm rinsing both at high speed, excellent results have been achieved.

Continuous String Washers

The first types built entirely of wood and using the tight strand system have been replaced by stainless steel machines such as the Laxloop continuous cloth washer, and the Cascade continuous cloth washer. The first one shown diagramatically (Fig. 12) consists of nine dolly washers arranged in a train. Most fabrics up to 16 oz/yd (500 g/m) can be scoured at 45–55 yd/pm and fabrics from 16 to 20 oz can normally be processed at 35–40 yd/pm and goods heavier than these at 25–30 yd/pm. The cloth is tensionless throughout the range, each unit kept in step by the special Free-Loop Electri-matic Compensators. When two loops of cloth complete a low energy electrical circuit by coming in contact with two metal bars, corresponding tubs are synchronized. Cloth is protected against damage by electrical stop motions. Two control panels—one for drive gear motors, and one for pump motors, make it possible for the Operator to control the whole range of each tub individually, from one spot.

One mill using this machine reported that it can process around 25,000 yards per day in two shifts using four men per shift; one for the sewing, two on the washer, and one on the scutcher. The overall length of the train including a soaper and scutcher is approximately 130 ft, normally set in a straight line, but may be set up in two parallel lines or at right angles to each other. For a full load, 3,000 yd of cloth are needed. Each tub holds 300 yd which are threaded between the two rubber rollers in 12 loops of 25 yd each, therefore, the total number of nips the cloth receives by passing through the entire machine is 108 (9×12). It has proven that a good washing procedure is to prepad the grey goods with detergent in a solution containing 3% detergent fortified with 0.5% soda. The first 3 tubs are the washing section, to the fourth tub it is advisable when soap is used as a detergent to add some phosphate such as calgon approximately 6 oz per piece using tubs 4 and 5 as a forerinse followed by warm water rinsing in tubs 6 and 7 and 8 and end up with a cold water rinse in the seventh tub. The average time for a piece of 60 yd to pass through the nine tubs is 40–50 min. For the same number of nips in a regular dolly washer, it would take 2 hr

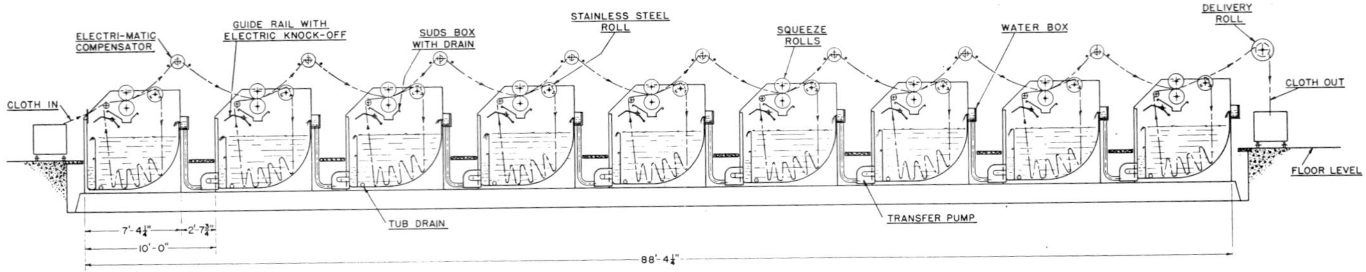

Fig. 12. Fleet Line Laxloop continuous cloth washer. (Courtesy of Riggs & Lombard, Inc., U.S.A.)

processing time, however, in a jet washer the time would be about equal to that in the continuous washer.

The Cascade continuous cloth washer (Fig. 13) is a variation of the 9 bowl washer wherein two bowls are set underneath a 3-roll squeeze roll stand. The machine is threaded so that the bottom nip handles the goods from the first bowl and the top nip handles the goods from the second bowl. This construction results in a considerable saving of floor space, and the fact that the counter-current is not done by pumps, but by actual overflow from one tank into the other with a 2 in. variation in head, allows a counter-flow system at lower cost. The machine is built as an eight compartment unit, and has all the necessary safety stop controls. The loops are also 25 yds controlled by a variable drive. The machine can be operated by two men per shift operating washer, scutcher, and extractor. With operating speeds anywhere from 15 to 45 yd/min

Fig. 13. Cascade Continuous Cloth Washer. (Courtesy of James Hunter Machine Co., U.S.A.)

daily production per 8-hr shifts of 75–230 pieces can be figured. The water requirements for both of these washers run between 5 and 10 gal of water per pound of cloth against up to 50 gal in a dolly washer. For straight worsteds, the water consumption is on the low end, whereas heavy fulled woolens, washed after fulling, it is on the high side of the range. With the water savings goes a potential saving in steam, since warm water is usually used.

Each bowl or compartment of the wash section must be charged with sufficient detergent at the start to attain a clean piece of cloth right from the beginning. The purity of the cloth is best checked by extraction with alcohol. The maximum extractable matter should not exceed 1.2%.

Broad washer

Certain fabrics that are liable to crease in the dolly washer must be washed in the full width in a broad washer. This applies especially to light dress fabrics and clear-finished worsteds. The cloth is run in full width over a series of guide rollers and between a pair of small, hard, rubber squeeze rolls, which can be adjusted for various pressures. In such machines six to eight pieces are sewed together in an endless band and run from one to four hours, following the same procedure as in the dolly washer.

The older forms of open width scouring machines are not very efficient and productivity is low. Therefore, as in the case of the old string washers a number of modern machines have been designed to operate as continuous washers. Examples of such machines are the Express Washer made by Messrs. F. Smith and Co., Ltd., of Roehdale, England described by Castle and Wilford (4), two German machines one built by Karl Hennecke (Fig. 14) and described by Kistenreich (9) and the Kontilana made by L. PH. Hemmer (Fig. 15), a Japanese washer by Sakurai and a Spanish machine built by P. Serracant Riba (19) (Fig. 16) and described by Carter (3).

The Smith machine consists of seven separate stainless steel becks, each one of which contains four double spray pipes. After each beck is a pair of squeeze rollers. The action of the double jets and the full width squeeze, together with a high liquor circulation provide the means of effective scouring.

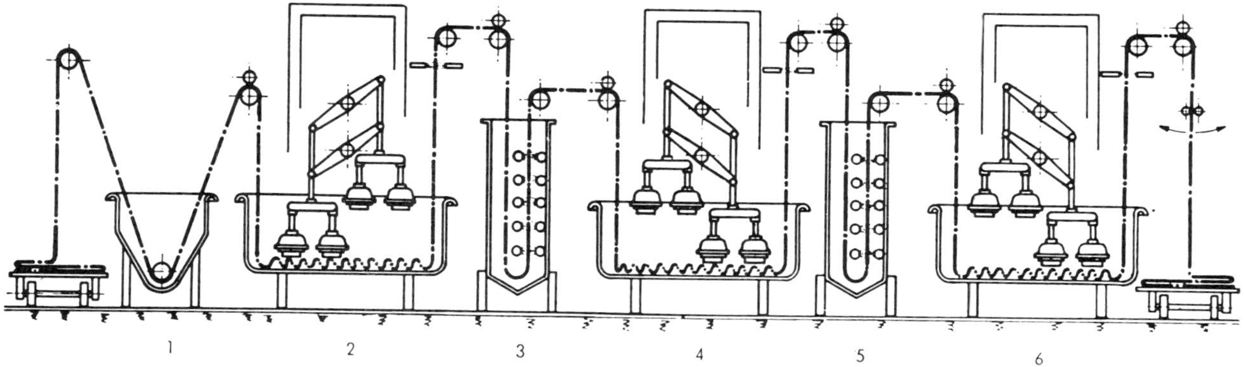

Fig. 14. Hennecke Continuous Cloth Washer. (Courtesy Karl Hennecke, Germany.)
1. Soaping Trough *2, 4 & 6.* Washing Tanks *3 & 5.* Spray Rinse

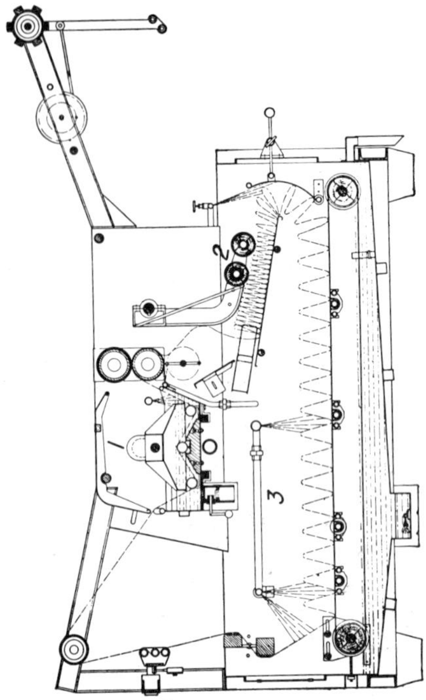

Fig. 15. Hemmer continuous broad washer. "KONTILANA"
1. Sucker *2.* Crimping chute *3.* Spray system

The Hennecke washer as seen in Fig. 14 consists of a soaping trough and three washing tanks with separated jet spray rinsing units. Each scouring tank is equipped with 22 diverbells which oscillate up and down and thereby work the wash liquor through the loosely tabled cloth by a suction and repelling action.

The Kontilana as seen in Fig. 15 is even a more complicated machine. The single unit has built in a combination of four different means of washing and rinsing. It contains (*1*) a "sucker," an inverted bowl or diverbell, dipping into a short bath where the passing cloth is held in position by three guide rolls, moves up and down 170 times, thus pushing and sucking the washing liquor through the fabric; (*2*) a "crimping chute" with a roller lid (Posser) producing a fulling effect as well as forces the liquor into

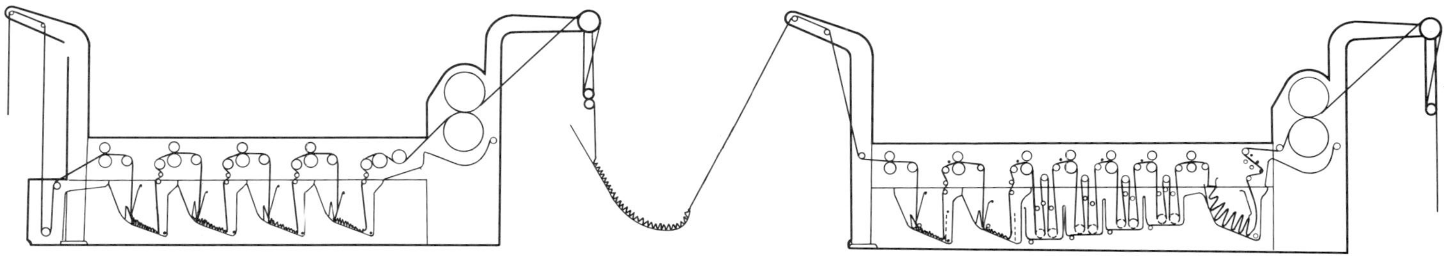

Fig. 16. Spanish continuous broad washer. (Courtesy of P. Serracant Riba.)

the folds; (*3*) a spray pipe system with circulation pumps intensifies the washing. The cloth travels and slides through the machine with practically no tension applied which is of particular importance for the finishing of stretch fabric such as Helanca.

The Riba machine is a construction which follows more the older forms (Fig. 16).

Cylinder washer

Very fine crepes and lightweight dress fabrics have been scoured to good advantage in laundry wheels. The individual pieces are put in bulk into each compartment, of which there are three to a machine. The perforated cylinder, enclosed in a hood, tumbles the pieces as it oscillates back and forth, performing the necessary mechanical action. Total scouring time is from 1 to 2 hr. This type of scouring produces a much better crepe than is possible in a string or a broad washer.

Fulling or Milling

The process of fulling or milling is peculiar to the woolen industry because of the characteristic felting quality of wool and certain hair fibers. The theoretical background of fulling was discussed in Volume I, Chapter 3, page 199. Newer information on the movement of wool fibers in fabrics during felting are reported by Makinson (12). Other studies were made by Blankenburg (2). Most of the recent observation on the mechanism of felting are summarized in *Wool Science Review*, No. 24 (27). Three factors are involved in fulling: moisture, friction, and heat. Moisture, the most important factor, causes the wool fiber to increase in elasticity and capacity for elongation. Under the influence of friction or pressure the wool fiber is forced to travel in the direction of its root (on account of scale formation) until it strikes against adjacent fibers and becomes entangled with them, resulting in a compact matting or felting. The heat produced by the constant friction of the pressure rollers and the trap assists greatly in the movement of fibers among themselves and increases their plasticity, which speed up the process. The best fulling temperature is between 90 and 100°F.

Moisture is provided in the form of a soap, detergent or acid solution. When soap is used, the pieces are soaped in preparation for fulling by running them through a machine designed for this purpose. It consists of a soap box with a rubber roller wringer that distributes the soap and removes any excess. The concentration of the soap solution, i.e., the amount of soap used, depends upon the felting quality of the wool and the shrinkage desired. For slight felting a 5% soap solution (70% strength) is sufficient, whereas for heavy uniform goods a 10% soap solution (70%) is required. The alkalinity of the soap depends upon whether the cloth is scoured or unscoured. When in a scoured condition a neutral soap is used. For gray goods, an alkaline soap containing approximately ½–1% of soda, which will aid the saponification of the oil, is preferred.

Another purpose of the soap is to reduce the friction of the piece in the mill and to prevent the formation of flocks and of fulling streaks or uneven places. Care must be taken that no excess of soap is present because of the possibility of the goods slipping, which would reduce the fiber movement.

To prevent bleeding of colors, weak acid solutions are preferred as milling agents for fabrics such as felts and white woolen blankets with colored borders. The quantity of acid used varies from 2 to 4%, depending upon the weight of goods to be fulled. The fulling time may vary from 5 min to 6 hr, depending upon the amount of felting which is desired. According to McPhee (15), in order to obtain optimum wearing properties with woolen fabrics, acid fulling is preferable to soap and soda fulling.

There are two kinds of fulling machines or mills; the stock or hammer fulling mill and rotary fulling mill.

Stock or hammer fulling mill

Into this machine the piece is put in bulk and two stocks or hammers work on or pound the cloth. Through this action the threads of the cloth are moved about and the fibers begin to felt. The amount of shrinkage of the cloth depends upon the time this action is continued. The shrinkage or felting takes place in length and width at the same time: there are no means to control separately the shrinkage in width and length. The machines are built in different sizes, and from one to four pieces can be milled at the same time, depending upon the weight of the goods. Stock fulling mills are sometimes preferred by finishers for woolen and worsted fabrics when a certain crepe or sporty effect is to be developed and a small amount of felting is required. Their main use is in the manufacturing of non woven felts. See Chapter 15, page 1075.

Rotary fulling mill

In this machine pieces travel in rope form in a manner similar to that found in a string washer. Two or more heavy rollers, with considerable pressure on the top roller, pull the cloth through the machine. Until recently these fulling rollers were made of hard wooden blocks or an iron cylinder, but now rubber and fiber composition rollers are used for this purpose. The latter have proved very successful because they require less turning down than wooden rollers and last longer. From the rollers the goods enter a wooden or stainless steel crimping box (also called a trap), into which they are crammed temporarily by a weighted lid or cover. This compression of the goods acts on the warp of the cloth and causes the piece to shrink in length. From the crimping box the goods slide down an incline into the bottom of the trough from where it is lifted and passed through a knocking-off device or stop motion. The stop motion serves to stop the machine when an entanglement occurs and at the same time aids in keeping the individual strings apart. The cloth now passes over a guide roller into a throat. The throat may consist of two adjustable side rollers or metal checks. From here the goods pass into the main fulling rollers, completing the cycle of continuous operation (Fig. 17).

The pressure given to the top rollers and its proper control is an important factor in good fulling. Up to recent years, pressure was provided through weights and springs, but a late development is the application of hydraulic pressure and compressed air cylinders. These devices assist greatly in definite control of pressure and ease of operation.

Modern fulling machines are equipped with roller bearings throughout, so adjustments for throats and top roller tensions can be made during operation (Fig. 18).

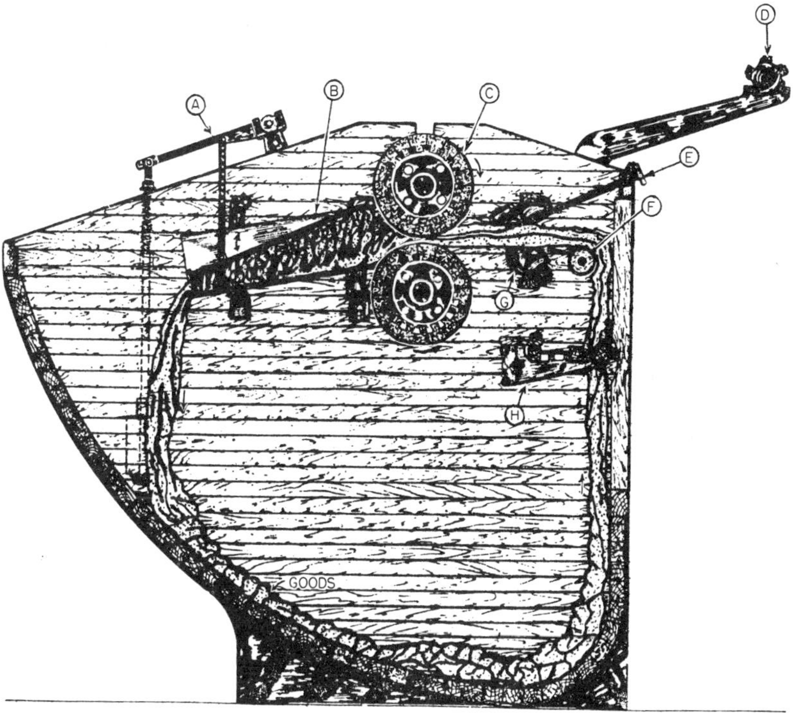

Fig. 17. Diagram of Rotary Fulling Mill. (Courtesy of James Hunter Machine Co.)

A — Compound Weighting
B — Crimping box or trap
C — Main rolls
D — Outboard Delivery roll
E — Throat Adjuster
F — Tension Whip Roll
G — Diamond Eye
H — Stop motion

In the last few years, more and more machine builders have begun to build fulling machines entirely of stainless steel. A new design has basic changes in the construction of the tub that enable the top part of the machine to be dismantled within a few minutes and the top and bottom rollers to be easily removed. Valuable, also, is the additional feature that this machine is supplied with pressure gauges to indicate the pressure not only of the top roller but also of the trap and the adjustable front guides.

Fig. 18. Rotary Fulling Mill. (Courtesy of James Hunter Machine Co., U.S.A.)

Continuous Rotary Fulling Mill

In the fall of 1946 the development of the first continuous fulling mill was announced. The machine was a combination of two rotary fulling mills facing each other, but built within one housing. Essentially, the mill consisted of two sets of rollers running continuously in opposite directions. The top rollers were mounted on a walking beam and the pressure was applied to them by means of pneumatic cylinders. This method of mounting the top rollers eliminated the necessity of reversing the rollers because the cloth would travel only in the direction imparted at the point where the two rollers were brought together; this mounting also allows the cloth to reverse after having been packed into the crimping box. The action on the cloth was exactly the same as in a regular rotary fulling mill.

The machine is still awaiting its further development from the pilot stage. The task is considerably more difficult to achieve than continuous scouring.

Combination scouring and fulling machine

Williams (26) and Peace in the middle forties developed a combination machine. The main advantage of such a machine is that the moving of pieces from string or broad washer to a fulling mill and back into washers is eliminated, thereby effecting considerable saving in time. A diagram of the machine is shown in Fig. 19. The two types built today have a capacity of 240 lb dry weight and 360 lb.

The Rodney Hunt Express fulling mill can also be converted into a combination washer and fulling mill by the addition of water supply pipe, sud box, and tube train. The machine is widely used in the United States for wool knit goods.

A new design is the Manerbio (13) scouring and fulling machine, which has a rather complicated construction.

Tacking

In preparation for fulling the cloth is generally tacked. The goods are doubled over, face in, and the two selvages sewed together by hand or, more commonly, by machine. The purpose of tacking is to prevent chafing of the face, and to permit ballooning of the goods,

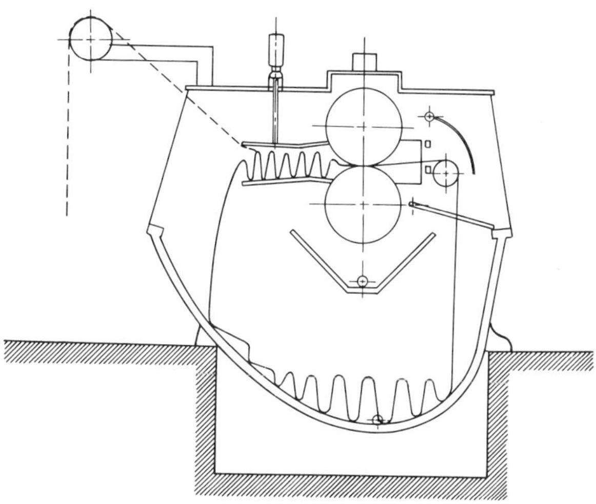

Fig. 19. Williams Combination Scouring and Fulling Machine. (Courtesy Sellers Huddersfield, England.)

in this way preventing fulling wrinkles and streaks. In cheaper grades of goods, tacking is resorted to for flocking, which means felting on shear flock for the purpose of adding weight to the goods.

Soaping

The bulk of goods is run through a soaping machine in open width or string form. By this method the soap is uniformly distributed and its amount controlled. Various types of soaps are used, such as tallow and palm oil soaps, the high-titer soaps preferred in New England mills. It has been proven that low-titer soaps, such as olive and oleic acid soaps, are just as suitable and have the added advantage of good rinsability. They are bought in solid or flake form of 70 or 97% actual soap content. Additional soap may be added as needed during the fulling to keep the pieces properly lubricated.

In recent years synthetic detergents, such as nonionic ethylene condensates, as well as anionic amido condensates, have been found very suitable as fulling agents for woolen and worsted fabrics. These products are especially useful for fulling and scouring of carbonized goods in acid.

The fulling mills can be loaded with goods by one of two methods, single draft or double draft. The former consists of running a piece through the fulling roller single strand, whereas in the latter the piece is doubled or run double strand. Single draft is used in qualities where the shrinkage takes place unevenly in width and length. In other words, when length shrinkage is desired the width shrinkage is retarded by providing less resistance in the throats or eye guide. Double draft is used when a quick felting action in length as well as in width is wanted. The amount of shrinkage given by fulling may vary from a few per cent to as high as 30% in length and 15–35% in width. Fulling shrinkage is controlled by stopping the machine at intervals and measuring width and length by hand or special measuring devices. The following example illustrates fulling shrinkage:

Army cloth	Before fulling	After fulling	Shrinkage, %
Width	80 in.	56 in.	30
Length	50 yd	38 yd	24

After fulling, the pieces containing the soap and other fulling agents are subjected to a washing process similar to that described under piece scouring. In acid fulling, the removal of the acid is accomplished by neutralizing with an alkali and thoroughly rinsing with water.

As the fulling process is essentially a shrinking process and is employed to regulate the weight of the fabric, the necessary calculations must be made before fulling. The weight of the goods will be materially lessened after the oil, as well as other foreign matter adhering to the fibers, is removed. In addition to the wear and tear of the fulling process and the later raising and shearing, a further loss in weight is sustained. There is considerable variation among the various fabrics in all of these weight losses and, therefore, a proper estimation of the loss can be attained only by actual trials. The percentage of loss varies with the stock or type of raw material used more than with anything else. The ultimate finish which the goods receives has a direct bearing on the weight loss.

A piece weighing 20 oz/yd raw should be fulled to 20 oz/yd finished weight. Through scouring and fulling the piece will lose 20%, which is equal to one-fifth, or 4 oz/yd. Therefore, with no shrinkage at all, the piece would weigh only 16 oz/yd. To make up this loss, the piece must be shrunk one-fifth in width and length. This means one-fifth of 1 yd, or 7.2 in. Therefore, the fuller has to shrink each yard of the goods 7.2 in. to obtain a finished weight of 20 oz/yd. This means that a 60-yd piece would be shrunk to 48 yd. For best fulling results it is essential that the contraction in width and length proceed simultaneously.

The fulling time depends on the speed the cloth travels through the machine as well as the pressures exerted by the roller, the trap of the crimping box and the throat. A study was made by von Bergen (22) on 5 Express Fulling Mills to establish the felting rate.

Main Dimension of Machines

Length 108 in.
Width 34 in.
Nip Height 77 in.
Rolls 20 in. diameter 14 in. width
Trap 35 in. × 14 in. × 8 in.

Throat 14 in. × 12 in.
Weight of top roller approximately 600 lb
Linear pressure on 14 in. width 43 lb
Trap pressure settings from 300 to 550 lb
Surface speed of rolls 530 ft/m; Cloth speed 500 ft/m

The cloth used for the study was a 17 oz Melton 69 in. wide from loom 0.1 in. thick. As previously stated, moisture, friction, and heat are the three essentials necessary for the felting process. The moisture was provided by running the pieces open width, through a soap solution, containing 10 oz of soap per gallon and 1.2 oz of soda ash. The dry pieces picked up 50%. They then entered the fulling mill in single strands, three pieces together. The average fulling time of the style used was roughly two hours. During the running of each individual lot, the pieces were measured every 20 min with an additional check test of every half hour. The width and length shrinkages were plotted and also the temperatures registered. The pH during the fulling was around 10.

Figure 20 shows the rate of felting shrinkages in width (filling) and length (warp) and the average temperature rise during the fulling of four lots run in machine No. 1. Table 2 illustrates the variation of the settings in the machines.

TABLE 2

Fulling Machine Settings

Machine no.	Trap weight, lb	Throat, in.	Roller pressure	No. of pieces
1	550	13	full	12
2	550	13	full	12
3	450	10	full	12
4	450	10	½	12
5	550	8	½	12

The difference between pieces within a lot is around 1%, as seen in Table 3, whereas between machines the maximum difference was 3% as shown in Table 4.

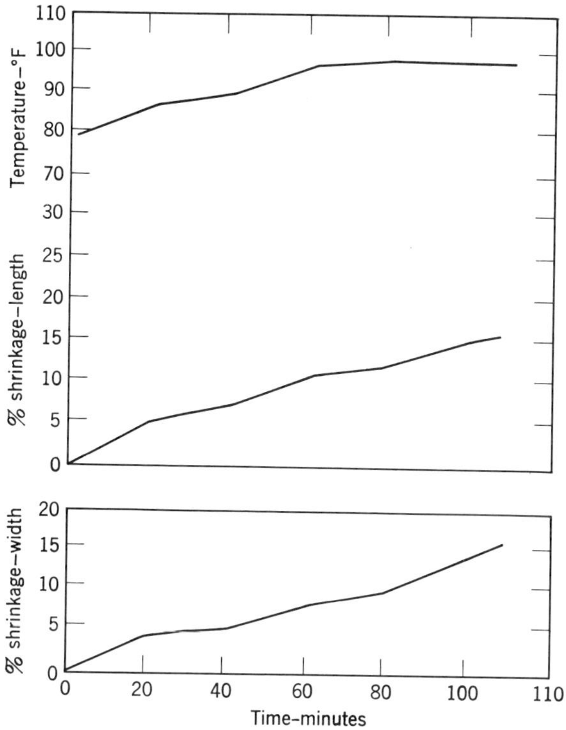

Fig. 20. Average fulling rate and temperature curves of 12 pieces of a 17 oz Melton in Machine No. 1.

TABLE 3

Shrinkage Range between Lots—Machine 1

Lot no.	Width range, %	Length range, %
1	17.1	17.0–17.9
2	16.0–17.7	17.0–17.6
3	17.5–18.6	16.3–17.9
4	17.6–18.6	15.6–17.7

The temperature started from room temperature of 80°F, advancing within an hour to 95° and slowly reaching a maximum average temperature of 102°. These graphs illustrate that it is possible to establish a standard setting for a certain style and reduce the variation within the pieces, caused by fulling, to a very low level.

TABLE 4

Shrinkage Range between Machines

Machine no.	Width, %	Length, %
1	17.5	17.0
2	18.3	18.2
3	17.4	17.0
4	15.4	15.0
5	17.1	17.1
60 pieces range	15.4–18.4	15.0–18.2

It is interesting to study the effect of fulling on the individual yarns. Depending on the length of fulling operation, the interlocking of the fiber first occurs within individual yarns and then, gradually, migrates to the adjoining yarns. The fabric discussed in the above experiments increased its thickness through the fulling operation from 0.1 in. to 0.14 in. which is equal to a 40% increase.

It is well known that considerable differences exist in the felting of wool from various origins. Also in stock dyed goods the dye procedure can affect the feltability. A record kept on the fulling time of the United States Army officers overcoating—Olive drab 26 oz Doeskin 56 in. wide (5 Harness Warp Satin Count 3) made from 64 merino fleece wool revealed the following facts: Changing the dye procedure from the after chrome method to the metachrome method reduced the average fulling time of 100 pieces from 6 hr 38 min to 5 hr 40 min (15%).

The blend change from Texas wools to Austral wools resulted in further reduction of the fulling time to 4 hr 19 min as seen in Table 5.

TABLE 5

Fulling Time in Express Fulling Mill
26 oz O.D. Doeskin 2 pieces per load

Blend	Dye method	Fulling instruction, g/meter	Average fulling time, hr/min	Fulling time range, hr/min
Texas	After chrome	1150	6:38	4:45–8:15
Texas	Meta chrome	1150	5:40	5:00–7:00
Austral	Meta chrome	1150	4:19	3:45–5:45

The following defects may be encountered in fulling:

1. Streaks along the edge of the piece, resulting from curling selvages.

2. Too much friction, from allowing the piece to run too dry or from excessive roller pressure.

3. Fulling rows or marks, from roping of pieces, one of the greatest troubles. With some goods, a constant shaking out and reversing of the direction of the pieces is necessary.

4. Irregular streaks in the cloth, from setting the piece too wide in the loom. This also entails prolonged fulling.

5. Transverse marks at the end of the piece, caused by uneven seams.

6. Fulling rows, from the use of very heavy soap, which retards the felting.

Carbonizing

The carbonizing process, the underlying principles of which were explained in Chapter 1, has for its principal object the destruction of vegetable matter by means of acids. The acid reduces the vegetable matter to carbon, which is removed by mechanical action and the excess acid then neutralized. There are four steps involved, namely, steeping in acid, drying and baking, dusting, and neutralizing.

As already pointed out, a much greater amount of wool is carbonized in the form of piece goods than in loose form. The reasons for this are twofold: first, because piece carbonizing is less costly and, second, to preserve the full fiber strength through the operations of carding and spinning.

The pieces may come from scouring, fulling, or dyeing, depending on the type of the goods. For example, polo cloths, flannels, and similar goods should be carbonized before fulling, because the removal of large burrs from the face of these goods is apt to leave marks, and spiral burrs are difficult to dust out, even though they may be well carbonized. Velours and most worsted fabrics are successfully carbonized after fulling. When carbonizing after dyeing, great care should be taken to avoid changing the shade.

Equipment for piece carbonizing has been primitive until recent years. The steeping was accomplished by running the goods in rope form over a winch, similar to a piece-dyeing machine. The pieces had to be opened, hydro-extracted, and again sewn end to end at full width and passed through the dryer.

Today, the steeping and the baking processes are accomplished in one continuous operation that has eliminated many of the troubles encountered in the old method. The continuous cloth carbonizing machine generally consists, in sequence, of a wetting-out-tank, an acid-soaking tank, a vacuum or squeeze-roller extractor and a carbonizing dryer. The European manufacturers of this type of machine prefer the vacuum extractor or a combination of the two, whereas the American builders have shown their preference for the squeeze-roller extractor. These two types are illustrated in Figs. 21 and 22.

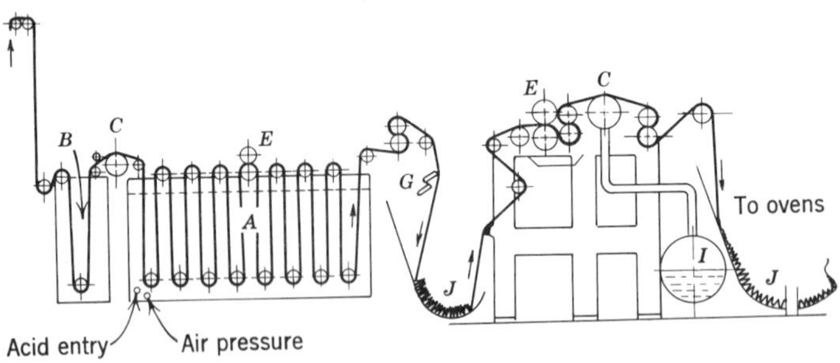

Fig. 21. Piece Carbonizing Range.

A — Acid Tank
B — Wetting our Vat
C — Vacuum Extractor
D — Dryer

The use of squeeze-roller extraction not only eliminates the often troublesome vacuum extractor, with its attendant high horse power requirements, but it offers advantages in the carbonizing operation itself.

Steeping in Acid

Assuming the goods are dry, a number of pieces are first sewn together, end to end, at full width to make a continuous chain that enters the wetting-out tank by means of two narrow rubber leaders, one at each side, which are already threaded through the entire machine. The object of wetting-out is to obtain uniform moisture content in the goods to permit it to absorb the acid in the next tank equally and uniformly. The goods pass over and under a set of top and bottom rollers and then are guided over a vacuum slot that draws out the excess moisture.

Whenever goods are delivered in a wet condition, the above procedure becomes unnecessary and the goods enter the acid-soaking tank directly. A recent development is the use of Fiberglas reinforced polyester resin tanks for the acid-soaking tank. These have always been lead lined and up until the last several years, when type #316 stainless steel has been utilized. Even the stainless steel has not been completely satisfactory, since absolute cleanliness is necessary to avoid corrosion. The polyester glass fiber tank seems to be the best answer to a standing sulfuric acid bath without corrosion. The pieces first pass over a guide roller and a spreader, into a tank filled with diluted acid, then over and under a series of six to eight top and bottom rollers. In the center of the tank a

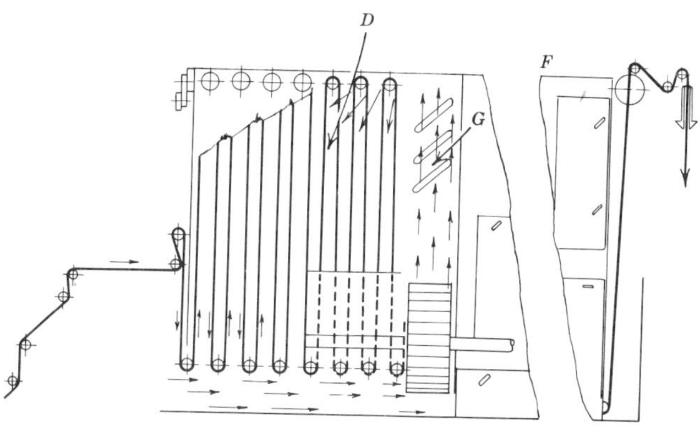

E — Squeeze Rollers
F — Baking Section

G — Hot air circulation
H — Waste Acid Tank

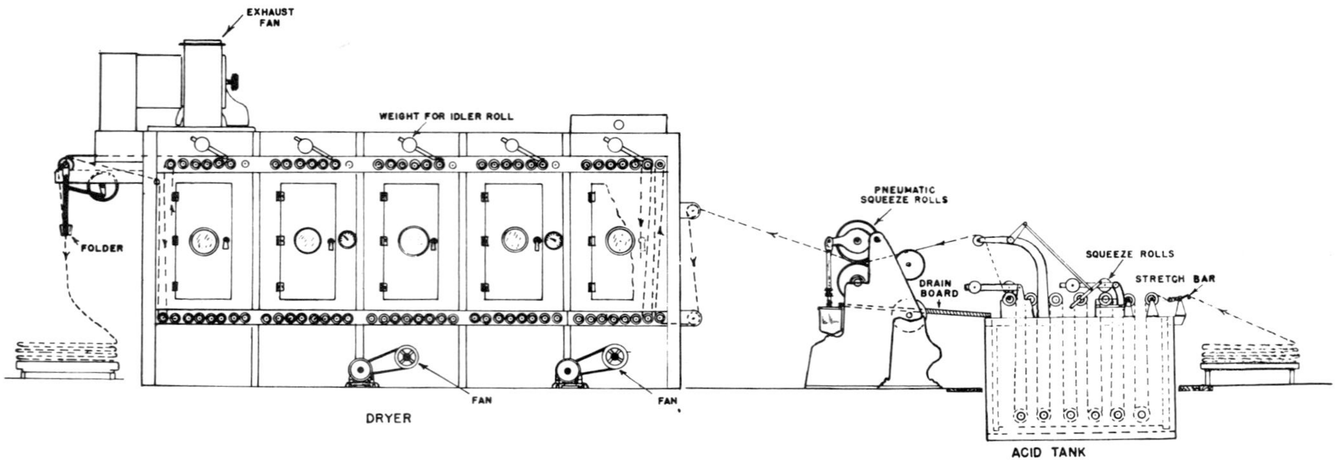

Fig. 22. Hunter continuous piece carbonizer.

pair of squeeze rollers serves to relieve strain as well as to assist the acid to penetrate the cloth by squeezing out the air. At the delivery end of the tank there are another spreader roller and a large pair of rubber squeeze rollers to remove the excess acid. In European machines the pieces pass over a vacuum slot for final extraction.

The speed of the goods varies. The slowest speed is approximately 5½–6 yd/min, whereas the highest speed may be 35 yd/min. For medium and heavy weights (from 16 oz up) the slow speed is required; for light goods a faster speed can be employed. The carbonizing agent is commonly sulfuric acid, which is best stored in a lead tank sunk below the level of the floor. Its normal strength is 66°Be.

For large users of sulfuric acid to reduce the problem of handling this acid the most economical way is to buy the acid in tank cars From these cars it is drawn off into a large open-hearth steel storage tank, holding about two tank carfuls. The acid is then pumped directly through open-hearth steel piping which delivers it directly to a lead-lined stock tank in the carbonizing department.

The acid is mixed directly with water to a concentration of 20°Be. From this stock tank the acid flows by gravity through an acid proportioning valve into the steeping tank.

The incoming acid is thoroughly mixed with the steeping liquor by compressed air, thus giving a uniform concentration throughout the tank. The actual steeping concentration is determined by the quality of the goods as well as by the amount of vegetable matter present. For light weight fabrics the acid concentration may be as low as 1.5°Be or, for 30-oz goods, it may be as high as 4°Be.

The most important task is the control of the concentration within closed limits. The Leeds & Northrup Speed Micromax system has given us excellent performance. The scale on the Micromax Recording Controller shows the ratio between the acid bath and the reference cell (Wheatstone Bridge). We found it very helpful to show the various effects which are normally encountered, such as: valve hunting, improper air mixing, contamination and cleaning of electrode. It is necessary to check the readings in the Micromax by direct titration of the acid bath; this is done hourly and is especially important if the acid bath is kept more than a week. The soda present in the pieces before steeping will accumulate in the steeping tanks in the form of sodium sulfate and will exert a buffering action which causes the speeder to drift. It also will give a warning when the electrode is being poisoned or contaminated. The actual immersion time at the low speed with eight pairs of rollers is approximately 3 min. The temperature is always room temperature.

Based on the studies by Mizell, Davis, and Oliver, as already referred to in Chapter 1, page 000 the possibilities are to work with higher acid concentration 7–7.5% in the presence of a nonionic type of wetting agent at low temperature 50°F and reduction of the immersion time to less than 1 min.

Through the use of heavy squeeze rollers, a uniform as well as a high degree of acid extraction is possible. Under this high pressure, a more thorough acid impregnation of the burr content is also accomplished. In order to reclaim the extracted acid from the cloth and return it to the acid tank for reuse, either a stainless-steel or lead-lined drip pan, which drains directly back into the acid tank by gravity, is mounted under the squeeze rollers, thus eliminating any necessity for pumps.

The extraction of excess acid also may be effected by a vacuum slot or mouthpiece made of acid-resisting metals. There are various types of mouthpieces such as single and triple slots, the most common being the single slot. The ends of the slots may be covered to match the width of the running piece that is electrically controlled at all times to prevent gaps.

Drying and Baking

Drying has two distinct functions: to dry the cloth itself and to char the vegetable matter. After the cloth leaves the extractor, it goes directly in open width, free from wrinkles, into the drying and baking section of the unit. To avoid destruction of the piece numbers it is customary during this process to treat the piece numbers (sewn on the head ends with cotton thread for identification) with an alkaline solution such as 1 part of sodium silicate and 10 parts of 6% soda solution.

The entire drying unit is made up of a series of sections or chambers, the number of which may vary from five to ten. In a six-section unit, the first four act as drying and the last two as baking chambers. Preheated air enters the chamber through circulating fans, blowing it across the width of the goods and then drawing it down through the fabric. An exhaust fan takes care of the removal of the nearly saturated air. After the cloth is thoroughly dried, it passes into the baking section, where the greatest heat is applied, usually 190°F or over. Special care is taken that the hot-air circulation maintains a uniform temperature throughout and at all times. The pieces leave the baking section at the top of the machine and are folded into a truck. The temperatures maintained in a unit of six sections are given in Table 6.

TABLE 6

Temperature Ranges in Carbonizing

Chamber	I	II	III	IV	V	VI
Degrees F. min.	120°	145°	160°	175°	190°	200°
Degrees F. max.	145°	160°	175°	190°	200°	210°

The rate at which the goods passes through the drying unit is dependent on the speed permissible in the immersion tank. The rate in the immersion tank is kept slightly lower than the rate in the drying unit to avoid the possibility of stretching or breaking the pieces between the sections.

A special compensating range drive that eliminates undue tension and building up of slackness in the dryer has been developed. This drive holds a predetermined tension on the cloth from the time it enters the unit until it leaves. The control is split into sections, and the tension in all sections may be set alike or it may be adjusted to give more or less tension in different sections. The time varies with the number of chambers; for a six-chamber oven it is approximately one-half hour, of which about one-third is devoted to the baking.

Infrared Drying

A new development is the application of infrared rays for the final baking. One type of infrared baking unit follows the last section of the dryer. The heating equipment consists of two banks of infrared lamps facing each other.

Cloth is fed from the dryer into the baker under a roller at the bottom and then vertically upward to the top, where it passes over a roller, then vertically downward to a third roller, after which the cloth repeats this cycle upward and downward over the fourth and fifth rollers and passes under a sixth roller and vertically upward and out of the carbonizer. By this means one side of the cloth is exposed to one bank of lamps and the other side of the cloth to the second bank of lamps. The baker has a capacity of 21 yards of cloth. At the normal speed of 25 yd/min each side of the cloth is exposed to the direct rays of the lamps for about twenty seconds.

The unit has not found general acceptance because of the danger of burns, and it is costly to run.

The main advantages of the use of infrared heat for the final baking are said to be two: a lower number of drying sections, with consequently less floor space required, and increased production as the speed can be boosted 30–40%. This is made possible by using the drying and baking section of the dryer for drying only and doing the baking with the infrared unit.

Crushing

The object of crushing or dusting is to crush and beat out the charred vegetable matter from the fabric. This is accomplished by mechanical means in a fulling mill. Instead of running the pieces with soap, they are run dry and subjected to the crushing of the fulling rollers. For economical reasons it is advisable to use a number of dry mills in series. The time of dusting varies from five to fifteen minutes, depending on the amount of charred matter to be removed. Loosely woven goods, lightweight pieces and qualities such as boucles, which are easily damaged, are not dusted at all. The process of dusting is most effective when the goods are dry and hot. The wool regains nearly all of its normal moisture in approximately one hour, hence no time should be lost in doing the dusting. The pieces leave the dry mill in rope form and are opened by running them through a mechanical opener or by hand.

Today the crushing is done on continuous machines installed as a part of full carbonizing ranges. Built by several manufacturers such as the Riggs and Lombard consisting of two units, set on tandem, each equipped with four sets of metal rolls, the bottom ones being knurled. The top roll pressure is applied through coil springs controlled by pneumatic cylinders.

The pieces travel in forward and reverse cycles adjusted for a desired production speed and synchronized with the speed of the carbonizing range. An even more compact unit is built by Hunter (Fig. 23).

The main advantage of the continuous units is that the fabrics are crushed as they come out of the baker section, which means that they get the crushing action while the charred burrs are still hot and very brittle and before they have had a chance to regain much moisture and become somewhat softened. This is especially vital in soft woolens where the burrs are buried in the core of the fabric.

Neutralizing

The object of neutralizing is to remove all traces of acid in order to eliminate difficulties in later processing as well as tendering of the goods. For this purpose the goods are treated either in a string

Fig. 23. Continuous Burr Crusher. (Courtesy of James Hunter Machine Co., U.S.A.)

washer or a broad washer with a solution of sodium carbonate or soda ash. The process is accomplished in three steps: preliminary rinsing, alkaline treatment, and final rinsing.

The purpose of preliminary rinsing is to reduce the acid content of the goods. This effects a saving in soda ash and eliminates any danger of damage to the material that might occur from the heat of reaction if soda ash were added directly. The acid content of the goods should be reduced to about 3% or less in this rinse, which requires about 20–30 min in a string washer (four pieces).

After this preliminary rinse the machine is drained and the proper quantity of soda solution added. The strength of the alkaline bath should be about 2½%. The goods are run in this bath until completely neutralized and alkaline to methyl orange. The running time depends on the acid content and on the weight of the goods, but is normally from 20 to 30 min.

The final rinse, which is necessary to get rid of all excess soda, is done by turning on the cold water full force and running the goods continuously for a period of 20–30 min. On stock- or yarn-dyed goods and color mixtures it is advisable after the final rinse ro run the goods for 10 min in a weak acetic acid bath for brightening of the colors that are apt to be dulled by the soda. A quart of acetic acid (28%) for each two pieces is sufficient. Data on neutralization of a coating material weighing 20 oz (56 in.) is given in Table 7.

TABLE 7
Neutralizing Process

Items	Carbonization	Pre-rinse	Soda	Final rinse
Running time	—	30 min.	20 min.	30 min.
pH of goods	1.8	2.4	10.6	7.5
Per cent H_2SO_4	6.0	2.8	—	—
Per cent soda	—	—	1.5	0.6

Continuous Neutralizing

The continuous washer lends itself to the successful application of continuous neutralization. In one plant, the soaper and bowls 1 and 2 of the wash section are used for preliminary rinsing, wash bowls 3 and 4 are used as neutralizers, and the reverse-rinse section is used for the final rinse. The sulfuric acid content of the carbonized cloths was reduced from 6 to 2.2% in the three-bowl, preliminary-rinse system employing separate overflows for waste in each bowl; this is equal to an acid reduction of 63%. A sodium carbonate solution of 2% strength was found to be necessary to neutralize qualities up to 20 oz, inactivating all of the residual sulfuric acid and leaving a surplus of approximately 2% alkali in the cloth. The alkali concentration must be controlled by titration. The total neutralizing time in the continuous washer, operating at a speed of 40 yd/min, is approximately 22 min, which is less than one-third of the normal neutralizing time necessary in the dolly washer. The preliminary rinsing time, with the inclusion of the soaper, amounts to 5 min, actual neutralizing time to 4 min, and final rinsing to 13 min. This is possible because there are six nips per minute in the continuous washer as compared to three nips per minute in the dolly washer. Figures 24*a* and *b* shows the titration curves of a four-bowl continuous neutralizing machine, as reported by von Bergen (22). The pieces which were treated according to curves in Fig. 24*a* still

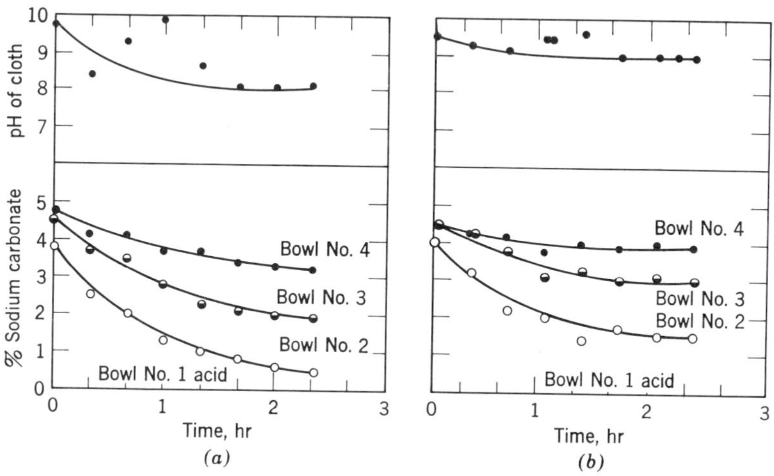

Fig. 24. Soda Concentration Changes in a four bowl continuous Neturalizing Machine.
(a) Rate of soda feed to bowl 4—Six gallons per minute 4% soda solution.
(b) Rate of soda feed to bowl 4—Seven gallons per minutes—bowl 3, two gallons per minute.

contained a small amount of sulfuric acid, after rinsing, whereas the pieces from the *b* curves ended up on the alkali side, with a pH between 8 and 9. In the final rinse the concentration of the alkali was reduced below 1% ending up with a pH of the cloth around 8.

Piece-dyed fabrics, which are carbonized after dyeing, are given only a water rinse. To avoid stripping of colors, soda has to be omitted. Aluminum chloride is today rarely used in piece carbonizing, although it may have its place for carbonizing of stock- or top-dyed fabrics on which dyestuffs sensitive to sulfuric acid are used.

Scutching

In most of the wet processes described so far, scouring, fulling, and neutralizing, the pieces are handled in string or rope form. In many instances the pieces are tacked. As the proper examination of pieces in rope form is not possible and because most of the operations that follow require that the fabric be in the full width, detacking and opening operations are necessary. In small mills these operations are performed by hand and in the larger mills by running the pieces through a scutcher. An arrangement for scutching, combined with detacking, is shown in Fig. 25. The continuous washer is generally connected directly with a scutcher.

Beaming, Wet Decating, and Blowing

Beaming

Face-finished goods coming from scouring, fulling or carbonizing, contain many creases and wrinkles that require removal before

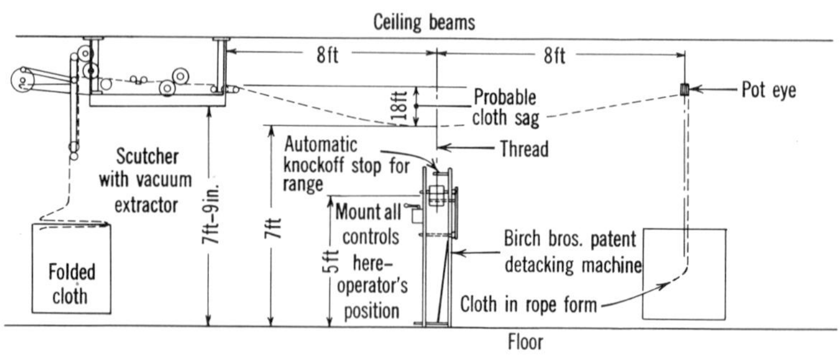

Fig. 25. Scutcher Installation. (Courtesy of Birch Brothers, Inc., U.S.A.)

napping or gigging. Creases can be removed by two methods, beaming or wet decating, choice of which depends on the type of the cloth and the severity of the creases. The easiest method is beaming, which is a simplified form of crabbing. The goods are run through warm or hot water depending on whether the fabrics are stock dyed, yarn dyed or piece dyed. The goods are wound tightly on a wooden roller with a heavy pressure roller on top, and then run for half an hour. Machines, as previously described, are used for this purpose.

Wet Decating

Where a more permanent setting of the fibers is required and at the same time a certain amount of luster has to be developed, the finisher will apply wet decating, also known as boiling, or when followed by a steam treatment, as wet blowing. Wet decating is accomplished by the combined application of heat, moisture and tension. The production of luster is partly attributable to the straightening and leveling out of the surface and partly to the natural luster of the stock itself. For example, mohair with a high original luster, needs little decating, whereas fine merino wools require considerable decating to produce a similar luster.

There are various types of wet decating machines, all of which have the following essential features: a metal trough containing several perforated metal cylinders to allow water or steam to circulate freely, a pump for this circulation, and a suitable tension arrangement.

In the preparation of the cloth for wet decating or blowing, cotton leaders are stitched to each end of the fabric, which is then wound in a suitable state of tension on a perforated metal cylinder. Thus, the cotton leaders become the interior and the exterior layers of the roll, notonly protecting the cloth during treatment, but also ensuring even penetration of the water and steam throughout the roll. After being bound with heavy cotton tape, the cylinder surrounded by the roll of cloth is placed in the trough. The piece is then treated for 5–20 min at temperatures ranging from 140°F to boiling with water circulating in both directions, or it is blown with steam from the inside of the roll for 5–10 min. The effects of the treatment can be considerably increased by the combined use of hot water and steam. However, such a treatment carries with it the danger of severe fiber damage. The cooling of the cloth is accomplished by drawing cold water or cold air through the roll by means of a vacuum pump (Fig. 26).

Sustmann and Zahn (20) made a systematic study of the effect wet decating has on the physical properties of the worsted fabric and the modification of the wool after the decating. They came to the following conclusion:

At water temperatures from 50 to 65°C the average loss in breaking strength is about 5–10% and at the higher temperatures from 80 to 95°C from 10–20% compared with the untreated cloth. Thus, higher temperatures than 70°C should be avoided.

An added steam treatment of 5 min has no further effect on the strength. The time of the treatment in the range of 5–40 min does not have any special effect on the breaking strength. The elongation at break increases in special cases up to 30%, but with increasing water temperatures from 50 to 95°C, the increased elongation drops and reaches only after a 5 min steam treatment, the break elongation of the untreated cloth. Also, the time of treatment does not seem to influence the elongation.

At warm water temperatures from 50 to 60°C, there is no change in the thickness of the fabric; only through treatments at 80 and 95°C and a following steam treatment the fabric gets thinner. In general it can be stated that wet decating treatments up to 65°C do hardly cause losses in breaking strength, but improve the breaking elongation and preserve the thickness which explains the full hand obtained with this method of decating.

The chemical analysis of the wool substance after the wet decating showed that the cystine content only after treatments in 95°C water was considerably lowered and a reduction in the urea-bisulfite solubility took place.

Raising

As the name implies, the raising process consists of lifting out from the body of the fabric to the surface a layer of fibers which has been developed in fulling or from the yarn itself. This layer of fibers is variously termed nap, pile, or cover. Raising is distinctly a mechanical operation. The action is simply to entangle the fibers with the sharp points of teasels or wire rollers and to bring the

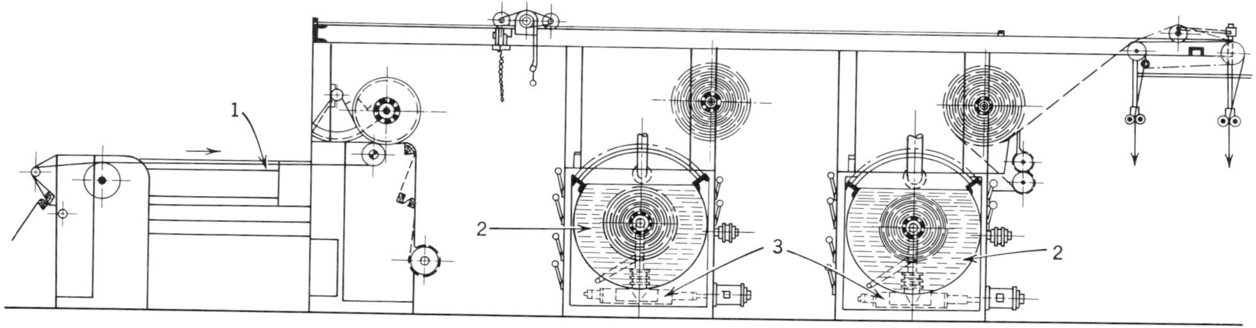

Fig. 26. Wet Decating Range. (Courtesy of Gessner.)

1. Tenter Frame *2.* Water Tank *3.* Water Pump

fibers to the surface of the fabric. Four objects of the raising process are:

1. To produce a nap, pile or cover on the surface of the cloth.

2. To secure softness and a lofty handle.

3. To conceal the weave or design, to soften the outline of the pattern and partially to blend the colors.

4. To promote a more thorough felting as a preliminary to the fulling process.

The raising operation may be accomplished by means of teasels (gigging or teaseling) and by wire-covered rollers (napping). There are four principal systems of raising the nap when submitted to the action of the teasels or wire-covered rollers; moist gigging and wet gigging, moist napping and dry napping.

The amount of napping or gigging applied to fabrics depends on the quality of the goods. The raw material used, the structure of the weave, and the weight and strength of the fabric are factors to be considered and the effect desired.

Gigging or teaseling

Gigging is done with teasels which are the dried flower heads of a species of thistle plant grown and cultivated specifically for this purpose in the state of Oregon and in the vicinity of Schenectady, N.Y. The largest cultivation of teasels is in southern France, where there is a large-size teasel, known as the king teasel, which is about 2 in. in length. The medium-size teasels are 1½ to 1¾ in. long and are the most expensive. Within the last few years, synthetic teasels have been made and are currently in use. (U.S. Patent No. 3,080,637 (March 12, 1963) by William H. Dutt, assigned to Albany Felt Company, Albany 1, New York.) These are more uniform than the natural ones, and have much longer life. This is another case where synthetics have improved over "nature." Note the difference in the photograph (Fig. 27). They are set into the frames as tightly as possible in two or three rows. For proper setting they are pretreated with hot water or steam for 5 min to soften them and then allowed to dry. These frames or iron slats are then mounted on a drum. Some finishers prefer the "rolling" teasel gig, where the teasels are mounted on rotable spindles fastened directly on the drum as seen in Fig. 20 of Chapter 15. This

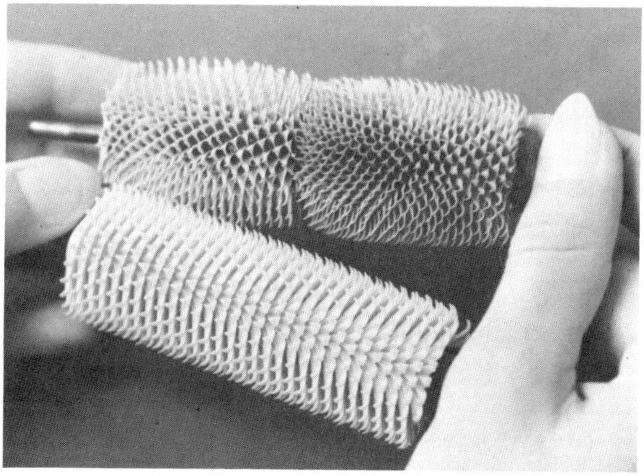

Fig. 27. Comparison of Natural and Synthetic Teasels. (Courtesy of Albany Felt Co., U.S.A.)

drum rotates at a speed of 100 rpm and the cloth is brought into direct contact with the teasel-filled slats. It is very important that the raising of the fibers be done very gradually so that the fibers are not torn out. They are merely untangled and lifted. Therefore, the teaseling process should begin with used teasels, the points of which have already been dulled and softened. Gradually new or sharp teasels are employed until the desired nap or pile is developed. As stated in Chapter 15, page 1109, synthetic teasels are made today which prove to be more uniform and much more durable.

Moist Gigging and Wet Gigging

Of these, moist gigging is the most general method. The raising of the fibers is done in the moist state followed by the process of wet gigging, in which the cloth runs saturated with water. Wet gigging really should not be considered a raising process. In this action the fibers that were raised in moist gigging are laid down in one direction on the surface of the cloth. In the moist and the wet gigging processes the fibers are swollen and their elasticity and rigidity are diminished. Consequently, they are more pliable to mechanical stress. This raising process is generally applied to broadcloth finishes, where an evenly laid down nap with a lustrous appearance is desired.

Figure 28 shows a one-cylinder, four-contact gig. Specially designed for wet gigging, it is made with a skeleton iron or a wooden cylinder into which the teasel slats are set in rows. The slats are staggered 2 in. by pairs and 20 or more are required for a cylinder. The number of teasels required to fill 20 slats varies from 2500 of the 2-in teasel to 4000 of 1½-in. teasel.

The goods are wound in full width on roller *A*, submerged in a trough of water from which they are unrolled by means of a leader, passed over the contact and the carrier rollers around the cylinder of the gig, and rerolled on roller *B*, also submerged in a trough of water. The cloth contacts the teasels at points *1*, *2*, *3*, and *4*.

The direction of rotation of the cylinder is the same as that of the cloth, but it runs at a higher speed. When all the goods have been unwound and the second leader comes up, the machine and the cloth are reversed, this being continued until a good nap has been produced. The contact rollers are adjustable and, if set closer to the cylinder, they give a greater gigging action, producing a greater amount of nap.

During the gigging process the teasels become ffilled with wool fibers and lose strength because they are wet. To restore them to use, they are dried and brushed clean by hand or with a brushing machine.

Double-cylinder gigs and also "rolling" teasel gigs have been built to overcome some of the drawbacks, such as low production

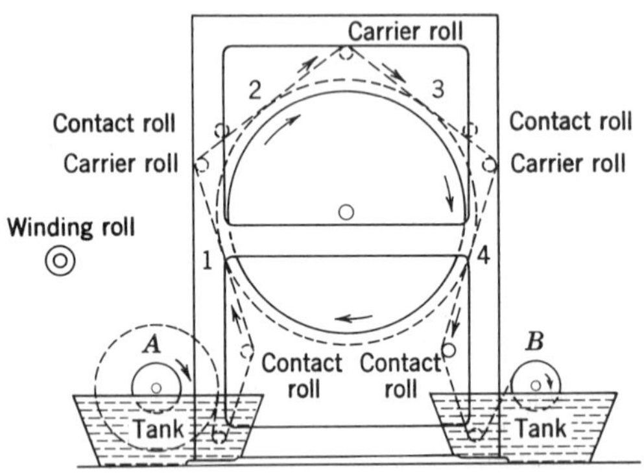

Fig. 28. One cylinder four contact gig. (Courtesy of Curtis & Marble Machine Co., U.S.A.)

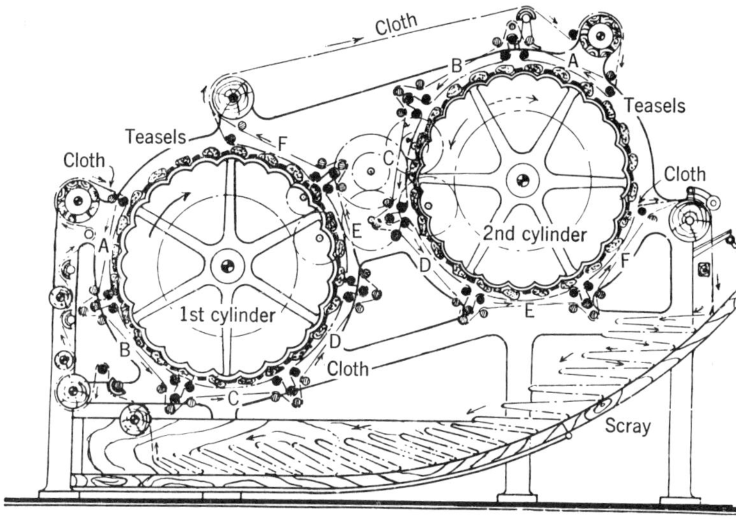

Fig. 29. Double cylinder—twelve contact Teasel Gig.

and reversing cylinders, of the old "up-and-down" gig. Also the number of contacts has been increased from four to six, which in a two-cylinder gig amounts to twelve contacts (Fig. 29). Instead of cloth rollers, scrays are used to enable continuous gigging of a greater number of pieces in a long chain. The cloth speed is 8–10 yd/min. The number of runs depends on the surface of the cloth and the character of the yarn.

Instead of teasel slats, the cylinder may be fitted with brush lags or rolls, set with any grade of bristle, hair, or nylon fiber and used for wet or damp brushing on face-finish goods.

Napping

It is now more than a century since attempts were made to employ fine steel wires in the raising process and today there are wire napping machines which give fast production and are so constructed that the raising of the wool fibers can be done very gradually. These processes are carried out by means of fine steel wires, called wire card clothing, that are generally applied to fabrics where a dense lofty nap is required. The fibers, in contrast to the laid-down nap, are lifted with the sharp points of the steel wire and kept in a nearly upright position. This produces a finish on woolen cloth similar to that on velvets and plushes, where the pile is formed by cutting the yarns (see Chapter 14).

Moist napping is usually a preliminary process to the succeeding moist gigging. In general, it is applied to fulled fabrics where the fibers have formed a dense layer of felt on the surface. The action of the wire points is more drastic in its nature than the teasel points and the entanglement of the fibers is more efficient.

Dry napping is more suitable for fabrics which are not fulled. The raising of the fibers must be done from the warp and the filling threads, and hence produces a spongy, lofty pile. Figure 30 shows three types of common naps.

There are two types of nappers, i.e., single acting and double acting. In the single-acting machine, also called the French or straight-away napper, the nap rises in one direction only. The cylinder runs in the direction of the goods at a fixed speed of 100 rpm. On its periphery are small wire rollers that rotate on their own axis at an independent and variable speed in the opposite direction of the cylinder.

The cloth runs through the machine in the same direction in which the cylinder rotates; the points of the card wire are carried

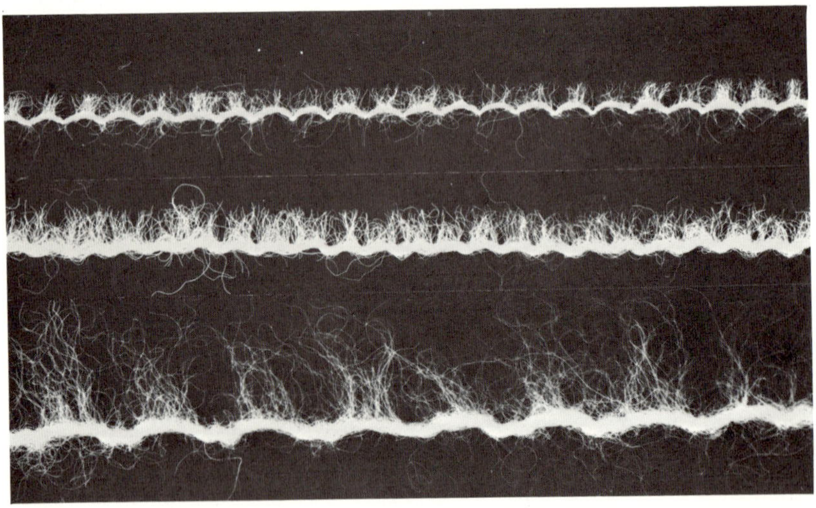

Fig. 30. Types of Naps: Suede, Duvetyn, Velour, top to bottom.

into the cloth by the forward rotation of the cylinder, the degree being determined by varying the speed of the wire rollers (Fig. 31).

On the drum of the double-acting napper there are two series of rollers, named *pile rollers* and *counter pile rollers*, which run counter-clockwise to the cylinder and the cloth. The points of the card wire are set in opposite direction to each other, the points of the pile roller in the direction with the cloth and the points of the counter pile roller against the cloth. By adjusting the speed of the two kinds of rollers, more or less raising effect can be obtained. The action of the pile rollers increases the effect by decreasing the speed; the action of the counter pile rollers increases the effect by increasing the speed (Fig. 32).

Here, too, the rising of the fibers must be done gradually so that most of the fibers are preserved as much as possible. The cloth speed in napping varies from 12 to 15 yd/min, which is 20–30% higher than the normal speed in gigging.

Another kind of a machine is a combination of wire card rollers and slats filled with wire clothing. The rollers and slats are mounted alternately on a drum. The wire card rollers do the raising and the stationary wire points lay down the nap. This kind of a machine is used on fabrics where a pile is laid at an angle in one direction, instead of standing upright.

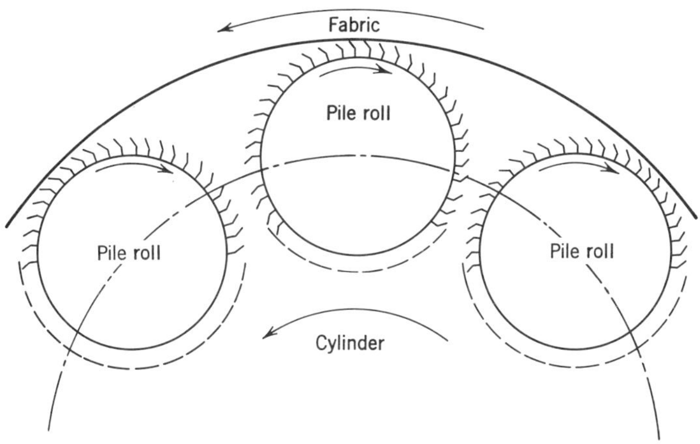

Fig. 31. Single Action Raising.

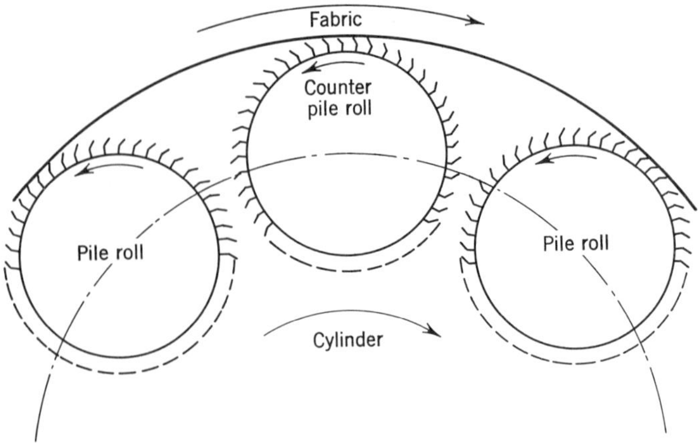

Fig. 32. Double Action Raising.

A considerable change in the driving mechanism and control of the napping action has taken place within recent years. In the earlier machines, both sets of rollers were driven by belts and the adjustment of napping action was obtained by means of cone pulleys. In the modern types vee-ropes have replaced the belts, and P.I.V. gear boxes the cone pulleys.

The operation of nappers requires considerable skill and experience if the best results are to be obtained. The napper has been helped by the introduction of control devices such as those on the auto-zero machine by Tomlinson (21) in which it is possible to locate the setting at which napping just commences to be effective. This automatically fixes the zero and it is possible then to regulate napping from this base.

A further improvement of this control by Tomlinson is the Electro-Zero Raising control which incorporates a new conception of the control of the working forces in a raising machine introduced in 1965.

The latest important development in the napping machine design has been the introduction of 24 roller single and double acting nappers such as built by Riley and Gessner (Hi-Torc) (Fig. 33) which have as their main feature unique and extremely variable hydraulic systems of controlling the worker rolls. In the

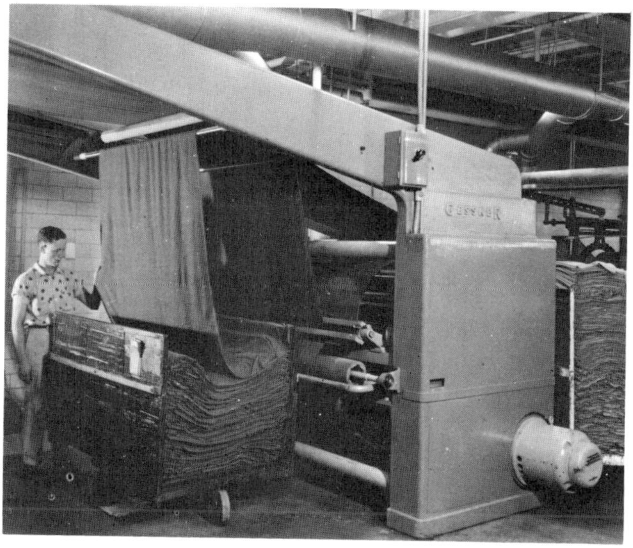

Fig. 33. Gessner Hi-Torc Napper. (Courtesy of J. P. Stevens & Co., Inc.)

double acting machine the hydraulic system consists of three units, one of which, chain driven from the main cylinder shaft acts as a rotary pump and provides a flow of oil to drive the rotors of the other two units which thus act as motors. One of these motors applies the energy to the pile rollers and the other operates the counterpile rollers. The cloth tension over the napping cylinder is also adjusted by a hydraulic unit. Figure 34 shows the control panel of the Gessner (Hi-Torc) napper.

By this method of driving the worker rolls, the complete range of work can be done on one machine including "breaking," "raising," and "felting." The "breaking" action is produced by adjusting the torque or fluid pressure drive to the necessary point to start raising the fibers out of the yarns at the desired rate.

Generally "raising" is produced by lowering the torque or fluid pressure drive to lesser raising action, or to a higher slip factor. This is accomplished by adjusting all pressure of the fluid drive to produce certain types of actions as indicated by the energy tachometers.

Fig. 34. Control Panel Gessner Hi-Torc Napper.

"Felting" is produced by the further adjusting of the fluid pressure drive by means of a "back pressure" or torque reversal control, to bring the worker roll effort to various degrees of negative, or felting action.

In addition to the simplified control and great flexibility the hydraulic system also has the advantage that it allows the resetting of pile and counterpile controls without stopping the machine. A simple control panel indicates napping energy in front of the machine and is in easy reach of the operator.

The cloth speed of the older machines generally can be changed with 3 to 5 settings from 5 to 18 yd/min. With the modern machines smooth changes in the cloth speed can be made up to 35 or even close to 50 yd/min. Generally it can be stated that two runs in an old machine are equal to one run in a modern machine. Actually, as much as 400% increase in production has been reported. The working widths are 60 in., 72 in., 80 in., 90 in., 96 in., 108 in., and 120 in.

For continuous operation two to three double action Hi-Torc nappers are used in tandem. This is especially suitable for dry napping of woolen blankets to produce a long fluffy nap at a production speed anywhere from two to four times that of the conventional napper.

As the space requirement for such napping ranges is large and the machinery too costly for small mills, one manufacturer, (Kettling and Braun) introduced a "non-stop" napping machine described by Kittel (11). The machine handles 3–5 pieces against the normal single piece napper making savings from 20 to 25% possible.

Factors in Napping

Whewell and Snowden (25) made a study of the factors influencing the results obtained in napping. The study was concerned mainly with changes in thickness and the tensile strength of the fabric and the yarns when various types of fabrics are napped. Also, some data about the fiber composition of the nap from various blends were obtained. The following information is gathered from this study:

Machine settings are perhaps the most important factors which determine the ease with which a nap is produced on a cloth. It is essential to note, however, that a high speed of nap production is not always desirable. In fact, if the nap is produced too quickly it tends to be wild and raggy. There is probably an optimum speed for napping each type of cloth and the experienced napper is well aware of this and has generally found out what it is. The following adjustments are of particular significance in determining the severity of the napping:

1. Settings of the pile and counterpile rollers
2. The number of passages through the machine
3. The speed at which the fabric goes through the machine

Settings of Pile and Counterpile Rollers

The speed at which nap is produced on a double action machine is determined by the speeds of the pile and counterpile rollers relative to the cloth. In practice it is necessary to obtain a correct balance between the two speeds, for excessive counterpile or pile action tends to produce a raggy and unattractive nap. The relative speeds of the two sets of rollers also affects the change in yarn strength which accompanies napping. Excessive pile or counterpile action has a tearing action and this causes the cloth to be greatly weakened. Data to illustrate this are shown in Table 8 which refer to a fabric napped on a Tomlinson double action Vee machine at various settings of the pile and counterpile rollers. Each fabric was passed through the machine twice and the tensile strength of the weft yarns was then determined and compared with that of the untreated yarn.

TABLE 8

Setting of machine		Tensile strength of napped fabric (lb)	Thickness of napped cloth (in.)
Pile	Counterpile		
Setting on dial	Setting on dial		
0	0	180	0.224
1	1	146	0.250
2	2	133	0.280
3	3	122	0.316
4	4	104	0.360
5	5	58	0.414
6	6	8	0.526

TABLE 9

Speed of cloth through machine (ft/min)	Tensile strength of napped cloth	Thickness of napped fabric (arbitrary units)
55	113	277
45	112	297
39	108	317
32	100	340
27	75	366

Number of Passages Through Machine

As would be expected the amount of pile obtained during napping increases with the number of passages through the machine until the weft strength is so low that napping merely detaches the fibers from the fabric. It is clear, therefore, that the weight of nap or the thickness of the raised cloth must approach a limiting value as the number of passages through the machine increases. In the early stages of raising, however, the relationship is linear (see Fig. 35) so that to a first approximation the greater the number of passages through the machine the more pile is produced. In a similar way, the tensile strength of the weft yarn decreases with number of passages until a limiting low value is reached but in the early stages of raising the relationship between tensile strength and number of passages is linear.

Speed of Cloth Through the Machine

In general the slower the cloth passes through the napping machine the greater is the loss in strength and the more nap is produced by each passage. This clearly arises from the longer time the wires are able to act on the cloth. Some indication of the magnitude of this effect is given in Table 9.

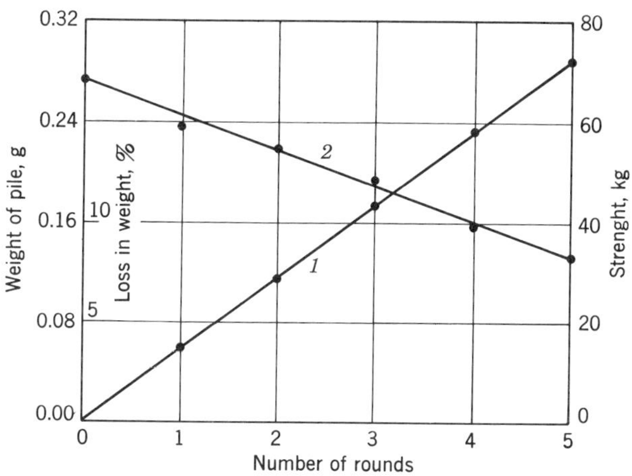

Fig. 35. Relationship between number of napping rounds and amount of pile and weft strength. (Whewell-Snowden.)

TABLE 10

Changes in Thickness and in Tensile Strength (Wet) on Napping

Composition of weft	Increase in thickness (arbitrary units) on napping	Tensile strength (oz)	
		Before napping	After napping
Nylon	269	115	60
Bright fibro	324	62	24
Matt fibro	411	57	14
Wool	443	34	16
Acetate Rayon	410	29	1

Composition of Fabric

All fabrics, whatever their composition, can be napped provided they are strong enough to withstand the action of the machine. The nature and appearance of the nap does, however, vary widely from one fabric to another. In wool cloths it would appear that density and length of pile are governed by the diameter and length of the fibers.

In most cases the blended fabrics respond to napping less readily than the corresponding all-wool cloths. This is most marked in fabrics containing Terylene, nylon or Fibroceta. The presence of

TABLE 11

Thickness of Napped Wool/Terylene Fabrics

% wool	Thickness (in.) of fabric		
	Before napping	After dry napping	After wet napping
100	0.077	0.105	0.215
80	0.075	0.105	0.205
60	0.075	0.100	0.195
40	0.072	0.080	0.210
20	0.074	0.080	0.180
0*	0.070	0.075	0.170

* 100% Terylene.

Note: The greater the percentage of Terylene in the fabric the more napping would the cloth stand.

Dynel on the other hand seems to facilitate napping, while the effect of Fibro depends on the type of wool and of Fibro in the blend. With fine wool and 3-den. Fibro the presence of Fibro hinders napping, while the reverse if true for cloths containing coarse wool and 8 den Fibro.

In several cases the nap produced on a blended cloth differs in composition from that of the yarn from which it has been produced. This is illustrated by the data given in Table 11. It is quite clear that the nap has essentially the same composition as the cloth from which it has been produced in the case of wool/Fibro and wool/Fibrolane cloth, but cloths containing wool and Terylene, nylon or Fibroceta yield a nap which contains more wool than the un-napped cloth.

Coth Structure

The part played by cloth structure in determining the response of a fabric to napping is illustrated by the data given in Table 12. The fabrics in each series were napped together so that truly comparable results could be obtained.

It can be seen (samples 18, 1, 2, and 3) that in cloths which are derivatives of the plain weave the susceptibility to raising increases with the length of float. Moreover, as the number of picks decreases the ease of raising increases. The change in ease of napping produced by decreasing the number of picks does, however, vary according to the particular weave. Considering the twill cloths 5, 6, and 9, of $\frac{1}{3}$, $\frac{1}{5}$ and 1/11 twill structures, increasing the length of the float again promotes napping, and all the twill structures are easier to nap than the plain weave material. Sateen cloths are more readily napped than either the plain or twill structures.

In a series of 2/2 twill cloths of different sets the amount of pile produced on napping was found to vary according to the number of picks and ends. In general, as the number of ends per inch increases, the weft set remaining constant, the thickness of the pile increases. Altering the warp set seems to affect napping to a greater extent than variation in weft set. This is presumably due to the fact that with a low warp set a greater length of weft is exposed to the action of the napping wires. Comparing cloths of different square sets it would appear that as the set increases, the thickness of the

TABLE 12

Effect of Cloth Structure on Napping

No. of fabric[a]*	Type of interlacing	Picks per in.	Increase in thickness (arbitrary units) on napping
1	2/2 Matt	24	205
2	3/3 Matt	24	317
3	6/6 Matt	24	395
4	2/2 Twill	20	197
5	1/3 Twill	21	291
6	1/5 Twill	20	379
7	2/4 Twill	26	337
8	3/3 Twill	26	303
9	1/11 Twill	24	460
10	2/10 Twill	24	427
11	3/9 Twill	24	383
12	4/8 Twill	24	419
13	5/7 Twill	24	440
14	6/6 Twill	24	481
15	4 end Sateen	20	351
16	6 end Sateen	21	493
17	12 end Sateen	26	582
18	Plain weave	16	99

[a] In all cases the warp was 2/16s Y.S.W. (2/120 Tex) with 20 ends-/in. The same yarn was used as weft.

nap obtained by a given napping treatment increases. Moreover, the plot of cover factor against pile thickness is linear.

Twist in the weft yarn is also important. For example, if cloths of identical structures, but made from weft yarns of different twists, are napped together the weight of pile produced from cloths with weft yarns with 4, 12 and 18 tpi are 0.18, 0.10 and 0.08, respectively. Folding the yarn also decreases the susceptibility to napping. This does not, of course, imply that yarns of considerable twist or twofold yarns are not used in the manufacture of napped cloths. They may be, for the total effect produced is obtained only by a correct choice of yarns so that the final product combines an adequate nap and strength.

Extent of Milling

In general the more the cloth has been milled the more difficult is it to nap. In spite of this, however, heavy milling is often necessary to obtain the desired result, e.g., as in Doeskins. It must be appreciated, however, that this consolidation will also increase the time required for napping.

With the advent of contractile fibers such as Rhovyl and Thermovyl, additional scope has been given to the designer of napped fabrics for these materials enable the compact ground which provides an excellent base for napping to be obtained by methods other than milling. It is, of course, essential that the contraction should take place uniformly, but if this is done the fabrics respond well to napping and attractive suedelike fabrics can be produced.

Moisture

Moisture plays an important part in determining the character of the nap produced. For example, only if the cloth is maintained really wet can the characteristic drawn finish be obtained. The amount of nap produced by a given set of conditions is generally increased by the presence of moisture, although this does not apply to tightly set Fibro cloths which often become boardy when wet. Moreover, the presence of moisture makes the Fibro weak and this results in too much fiber being detached from the cloth during napping. Cloths made from hydrophobic fibers such as Terylene, are little affected by moisture.

Other factors investigated by Whewell (23) were the presence of acid and alkalis which facilitates napping, whereas neutral salts have the opposite effect. Lubricants in many cases improve the efficiency of napping, however, care must be exercised on the particular lubricant and on the amount applied for if an erect nap is required, too much lubricant will cause the nap to be laid. Dyeing in general lowers the napping effectiveness; the real differences are associated with the method of dyeing which has been used. Pieces dyed with acid colors nap better than chrome dyes.

In conclusion Whewell states: It is evident that many factors influence the napping of textile fibers, and that careful control over these is necessary to obtain results which are acceptable commercially. This is normally exercised by the napper himself and napping is still regarded as something of an art. It is possible, however, by carefully planned experimentation to develop a more reasoned approach to the process and the data given previously indicate that a rational explanation is possible for many of the apparently illogical observations which form part of the "folklore" of nappers.

References

1. Bell, J. W., *Textile Inst. Rev. Textile Progr.*, **14**, 426 (1962); **15**, 412 (1963).
2. Blankenburg, G., Dissertation Techn. University of Aachen, 1961; *Textile Res. J.*, **32**, 986 (1962).
3. Carter, E., *Melliand Textilber.*, **46**, 1323 (1965).
4. Castle, H., and Wilford, E., *J. Textile Inst.*, **52**, P634 (1961).
5. Crowder, N. F., and White, W. A. S., *Textile Inst. Ind.*, **1**, 11 (1963).
6. Derby, R. E., *A. D. R.*, **18**, 28 (1939).
7. Elsworth, F. F., *Textile Inst. Rev. Textile Progr.*, **16**, 326 (1964).
8. Hemmer, L. Ph., *Intern. Textile Bull., Dyeing Finishing*, No. 3/1967—332, 40.
9. Kistenreich, H., *Z. Textilind.*, **62**, 608 (1960).
10. Kittan, A., *Melliand Textilber.*, **47**, 173 (1966).
11. Kittel, H., *Melliand Textilber.*, **47**, 800–803 (1966).
12. Makinson, K. R., *Textile Res. J.*, **32**, 364 (1962).
13. Manerbio, M.A.T., *Melliand Textilber.*, **48**, 1498 (1967).
14. McPhee, J. R., *Textile Inst. Ind.*, **1**, 10 (1963).
15. McPhee, J. R., *Textile Mfr.*, (Dec. 1965).
16. Mizell, L. R., Davis, A. E., and Oliva, E. C., *Textile Res. J.*, **32**, 497 (1962).
17. Poole, E. J. D., and Turner, J. H., *Textile Inst. Ind.*, **3**, 304 (1965).
18. Scibior, J., and von Bergen, W., Forstmann Woolen Co., 1959.
19. Serracant Riba, P., *Intern. Textile Bull. Dyeing-Finishing*, **2**, 42 (1967).
20. Sustmann, C. L., and Zahn, H., *Textilind.*, **65**, 286 (1963).
21. Tomlinson, J. M. D., *Textile Inst. Ind.*, **2**, 7 (1964).
    Tomlinson, J. M. D., *Textile Inst. Ind.*, **3**, 217 (1965).
22. Von Bergen, W., 2nd Canadian Textile Seminar, 1950, *T.T.F.C.*, p. 171.
23. Whewell, C. S., *Ann. Textiles*, **3**, 9 (1957).
24. Whewell, C. S., and Snowden, D. C., 21-28, 1961.
25. Whewell, C. S., 7th Canadian Textile Seminar, III, 1960.
26. Williams, D. R. H., *J. Textile Inst.*, *A***215** (1958).
27. *Wool Science Review*, No. 23, July 1963; No. 24, Feb. 1964; No. 34, July 1968.

CHAPTER 13

Part I: Dry Finishing Operations

WERNER VON BERGEN

Dry finishing operations are carried out on the cloth in the dry condition and include the following: extracting, drying or tentering, shearing, brushing, dry decating, damping, pressing, steaming, sponging, examining, stamping, measuring, weighing, folding, inspecting, and shipping. These processes are listed in a general order only and do not necessarily follow each other in any definite sequence.

Cloth drying is accomplished in two steps: removing the excess, loosely held water from the piece by mechanical means, i.e. squeezing, suction and centrifuging, and removing the remainder of the water by evaporation. The first of these steps is spoken of as *extracting*, the second as *drying*.

Extracting

A woolen fabric at room temperature at 100% relative humidity can absorb approximately 26% water vapor calculated on a wet basis (moisture content) or 35% on a dry basis (moisture regain). Any water above this figure is present as fine droplets in the fiber interstices and can, therefore, be removed by mechanical extraction. Three types of machines are used:

Squeezer or Wringer

Until recently the squeezer was the least effective of the three machines. However, the newest type of squeeze rolls can reduce the moisture content to as low as 38%, or 61% moisture regain, almost the maximum in efficiency. The pressure of this machine is hydraulically or air controlled. The squeezer illustrated in Fig. 1 is air controlled and, in addition, is equipped with a dial gauge and an automatic safety valve. The dial regulating valve shows the exact roll pressure at all times. thus ensuring proper control during this operation. Pneumatic controls cushioned against momentarily greatly reduced pressure when the seams or bunches pass through the rolls keep the pressure on the material uniform at all times and at all points across the width of the roll. Today there is a tendency toward replacing centrifuges and vacuum extractors with the new types of squeezer or roller, because the latter are highly efficient.

Fig. 1. "Wring Master" Extraction Squeezer (Courtesy of Rodney Hunt).

Vacuum Extractor

In the vacuum extractor the cloth is treated in full width. The machine consists of a cylindrically shaped housing with a narrow slot on the top, extending from end to end, over which the cloth passes. Connected with the housing is a vacuum pump that draws air from the atmosphere through the wet cloth as it is drawn over the slot, thus forcing the water out. Of the three drying machines the vacuum extractor has the lowest efficiency. Generally the moisture content of the cloth after vacuum extraction is about 40–50%, corresponding to 67% moisture regain. With lightweight fabrics, a heavy felt or rubber cover is placed on top of the cloth while it passes over the vacuum slot. This helps to create sufficient vacuum. With thick and heavy qualities, covering of the slot is unnecessary as the tightness of the fabric gives enough resistance to the vacuum. The advantage of this machine is the minimum of disturbance which the cloth receives. The disadvantage is the danger of clogging, necessitating frequent cleaning.

Hydroextractor, or Centrifuge

The excess water is removed from the cloth by centrifugal force. The machine consists primarily of a perforated copper or galvanized iron cage or basket built around a central spindle, the whole being encased in strong cast iron or steel. The spindle is rotated at a speed up to 1000 rpm, developing centrifugal force which is resisted by the inner surface of the cage. This causes a pressure on the cloth in the cage and forces the excess water through the perforations in the sides to the casing, where it is drained away. There are various methods of driving the drum; belt drive, friction drive, and direct drive with a built-in electric motor. The speed of the basket varies with its diameter:

36-in. diameter:	1,000 rpm
48-in. diameter:	900 rpm
60-in. diameter:	750 rpm
72-in. diameter:	600 rpm

The time required for hydroextracting depends upon the speed of the spindle and the diameter of the drum (which give the centrifugal acceleration of the machine), and upon the length, weight, and thickness of the cloth; the latter three factors are responsible for any resistance to the centrifugal action. Maximum extraction with this method is about 30% moisture content, a figure which is only slightly higher than the theoretical limit of 26% moisture content. This limit is generally reached after two to ten minutes, depending on the type of cloth and the load of the machine. The advantage of the hydroextractor over the squeezing machine is that the material is placed in the machine in a loose and free condition without pressure, and there is less danger of wrinkling or of pulling the cloth out of shape. Its disadvantage is that it is a batch process which makes production time-consuming and therefore costly.

Drying or Tentering

Drying plays an important part in the finishing of woolen fabrics. The number of drying operations necessary at intermediate stages varies with the quality requirement. It is an expensive factor as it involves costly machinery with high steam consumption.

There are two principles used, namely, tensionless drying, and drying on a frame. In the finishing of woolen and worsted fabrics, tensionless drying is mainly used in the sponging operation. Bergmann (1) wrote an informative survey on the tensionless dryers built today.

The important position in the drying of woolen fabric is held by the tenter frame, because it enables the cloth to be guided both in the width and length, with a certain degree of shrinkage control.

Drying was originally done in the open air. The cloth was stretched by hand between two sets of parallel bars fixed on vertical posts, and held in position by tenter pins. A further development was the "tenter house," in which the cloth was dried indoors by hot air. The tenter frames were placed in long narrow rooms heated by steam pipes. Today cloth drying is an automatic process carried out by means of various tentering machines which perform the double purpose of drying and stretching. The principal feature of a tentering machine is a pair of endless traveling chains, carried on tracks, on which fine pins are mounted. As the cloth enters the machine, it is engaged in the selvages by these pins. Frame dryers are subdivided into storied frames or multiple pass dryers and flat frames or single pass dryers. The machine built in stories (Fig. 2), the cloth-carrying chains passing over sprocket wheels at the end of each story to the next lower level. The piece enters the machine from a platform, passing between tension bars to the pins. A small brush wheel placed over each chain presses the selvages onto the pins and the cloth is carried from the top story through the whole series to the lower story, where it is released from the pins and delivered by ordinary folding motion. Tin cylinders support the cloth at each end of the stories. The distance between the two chains is adjustable so that the cloth can be brought to the desired width. Wrinkles and creases are pulled out evenly and the warp and the filling are put into the proper geometric relationship.

The latest devices for automatically feeding the fabric onto the pins of the chains, and at the same time controlling the width are electrical. The cloth is picked up by two feelers, and, according to the contact made, moves the chain arms in or out to accommodate its width. This provides for accurate and consistent feeding to the pins which hold the cloth at the selvages (Fig. 3).

Advances in trying techniques have been considerable in the last 20 years. The main change is in the direction of the hot air which was previously passed over the cloth from the sides as seen in Fig. 2. It is blown vertically on to the top and bottom sides of the cloth at higher flow velocities through a system of nozzles. The system of nozzle or jet drying by vertical discharge on to the cloth is used today by practically all builders of dryers.

Examples of some of the recent designs are the multitier jet tenters of H. Krantz Sohne, Aachen, Germany described by Zanke (18). The unit is built as part of a range which consists of a suction hydro-extractor, a two bowl padder and the jet tenter for drying and heat setting. This makes such a range a truly universal one because a modern tenter must not only be suitable for drying all types of wool and wool mixture fabrics, but it must also be possible to use it for other heat treatment processes, e.g. heat setting of wool and synthetic mixtures. The heating in the heat setting zone is done with saturated steam at a pressure of approximately 17 at (240 psi) maintaining the setting temperature of 180°C (367°F) within an accuracy of 1°C (2°F).

The three tiers or six pass dryers is approximately 10½ feet high and the length of the bay 10 ft. The speed range is from 9 to 52 yd/min and the working width can be set from 30 in. to approxi-

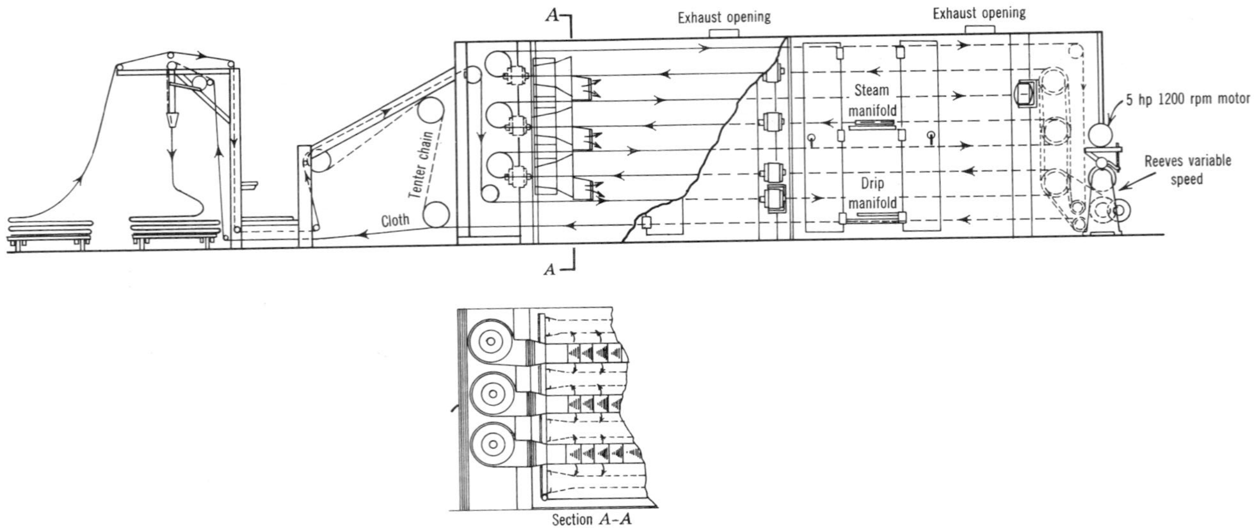

Fig. 2. Diagram of Uniflow Tenter (Courtesy of James Hunter Machine Co.).

Fig. 3. Entrance Bar with Pin Chain Vertical Return (Courtesy of H. Krantz Soehne).

mately 71 in. The drying rate is approximately 660 lb water evaporation per hour, corresponding to an output of some 22 yd/min for worsted and woolen fabric with a weight of 15 oz per linear yard. The heat setting output is approximately 15 yd/min for a 30 sec exposure. An overall view of the three tiers tenter is shown in Fig. 4. The plan of the fabric run can be seen in Fig. 5 whereas that of the air stream is shown in Fig. 6. Above and below the six fabric passes four jet-boxes are fitted in each bay and each jet-box blows from two rows of jets onto both fabric surfaces. To insure that the fabric remains stable, the jets working in pairs are so arranged that the points of contact with the upper and lower jet streams alternated causing the fabric to billow and sag alternately, with a quilted effect, so that the slack in the loosely fed-in cloth is taken up. In Fig. 7 a view of a nozzle set is shown. Krantz has also developed a single tier tenter and for processing knitted goods a single pass thermo setting tenter. Other recent designs are the James Hunter (6) High Speed Jet Dryer called the Uniflow; the high production tier tenter of Famatex. Riedel (8) described a

Fig. 4. Cloth Accumulator in Front of Krantz Jet Dryer.

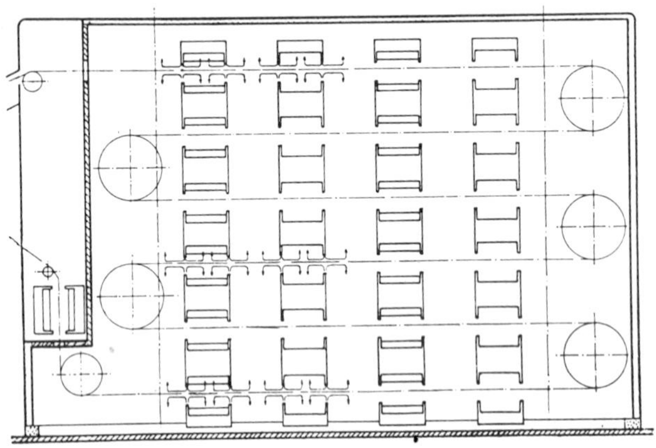

Fig. 5. Plan of Fabric Run—Krantz Jet Dryer.

multipass dryer with turnable blow jets especially suitable for laydown velours and fleeces.

A rather complete list of piece goods drying machine was published in the *International Textile Bulletin: Dyeing, Printing and Finishing 2/67* (19).

Other systems of drying piece goods are perforated drum dryers and cylinder dryers. The suction nozzle drum dryer of Fleissner is of special importance for drying tricot. The capacity of this design

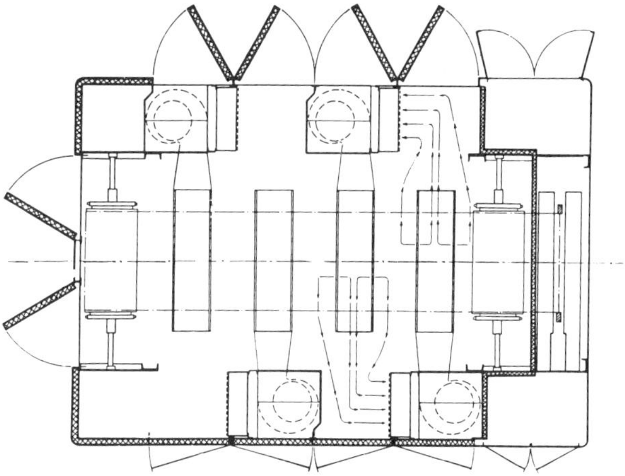

Fig. 6. Plan of the Air Stream—Krantz Jet Dryer.

Fig. 7. View of Nozzle Sets—Krantz Jet Dryer.

can be seen from the drying performance of 2,645 lb (1200 yd) of woolen tricot in 8 hr dried from 50–60% moisture to 16% at 176°F. Cylinder dryers on which the cloth is led over heated cylinders are rarely found in woolen finishing plants.

Gottschalk and Weiss (5) published a thorough study of the points which must be considered in the installation of a new tenter dryer.

Steam Consumption

In the newer drying equipments, steam consumption lies in the region of 1.5–1.6 lb for the evaporation of 1 lb of water present in the fabric. The drying process is mainly depending on the structure and weight of the fabric and not by the fiber. The duration of the drying is decided by the nature of the cloth especially its surface. Raised and heavy felted fabrics dry slower than those with a smooth finish. To calculate the steam consumption per piece, reference can be made to the chart, Fig. 8, which shows the number of pounds of water which must be evaporated by drying from fabrics having an initial moisture content of between 35 and 48% in order that they may leave the dryer with a moisture content of 8%. For example, a 60-yd piece weighing 16 oz/yd and having a moisture content of 48% would contain an excess of 48 lb of water. As 1.6 lb of steam are required to evaporate 1 lb of water, the total amount of steam required per piece would be 1.6 × 48, or 76.8 lb. If the same piece should enter the dryer with a moisture content of 35%, only 25 lb of water per piece would have to be evaporated. The steam consumption would be 1.6 × 25 or 40 lb, a saving of 47.9%.

Contrary to popular opinion, research has shown that when cloth is exposed to a high temperature when wet, more damage is done than when a cloth is exposed to a high temperature when it is nearly dry. The latter method of drying has been used for years in the carbonizing dryer. To cool the cloth and to provide it with the necessary amount of moisture, damp air is drawn through the bottom chambers, which have a relative humidity sufficiently high to provide a final moisture content of between 8 and 10%. This method of drying unquestionably simulates outside drying conditions more closely than any other. A few manufacturers have led the way in incorporating this principle in their regular dryers (Fig. 9). Three factors must be considered in drying:

1. The moisture-absorbing property of air is increased by raising the temperature.

2. The dry air must be brought into contact with the material.

3. After acting on the cloth, the moisture-laden air must be removed continuously in order to admit fresh, dry air.

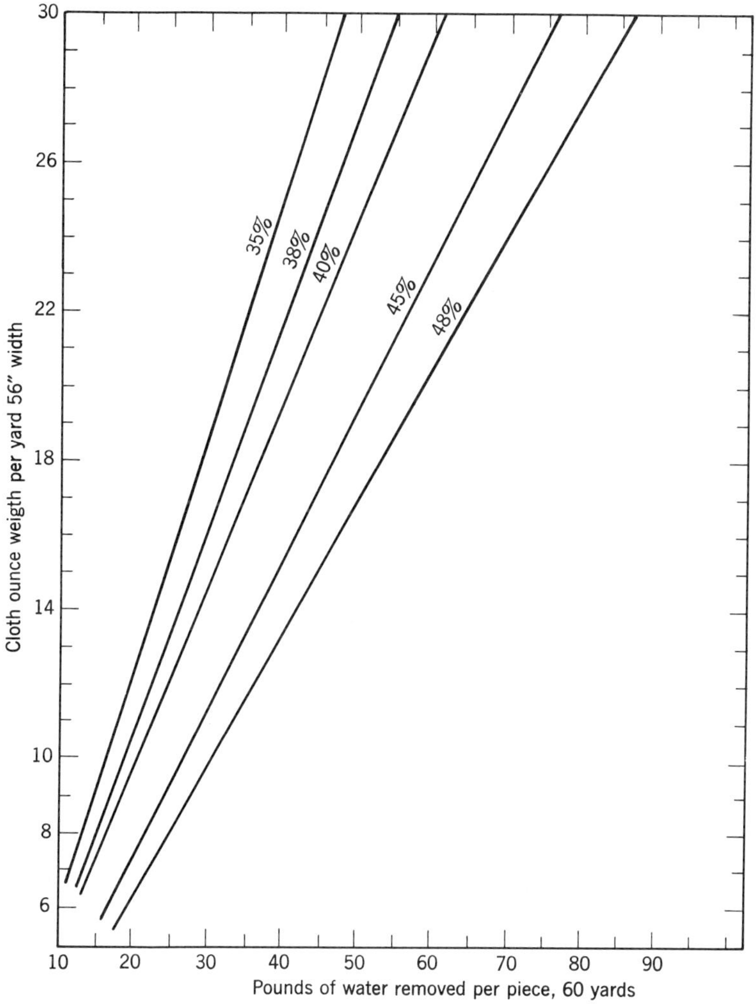

Fig. 8. Moisture Evaporation Chart.

Cooling after Drying

In the modern dyers, cooling and conditioning zones have been made part of the machines. In the Kranz machine cooling of the fabric takes place in the vertically rising fabric pass after it leaves the drying chamber. The cooling air, sucked in by means of a fan

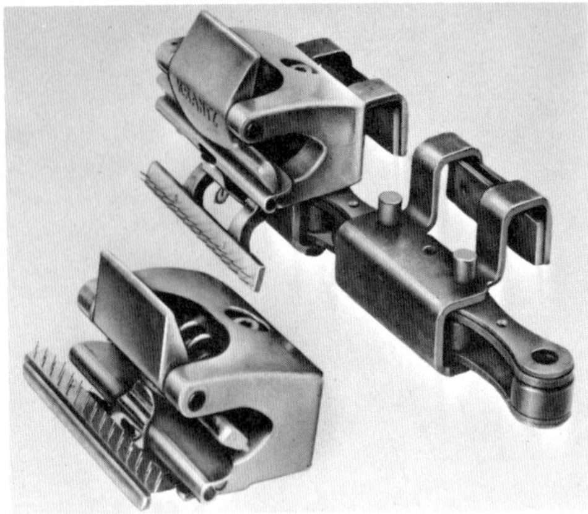

Fig. 9. Combined Pin and Clip Chain Links with Depressable Pin Plates (Courtesy of H. Krantz Soehne).

is blown on to the fabric by means of jets. The ideal condition of the fabric after drying would be a residual moisture content between 8 and 10%. To prevent overdrying, hygrometer and various electrical instruments are employed which enables the speed of the dryer to be changed automatically via relays controlling the moisture content desired. Such instruments are not only of decisive importance for the quality of the cloth, but also make vital contribution toward improving the economics of the drying process. Examples of moisture measuring and controlling installation are the Fielden described by Bunch (3) and the Mahlo.

Needle Design

With the high air velocities there exists the danger of the cloth being blown from the needles on the tenter frame. Hunter overcomes this problem with a swinging type pinplate, as the cloth enters the drying section, the pins swivel to a 40° angle from the vertical, which makes it impossible for the cloth to come off the pins. Krantz supplies special designs such as pinplates with Z shaped pins and bridged pin plates with locking system. Figure 9 shows a combination of clip and pinplate.

Warp and Filling Shrinkage

To enable the cloth to shrink in the direction of the warp during the drying process, overfeeding devices are fitted at the entrance field of all tenters nowadays. The contraption at the feed to the needle chain has a higher speed than the chains which take over the fabric, so that the latter is pressed on to the needles by the selvages with small transverse folds, enabling it to become shorter in the course of the drying process. Settings of 5–10% overfeed are normal. To obtain shrinkage in the direction of the filling, the two guide rails which normally run parallel can be adjusted so that they taper towards the delivery end of the machine.

The feeding mechanism for automatically feeding the fabric onto the pins of the chain, and at the same time guiding the width are controlled either mechanically or electrically. Part of this arrangement are also often selvage uncurlers to the control or removal of double or turned selvages which can be a serious problem in fabric processing. Mount Hope Model R and S are such devices.

Weft Straightening

All finishers are aware of the difficulties they face in correcting fabrics such as checked cloth spoiled by distorted filling yarns. This problem has been mainly overcome by the introduction of Weft thread straightener such as manufactured by Mount Hope (Fig. 10*a* and 10*b*) Dornier (4) and Krantz. The proper placement is ahead of the tenter frame as it is imperative that the material enter the setting point of its finishing process already straightened, which means that slanting or curved weft threads are aligned at right angles. Any bow or skew present at this point will often be permanently imparted to the fabric by the setting process. This applies not only to the synthetics but also to wool and its blends. A good summary on the measurement and regulation of moving fabrics was published in *International Textile Bulletin* (20).

Shearing or Cropping

The object of shearing is either to remove the protruding fibers from the cloth surface or to even out or level the fiber pile or nap

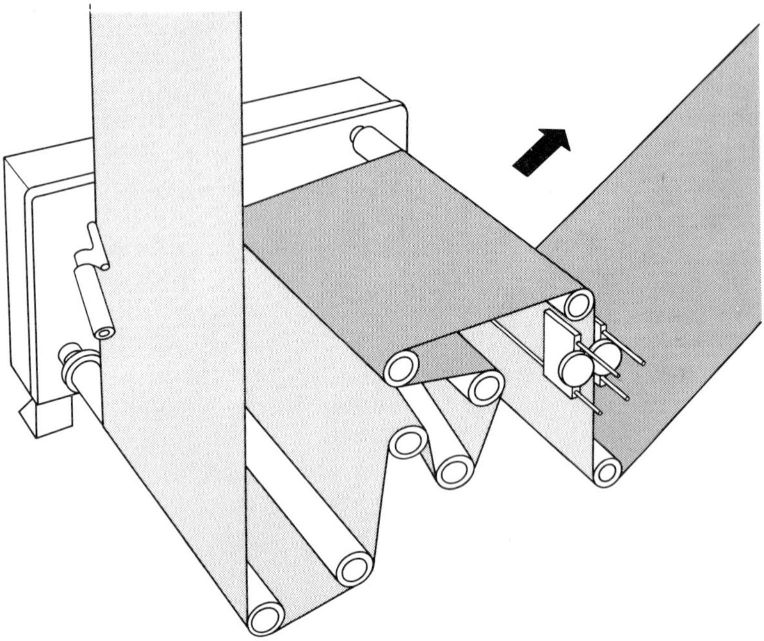

(b)

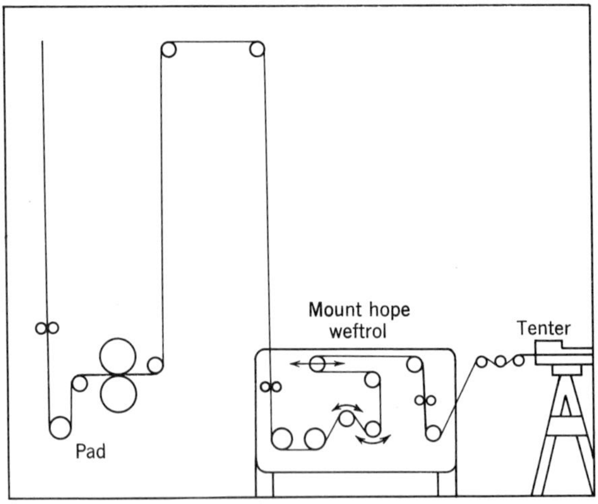

Floor mounted prior to tentering

(a)

Fig. 10. Mount Hope "Weftrol" Bowed and Skewed Weft Straightener.

which has previously been raised. The amount of fibers to be cut from the surface of the cloth depends on the type of finish desired. This dry-finishing operation is used on all woolen and worsted goods.

In singeing, the hairy surface was burned off, but in subsequent wet processes, additional fibers which were raised through manipulation can now be removed only by shearing. Another singeing would affect the handle of the goods and create an odor which would have to be removed by another wet treatment.

The general aim of shearing is to clear the surface of the cloth of surplus fibers and thereby bring out certain characteristics such as the design or color pattern. The most drastic shearing is done on fabrics which require a clear face, such as gabardines and serges. The amount of shearing done on napped fabrics, such as velours and broadcloths, depends entirely on the length of the desired nap. On velours sufficient cutting or cropping is done to produce a plushlike effect, whereas on duvetyns and suedes the nap is shortened considerably (see napping). In pile fabrics shearing is done before and after dyeing. Broadcloths require the most "cropping." To produce the high luster of a broadcloth, a laid down short nap, the cover has to be shorn off very gradually.

The earliest mechanical shearing machine was constructed as a cross-cutting machine. The piece was fastened on pins at the selvages and the cutting device moved over the length of the piece and so length for length was sheared.

Modern shearing is a continuous operation. The ends of the piece are sewed together and the cloth runs through the machine in an endless band over stretching bars, guide rollers, and brushes at an average speed of 8–20 yd/min. A four blade automatic shearing machine illustrated in Fig. 11 and a three blade machine shown diagramatically in Fig. 12 which have been designed especially as production machines can be run at speeds of from 30–40 yd/min or higher if necessary. Flannels and worsted suitings which need to be sheared once on the back and twice on the face can be processed on the newest three-blade shears at a rate of up to 60–70 yd/min.

The cutting of the nap has to be done gradually, meaning that the setting of the cutting device has to be higher in the beginning and gradually lowered to the desired point. This method gives a more evenly sheared surface—in trade terms, a round sheared nap.

Fig. 11. Four Plate Automatic Shearing Machine (Courtesy of J. P. Stevens & Co. Inc.).

If, in the beginning the setting is too low, the cutting device cannot cut the immense amount of fibers at once and the fabric tears.

The critical adjustment in shearing lies in the mutual relations of the ledger blade and the cylinder and of these two with the bed. Before commencing to shear a piece, the cutting device is adjusted to the bed by inserting a thin paper between the bed and the cloth as it rests on the ridge of the bed. By doing this across the width of the machine and feeling the amount of resistance to the withdrawal of the paper, a judgement is formed as to the pressure between the cloth and the cutting edge. The keenness of the cutting depends on the sharpness of the knives. If the machine, when the cylinder is turned by hand, will shear a fine paper with a clean cut, it is concluded that the cutting device is in proper adjustment. With the proper adjustment of the cylinder with the blade and the adjustment of this combined system with the bed, the spirals are able to reach all projecting fibers as they stand up from the cloth

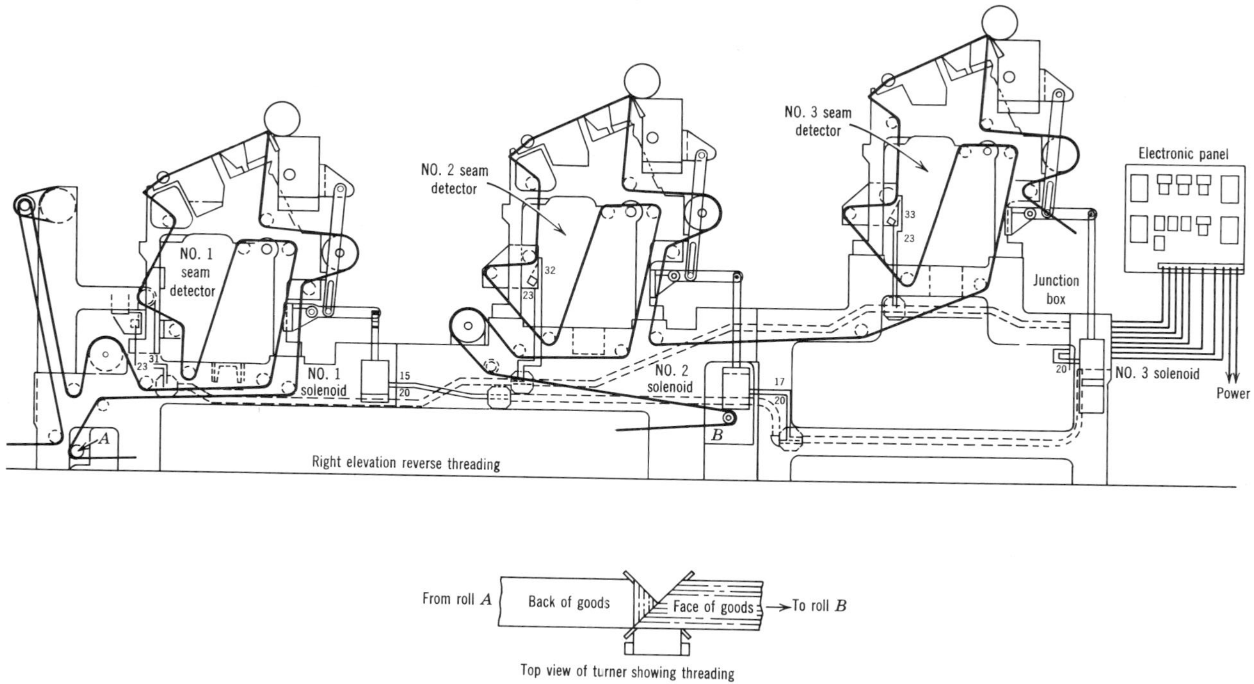

Fig. 12. Diagram of Three Plate Automatic Shearing Machine (Courtesy of Parks and Woolson Machine Company).

when it passes over the ridge of the bed. These fibers are then dragged by the revolving spirals against the keen edge of the ledger blade and there cut off (Fig. 12).

In the Raxhon shearing machine the shearing system is equipped with a suction cooling device. As the suction is concentrated at the cutting place, it raises the surface fibers of the cloth to the best possible erect position for efficient cutting. The suction realized in this manner removes at the same time the shear flocks and automatically cleaning the spiral blades of the cylinder. In the Gessner shear the flock removal ducts are integrated in the design of the machine in such a manner that they remove the flocks from all cutting parts as well as from the cleaning brushes, the back brush or beater, and the laying brush.

When the seam of the cloth approaches the cutting point, the cylinder and blade, which is the cutting device, has to be raised by pressing down a lever or foot rest to allow the passage of the seam. To this movement the operator of the shearing machine has to give special attention. If the seam passes into the cutting device without being disengaged, considerable damage can be done to the cloth, the ledger blade, and the spirals. In modern machines the lifting is done automatically. In the Gessner shear instead of lifting the cutting unit, an electrical and pneumatic reciprocating movement retracts the cloth rest to pass seams. The raising and lowering of the rest is done by cam motion.

Beds

There are several types of beds or rests available over which the fabric travels while cutting takes place; solid or hard rests, hollow or gap rests, sliding spring, rubber tube and cushion beds. For the shearing of woolens and worsteds, the first two are most widely used. Smooth-surfaced qualities with nap, such as meltons and velours, can always be sheared on a bed with one ridge or hard rest.

For the cutting of worsted goods, especially clear finished ones, and for weaves made of yarns with fancy knots and curls, it is safer to use a hollow bed, which has two ridges about 4 in. apart, known as the "gap rest" (Fig. 13). The cutting device works on the gap between the two ridges and because there is no solid object underneath the cloth, the projecting curls and knots will not be cut off.

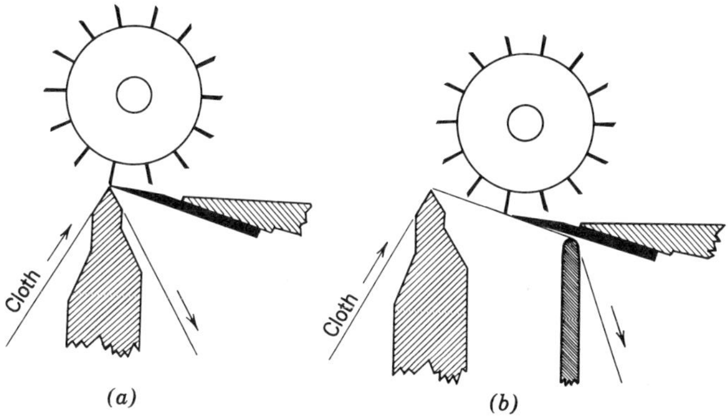

Fig. 13. Two Methods of Shearing Fabrics: (*a*) "Hard Rest" Shearing. (*b*) "Gap Rest" Shearing.

There are three main advantages to the "gap rest." First, it reduces considerably the amount of burling necessary, since most of the knots do not have to be pulled through to the back and cut before shearing. Second, it eliminates all knot holes in finished goods, as it is impossible to produce them with this type of rest. On some goods, where the knots are difficult to locate, this is a decided advantage. Third, with the new high-speed machine the gap rest allows for continuous production on numerous types of fabrics varying in thickness, without the necessity of resetting the cutting knives.

The production with gap-rest shearing is about 20% lower than on the shearing machine equipped with hard rests. In the new automatic cloth-shearing machine using hard rests, speeds of up to 50 yd/min can be obtained, whereas using the gap rest satisfactory shearing is possible only at a speed of about 38 yd/min. A further disadvantage of the gap rest machine is that fact that the plates must have a flat ledger, which is considerably harder to grind.

The Sellers shear is built with dual purpose solid-hollow beds as seen in Fig. 14. For cropping of carpets roller and bull-nose beds are used.

The older cutting cylinders were made with 10 spirals, whereas the newer have 17–26 spirals. The cylinders run at speeds between approximately 900–1800 rpm. The efficiency of the shearing

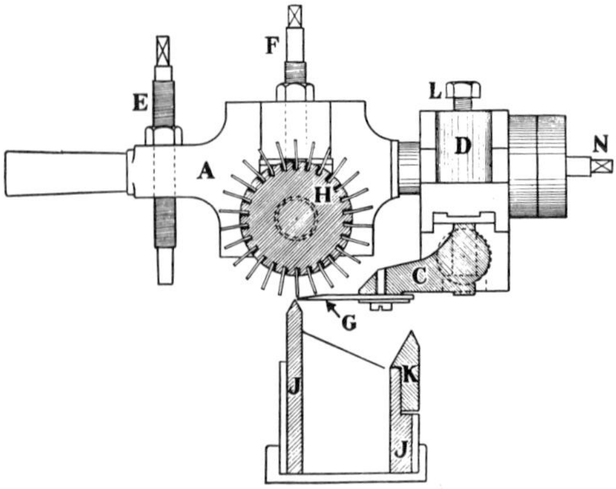

Fig. 14. Head Stock Detail of Robot Shear Dual Purpose Solid Hollow Head (J–K); Spiral Cylinder (H); Ledger Blade (G); Front Pin for Depth Adjustment (E); Middle Pin Adjusts Cylinder into Ledger Plate (F). (Courtesy of Sellers & Co.).

depends on the cutting contacts made per inch of cloth which can range from 15 to 40. The contacts are calculated from the following formula:

$$\text{Contacts per inch} \times \frac{\text{cylinder speed rpm} \times \text{number of spirals}}{\text{cloth speed } Y/M \times 36}$$

Whereas the average speed of 20 yd/min for worsted and 12 yd/min for woolen fabric was normal several years back, today it is about twice that. Table 1 gives some shearing machine details as they existed around 1960 in one European country.

TABLE 1
Shearing Machines Details

	Cylinder speed, rpm	No. of spirals	Cloth speed, yd/min	Cutting contacts per inch
Worsted	1,200	17	38	15
	1,200	10	20	17
	1,000	26	28	26
	1,150	10	8.5	38
Woolen	1,440	10	10.5	38
	1,000	28	28	26

To obtain the desired height of cut, the newer shearing systems are provided with micrometric settings which afford rapid and accurate adjustment of the shearing headstock to the fabric. The speed of the raising brushes installed before the shearing system play also an important part in freeing and raising the surface fibers of the cloth for best presentation to the cutting device. The brushes normally run opposite the direction of the cloth, but it was found in many instances that to brush with the cloth a better cut is achieved. The speed range of the raising brush should be sufficient to brush with or against the cloth.

Single control panels generally carry stop and start buttons, and seam passing controls. Indicators show cloth speed, surface speed of the raising brushes and the degree of cloth tension. Control panel makes it easy for the operator to make and reproduce all settings quickly and accurately.

Regardless of what type of machine is used, to secure a clean even cut, it is very important that the cutting device be kept in good condition. This necessitates regrinding the spirals and the blades from time to time. The sharpness of the blades and the spirals, when properly ground, should last from four to six months, depending on their use.

In normal hard rest shearing, it is essential that the back of the cloth be free from weaver's knots and slubs in the yarn and that no flocks get between the bed and the cloth since any of these would lift the material at that point. Any lifting of the material results in an increased cutting action on the upper surface which might even be great enough to cause a hole to be made in the cloth.

Flock Removal

In the modern shearing room, the shear flocks are removed from the machine as they are produced: In addition to the danger of the flocks adhering to the cloth and causing cut holes and choking running parts, they may soak up as much as 6% of their own weight of lubricating oils. The suction system fitted over the cutting cylinders conveys the flocks from the machines to a central separator.

Fancy Shearing

A special development in the art of shearing cloth is fancy cutting, which is often applied to ladies' coats and jackets of the velour type, sometimes called imitation Jacquards. This is done by using sector beds and lifting the cutting device at determined intervals. According to the design cut on the sector bed, a design will be produced on the surface of the cloth. Fabrics of this kind have straight-line designs, either in the warp or the filling direction or running diagonally, simulating stripes and checks or, in more elaborate examples, floral designs may be imitated. A more universal system of producing patterns, curvilinear or otherwise, is to replace the ordinary bed by an engraved metal roller.

Chinchilla or Ratine Finish

This kind of finish is applied to heavy woolen cloth used for women's coats and men's overcoats; its chief characteristic is its nub or frizzy finish on the face of the fabric. The nubs or beads are made by the Chinchilla or Freizing Machine.

The finish can be created only on a pile nap. The cloth is drawn at a speed of 3–4 yd/min over a stationary, plush-covered table. Above this stationary table is placed a swinging table covered with a rubber pad, which rests with its own weight on the surface of the cloth. This table is driven by a beveled gear from the side of the machine and its reciprocating motion is controlled by eccentric cams on both sides. Five effects can be made by changing the setting of these cams, namely:

1. A nub effect.
2. A wave effect in the direction of the warp.
3. A wave effect in the direction of the filling.
4. A wave effect diagonal from left to right.
5. A wave effect diagonal from right to left.

To produce a nub effect the upper table moves at a fast speed in a short circular motion. The size of the nub depends on the length of the nap; a short nap will give a small nub and a long nap a larger nub. To produce all other effects the table moves in a reciprocating motion in the opposite direction of the wave. In all wave effects the size of the wave is determined by the length of the stroke the table makes in its reciprocating motion.

Brushing and Steam Brushing

This is a process which often follows shearing and has for its object the removal of the loose, short-cut fibers which have dropped into the face or nap. It is often combined with steaming; then it is called steam brushing. The principal features of such a machine are shown in Fig. 15. One of its important tasks is the laying of the nap in one direction and fixing it in that position with steam. The machine is built normally with two brush cylinders about 13 in. in diameter. For cleaning the back of the goods, there is a flock brush with flock pans which catch the flocks and lint removed by ghe brushes. The cloth strikes each brush twice, giving a large brushing capacity. The amount of contact on the brushes is governed by hand wheels acting through a simple cam arrangement. The operator can adjust the machine instantly so that both cylinders brush the goods alike or so that one cylinder brushes the goods more heavily, as may be required.

The machines are most commonly arranged with a back folding attachment and a cloth scray for guiding the cloth underneath, as shown in the figure. The cloth scray is made with a hinged table at the rear on which the goods are folded after being brushed, and at the same time a new piece may be run into the machine. The machines are also built with a top frame by which the goods are

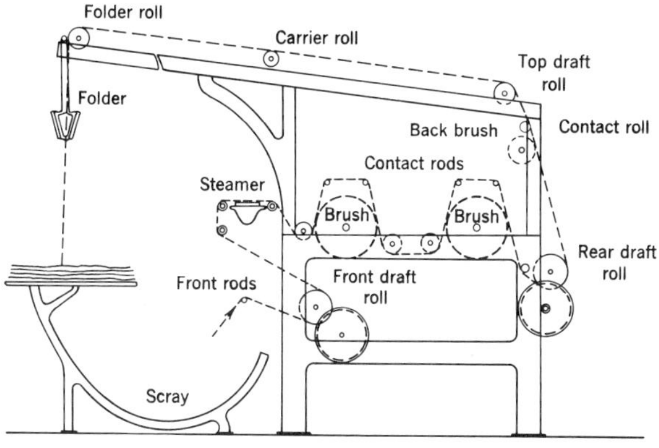

Fig. 15. Two Cylinder, Double-Acting Brushing Machine (Courtesy of Curtis & Marble Machine Co.).

carried overhead and dropped into a cloth scray in front, so as to run endlessly, or folded neatly on a platform.

A steaming apparatus is set high on the frames so there will be less liability of the steam affecting the brushes.

Dry Decating

In the finishing routine, dry decating or dry blowing generally follows shearing but for a high luster finish it follows pressing. Some reference has already been made to the decating process in the sections on crabbing and wet decating in the chapter on wet-finishing operations. The essential feature of the process is the application of dry steam to the enclosed dry cloth while in a state of stress. The subsequent cooling off produces a permanent setting of the fabric. The dry cloth is wound under tension on a perforated decating roller, the general working arrangement being the same as that described in the wet-decating section. Under the influence of the hot steam, which provides heat as well as moisture, the wool becomes plastic, causing the tension to be relieved and with it all additional strains present in the form of creases and folds. The fibers of the yarns and the yarns themselves that make up the fabric reach an equilibrium resulting in a balanced, unstressed texture. The final fixing occurs during the cooling-off period, in which a current of cold air is drawn through the roll. Dry decating can be accomplished in either of two ways, by the pressure-blowing or by the open-blowing method, depending upon the permanency of the setting desired.

Pressure blowing

The general features of a pressure-blowing machine, also known as a vacuum steamer, are illustrated in Fig. 16. This outfit consists of a winding machine, a cooling frame, a Kier and a vacuum pump. The cloth is wound tightly with a fine-faced cotton leader on a perforated roll. Two pieces can be put on one cylinder. A heavy leader cloth is wound on top of the finished roll to give the outside end of the cloth sufficient pressure and protection. This roll is put into the preheated boiler which is sealed air-tight with a cast-iron cover. Before opening the steam valve a vacuum in the boiler is

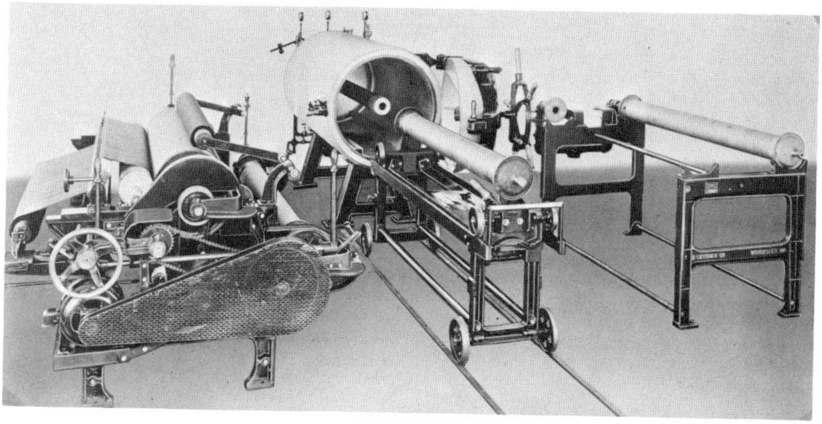

Fig. 16. Vacuum Steamer with Winder and Cooling Frame (Courtesy of David Gessner & Co.).

created by means of a vacuum pump. Then the steam enters the boiler, first through the inside of the cylinder on which the cloth is wound, and forces its way through the cloth for two or three minutes. Then the steaming is reversed. The steam enters the boiler and forces its way through the cloth from the outside to the inside for another 2 or 3 min. During this process the cylinder does not revolve. The pressure created in the boiler is usually not higher than 20 lb/sq in. The roll is then taken out of the boiler and carried to the cooling frame, where a vacuum pump sucks the steam out of the cloth and also cools it with fresh air. This takes from 5 to 10 min according to the weight of the cloth. The cloth is then unwound. By means of this decating process the cloth is "set" in width and length and the handle of the cloth is made firmer and thinner. The pressure of the leader which is wound with the cloth on the roller and the steam pressure in the boiler during the decating process develop plastic setting and leveling of the surface of the cloth. The newest machines are equipped with full automatic controls. Figure 17 shows a diagram of such a machine.

Dry decating, as well as wet decating, gives a marked and variable character, such as brightness of appearance or luster in varying degrees, to the woolen cloth which cannot be produced by any other finishing routine. Increased luster, which is caused by a better reflection of light, is the result of the improved smoothness of

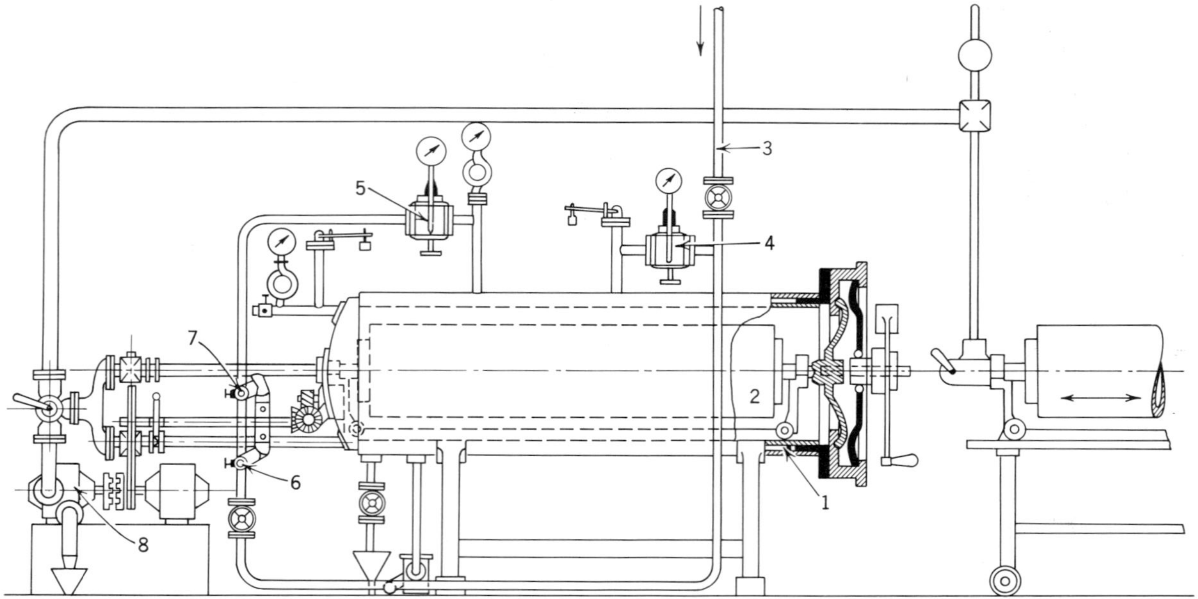

Fig. 17. Pressure Decatizing Unit—David Gessner & Co. 1. Chamber. 2. Cloth Roll. 3. Steam Line. 4. Reduction Valve. 5. Reduction Valve for Chamber. 6. Steam Valve Out to Inside. 7. Steam Valve Inside-Out. 8. Suction Pump.

the surface. In some goods it is essential and is developed to a very high degree. A high-luster broadcloth is an example. Such a luster, when permanent, is unaffected by the weather and is impervious to rain.

As referred previously in Chapter 12, page 844, Sustmann and Zahn (16) made a systematic study of the wet and dry decating processes.

The working routines which they found in various finishing plants as far as dry decating was concerned varied for the steam pressures from 0.5 to 2 atm (7 to 30 psi) and steaming times between 5 and 10 min. Therefore, they use these figures as the basis of their experiments.

To establish the temperature changes which take place where the steam penetrates the various layers of the roll of cloth on the perforated cylinder thermocouples were inserted into the outer and inner part of the rolls. A method which was successfully used in 1950 by Macaluso (7) in the Forstmann Woolen Company Plant. The temperatures were recorded by an electronic compensator and curves obtained as seen in Fig. 18 with a steam pressure of 2 atm. These curves are very similar to the ones obtained by Macaluso (Fig. 19). The results prove clearly that the steam pressure in a

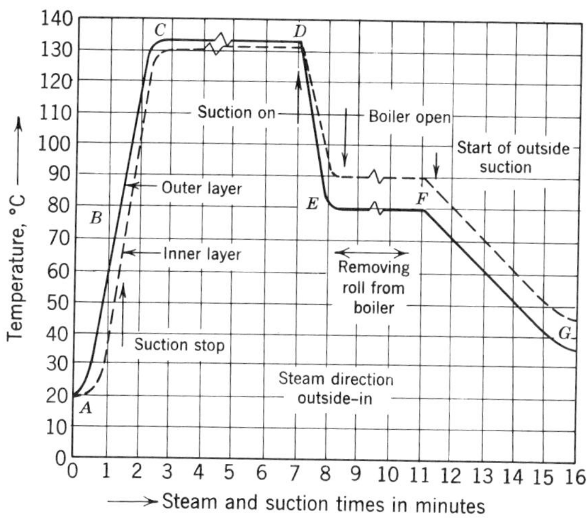

Fig. 18. Temperatures during Steaming and Suction Operation in Outer and Inner Layer of Cloth Roll with a Decatizing Steam Pressure of 2 atu—Sustmann.

closed steamer should not exceed 1 atm (15 lb) and the steaming time not longer than 6 min. Because already at 1 atm there is loss in breaking strength of around 8 percent and a reduction in the yarn elongation of 20%. As a bordercase a pressure of 1.5 atm and 2 min steaming time could be allowed.

Any increase in the decating effect above the stated limits will result in severe damages.

Dry decating affects the thickness and compressability of the fabric reducing the thickness around 30% and a reduction in the compressability from 27% for the untreated to 10–15% for the steamed.

The expression used by the finisher for the thinner fabric is "Paper Touch." There is also a yellowing effect which becomes stronger as the pressure and the decating time increases.

The chemical analysis showed a slight reduction in the pH from 7 to 6 with the inner layer slightly lower. The cystine content was

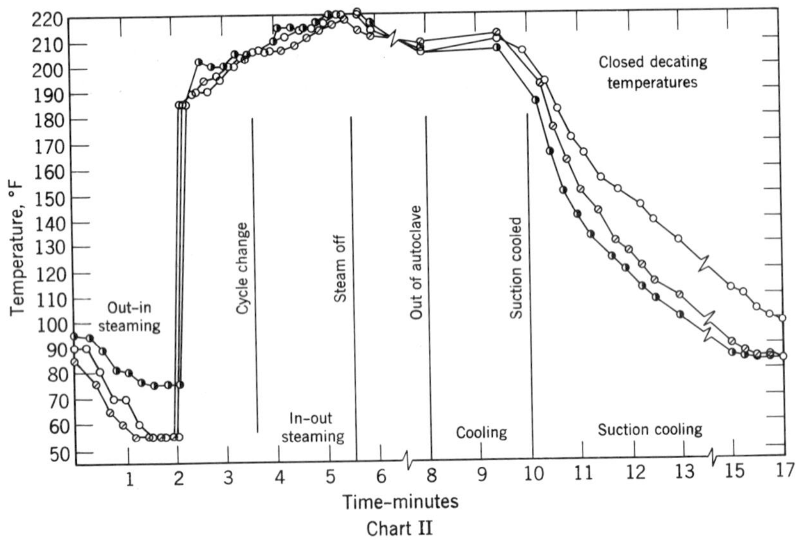

Fig. 19. Temperatures in Gessner Decator during Steaming cycle—Macaluso.

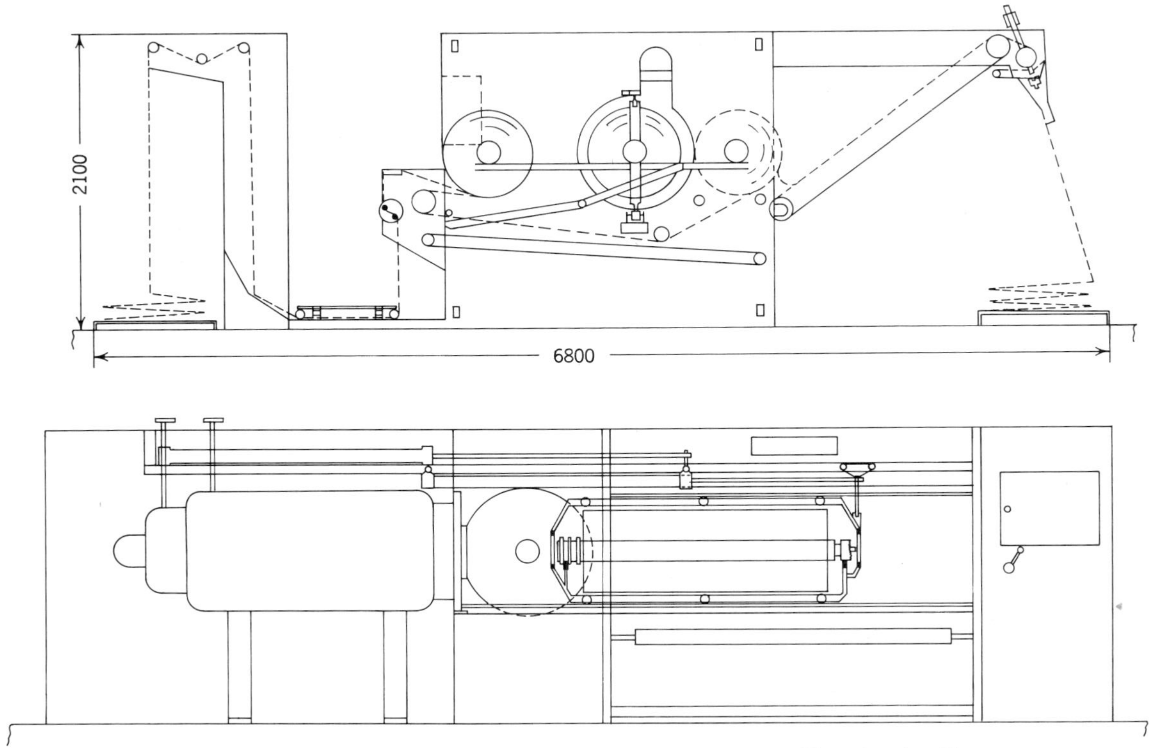

Dimensions in millimeters

Fig. 20. Permanent Finishing and Setting Machine—Biella Shrunk Process.

already reduced after the mild treatment of 0.5 atm and 2 min steaming and after 14 min of steaming at 2 atm, 25% of the original cystine was destroyed. The ureabisulfite solubility number dropped from 26% down to 13% after 14 min at 0.5 atm and to a low of 5.8% at 2.0 atm. There is also a considerable increase in the speed of absorption of dyeacid after 1.5 atm and 2 atm steaming.

For permanent setting and stabilization of woolen fabrics by mechanical and physical means the process of Kier decatising is still the best. Due to the demand for more luster finishes, there has been a revival in this technique which in turn led to considerable modification and improvement of machine designs. Two outstanding examples are the Biella Shrunk process (2) and the Kettling and "Decoclave" method as described by Riedel (9).

The Biella machine Model KD-Matic 66 is completely automatic shown in diagrams (Fig. 20). With a few controls from push buttons the whole decatising process is operated by one person. The machine consists of three interconnected units namely:

1. Preparation and batching unit. This serves to roll the cloth on to the perforated stainless steel cylinders and unwinding it from the cylinder after treatment. Transfer the carriage from the preparation machine into the setting chamber and out of it without manual intervention.

2. The setting chamber in which the actual process is carried out is completely automatically controlled and contains 6 magnetic

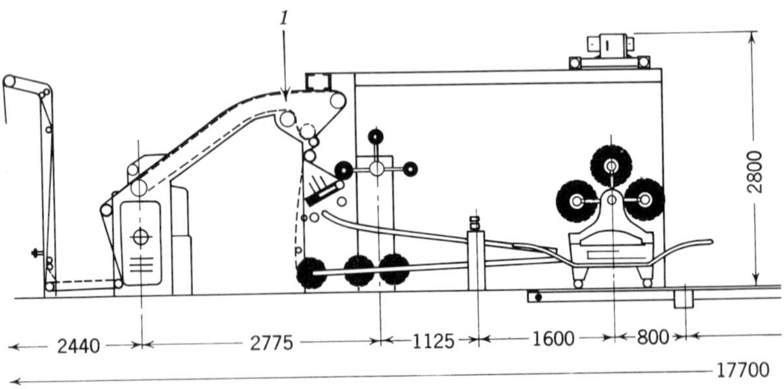

Fig. 21. Decoclave Dry Decatizing Installation—Kettling and Braun. (1)

valves for steam setting from 0 to 3.5 atm as well as the timing and temperature controls.

3. Electrical control panel which supervises the operation of the setting chamber as well as the movement of the preparation and batching unit. The hourly output of the machine is for worsted fabrics 9–10 charges of 130–160 yd each, and for heavy woolen goods 10 charges of 60–70 yd each. This limits the treatment in the Kier to 2–3 min. For this production the steam consumption is put at 65 kg/hr and the current at 8/9 kw per hour. The finish either dull or glossy, crisp or soft, the claim is made that it will be absolutely permanent against pressing, steaming, ironing or wetting.

The Decoclave process involves three separate steps namely: Preparation of rolls-winding, steaming and unwinding, each operation performed by a separate machine. The basic contrivances allow the decatising of several rolls together as one unit. For this purpose so-called "Wickelsterne" are used, which on three turnable arms carry the perforated cylinders. The overall installation is shown in Fig. 21 consisting of (*1*) Equalizing Machine, (*2*) Large Chamber Autoclave, and (*3*) Cooling and Conditioning Machine.

At the equalizing machine the cloth is automatically fed into two needle chains which carry the cloth at a fixed width into the cylinder where it is wrapped with the control of a pressure roller to obtain even hardness and even tension of the cloth layers be-

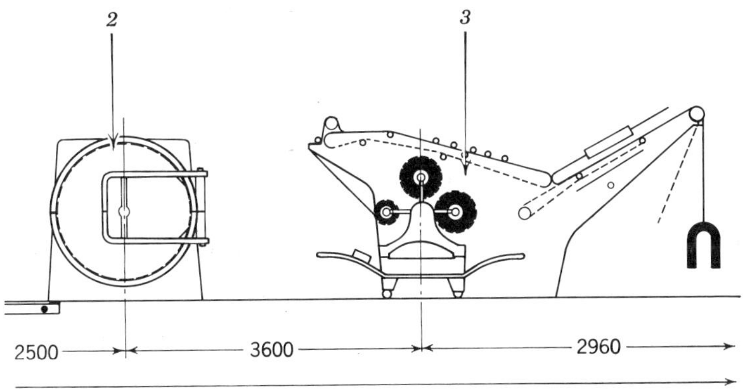

Equalizing Machine. (2) Large Chamber Autoclave. (3) Conditioning Machine.

tween the start and end of the roll. After all three cylinders on the Star frame are wrapped, it is lifted by a crane onto a carriage and moved in front of the autoclave.

The loading and unloading of the autoclave as well as the steaming cycle are ordered by a program unit. After the steaming is finished the star frame with the three rolls is moved to machine 3 and connected with the unrolling and vacuum cooling arrangement. In addition to the cooling on the rolls the machine is equipped to further cool and remoisten the cloth by running the cloth between the unrolling and folding over two transporting aprons. Above the aprons are a series of blow-jets installed from which a stream of moist and cool air hits the cloth. Under the apron are several suction nozzle boxes located. Both these aggregates form together a waterfog cooling cycle.

The hourly output for the Decoclave process is from 6 to 7 charges of one piece per roll (70–90 yd long) 18 to 21 pieces and with two pieces (140 yd) per roll 36 to 42 pieces per hour.

Open blowing

The mantle-steaming and air-cooling machine, also called an open-blowing machine (Fig. 22), is often used in place of the closed-blowing machine. Its advantage lies in the fact that the danger of tendering the goods is eliminated and the colors are not affected. Open blowing is as a whole a milder treatment and less effective. The process is as follows: The fabric to be processed is wound in a specially woven, smooth-faced, cotton mantle or wrapper upon the steaming cylinder. In winding the goods onto the steaming cylinder, compression may or may not be used as required. Tension is applied to the mantle or wrapper by means of a friction device on each end of the roll on which the mantle is wound. The fabric is subjected to the action of dry steam while enclosed in the mantle on the steaming cylinder. A separator ensures "dry" steam.

The exact amount of steaming on the cylinder varies from one to two minutes, depending on the particular fabric. The steam is then shut off and, by opening a quick-acting valve, air is drawn through the cloth in order to cool it. The amount of extracting or cooling also depends on the type of finish desired. Ordinarily three to

Fig. 22. Mantle Steaming or Open Blowing Machine.

five minutes will be sufficient. After this treatment, the goods are unwound from the cylinder, rolled up or folded.

Increased cloth output is achieved by a double blowing roller arrangement (see Fig. 23). The rollers, working side by side are individually driven and alternatively work with a common wrapper as well as a common exhaust pump. While length of cloth leaves the one roller decatized, another length can be wrapped on the second blowing roller, the wrapper being simultaneously supplied from the first roller. With the Sellers double decatizing machine wrapping and unrolling of the cloth can be carried out at steplessly adjustable speeds ranging from 5 to 36 m/min.

All processing cycles can be set into operation by push buttons arranged on a control panel located on one side of the machine. The decatizing process takes place fully automatically.

Other types of semiautomatic decating machines are the so-called Finish Decators such as built by Kettling and Braun and Monforts which could be classed between the full decator and semidecator types. Figure 24 shows a diagram of the K.&B. Planet-Automat. The steaming cylinder sits within a metal housing with a sliding

Fig. 23. Double Roller Blowing Machine (Courtesy of Sellers & Co.).

door. The working procedure can be set fully automatic and programmed for steaming time, steam condensation time, cooling period and unrolling. All operations can also be controlled manually.

Sustmann (11) in his study of the effects this machine produced on various fabrics gave the following data:

Moisture takeup 1–3%

Steaming time from 2 to 8 min linear increase

Breaking strength and elongation of fabrics are not affected with steam pressures between 4 and 8 atm and steaming times from 2 to 8 min. Thickness change: 18% decrease in thickness against untreated cloth. A reduction in resilience from 27% for the untreated to 16% for the treated cloth. Shrinking effect very small, by all steam pressures without exhausting maximum 0.3%

Continuous decating machines

Modern working methods demand continuous working machines, but only within the last 10 years have several machine builders offered continuous decating machines. The first ones were

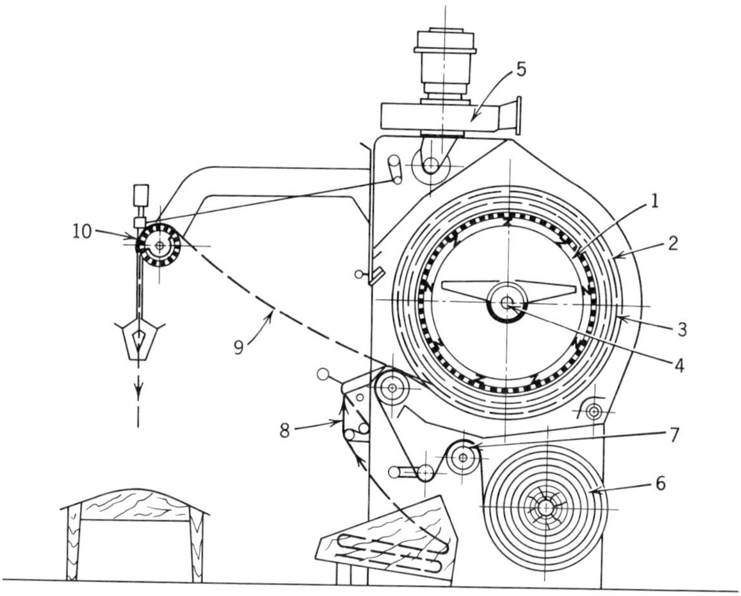

Fig. 24. Planet Automat Finish Decator—Kettling and Braun. 1. Perforated Cylinder. 2. Melton Leader Cloth. 3. Fabriq. 4. Steam Entrance. 5. Suction Fan. 6. Leader Cloth Roll. 7. Leader Cloth Dryer Roll. 8. Fabric Feed In. 9. Fabric Run Out. 10. Cooling Cylinder.

built by Sperotto/Italy and Raxhon/Belgium; Gessner, United States and Monforts, Germany followed later.

The Sperotto machine as seen in Fig. 25 consists of three perforated steam cylinders of 26 cm (10 in.) diameter, a large suction cooling drum of 90 cm (35 in.) and the cloth is carried through the machine on an endless cotton apron, which is continuously kept dry by passing it over a drying drum.

The Raxon machine as seen in diagram Fig. 26 consists of a large perforated steam drum of 120 cm (47 in.). An endless apron presses the cloth against the drum. The drum is divided into two sections—the steaming section—lower part and the cooling section, upper part.

In the Gessner machine (Fig. 27) an endless cotton apron, without seam, carries the fabric through the machine, first over a 36 in. steaming cylinder and then over a 24 in. cooling drum. Directly in

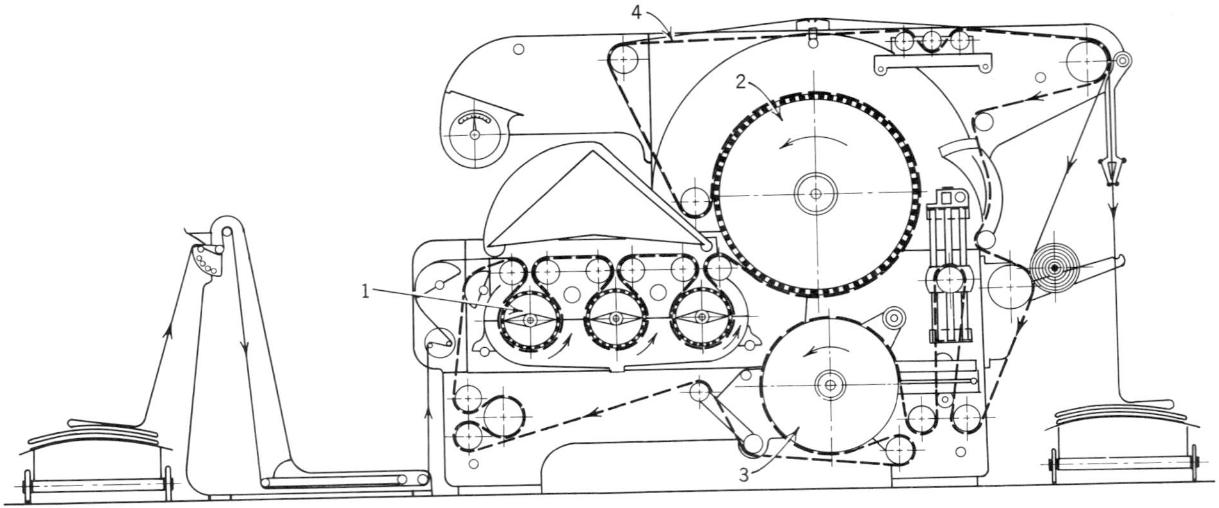

Fig. 25. Sperotto Continuous Decatizing Machine. 1. Three Decating Cylinders. 2. Suction Cooling Cylinder. 3. Leader Cloth Drying Cylinder. 4. Endless Leader Cloth.

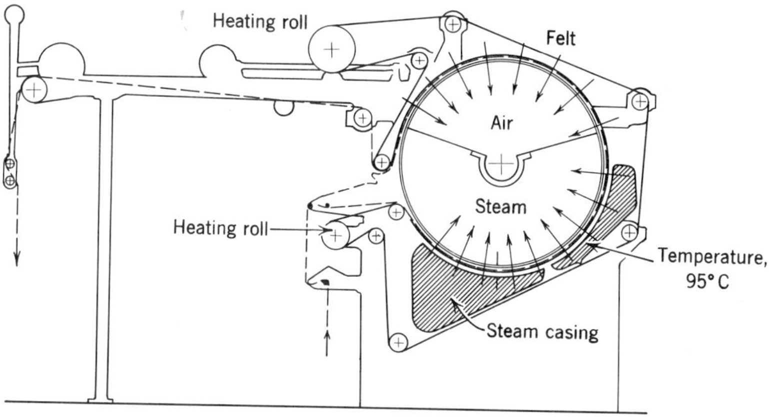

Fig. 26. Raxhon Continuous Decatizing Machine (Courtesy by Ateliers Raxhon, Belgium).

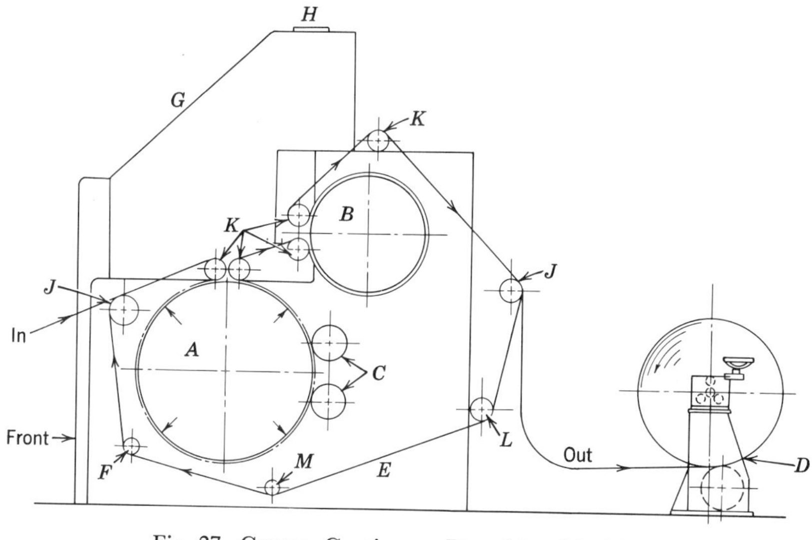

Fig. 27. Gessner Continuous Decatizing Machine.

back of the steaming cylinder are two 8 in. pressure rolls. These are steam heated and are applied to the cloth after it traveled more than 180° around the cylinder. The center wind batching unit on the back of the machine enables the taking off of large rolls at a controlled tension. The operator can vary the degree of finish on a wide variety of cloth by regulating the cloth pressure from a central panel while the machine is in operation.

The Monforts machine works with two steaming cylinders, one on top of the other to which the cloth is fed alternately and then run over a cooling cylinder. The cloth runs between two endless blankets.

The cloth is decatized twice over, because of the automatic reversal of the cloth wind during the decating process. The cloth speed on this machine is adjustable between 15 and 40 m/min. Working time from 3 to 8 min per 60 m piece with a winding length of 120 m.

The Sperotto and Raxhon machines were tested by Sustmann (12) and compared in respect of their speed of working, their steaming time, the moisture ratio of the material, its shrinkage, the change in thickness and its relative compressability. The results are given in Table 2.

TABLE 2

Performance of Continuous Decating Machines

	Sperotto	Raxhon	Gessner	Kettling and Braun, Discontinuous
Cloth speed m/min	10	10	10	
Passage time through machine 50 meter piece min	5	5	3	
Steaming time sec	12	8	18	120
Moisture pickup, %	1–2	0.5–1	2	3
Shrinkage, %	0.5–1.5	1.1–1.5	0.5–3.0	0.3
Thickness change, % Worsteds minus	7–14	10–15	—	18
Compressibility, % minus	0–5	2–10	—	

Sustmann's (12) conclusion was as follows:

The advantages of a continuous machine apart from the continuous working are: No wrapper needed and uniform steaming from start to end of pieces. In the steaming action, moisture uptake, thickness change of fabric and shrinkage effect they are similar to the Finish decators. Both these machines act like a Hoffman press with a one time 10 sec steaming and suction cooling.

In a narrow sense the continuous decating machine as well as the Finish decator are not decating machines at all. Because the machines produce not the stabilization effects such as shrinkage permanency, luster fixation, and the fixation of the nap as they brought about to a certain degree by luster press decator and to a full degree by the vat type wet decator and the dry full decator. All that these machines produce is fundamentally a steam iron effect.

The highest and most permanent luster in woolen fabrics is obtained by repeated wet and dry decating. In many instances, luster is not desired but, since decating is necessary to set the cloth, the excessive lustering must be removed by open steaming. All decating processes, if not carefully applied, can produce serious defects. Such defects are harshness of feel, paper handle, water marks, and change of shade. The harsh and papery feel is a result of loss of moisture when the pieces were subjected to high temperatures. Normally, the pieces are able to regain the lost moisture on standing, thus becoming reconditioned, but in case the harshness does not disappear the pieces must pass through another wet process.

Damping or Dewing

Woolen and worsted products are often subjected to a damping or dewing process before being hot pressed. Saxonys and some tweeds, which can be cold pressed, are the only exceptions. Most woolen and worsted products are subjected to hot pressing. In hot pressing, the wool in order to obtain full benefit of the operation must pass into the plastic state. Not only heat but also moisture is necessary to bring this about; therefore, the fabric must be properly conditioned.

The cloths as they come from the tenter frame and from the blowing processes generally are overdryed, often containing less than 5% moisture. Such pieces will not take a press finish properly and will have a thin, papery handle. It is, therefore, good practice before pressing, to bring the fabric into the required condition by dewing or by storing it sufficiently long in a damp atmosphere.

There are various types of machines employed in the dewing operation, they include the brush type, the air nozzle type, and more recent the atomizer type.

In the brush machine, the cloth is drawn over an open trough containing water in which a grooved, wooden roller revolves. Over this roller, revolving at a high speed, is a circular brush which takes the water from the wooden roller and distributes it in the form of a fine dew or mist on the cloth as it passes over it.

An example of the turbo atomizer type is the Monforts fog dewing machine. As seen from the diagram (Fig. 28) the cloth enters the top of the fog chamber and is guided without length tension into a 70 in. sieve drum, then carried by the rotating drum close to a full cycle and then leaving the drum and the chamber above the incoming cloth. Five atomizers spray the foglike moisture into the back side of the cloth on the drum. High velocity fans suck the moisture filled air through the cloth and the perforation of the drum thereby restoring the proper moisture content of the cloth.

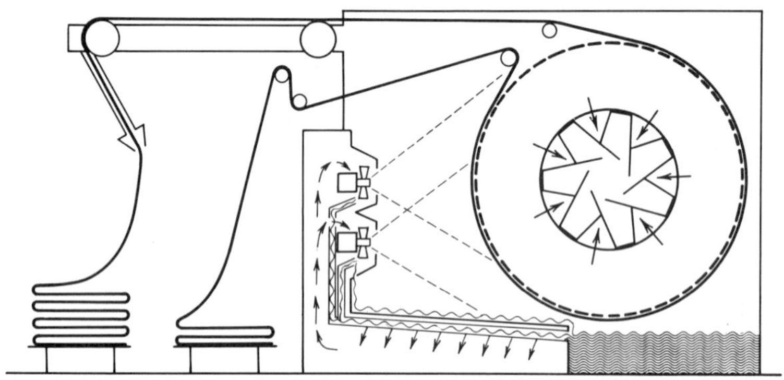

Fig. 28. Monforts Fog-Moistening Machine.

Pressing

As a rule, the process of pressing is the last of the finishing treatments and therefore demands the utmost attention. Its object is the last leveling and flattening of the texture and, at the same time, pressing brings out the handle, luster, and appearance desired in the finished fabric. The basic factors controlling pressing are heat, moisture, pressure, and time. Since all four of these factors can be varied, there are infinite combinations which can be used. The principle of the various kinds of machines used for pressing woolen and worsted goods differs in the way use is made of these four physical factors.

The old fixed hydraulic paper presses used a moderate temperature over a longer period of time. The new intermittent and rotary presses employ higher temperatures over a very short time, thus giving increased output. There is no question that in point of fine finish, soft handle, and appearance the old method is superior.

There are two distinct systems of pressing: in one, pressure is applied vertically and, in the other, longitudinally. In both systems the machines used employ the hydraulic principle to secure the necessary pressure. There are at present three types of pressing machines in use: the vertical paper press, the intermittent vertical paper press, and the rotary press. In the vertical paper press the cloth is folded between heated press papers, piled into a stationary hydraulic press, and compressed for several hours. In the intermittent vertical paper press the cloth is fed intermittently between papers held in heated gaps, and compressed for a few seconds. A new portion of cloth, equal to the width of the plates, is then fed forward and the process repeated. In the rotary press the cloth travels between two heated metallic surfaces, forced together either by a hydraulic ram or by mechanical means, thus giving continuous action.

Vertical Paper Press

This installation, shown in Fig. 29, is usually erected on two levels. The papering-in, the turning, and the papering-out processes are carried out on the upper floor, whereas the heating up and the actual hydraulic pressing is carried out on the lower floor. In papering in (Fig. 29*a*), the cloth is drawn forward over suitable guide rollers onto the table of the middle lift, a stack of papers being in position on each of the end lifts. As the cloth is fed to the middle table, each of two operators standing on opposite sides inserts press paper made of glazed cardboard between each layer of cloth. Each twelfth cardboard has a metal lining with exposed metal tabs which serve as electrical contacts for heating. As the piles alter in

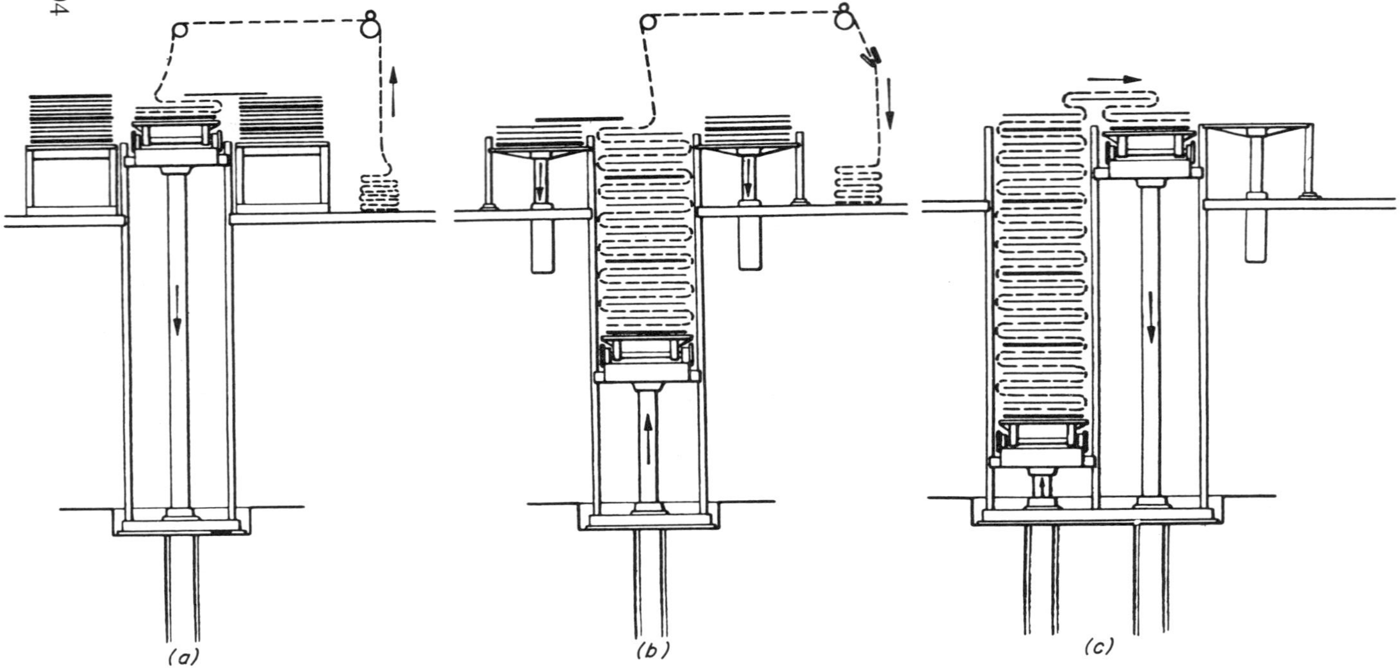

Fig. 29. Automatic Cloth Handling in a Vertical Press.

height, the levels can be adjusted by operating levers controlling the three hydraulic lifting presses. As the middle one is slowly lowered, the outer two, which carry press papers, are gradually raised. When about twenty pieces are thus papered, the middle lift is lowered to the floor below. A movable platform is then brought opposite the lift and the full load run onto it. Small trucks are used to transfer the load of cloth with the paper from the papering lift to the hydraulic press. It is then pushed in front of the press and run into position.

This press consists of a lower and an upper heavy metal plate, fastened together by four steel pillars. A hydraulic pressure arrangement presses the lower plate carrying the load of cloth folded between the press papers and metal plates against the top. Up to 35 pieces can be pressed in a machine at one time. The metal tabs mentioned above are then connected to the electric current which distributes the heat evenly through all layers of the cloth (Fig. 30). A hydraulic pressure of 450–500 lb/sq in. is sufficient to produce the desired effect. The cloth remains under this pressure from three to four hours. It is then taken out and transported back to the lift arrangement for turning or refolding (Fig. 29*c*). In the turning process the cloth is refolded between press paper in a manner similar to that of "papering in" but the edges of the cloth which overlapped the press papers in the first pressing will be placed in the middle as the pressing process is repeated. The papering out which follows the second pressing is illustrated in Fig. 29*b*. In papering out, the load is taken to the middle lift as at first, the cloth is drawn out by the roller mechanism, and the papers withdrawn and delivered to the side lifts ready for the next batch of pieces.

It is thus apparent that the operations are largely automatic, the actual work of papering being unskilled. Only one competent workman, usually the foreman finisher, is required to decide the time, heating, and pressure necessary to obtain the requisite finish for particular fabrics.

Intermittent Vertical Paper Press

This machine is also known as the automatic, flat-plate press. It was designed to carry out automatically the function of the stationary paper press. The principle features of this machine are seen in Fig. 31. This machine consists of a top plate *A* and a lower

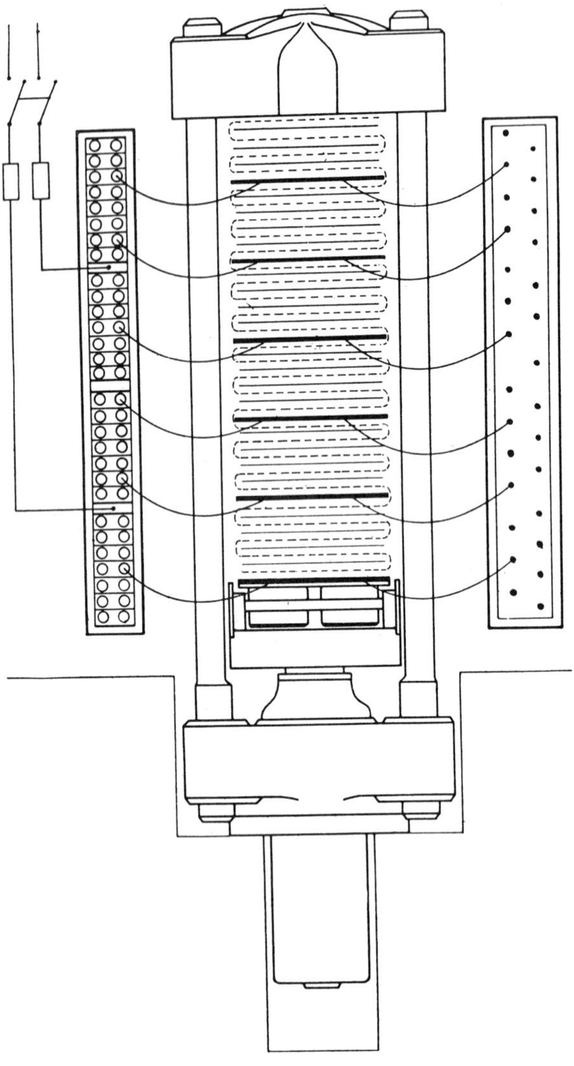

Fig. 30. Fixed Vertical Paper Press.

plate *B*, which is connected to the double hydraulic ram *C*. Between upper and lower plates is an intermediate plate *D*, which is capable of sliding up and down between guides. All of the plates are hollow and connected by flexible pipes to a steam supply.

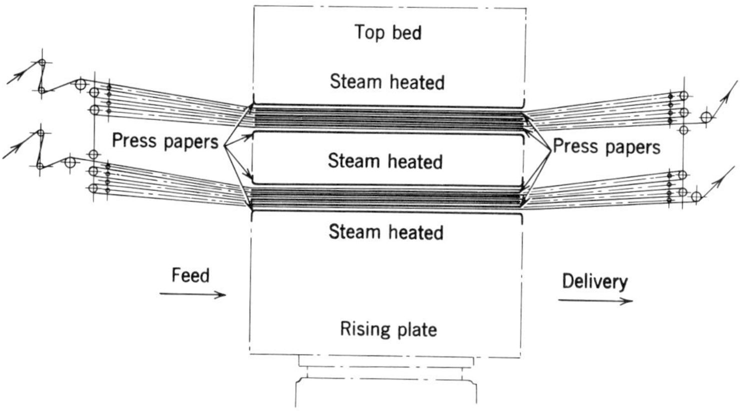

Fig. 31. Automatic Hot Plate Pressing Machine (Courtesy of Hattersley Pickard & Co. Ltd.).

Owing to the short contacting time, the temperature necessary is much higher than that used with the nonautomatic vertical press. The openings between upper and middle plates, and middle and lower plates are taken up by the cloth and the press paper. The pressure is created by the two hydraulic rams. The smaller ram lifts the lower plate *B* and hence, a little later, the middle plate *D*, subsequently closing the two gaps containing the papers and the lengths of cloth. At this point, the larger ram comes into action, applying the requisite full hydraulic pressure.

The machine just described is known as the double machine and will press two full-width pieces or four narrow-width pieces at one time, at rates of from 250 to 700 yd/hr. The newest machines are built to handle three or more pieces, for example a six-piece machine has an output of 2700 yd/hr (2470 m/hr).

By arranging that some of the plates are heated and others cooled, it is possible to use the machine for hot and cold pressing in one operation.

Rotary Presses

For high production requirement, there is considerable opportunity for the rotary press to be successfully used especially on medium priced cloth (Fig. 32).

Fig. 32. Rotary Press (Courtesy of David Gessner & Co.).

With the high speed type machines, speeds from 18 up to 50 yards per minute have given satisfactory results depending upon the fabric and the finish desired.

The main working parts of a rotary press are the cylinder and the beds which are constructed in different sizes. Both the cylinder and bed are generally heated with steam at 50–60 lb pressure. The diameter of the rotating cylinders range from 15 to 25 in.

There are two types of beds made, the single and double bed type. In the first type, the single bed is directly underneath the cylinder as shown in Fig. 33. The pressure is applied either pneumatically or hydraulically by lowering the cylinder down onto a stationary bed. In the double bed type, on each side of the cylinder is a half round shoe or bed. As the cloth passes between the half round beds and the cylinder, the beds are pressed against the cylinder to give the cloth an ironed effect.

As it is being pulled from a truck or roll, the cloth passes under the operator platform, then between a stretcher to smooth the fabric and remove any wrinkles. Then rotary brushes brush bath the face and back of the cloth, the brushed cloth runs over a steamer,

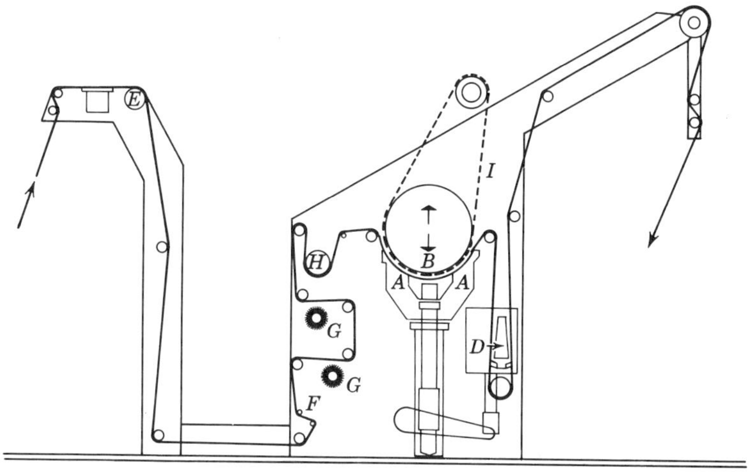

Fig. 33. Diagram of Single Bed Press (Courtesy of A. Monforts Maschinen fabrik). (A) Bed. (B) Press Cylinder. (C) Counter Press Device. (D) Cooling. (E) Brake Roll. (F) Stretcher. (G) Rotary Brushes. (H) Tension Release Roll. (I) Felt Apron.

tension release roll and guide rollers into the pressing unit. The machines may be built with a cooling device through which the pressed cloth passes before it is folded on a truck or rolled up. The counter pressure device in the Monforts machine allows to regulate the counter pressure in the center of the bed precisely with the cylinder pressure that the distance between bed and pressure cylinder are exactly even under all pressure conditions over the whole width.

Some press models can be ordered with an endless felt apron to convert it into an Apron Press. In such a machine the cloth is carried through on an endless woven felt, Fig. 33*d* between the cylinder and the bed. The Apron Press has two distinct advantages over the conventional rotary press. First it does not flatten the fabric when the roundness and softness must be maintained. Secondly, fabrics are not stretched.

Shrinking and Sponging

During the process of finishing woolens and worsted, certain strains are set up which stretch the cloth lengthwise (warp), or widthwise (filling). Before such a fabric is ready for the needle, respectively can be cut and sewn into a well tailored garment, these stresses must be relieved; otherwise the garment will shrink, pucker or go out of shape.

Many mills do this shrinking and sponging, (moistening) as part of their finishing operation. However, a still greater number of mills especially in the United States sell their cloth with only a so-called mill finish and rely on either a public sponger or a clothing house equipped to do their sponging and shrinking.

The problems in tailoring since World War II have been increased tremendously through the demand of lighter weight fabrics but also through the wool-synthetic fiber blends which are difficult to tailor.

New ways of finishing had to be found and better shrinking equipment developed within the framework of higher efficiency and productivity.

The two methods open to reduce the relaxation shrinkage down to a one percent level for fabrics without a nap are cold water shrinking and steam shrinking.

Cold Water Shrinking

This process also known as London shrinking is used primarily in the United States for men's wear worsteds and mainly done by spongers and cloth tailoring houses. Mills generally shy away from it due to its slow productivity. Not much can be done to speed it up, therefore, hardly any changes have been made with this method over the years. The detail description is found under the section of conditioning before cutting up.

Steam Sponging and Shrinking

Steam sponging was found to be sufficient for a ready for the needle finish for most woolen and worsted fabrics. But this became true only in recent years through the combined efforts of science and industry, which on one hand led to a clear understanding of the fundamentals involved and on the other hand resulted in the use of this knowledge by the machine builder to give the finisher the proper tools to work with.

There are two types of machines available, namely, the table Steam Sponging, and the Cylinder Steam Sponging Machines.

As the table type is a continuous, the main effort by the machine builders have been concentrated on the perfection of this type. Two recent machines are built by Drabert-Kettling & Braun (Fig. 34), and by A. Monforts (Figs. 36 and 37). In both of these machines, the cloth travels through three zones, namely, steaming, heating and cooling.

The action of the K&B, London, Shrunk Machine as described by Riedel (10) is as follows:

The shrinkage and the stabilization of the cloth in warp and filling direction is achieved by means of a double intensive shock treatment, a change being made twice between extremes of conditions: steam damping/hot iron and hot iron/cold humid air. The changes in temperature and moisture of the fabric taking place in the three zones are shown in Fig. 35.

The studies which are the basis of Fig. 35, were made on wool fabrics ranging in weight from 400–600 g/m (12–20 oz/yd) run continuously at a speed of 8 m/min. The diagram (*a*) shows a rapid increase of the cloth temperature through steaming from 25 to 83°C (77–180°F). In the connected heat zone a slight further increase occurs but then cooling sets in through evaporation down to approximately 58–50°C (136–122°F). The shock type cooling through moist air which follows, results in further shrinking of the cloth, which cools down to 17–21°C (63–70), respectively,

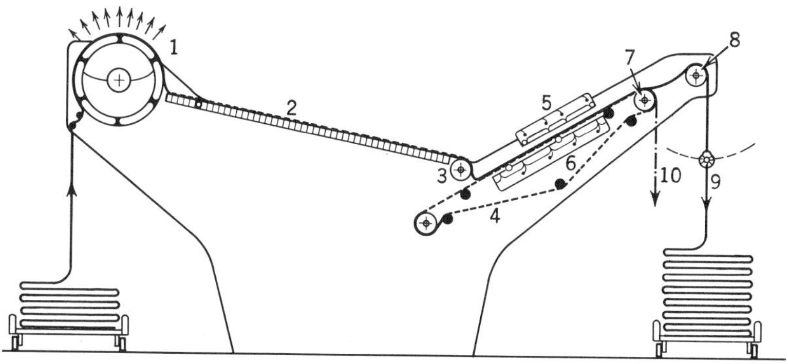

Fig. 34. Diagram of the K & B "London-Shrunk" Shrinking Machine (Courtesy of Drabert—Kettling and Braun, Appreturmaschinenfabrik). 1. Steam Cylinder. 2. Shrink Plate. 3. Take-off Roll. 4. Transport Apron. 5. Dewing Unit. 6. Cooling Unit. 7. Transport Roll. 8. Take-off Roll. 9. Folder.

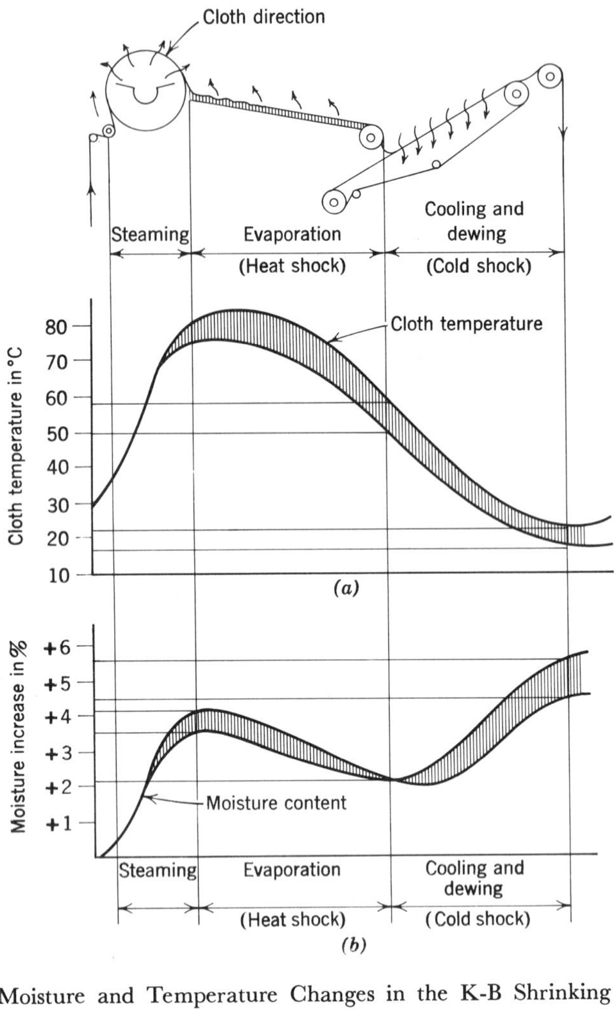

Fig. 35. Moisture and Temperature Changes in the K-B Shrinking Machine.

room temperature. Diagram (*b*) shows the changes in the moisture content of the cloth during the whole shrinking operation ending up with a 4.5–5.2% moisture takeup. Another major factor affecting the moisture and heat absorption of the cloth is the working speed. With increased cloth speed the moisture take up and temperatures are lower, but the final temperature of the cloth will be

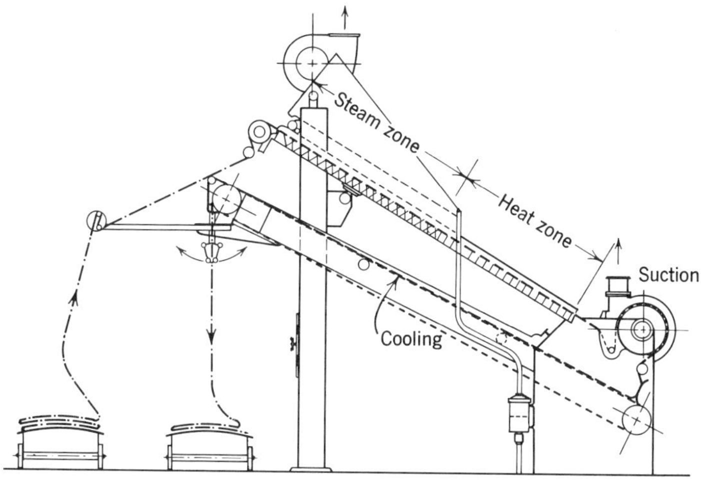

Fig. 36. Diagram of Monforts Steaming and Shrinking Machine.

Fig. 37. View of Monforts Steaming and Shrinking Machine (Courtesy of A. Monforts Maschinenfabrik).

higher. Based on test runs and many measurements the relationship between the working speed and the attainable residual shrinkage for this machine is shown in graph form in Fig. 38. To attain a residual shrinkage of 1% the maximum cloth speed can be in the range of 12–15 m/min.

Sustmann (13) made a study of the Monforts steam and shrinking machine and reported the following results:

Analyzing separately the effects of steaming, heating and cooling on the shrinking of the cloth from Fig. 39 it can be seen that with increasing steaming time the shrinkage increases. It is also apparent that the main part of the shrinkage is caused by the moisture of the steam and the heating of the moisture present in the cloth. The heating grate following adding more heat produces a small additional shrink effect, indicating that the shrinkage also in narrow limits depends on the time factor. The influence of the suction and cooling is slight and can be disregarded. Its only advantage is that the cloth is cooled and can be folded without the danger of forming heat wrinkles. The optimum of a good shrink effect on the Monforts machine is around a cloth speed of 11 m/min. In Fig. 39 are shown the shrinkage diagrams of three fabrics with the 1% tolerance line. The data indicates that the residual shrinkage of all three fabrics will be on or under the 1% limit with 6 sec steaming at the corresponding cloth speed of 15 m/min. The tests

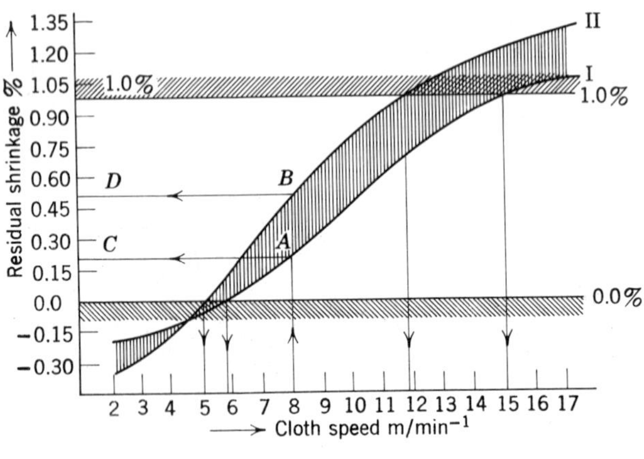

Fig. 38. Cloth Speed and Residual Shrinkage.

for the residual shrinkage were performed on a Hoffman press 5 × 10 sec steaming.

The softer hand attained with this type of sponging is due to the increase in thickness of the fabric and a higher compressability. This increase in thickness varies with the fabric type and all the factors involved in the processing.

An English A.M.P. (13) continuous cloth shrinking machine was introduced in 1951. This machine used infrared heat in the drying zone, and it was stated that the total shrinkage obtained occurred while the material was passing through the chamber which housed a Metrovick infrared projector. The cloth was first overfed at a calculated differential into a steaming conveyer, then moved into the heating chamber. After emerging from the infrared unit, it was then cold set by high velocity air impact. According to information from the I.W.S. the machine is not made anymore.

One of the newest machines in this field is the Italian Thermofrigofinish machine which is a combination of table and cylinder

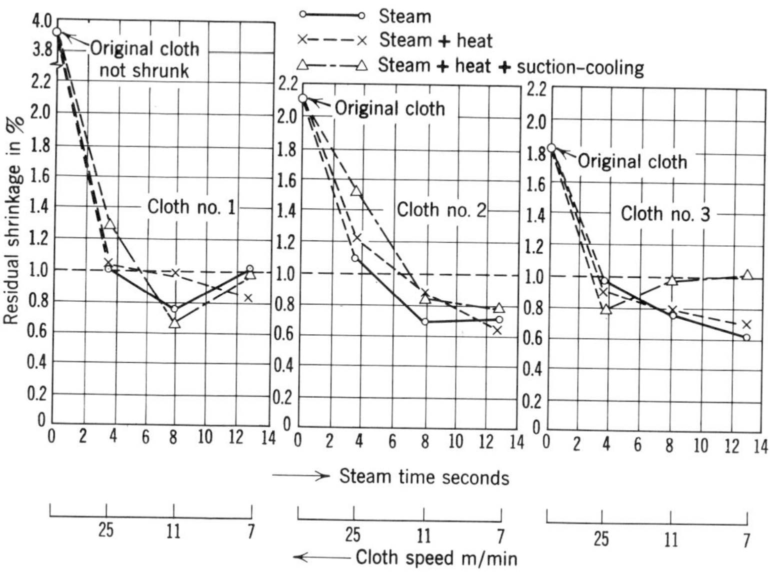

Fig. 39. Residual Shrinkage Diagrams of Three Fabrics in Hoffman Pressing by Sustmann.

shrinking. The cloth travels first over a short steam table (vaporizer) where it is exposed to wet steam, then is picked up by an endless woolfelt and carried around a 60 in. steam cylinder and then shock cooled by passing through a refrigerator unit, which has a temperature range from 30° to −5°.

In Germany, one of the better known machines for cloth shrinking is built by Kannegiesser.

Examining

This operation is performed several times during the finishing routines to detect imperfections, stains, and damages which may have occurred during the finishing operations. As a rule the cloth passes through two inspections in wet finishing and three after dyeing. The number of inspections necessary depends largely on the equipment of the plant and the skill of the workers on the machines.

In wet finishing one examination is practical after scouring, fulling, and carbonizing and one at the last stage before dyeing. After dyeing and drying, the cloth has to be inspected for color and for any wrinkles which may have occurred. An inspection of the cloth is necessary after shearing to detect small knot holes, oil stains, and cuts.

The most important of all examinations is the one at the end of the finishing routine when the cloth is thoroughly inspected for all imperfections and unevenness in color. Here every noticeable imperfection is marked with a small string at the selvage and the customer is given an allowance according to the size of the imperfection (Fig. 40).

An electrically operated and electronically controlled automatic marking device has been developed under the trade name Mark-O-Mat, which affixes the string marking a fault in less than one second. The apparatus forms a self-containing unit and can be mounted on the cloth inspection machine. Two types are available, one for a single color marking and the other for two color markings. For the effective removal of oil stains various solvent pistols have found wide use.

The final examiner must have good eyesight to detect all imperfections and good practical knowledge and judgment as to how

Fig. 40. Final Examining (Courtesy J. P. Stevens & Co., Inc.).

far he can go in passing or marking imperfections and variations in color. He must also be able to correct imperfections which do not affect the wearing quality of a fabric. For the final examination the machines used should be placed in good light, preferably with northern exposure.

Stamping, Measuring, Weighing, Folding, and Rolling

Stamping, measuring, weighing, double folding, also known as rigging and cuttling, and rolling, in other terms, "Making up for shipping," are operations through which the cloth passes after the final examination.

Stamping

Today stamping or marking of the cloth is a common practice in the woolen industry. The method most used is dry transfer marking. This process was invented in the mid-1870s in England and introduced to the United States in 1902 by the Kaumagraph Company.

The dry transfer consists of a strip of tissue paper on which a design, trade-mark, or description has been printed in fusible ink from a metal engraving. When laid face downward on a piece of fabric and pressed with a heated iron, the fusible ink adheres securely to the fabric. When the paper is lifted the design or printing has been transferred to the fabric.

Stamping was originally done as a protective function, known as "truth markings," applied to the inside and the outside end of pieces. The ease of marking with dry transfers has caused stamping to become much broader and of more informative significance. In many mills stamping is the standard means of numbering, truth marking, and of marking for shrinkage control and other identification purposes essential to the processing of woolen goods. The more informative significance was brought about by the use of stamping to apply the trade-mark of the manufacturer on the back of the cloth or along the selvages. On woolens stamping became a guarantee to the cutter against impurities in content, shrinkage, and incorrect shade; a proof to the wholesaler of authenticity of manufacture; and a guide to the consumer for quality purchasing.

Individual digits to make up any required number can be reproduced by means of a hot iron applied manually or automatically to a dry transfer specially developed for this numbering process. The Transferotor is a device which can be adapted to many cloth-handling machines or which can be built as a unit. It is used for applying transfers at regular intervals on a piece of material. The transfer in roll form is situated in the Transferotor. At predetermined intervals a self-contained electric iron applies the transfer to the web of cloth moving at speeds of 80–100 yd/min. Frequently this is combined with measuring, folding, winding, and other cloth-handling operations. The Transferotor offers the advantage of close printing on the tissue with reproduction on the cloth at wide intervals, usually 1, 2, or 2½ yd.

Transfers can be applied to the back or selvage of cloth by leading the paper from the transfer roll into the calender stack or pressing rollers. This allows the natural heat of the operation to apply the transfer to the cloth. This method is more expensive than that of the Transferotor since transfer printing is spaced on the paper to coincide with the required spacing on the fabric, usually 18 or 36 in.

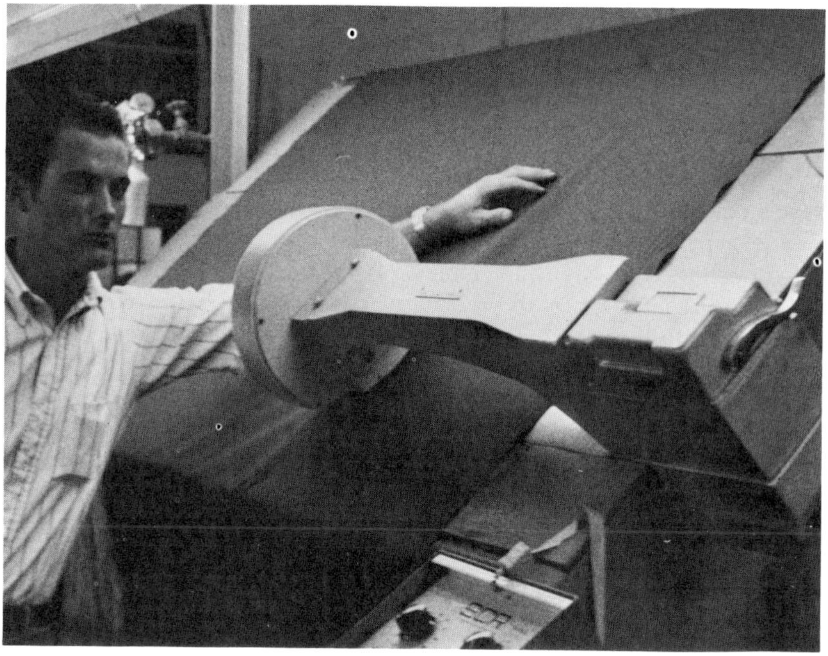

Fig. 41. Trumeter on Final Inspecting Perch (Courtesy J. P. Stevens & Co., Inc.).

Measuring

In small finishing plants, the measuring is done by pulling the pieces over a flat measuring table with yards or meters markings, also used in large mills for sample cutting. For continuous measuring, for years special machines equipped with a rough surface measuring roll of 2 yards or meters circumference with counter or dial were standard equipment. Most of these machines are built as combination measuring, rolling or folding winders or tubing machines. In recent years a British measuring device known as the Trumeter has been widely accepted as the most accurate and reliable textile measuring machine. It keeps a constant check on every yard and fraction of a yard of material passing through the machine and produces a printed voucher at any point clearly stating what measure has been run. The Trumeter is fitted easily to most types of existing machines such as rolling, winding, tentering and perch (Fig. 41). Figure 42 shows the Woolco model of the Trumeter specially recommended for the measuring of woolen cloth, finished or unfinished from 8 to 35 oz.

Weighing

Woolen fabrics are sold by the length, the cloth has to meet a specified weight per linear yard or meter, generally with a half ounce range per example 10 to 10½, or for government cloth a minimum weight has to be met. Various models of scales mostly of the platform type are used. The use of computers is possible to quick calculate the weight per linear yard and print it on the same ticket with the length.

Folding and Rolling

The packaging of the finished material is done either by double folding, rigging and cuttling or rolling. The rigging machine folds the material down the middle, with the face inside, this operation being known as rigging, and also cuttles or folds the piece into a neat compact form. The doubling winder folds the cloth and winds it on a cardboard.

Rolling

The rolling or tubing machine rolls the material full width around a cardboard tube Mandrel. One of the recent machines is the Somatex loop-controlled examiner tuber for stretch, knitted and woven fabrics. The machine gives accurate relaxed measurement and straight-sided tube package.

The final step in packaging is the wrapping of the folded or rolled fabrics with a strong paper for protection in shipment and storage.

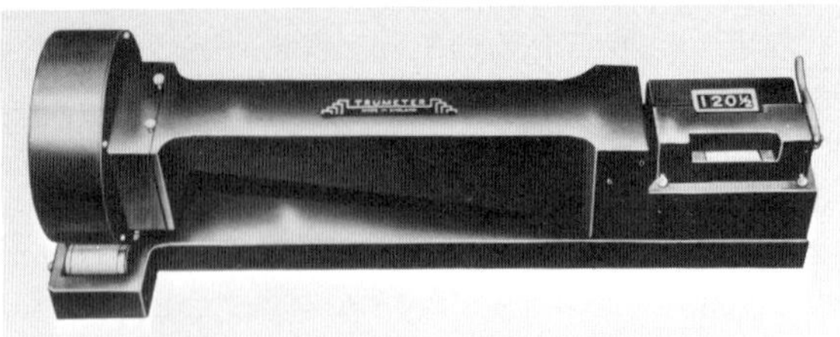

Fig. 42. Trumeter—Woolco Model.

In all the wet and dry finishing processes the finisher is faced with the problems of fabric guidance. The various devices and cloth guides are analyzed in three articles of the *International Textile Bulletin* (21).

Conditioning for Tailoring

As previously stated, the final sponging and shrinking of the fabric to bring it to the best condition for tailoring ready for the needle is either done as part of the finishing operation in the Woolen Mills or by commission spongers or by the tailoring houses.

For unsponged goods as delivered by the mills to be conditioned for cutting up the normal allowance made by the trade are up to 5% warp shrinkages and up to 2 in. (5 cm) width shrinkage. The United States Army Quartermaster per example specifies the following maximum warp shrinkages:

	Serge 18 oz	Elastique 18 oz	Melton 16 oz	Kersey 30 oz
Warp %	4	8	3½	2½
Filling	Above 56 in.	Above 56 in.	3½	2½

	Wool crepe 7–8 oz	Tropical worsted 10 oz	Flannel shirting 10½
Warp %	10	4	5
Filling	6	3	4

The work of the commission spongers and tailoring houses preparatory to the cutting up is twofold mainly examination and a thorough preshrinking.

Examination

The examination corresponds to the previous inspection at the mill acting as a double check for the cloth merchant or the tailor house which bought the goods. Generally, all the pieces of a lot or order are passed over a perch equipped with a measuring device like the Trumeter and weighed. The yardage and weight compared with the one given on the piece ticket or stated on the invoice. The responsibility of the inspector is also to carefully view the imperfections marked by the mill examiner and to catch overlooked ones. For perfect goods a maximum number of strings per piece may be set. Depending on the defect 1/8 to 1/4 yd allowance is given to the customer.

The examination defects fall into two principal classes: apparent or visible defects and latent defects. In the first class come various mispicks, bars, roweyness, uneven filling, warp runners, holes, stains, shadiness of color, and tenderness. The last two defects often can render the goods unmerchantable. In the class of latent defects are included streaks, which appear after pressing and in the case of napped goods, pilling after less than reasonable wear, and undue fading of colors. Washer streaks and lay wrinkles do not always appear in the goods when they are perched or examined and often appear afterward as a latent defect.

Shrinking

Passable goods are then subjected to different preshrinking operation similar to the ones previously discussed. The two print ciple methods are cold water shrinking and shrinking by treatmen- with steam.

London Shrinking

The process, also known as cold-water shrinking is used primarily in the United States for fine men's-wear worsteds. It calls for three principal operations, namely, dampening, drying, and refinishing.

Dampening

There are two ways of dampening or moistening the fabric, the old patch process using wet blankets and the continuous process which combines wetting in a padder and drying. Blanket dampening consists of wetting the goods in the following manner. A raised platform from 3 to 6 yd in length is used. Cotton or woolen blankets are thoroughly wetted and constitute the medium for dampening. A wet blanket is first laid on the platform and the first layer of the cloth is placed on top spread in full. Wet blankets and cloth are alternately spread out on the platform. The number of layers of cloth as compared to the number of blankets used between them varies according to the weight of the fabric and is governed by practical experience. When the requisite number of pieces is laid in between the blankets, sufficient weight is put on top of the entire pile to compress the various layers and so that the wet blankets come in contact thoroughly and evenly with the cloth being processed. The cloth is allowed to remain in this condition for about 12 hr, and then the dampened cloth is removed from the platform and festooned dried.

Continuous Process

This is done on cold water shrinking units such as developed by Gessner and shown in Fig. 43. The cloth first passes over a steam box to open up the fibers, then enters a stainless steel dip tank which may be divided into two sections for warm and cold water. After passing through the dip tank, the fabric runs through two pneumatically operated squeeze rolls. The squeeze rolls deliver the fabric into a traveling scray. Here the fabric is accumulated and allowed to sweat for a period of roughly 20 min before it enters the drying chamber allowing the cloth to fully relax.

The sponger may also be requested to waterproof or moth proof the fabric in such instances instead of water alone in the dip tank water repellent and moth proofing compounds are added. As these compounds are applied hot, special tanks for mixing have to be provided.

Drying

Drying is accomplished by hanging the dampened goods on sticks and in folds until they are thoroughly dried by natural

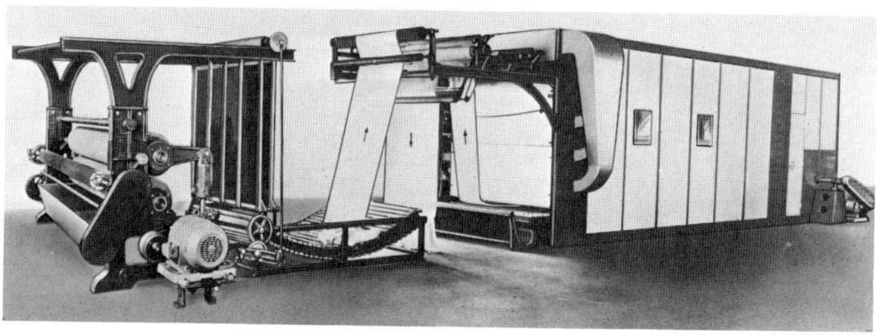

Fig. 43. Continuous London Shrinking Machine (Courtesy of David Gessner & Co.).

room air, or through the medium of forced air currents in regular drying machines or automatic festoon dryers equipped with steam coils.

Drying on hanging devices in free atmospheric air according to Bergmann (11) is only practiced nowadays in Britain and Italy for drying of high grade finished articles of pure wool. It is a low productivity method but exerts a highly beneficial effect on the hand of the cloth.

Suspension or festoon dryers with mechanical devices and hot air circulation working at moderate temperatures around 40°C (104°F) are still used. Figure 44 shows a diagram of a mechanical hanging dryer.

The Gessner dryer specially designed to shrink cloth is a multi-pass apron dryer. The traveling horizontal apron is composed of slats about four inches apart. The cloth lays in scallops on the slats while warm air keeps it vibrating constantly. When the cloth emerges from the dryer it may be folded on a truck or rolled up on a roll (Fig. 43).

Refinishing

After shrinking and drying, it is necessary to refinish the cloth so that it will regain its finish and lie perfectly flat on the cutting table. Depending on the fabric this is done either in an open blowing machine such as a semidecator or is subjected to hydraulic pressing.

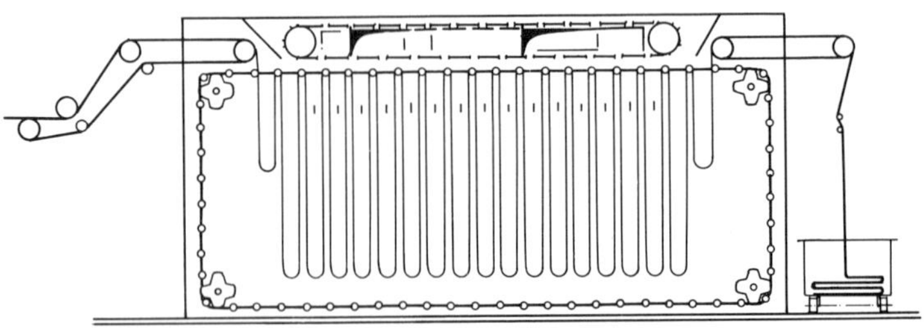

Fig. 44. Festoon Dryer.

Steam Shrinking

Steam shrinking is carried out by means of one of three machines, the choice of which depends on the character of the goods and its fiber contents.

1. The table steaming and cooling machine is used chiefly for coatings, overcoatings, and similar heavy goods, and for napped and pile fabrics. Minimum tension on the cloth and a proper cooling with suitable folding down attachments are essential in obtaining satisfactory results with this method.

2. The cylinder steaming machine is used chiefly for men's suitings, tightly woven tweeds, and similar fabrics which require stationary and double steaming. These cloths are properly cooled by being left for a sufficient length of time on wooden poles or rods on a cooling frame.

3. The semidecating machine contains a perforated cylinder (or a pair of cylinders) around which is wound a cotton leader of soft or hard character depending upon the finish desired on the goods. Lightweight dress-goods are "set" by this process with a minimum shrinkage and a retention of their original appearance. Coatings, requiring a high face-finish, particularly broadcloths, are also finished by this process.

After the cold water or steam shrinking, the cloth is wound on wooden spindles or cores and brought back to the examiner for re-examination and detection of any possible latent defects that might have developed during the process. Finally, the cloth is rolled and remeasured, after which it is ready for the cutting table.

Unshrunk or improperly preshrunk wool fabrics are unsatisfactory to the cutter-up and the consumer because:

1. The fabric will not react properly under the steam press or hot iron when the garments are manufactured.

2. The garment will shrink when the consumer has it dry cleaned.

3. The garment will shrink or expand when subjected to rain or excessive humidity.

Ready for the Needle Testing

Sustmann (14, 15) investigated the effect of steaming on the Hoffman press on moisture and shrinkage of wool fabrics under different testing conditions. The moisture increase of the fabric under steam pressures from 45 to 75 lb starts around five seconds and increases sharply between 10 and 15 sec, reaches the maximum at 20–30 sec and then slowly reverses dropping into the negative. Working with suction cooling a loss of the natural fabric moisture will occur.

The smoothing and forming of a garment under the Hoffman press is due to the moisture in the fabric in combination with the heat from the steam.

Without suction the shrinkage is higher than with suction. Figure 45 shows the shrinking action of the Hoffman press after several steaming operations as they occur in garment making. The shrinkage increases stepwise from steaming to steaming reaching its maximum after the internal stresses are relaxed (23).

Ready for the needle or ideally sponged cloth should have approximately two percent relaxation shrinkage left in both warp and filling direction. However, for specialty finishes such as Broadcloth or other napped and brushed finishes it is extremely difficult to maintain this level without impairing the finish by presently known sponging techniques. The garment manufacturer is, therefore, called upon to make allowances in his patterns. The subject of shrinkage is still one of the most misunderstood topics in the trade. Major difficulties experienced in tailoring are not so much the cause of too high or too low shrinkage, but are rather due by inconsistent levels of shrinkage. A garment manufacturer

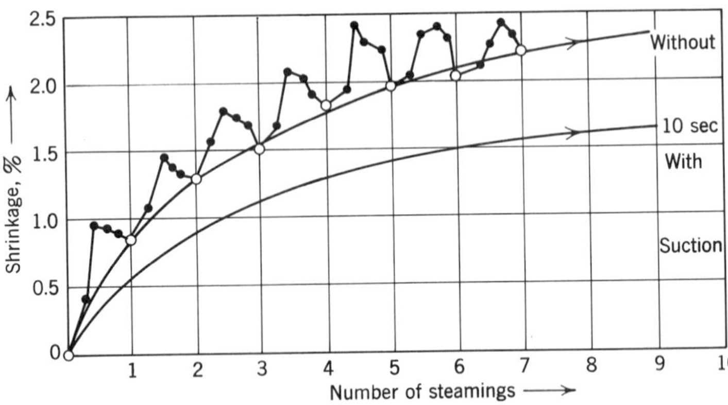

Fig. 45. Progress of Shrinkage after Multiple Steamings on a Hoffman Press—Sustmann.

dealing with a reputable sponger may find that the consistency of handling of all kinds of materials from various mills is a valuable asset for controlling garment sizes.

An extensive study to determine the tailoring characteristics, including appearance and dimensions of the garments made from variously finished men's wear fabrics and to correlate the cloth movement, during tailoring to the laboratory tests of the cloth was made over a period of several years by von Bergen, Clutz, and Macaluso (17).

It was found that there is a direct correlation between the cloth movement and the laboratory tests as can be seen from the results in Table 3.

TABLE 3

Correlation of Cloth Movement in Garments to Laboratory Tests

	Mill finish		Cold water sponged		Garment specifications
	Warp	Filling	Warp	Filling	
			Worsted Suiting		
Laboratory tests					
Steam Test	−1.5	−1.1	−0.2	0.0	
Sponge Test	−1.1	+0.1	+0.1	+1.1	
Relaxation	−3.6	−0.6	−0.7	+1.4	
Stretch	+3.0	+7.3	+5.4	+8.7	
Garment test					
Length	−1.3		+0.4		31¼
Waist	–	+1.0		+2.4	48⅛–½
Seat	–	−2.0		−0.7	51¼–½
			Flannel		
Laboratory tests					
Steam	−1.6	−1.5	−1.0	−0.6	
Sponge	−0.8	+0.3	−0.5	+1.5	
Relaxation	−3.1	−1.2	−1.5	+1.0	
Stretch	+3.7	+9.5	+5.1	+11.4	
Garment test					
Length	−0.9		−0.3		31
Waist		+1.9		+3.0	48⅛
Seat		−2.5		−0.6	51¼

Garment Model Dunn Size 40 Regular

The mill finished pieces showed higher steam and sponging test shrinkages and lower stretch than the cold water shrunk fabrics. In a trial covering 28 suits, the garments made from the sponged pieces had a clear tendency to be on the large and long side, indicating over sponging.

In a three-month study on 892 suits made of Forstmann fabrics as delivered ready for the needle and 389 suits made of various mill fabrics mostly imported shrunk by commercial spongers gave the following results: In comparing the coats sizes approximately 70% of all the coats in both groups came up to size which includes coats which were only ¼ in. off size, but the overall figures actually hide and do not show the difficulties which are present in specific fabrics. For proper comparison, it is necessary to single out indi-

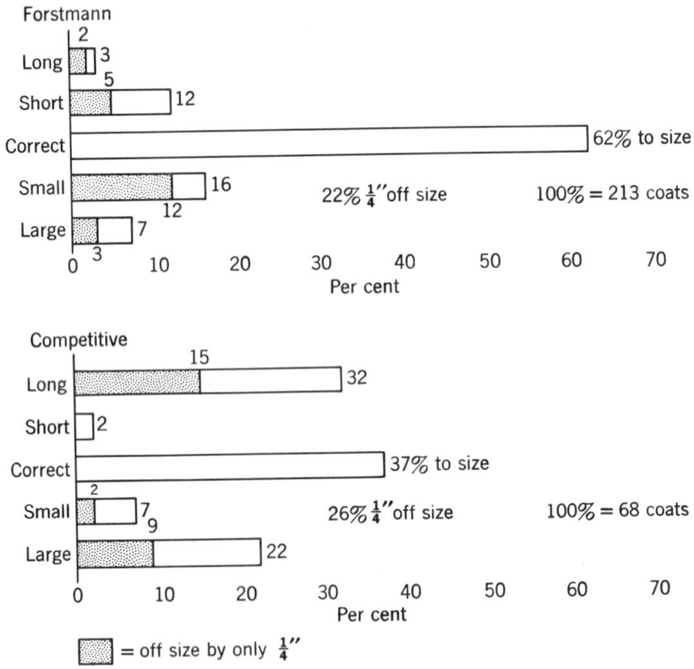

Fig. 46. Variation in Sizes of Tailored Men's Coats.

vidual styles comparable to each other. As examples out of 25 styles analyzed, two comparisons are given in graph form Figs. 46 and 47. The coats were classified as correct when the length and waist measurement were all within ⅛ of an inch of the specification when a ¼ in. or more off they were classified whether as long or short for the length, or small or large referring to the waist.

In Fig. 46 are compared two clear finished worsteds, Forstmann 10½–11 oz, and competitive 11–11½. Whereas, in the Forstmann coats the off sizes are highest in the short and small classes amounting to 28% of the total in the competitive style, the long and large amounting to 54% comprise the main percentages of the off sizes. Including the coats which are only ¼ inch off size, the overall correct sizes amount to 72% and 63% respectively.

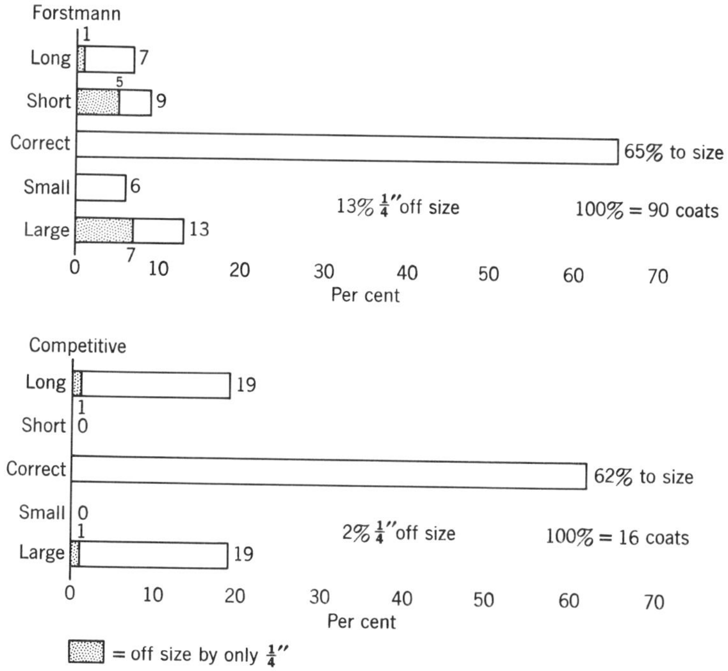

Fig. 47. Variation in Sizes of Tailored Men's Coats

In Fig. 47 are compared two mill finished worsted fabrics 12 oz each. The tailoring performance of both styles may be considered as very good and about equal. But again we have the phenomena that in the off sizes the competitive fabrics shows 38% of long and large and no short and small, whereas, in the Forstmann cloth the off sizes are quite uniformly divided between long, short, large and small.

The Forstmann fabric in general produces an even distribution of the off sizes, which was taken as a sign of a well balanced fabric ready for the needle finish.

Eleven competitive fabrics showed the off size groups of the coats to be mainly limited to long and large, which indicates an over sponging, leading to growth of the fabrics in tailoring in both directions, warp as well as filling.

In soft fabrics such as woolen flannels and soft loose sport fabrics, off sizes occur often in the large group. This is generally not caused by over sponging but rather by excessive stretching of the fabric in the filling direction during the tailoring operation.

The tailoring houses are not too much concerned about a small percent of off sizes in the long and large groups because alteration can be made, whereas they claim they can not do much when the sizes turn out too short or small.

Bonded and Laminated Fabrics

Jackson Bauer (24) discussed one of the major innovations that has taken place in the textile industry in recent years, namely the production of bonded and laminated fabrics, particularly for the ladies' apparel trade. A phenomenal growth took place beginning about 1962 in the bonding of various types of face fabrics, applied either directly to a backing fabric or first to a thin layer of polyurethane foam, which in turn was joined on the back to a suitable backing fabric. A large proportion of the face fabrics have been circular knits, although woven fabrics have also been widely used; and, to some extent, tricot face fabrics have been employed. The fiber content has varied from 100% wool through wool/synthetic blends, 100% synthetics, as well as cottons and cotton blends. The backing fabrics have been almost exclusively acetate tricot. The reasons for the nearly universal usage of acetate tricot have been its relatively low price, the available color range, its ability to conform to the face fabric, and the pleasant feeling when this fabric touches the skin. Urethane foams have been used to obtain increased bulk as well as insulation. Originally, fabrics joined to urethane foam were too stiff and did not drape well, but the use of thinner foams and improved bonding techniques have overcome this limitation.

The use of bonding and laminating techniques have resulted in the ability to use light construction fabrics which, in themselves, would not be suitable for apparel purposes. Thus a large volume of lightweight single knit fabrics produce the general characteristics of normally much heavier fabrics when they are bonded or laminated.

The term "bonded" is used to refer to fabrics which are directly joined with an adhesive to a backing fabric. On the other hand, the term "laminated fabrics" sometimes is used to refer to fabrics where an intermediate layer of urethane foam is used.

There are two main processes used throughout the trade. One of these is flame lamination. In this process, a urethane foam is introduced into a machine which applies a gas flame to the surface of the foam just before it is joined under nip pressure to a fabric. The flame melts the foam, thereby making the surface of the foam into an adhesive which can be joined by pressure to a fabric. The process is then repeated if a backing fabric is to be joined to the other side of the foam. The other process involves the use of an adhesive which is applied by an applicator roll to the back of the face fabric, generally in a discontinuous pattern to avoid subsequent stiffness, and then this is joined to a backing fabric such as tricot. Water-based adhesives are widely used, and these are generally acrylic types. For good durability, the acrylics must be cross-linked so that they are durable to laundering and dry cleaning. More recently, there has been increased interest in the use of solvent-based urethane adhesives. These require either the use of inflammable or toxic solvents, and care must be taken to avoid any disastrous effects in the mill from their use. While the water-based acrylics require drying to remove the water and subsequent curing to cross-link the resin, urethanes generally cure at relatively low temperatures although the process is sometimes extended into several days. The urethanes have the advantage of being elastomeric in nature and are thought by some to produce a more durable bond. However, properly applied and cured acrylics have also been shown to be completely durable to the normal fabric cleaning processes.

Bonded and laminated fabrics have been produced and marketed without adequate bond in some cases, and, as a result, various segments of the industry have introduced guarantee programs whereby quality controls are established and the fabrics marketed must pass certain tests and meet certain test specifications before permission is given to market these under the guarantee programs.

By joining together the various substraits, it has been possible to introduce a complete new range of fabrics for the use of the garment manufacturer. In many instances, the retailer has been able to offer completely lined garments at about the same price range as that existing previously for unlined garments. In a number of instances, the bonding has contributed to dimensional stability, crease resistance, and surprisingly good drapability. With the advent of bonding, wool fabrics could be made in lighter weight and bonded to tricot at a much lower price than previously possible with a heavier, approximately equivalent quality, wool fabric.

Although the large volume of backing fabric has been acetate tricot, other tricot fabrics have been used such as nylon and triacetate. These fibers have the ability to be heat-set and thus improve dimensional stability; at the same time, under some conditions they can be more washfast when properly dyed. On the other hand, acetate is lower in price and has excellent drape, particularly in the jersey stitch which is widely used. Other advantages of nylon lie in its higher tensile strength and superior abrasion resistance when this factor is of importance. Circular knit backings, particularly in cotton, with or without hot melt adhesives, are useful for certain segments of the market. Nonwovens, at the present time, have limited use, such as on wool interfacings. Of course, there are limitless possibilities for combining various fabrics, and many of these have come into use. For example, lace has been bonded to tricot in contrasting colors; and in some instances, nylon scrims have been bonded to open knits which stabilize and yet retain the open look of the knit.

Generally, the face fabrics and backing fabrics are separately dyed before they are bonded or laminated. However, particularly in the case of wool face fabrics, undyed face fabrics have been bonded to undyed nylon tricot backings; this combination then is piece-dyed using selected dyestuffs to union shades.

Although penetration of bonded fabrics into menswear markets has been limited, it is expected that there will be increased acceptance in the area. Continued growth is also expected in children's wear as well as in the established area of women's wear. In many cases, bonded and laminated fabrics have not displaced regular fabrics but have created new markets and new concepts and therefore have taken their own place as new and different products.

References

1. Bergmann, R., *I.T.S. Textile Technology*, 6002, 6/5.
2. Biella Shrunk Process, *I.T.S. Dyeing & Finishing* 2, **12**, 349 (1965).
3. Bunch, D., Jr., *Am. Dyestuff Reptr.*, 369 (1950).
4. Dornier Gesellschaft Mbh., *I.T.S. Dyeing & Finishing* 2, 6-1 223 (1964).
5. Gottschalk, K. H., and Weiss, A., *Melliand Textileber.*, **47**, 924 (1966).
6. Hunter, J. H., *A.D.R.*, **43**, 236 (1954).
7. Macaluso, Ch., F. W. Co., *Report Pr.*, **41** (1951).
8. Riedel, D., *Melliand Textilber.*, **45**, 303 (1964); **43**, 173 (1962).
9. Riedel, D., *Melliand Textilber.*, **47**, 1403 (1966).
10. Riedel, D., *Melliand Textilber.*, **45**, 1271 (1964).
11. Sustmann, Cl., *Z. Ges. Testilind.*, **65**, 368-376 (1963).
12. Sustmann, C., *Z. Ges. Textilind.*, **65**, 622-627 (1963).
13. Sustmann, Cl., *Z. Textilind.*, **67**, 901 (1965).
14. Sustmann, Cl., *Melliand Textilber.*, **38**, No. 4 452 (1957).
15. Sustmann, Cl., *Textilveredlung*, **13**, 387 (1958).
16. Sustmann, Cl., and Zahn, H., *Z. Ges. Textilind.*, **65**, 286 (1963).
17. von Bergen, W., Clutz, C. S., and Macaluso, C., Forstmann Woolen Co. Laboratory files.
18. Zanke, P., *Melliand Textilber.*, English Ed., **41**, 138 (1960).
19. *International Textile Bulletin*, Dyeing/Pr./Finishing, **2** (1967).
20. *International Textile Bulletin, Dyeing & Finishing*, **3**, 263 (1967), Part I, p. 14; **1**, 263, Part II (1968).
21. *International Textile Bulletin, Dyeing & Finishing*, **1**, 104; **2**, 207; **3**, 207 (1967).
22. *Wool Science Review*, No. 8, 61 (1952).
23. Sustmann, Cl., *Wool Science Review*, No. **19**, June 1961.
24. Bauer, J., Personal Communication, 1968.

Part II: Special Wool Finishes

Introduction

The most important developments that have occurred since 1950 are in the field of improved serviceability such as Wash-and-Wear—Durable Creasing and Permanent Press. The advent of wool and synthetic fiber blends in fabrics which could be permanently creased and pleated was the stimulus for the research work on finding methods whereby all wool fabrics could be given similar properties.

Treatments were developed for wool such that pleating or other fabric deformation could be obtained of a stability equal to that of the thermoplastic synthetic fibers. Similar processes were applied as a flat-setting procedure yielding fabrics of improved crease resistance.

In all these treatments, recognition has been given to the important role of fabric structure on the ultimate performance of the finished garment. In addition, the treatments have been utilized for special finishes, such as lustering, embossing, and permanent stretch, which have resulted in a range of fabrics not only of new design but of higher utility and beauty.

The further successful development and application of fiber utilization, fabric design, and finishing processes to the production of wool garments that are of improved wearability has become the vital goal of the wool industry.

Another area which has proved fertile for development is that of multi-purpose treatments. Sironizing is such a process where flat setting is carried out on shrinkproofed wool fabrics. The incorporation of moth proofing agents in either aqueous or solvent bath of a shrinkproofing treatment will convey durable moth resistance to the treated wool. Similarly the combination of flat setting, water and oil repellent as well as shrink resistance improve further the serviceability. The incorporation of insecticides in a solution of a water repellent can be successfully done.

Permanent Setting of Wool

The first investigation of the chemical reactions associated with imparting permanent set to wool was made by Speakman (25,26) as referred to in *Wool Handbook*, Vol. 1, pp. 252-253. But it took over 20 years before the value of imparting a permanent set to wool was recognized and practical processes were worked out.

The main fundamental studies of the chemistry involved were made by the members of the staff of the C.S.I.R.O. Wool Laboratories, Division of Protein Chemistry—Maclaren (18), Caldwell (3), and others—and the basic studies of practical processes by members of C.S.I.R.O. Wool Laboratories, Division of Textile Industry—Farnworth (10), McPhee (19), Lipson (17), Delmonico (8), and others. What takes place in the setting and creasing of wool is summarized in the C.S.I.R.O. Wool Textile News as follows.

Wool fibers are naturally elastic, but can be made plastic. Elasticity is desirable so that the wrinkles and crumples of garments will quickly disappear after wear. Plasticity is needed when setting pleats or creases in a garment. What is needed is controllable plasticity. This enables wool materials to be shaped in the desired way. Afterward the new shape should remain when the garment is worn. The result can be more or less permanent depending on how the garment has been processed. The forces that oppose alterations of shape are the chemical bonds that hold the protein chains together. The elastic character of wool is determined mainly by chemical bonds of several kinds holding the protein chains together. The two most important in setting wool are a large number of weak bonds known as hydrogen bonds and a small number of strong disulfide crosslinks.

Swelling wool fibers in water breaks hydrogen bonds. Any latent stresses present are released and the fibers tend to return to their original shapes. Raising the temperature provides more energy for bond-breaking, so that hot water is even more effective than cold water in breaking hydrogen bonds. This is the mechanism that allows damp wool to be given a temporary finish with an iron or press. Hydrogen bonds are broken by the moisture and heat so that the wool structure can be reshaped by the mechanical action of the iron or press. Simultaneously, the heat dries the wool and new hydrogen bonds are formed in the wool structure as the water escapes as steam. These new hydrogen bonds maintain the wool in the shape so long as the humidity remains low. In high humidity, or if the wool is damped with water, the new hydrogen bonds are broken and the molecular structure of the wool reverts to its former shape for the reason indicated above. This is why garments shaped by ironing lose their creases and flatness, and show relaxation shrinkage on wetting. To obtain a permanent set or finish it is necessary to rearrange the disulfide crosslinks contributed by the cystine in wool to maintain the desired shape even when wet. To do this the original disulfide crosslinks must be broken and reformed in a new configuration. Basically it depends on a series of bond interchanges each taking place between a disulfide crosslink and a free thiol group which must be in the ionized state.

Virgin wool does not contain many thiol groups so there is not much opportunity for the thiol–disulfide interchange to take place. The interchange is facilitated by generating thiol groups by splitting disulfide crosslinks between the two sulfur atoms. This can be done by a reducing agent such as ammonium thioglycollate. Si–Ro–Set is based on this reaction. The formation of thiol groups and the thiol–disulfide interchange is illustrated diagrammatically.

The more thiol groups there are available, the more easily the interchange takes place in either direction. Thus, not only is set facilitated by increasing the number of thiol groups, but it can be made more permanent by destroying the unwanted thiol groups after the wool material has been molded and set into the desired shape. This can be done by oxidation with, say, peroxide. It converts the thiol groups to oxidized forms, rendering them inoperative.

Permanent Creasing and Pleating

McPhee (19) stated that the interest in permanent creasing and pleating of 100% wool garments was greatly stimulated a few years ago by the introduction of blends of wool with synthetic fibers. Such fabrics can be given permanent creases or pleats by steaming

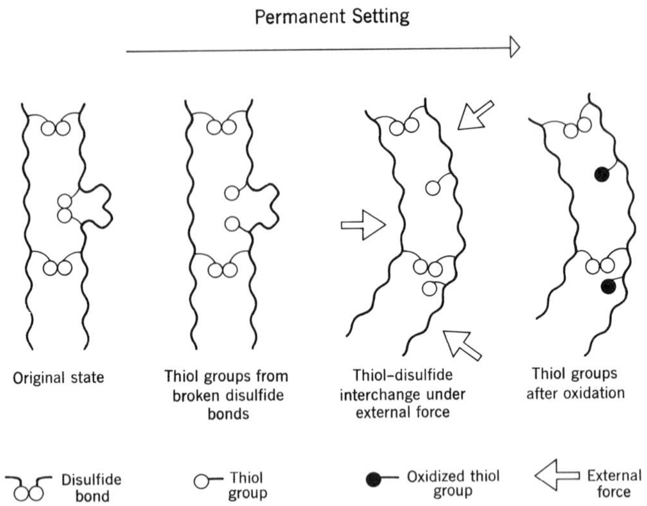

Diagram illustrating the thiol–disulfide interchange in the setting of wool. For simplicity, short segments only of two keratin chains are shown and only one interchange.

techniques very similar to those used conventionally for pressing trousers or pleating skirts. A challenge was thus presented to wool interests to devise, and introduce on a large scale, methods for producing similar effects in all-wool garments.

The first industrial process for producing these effects was the Immacula finish invented by Speakman (27) (bp 775,486) and described by Wolfram and Speakman (32). It is based on the use of sodium bisulfite, a process of presensitizing the wool fabric by treatment for 15 min in a cold solution of 2–3% sodium bisulfite and the permanent creases desired are obtained when the garment is steam pressed in tailoring.

The Si–Ro–Set process, an Australian development by Commonwealth Scientific and Industrial Research Organization CSIRO, was introduced in 1959 (33,34). The principle of permanent creasing or pleating is the wool, while held in the required form, is given a permanent set by the simultaneous application of heat, water and a chemical reducing agent. In the Si–Ro–Set process, an aqueous solution of a reducing agent is sprayed on to the appropriate areas of the finished garment and steaming carried out immediately while the fabric is still wet. The reducing agent can be ammonium thioglycollate, sodium bisulfite or monoethanolamine sulfite (MEAS) and, from the technical point of view, it is immaterial which reagent is used. Pardo and Foster (20) suggested the use of MBA (*N,N′* methylene bisacrylamide) (U.S. Patent 3,216,781). In practice, the choice of chemical depends on the availability of a suitable, stable concentrate, the price, and the service given by the chemical supplier. Because of these factors, different reducing agents are used in different countries, and even in different areas within a particular country.

Other studies covering the development of methods for durable creases were made by Davidson and Howitt (7). They found that the addition of Urea in concentration of 6–12% increased substantially the crease retentive properties imparted by the bisulfite, Cednäs and Karrholm (4) made a more fundamental study of different stages in the setting process using concentrated solutions of urea–bisulfite. After bisulfite or MEAS (low concentration) steam setting, no careful rinsing is required. By urea–bisulfite or MEAS (high conc.) setting, rinsing is necessary, and the cloth must be kept in a flat state.

The Si–Ro–Set process was first used on a large scale in Australia and, on a population basis, more permanent creasing and pleating of wool garments (more than 1 million annually) takes place there than in any other country. The use of the technique is however increasing continually in USA, Japan, and Europe.

The details of the technique as used in the USA by manufacturers of men's slacks and suit trousers are given in the WB-4 report of the Wool Bureau Inc. (35). In the United States in 1967, approximately 5½ million slacks and trousers have been treated.

An alternative method to the Si–Ro–Set process is the technique of "presensitizing" developed in the USA by the Wool Bureau Inc. The reducing agent is applied to the fabric in the finishing mill and the permanent creases or pleats are then produced, after the garment has been made, by adding water just before steaming. The relative merits of the two procedures are best compared by considering separately the manufacture of trousers and skirts.

For trousers, the Si–Ro–Set technique appears to be a more useful way of producing permanent creases than the use of presensitized fabrics. Here, the extra cost of presensitized suit fabrics is wasted for the cloth which is made into jackets, while the suggested use of damp cloths on the press as a means of adding the requisite amount of water to the trousers is very uncertain in practice. In order to obtain the best results consistently with presensitized fabric, it is necessary to spray a minimum of 20% water on to the trousers immediately before pressing. The men's clothing industry has, therefore, generally preferred the Si–Ro–Set process.

For permanent pleating of all-wool skirts presensitized fabrics are preferred. The required pleating technique is described in WB-5A report of the Wool Bureau Inc. (36).

The actual technique of presensitizing is very simple. Generally, the fabric is impregnated with a 2% solution of sodium bisulphite, then tentered at the usual temperatures and given a light open decating in the normal way. Other procedures used involve impregnation with monoethanolamine sulfite (MEAS) (WB-5 report of the Wool Bureau Inc.), or the treatment of cloth with either of the reducing agents in a dolly washer or winch-dyeing machine, then centrifuging or squeezing before drying.

The disadvantage of these particular creasing and pleating methods is the necessity for free water to be present during steaming. To overcome this disadvantage Cook and Delmenico (5) found a dry setting technique based on the use of a reducing agent/urea combination.

Pretreatment consists in padding the otherwise finished fabric through a solution containing the reagents and a wetting agent so that the wool retains 5% urea and 2½–4% diethanolamine as the carbonate. This is followed by tenter drying and, preferably, hydraulic pressing. At a later stage, the fabric of normal regain can be creased, pleated or otherwise set simply by steaming.

Another technique is the setting of creases or pleats in all wool fabrics by high pressure steam without the use of reducing chemicals. Davidson (6) found that durable creases in all wool trousers can be produced by steaming the neutral fabric for 10 min at 120°C (248°F). The loss in breaking strength is around 4% Kopke and Lindberg (14) developed a method for producing permanent pleats by steaming the dry fabrics in an autoclave at 132°C (270°F) for 25 min. The degree of setting is influenced only slightly by variations in cloth construction, finishing routine, pH and moisture regain. However, cockling in the pleated material is found to be dependent on the moisture regain of the cloth measured before pleating and setting. The cockling is caused by hygral expansion—as higher the moisture regain, the less cockling. Most fabrics were found to be free of cockling when conditioned at 65% RH.

The high temperature setting causes a yellowing of the wool, which may change the shade of pastel and bright blue colors.

The reduction in the breaking strength found in eight panels was 11% for the dry strength, and 25% for the wet strength.

Further detailed information on the subject of hygral expansion and cockling are given in Chapter 16.

Flat Setting

The principle used for setting wool fabrics in creases or pleats has been applied to the setting of full pieces of fabric as a new finishing operation. This new finish is called "flat-setting" and by now has been given about a dozen different trade names round the world.

Lipson and McPheen (17) in recent papers give the following details on this wool finishing procedure. Flat-setting of wool fabrics with sodium bisulfite solution results in an improved wet wrinkle-recovery, which helps to maintain a smooth surface appearance after wetting and air-drying. Flat-setting combined with shrink-resist treatments therefore gives a method of producing washable, minimum-iron fabrics. Flat-setting also gives a luster and smooth handle to the fabric, and flat-set fabrics are easier to tailor than normal fabrics. For these last-named reasons, flat-setting is becoming a common finishing procedure in many countries.

As with all chemical treatments of wool, flat-setting must be carefully controlled for satisfactory results, free from undesirable side effects, to be obtained. The recommended method for flat-setting is that the fabric is impregnated with a 0.5–1.0% solution of sodium bisulfite and steamed for 2–5 min. The pH of setting is important. Below pH 4.5, the degree of setting is usually inadequate, and, above pH 6.0, stiffening of fabrics and yellowing of the wool can occur. The pH of an aqueous extract of the fabric should be in the range 4.5–6.0 before impregnation with bisulfite takes place. This is necessary, even though the pH of the bisulfite solution is about 4.5, since a simple impregnation does not necessarily bring the pH of an aqueous extract of the fabric to this value. For example, the pH of an aqueous extract of a fabric may be lowered only from 6.8 to 6.2 by flat-setting, whereas a fabric whose aqueous extract is pH 9.2 can still have an extract of pH 8.0 after setting with sodium bisulfite.

Setting with Monoethanolamine Sulfite Solutions

When monoethanolamine "sulfite" or "bisulfite" solutions are used industrially for flat-setting, adjustment of fabric pH prior to impregnation is particularly important in this case, since the "sulfite" solutions are prepared with pH values of 6.0–7.0.

If concentrations of reagents and pH adjustments of fabric and solutions are accurately controlled, results obtained with monoethanolamine sulfite–bisulfite solutions and sodium sulfite–bisulfite solutions are identical.

In practice, the technique can cause difficulties unless the mill is reorganized to carry out the process on a reasonable scale, when it can become very cheap and simple. The only technical problem in industrial application is the steaming of the wet fabric, since this means that the wrappers on the blowing machines quickly become wet. Most mills carrying out a lot of flat-setting have at least two wrappers available for each machine, scouring and drying them at regular intervals, while some use specially woven lightweight wrappers made from synthetic fibers.

Flat-setting was first used industrially as part of the Sironizing processing. In this case, flat-setting is carried out on shrinkproofed wool fabrics to give the cloth the ability to retain a smooth appearance during washing and, in Australia, this is still the main use of flat-setting. Flat-setting is a very effective way of crabbing and can be used to simplify and shorten finishing routines. Some European mills now flat-set fabrics straight from the loom, then scour, tenter and give a light steaming. This gives fabrics that are quite acceptable for many uses and, when the necessary reorganization has been carried out, can significantly lower costs.

Flat-setting is now a common finishing technique in the U.K. and Europe, being used mainly because of the luster and improved handle which is given to worsted fabrics.

Sutcliffe (30) found that the combination of flat-press characteristics with water repellency assists the resistance to wrinkling.

In combination with the Cravenette Corp. a silicon compatible with monoethanolamine sulfite (Oncet BW) was developed. The final formula consisted of

5%	Silicone 300
2.5%	Catalyst 54
5%	Oncet BW
0.5%	monoethylamine boron trifluoride
50	gal volume

The padded fabric with a 70% pickup was steamed in a semidecator for 5 min (steam pressure 80 lb) and vacuum extracted for 7 min and then dried. After drying the cloth was finished normally ending up with a treatment on the semidecator for 3 min steam, 5 min vacuum.

Permanent Press

The ability of a garment to remain pressed or in other words has a lasting good appearance is a very desirable attribute. This has become an important feature. Today's concept of permanent press in garments involves the smooth pressed condition which, together with sharp creases and pleats, last through repeated washings and tumble dryings so that the garments may be worn without ironing. Other terms used for this feature include durable press, press free, washable non-iron, wash-and-wear, and minimum care.

Permanent pressed wool fabrics as reported by Farnworth (10) and McPhee (19) can be produced by combining any chemical shrinkproofing treatment with a flat setting. Such fabrics were first introduced in Australia, licensed under the trade name Sironized. Today such fabrics are produced on an increasing scale worldwide based on continuous processes as described under shrinkproofing and flat setting.

Lustering of Wool Fabrics

The demand for wool fabrics that have a high luster is not new. It is well established that careful selection of raw materials, such as luster wools and specially hair fibers such as Mohair and Alpaca still play an important part in this field. Of the blends the oldest to achieve a permanent luster are the silk-wool combinations. In many of the modern wool-type fabrics good luster effects are obtained by the addition of lustrous man-made fibers.

In recent years as brought out in the literature by Whewell (31), Scott (23), and many others, there has been an increasing amount of attention devoted to the production of lustrous wool fabrics. The particular emphasis at the present time is placed on the luster produced during the wool finishing, with the requirement that these finishes should also be permanent. The latter feature has posed, and still does, particularly tough problems for the cloth finisher. High luster finishes on woolens and worsteds have been readily attainable in a variety of ways, but to make such finishes stable to heat and steam treatments, such as they have to undergo at the hands of the spongers and garment manufacturers, has perplexed many wool manufacturers.

Luster on Pile Fabrics

Luster on pile fabrics in finishing can be obtained in two ways: Mechanical process such as teazle raising and laying the surface fibers in one direction and specific methods to flatten the cloth such as wet decating, blowing, rotary and hydraulic pressing, previously discussed, are ways to obtain appreciable luster. Chemical processes in the main, are related to mechanical processes such as raising, wet decating and blowing but combined with a pretreatment of the cloth with chemical setting agents such as sodium sulfide or monoethanolamine bisulfite.

The biggest change in the finishing of luster pile wool coating fabrics has come about through the adaption of the method and machinery developed years ago by the fur industry to convert sheepskins into the fur known as mouton, as stated by Frishman (11). The firm which pioneered in the development of the machinery and a new knitted deep pile fabric of man made fibers was the Borg Electronics Corporation. The wool finisher was very slow in adopting this method simply because of the low permanency of the pile. But today there is hardly any wool finishing plant without such a machine. The process is known under various names such as "electrifying," "polishing," "ironing," "electropolishing," "glossing," or "lustering."

Machines are built by many machine manufacturers such as: The Electro Finisher by Turbo Machine Co., and the Egan Electric Ironing Machine in the United States, the Pol-Rotor by Heinz Hergert of Frankfurt, Germany, the glossing machine by T. Mortamet, Villeurbaune, France and the unit by Comet, Prato, Italy.

A schematic drawing of one type of electrifying machine is shown in Fig. 48. Frishman (11) gives the following general description:

"The fabric is carried through the machine by a series of rolls designed to maintain constant fabric tension and to eliminate wrinkles. An apron brings into contact the pile fabric and a heated cylinder rotating at about 1000 rpm. The surface of the cylinder or roll has long diagonal grooves. Each groove has on its leading edge a strip of metal ('knives' in Fig. 48) which is serrated or 'file cut.' "

Figure 49 shows an electrically heated machine. Actually, there are several variations of this machine and in practice they are used in many ways. Cylinders are heated with gas or electricity; grooves extend the full width of the cylinder or they are shorter and placed at steeper angles; grooves are shallow or deep; the cylinder rotates

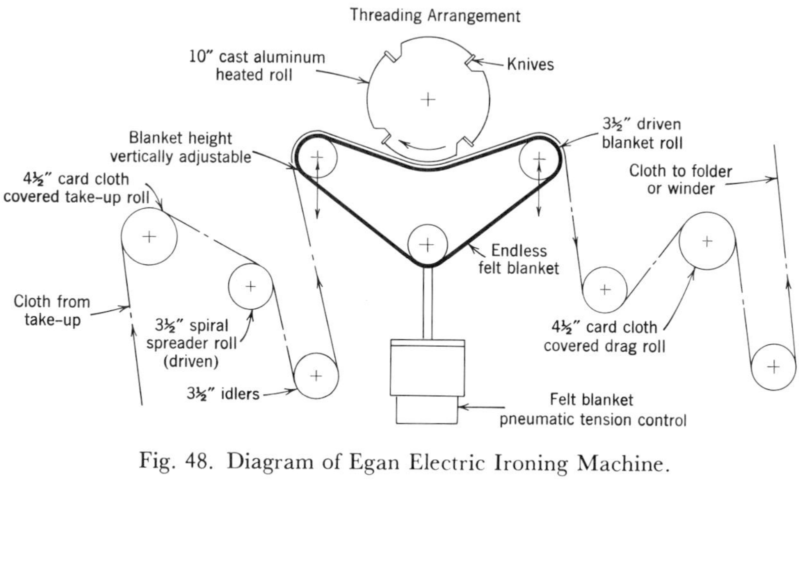

Fig. 48. Diagram of Egan Electric Ironing Machine.

Fig. 49. Photograph of the Electrifying Machine.

either in the direction of the cloth travel or in the opposite direction, with the grooves appropriately placed in either case. Cylinders range from 10–20 in. in diameter; cylinders are made of steel or aluminum but in the latter case the edges of the grooves are steel. The apron which supports the fabric may be free turning or driven; fabric speed is varied from 2–15 yd/min; temperatures are varied from 180–500°F, depending on the type of fiber, speed of the fabric, and the effect which one wants to achieve. The apron is either maintained at a fixed pressure against the cylinder or at a fixed spacing. Cylinder speed varies from 800 to 1,500 rpm.

Lustering machines are designed to remove fiber crimp and to disentangle the fibers from each other. This is done by the rubbing and beating action of a heated, rotating, grooved, metal cylinder which is brought to bear against the pile surface. The friction between the smooth sections of the cylinder and the pile surface causes the fibers to be pulled straight while the alternating grooves (assisted by serrated edges) on the surface of the cylinder have a combing action, untangle the fibers from each other, and tend to make the pile fibers stand perpendicular to the base. Figure

50 shows the change in the position of the wool fibers forming the pile before and after the passing under the cylinder.

The grooves on the surface of the cylinder are angled so that they meet the fabric surface alternately at an angle of about 15° and then at about 165°, with respect to an imaginary line drawn across the width of the fabric. The alternating, angled grooves obviously throw fibers from side to side, and this apparently assists in the overall action of the grooves. While the mechanical action is the same on all types of fiber, the mechanism whereby the fibers are fixed in the uncrimped condition varies, depending on the chemical and physical properties of the fiber.

Fig. 50. Change in the Position of the Wool Fibers before and after Passing through a Lustering Machine.

Mechanism

In the case of wool and other animal fibers, straightening takes place in the presence of formaldehyde which, when heated (the cylinder is 400–500°F) it was presumed reacts to permanently set the fibers in their uncrimped state. Formaldehyde forms molecular crosslinks and reacts with wool protein to reduce the tendency of the fibers to swell in the presence of moisture and thereby inhibits the reversion to a crimped condition. Following the "secret" formula (2) of the mouton fur finisher, alcohol (methyl, ethyl, or isopropyl) was added to assist in the penetration of the formaldehyde, and a strong organic acid, formic acid, as a catalyst. Today lustering agents such as the Pol-rotor Fix agent and Wool lustering agent Ciba described by Bitterli, are available for that specific purpose. The testing of furs and woolen coatings revealed that premoistening with lustering agents applied through rotating brushes was not sufficient to make such a luster stable to heat and steam treatments. Not only is it essential to make use of chemical setting agents such as sodium bisulfite or monoethanolamine bisulfite, or thioglycolic acid, but the cloth must be subsequently well blown while still damp.

Two recent patents in this field are French Pat. 1,123,848 (1956) which stipulates pretreatment of the cloth in an aqueous solution of 0.1–5% ammonium thioglycolate, extracting and fixing the fibers by ironing. The ironed and dried fabric is then treated with a water proofing agent which is soluble in volatile solvents. The U.S. Pat. 3,051,544 (1962) achieves a highly lustrous finish, stable to steam processing on raised cloth by brushing with 1.5% a monothioglycerol solution at a pH of 7, followed by treatment on a lustering machine. The cloth is then treated with 3% formaldehyde solution at pH 3.8 and passed again through the lustering machine.

Luster on clear finished worsteds

Permanent luster finishes on worsted are best achieved by the combination of plate pressing and full decating which as Scott (23) claims has brought about a revolution in worsted finishing standards. Basically, the plate press produces a beautiful luster, and the full decator renders it permanent.

The effectiveness of the full decator is considerably improved by the use of high pressure steam.

This led to a change in the construction of autoclaves which are able to stand pressures as high as 60 psi. The first American built machine in the early 1960's was the Pressure-Matic Full Decator built by Gessner and described by Herard (12), shown in Fig. 17.

Fabric evaluation showed that process cycles set in the 20–30 psi range produced a finish that had long been desired together with the highest possible dimensional stability, 1% or less residual shrinkage remaining.

Tensile strength will be sacrificed in the high pressure decating and also some dyestuffs on fabric are prone to change shade.

In order to speed up the winding and rewinding of the rolls, Gessner designed a Turret Winding Machine (Fig. 51).

The Gessner Turret Winding Machine operates as follows:

Cloth to be processed runs onto a perforated cylinder located at station Number 1. As the cloth to be processed is being rolled at station Number 1, cloth that has already been processed is simultaneously being unwound and cooled at station Number 2. The apron winding onto the cylinder at station Number 1, is coming from the cylinder being unwound at station Number 2. When this cylinder comes from the chamber and is placed in station Number 3, the turret is indexed one station, and the process is started all over again.

The floor carriage that transports the cylinders from the winder to the chamber is composed of a bottom section that remains on the floor and a top section which accompanies the cylinder into the chamber.

The chamber is built in two sizes, PF-33 in. and PF-42 in. PF-33 in. will take wound cylinders measuring 28 in. in diameter. The chamber has a power lock door with safety interlock which safeguards the starting cycle, and the opening of the door at the end of the cycle. The processing steam can be varied from below atmospheric pressure to a pressure as high as 50 psi.

Versatility of finish extends from very light effects to permanent set which neither steam, nor pressing nor water will change.

Production of 1000–1500 yd/hr are obtainable with a single operator.

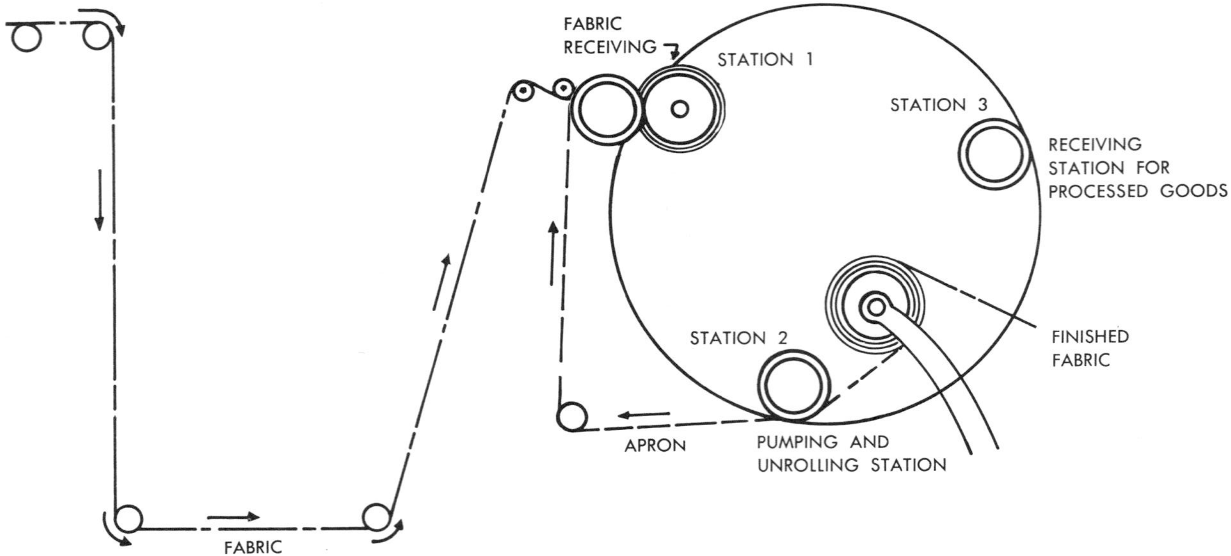

Fig. 51. Diagram of Gessner Turret Winding Machine.

Several machine manufacturers developed single luster decatizing machines, such as the one seen in Fig. 52. Sustmann (29) reported with steam pressure of 0.4 atm (6 psi) in the 12 in. decating cylinder different temperatures for both ends of piece, namely 108 and 100°C resulting in a difference of the press effect, specially noticeable in the thickness change, the outer layer was 18% thinner whereas the inner layer was 30% thinner than the untreated fabric. A satin leader cloth is essential for a good luster effect.

Lustering with chemical agents can be obtained by impregnating fabrics with a solution containing 1% sodium bisulfite, 37% ethyleneglycol and water before running it through a hot press or calender as suggested by Speakman and Wolfram (28). Another suitable bath is 40% ethyleneglycol, 2% thioglycolic acid adjusted to pH 9 with ammonia. The fabrics are treated full width for 20 min then squeezed leaving approximately 60% moisture in the fabric. The moist fabric passes between the heated calender rolls (180°) at a speed of 3 yd/min. Precised aligned roller pressure and temperature across the full width of the cloth will result in a uniform luster effect.

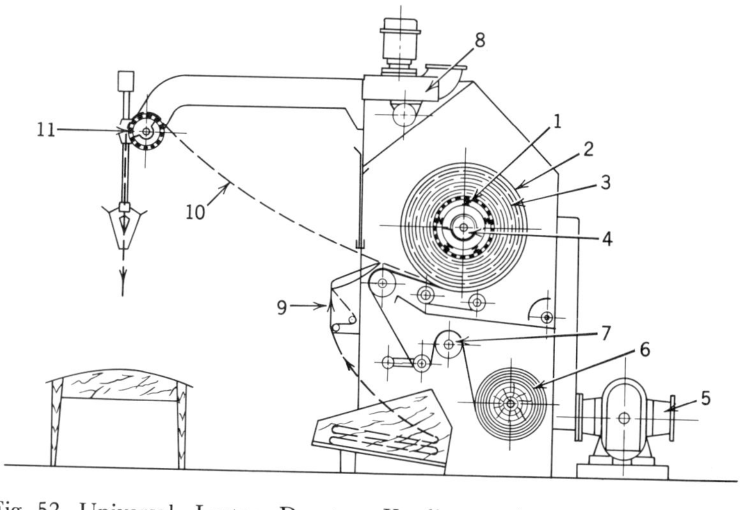

Fig. 52. Universal Luster Decator—Kettling and Braun. 1. Perforated Cylinder. 2. Satin Leader. 3. Fabric. 4. Steam Entry. 5. Suction Pump for Cylinder. 6. Leader Cloth Roll. 7. Leader Cloth Dryer Roll. 8. Suction Cooling Fan. 9. Fabric Feed In. 10. Fabric Unrolling. 11. Cooling Cylinder.

The principle of the above method has found further application in the production of washable seersucker effects in an all wool worsted fabric as reported by Sinus and Speakman (24). The impregnated fabric was fed to a heavy-duty compressive-shrinkage machine having a heated top roller grooved at right angles to the axis. The seersucker effect was produced by compressive shrinkage (12–13%) of the fabric where it made contact with the grooved roller and permanently set at the same time.

Stretch Fabrics

The need for form fitting garments from stretchable fabrics traditionally has been met more or less satisfactorily by Knitwear or by tailoring woven fabric on the bias.

As brought out by Keaton and Stuart-Smith (13) during the 1950's techniques were developed whereby normally flat rigid yarns of a thermoplastic nature, such as nylon and terylene could be bulked or crimped, thus imparting stretch properties to the yarns. These physically modified yarns have created a new fashion trend in both women's and men's wear fitting perfectly in the current vogue for casual or leisure wear fabrics.

The greatest success with such yarns in the woolen industry was achieved in the production of warp stretch ski trousers around 1960 from Helanca ski cloth. One of the leading fabrics consisted of a 100% nylon stretch yarn in the warp—2 ply 70s denier Helanca SPZ300 and 2/40s worsted in the filling weighing 19 oz per linear yard 60 in. wide. The warp stretch in the finished fabric ranging from 25 to 30%.

It was only natural that this created a demand for the development of all wool stretch woven fabrics. It is possible to produce wool stretch fabrics by various methods. Bilbie (1) outlined three of the basic methods which can be used.

Stretch Yarns

The best known technique is probably the use in weaving of a yarn which already has the power to stretch and recover. Although there are various ways of producing a stretch yarn, the most common method employed is that of twisting, setting and untwisting a twofold yarn. This method is similar to those which are used for thermoplastic synthetic fibers. The amount of stretch developed in the fabric is dependent upon the amount of stretch produced in the yarn, and also the freedom of that yarn within the fabric. To take as an example a fabric where this type of yarn is used in the warp; the yarn is woven under considerable tension which removes the crimp to a large extent, and the construction of the fabric has to be such that the crimp in the yarn can reestablish itself during scouring. The two main factors which influence the degree of stretch in the fabric are the length of float in the warp and the density of picking. The amount of stretch increases as the warp float increases and the number of picks is reduced, other factors being kept constant.

Inducing Yarn Crimp during Weaving

This method is used particularly on the continent in the production of warp stretch ski and "apres ski" trousers. One such method is to weave from two warp beams, the first warp being composed of worsted yarn and the second cotton (Vernier Patents: French 1,236,301 (1960), U.S. 3,189,972 (1965)). The yarn from the second beam is fed to the loom under considerably greater tension than the first.

After weaving, the fabric has to be set before the cotton yarn is removed. Setting of the wool component, in scouring and dyeing take place before carbonizing out the cotton, and only after carbonizing does the fabric exhibit the stretch quality.

Permanency of stretch is not as good as the Helenca type and weaving more difficult.

Yarn Crimp Development, Interchange and Chemical Setting

This method has been developed from experience gained by the chemical flat setting of conventional wool fabrics. This method of producing stretch fabrics is certainly cheaper than either of the other techniques and, as the process can be carried out as one of the last operations in fabric finishing, it is much easier to produce a fabric with consistent, stable dimensions, which in turn gives a consistent degree of stretch throughout the fabric.

Again, in this method of producing stretch fabrics there are many possible variations and refinements, Stevens Patent U.S. 3,077,655 (1963); U.S. 3,281,205 (1966), but basically an all-wool weft stretch fabric can be produced simply by taking the cloth in the scoured, dyed and dried stage, padding through a solution containing reducing agent, e.g., monoethanolamine sulfite, and feeding onto a decatizing machine (Fig. 53) or through an autoclave under warp tension. The fabric is steamed to complete the setting, cooled, removed and dried. Under these conditions there will be an interchange of crimp from warp to weft.

Cloth construction influences the degree and direction of the stretch which can be produced by this method, for example, in a cloth made with the same yarn in both warp and weft, if there are more ends per inch than picks per inch, warp stretch will develop, while if there are more picks per inch than ends per inch, weft stretch will develop. This stretch can be assisted and increased in magnitude by extending the fabric in the appropriate direction prior to setting.

Another arrangement to control warp tensions and stretch was developed by Krantz (15). It consists of two sets of three rolls each at two points over which the fabric passes and it stretched in the warp direction.

The Stevens-Runton process is a continuous operation, the machinery as seen in Fig. 53 consists of an immersion tank followed by squeeze rolls, an autoclave for steaming followed by a rinsing unit with squeeze rolls. The following example given in the patent illustrates the working of the process. Four pieces of all wool fabric were immersed in a 3% aqueous solution of monoethanolamine sulfite at pH 7.1 containing a wetting agent, squeezed and then steamed for 2½ min in the autoclave at temperatures between 212–218°F (100–103°C) then passed through a 0.2% formic acid bath. The autoclave is kept under slight pressure above atmospheric which in conjunction with seal prevents the entrance of air. The fabric is unconfined laterally so that it is free to shrink in the width direction. The warp tension is kept constant by the control roller units on both ends of the autoclave. The amount of filling stretch imparted to the fabric is preferably from 6% to about 25%.

Finishing Helanca Ski Cloth

Finishing procedures for Helanca type warp stretch ski cloth vary considerably from mill to mill. The two examples given below illustrate this.

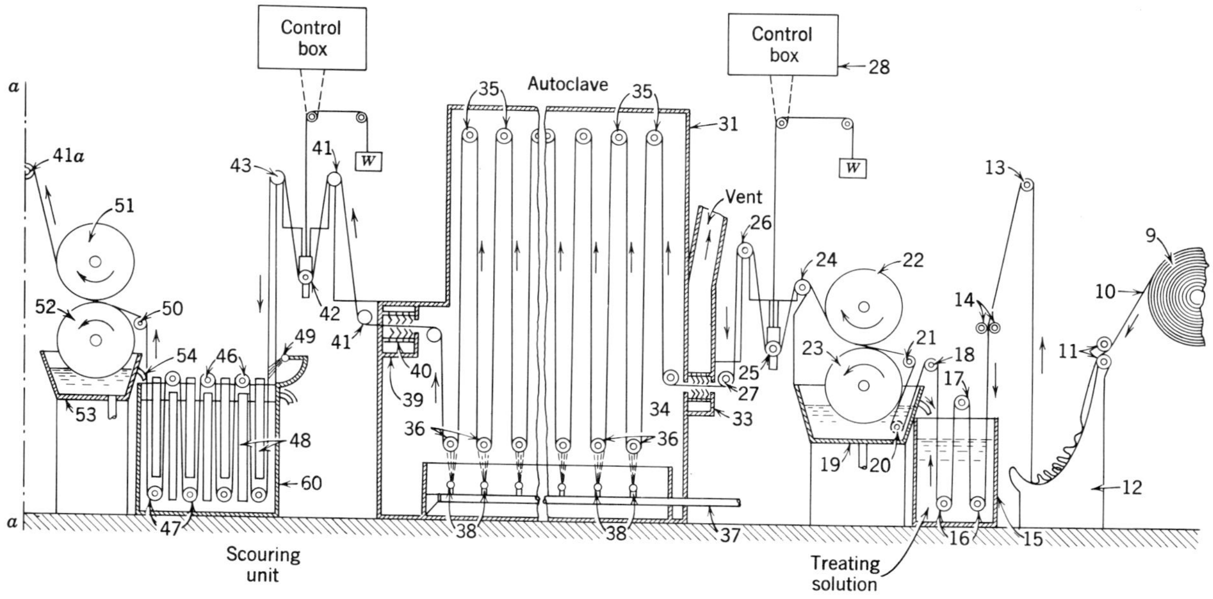

Fig. 53. Processing Range for Filling Stretch Wool Fabrics.

Mill 1. Five hours broadwashing in special machine, squeezed and rolled. Dried on tenter frame with 20% overfeed. Leader decated in high luster decator two minutes full steam, cooled, tacked and beck dyed, broadwashed combined with water repellent treatment, squeezed, dried, brushed and steamed, sheared and leader decated.

Mill 2. Continuous crabbing, 140–180°F slack dried, crabbing at boil, cooled, slack dried, tacked and dyed in beck, dried with 18% overfeed, sheared, water repellent treatment dried and cured 270°F open decating. A good summary of methods for producing wool stretch fabrics is given in *Wool Science Review* No. 31(39).

Water Repellent, Stain Resistant Treatments for Worsted and Woolen Fabrics

There are three distinct classes of water repellent compounds. The renewable, semidurable and durable.

The *renewable* water repellents are aqueous dispersion containing aluminum acetate or formate emulsifying agents combined with waxes such as paraffin and Japan wax. They do not stand dry cleaning.

Semidurable water repellents are supplied by chemical manufacturers in the form of two separate materials which are mixed together immediately before use. One of the materials usually consists of a combination of water soluble soaps and the other solution of zirconium salts. This type of product generally lasts through 2–3 dry cleanings.

Durable water repellents include the silicones certain chromium compounds and fluoro chemicals. Even though they are called "durable" they generally break after 6–7 dry cleanings, occasionally may last 10 dry cleanings.

Silicones

For waterproofing of woolen and worsted goods, the silicones of Dow Corning and General Electric have found the most use. For the two methods of application available, there is (*1*) the exhaust method, for which Dow Corning's No. 61 is a suitable product, and (*2*) the pad application, with Dow Corning's No. 71.

One of the interesting developments in the last years has been a combination treatment of mothproofing and waterproofing. Dow

Corning's Bulletin No. U8-247 of April, 1960, gives details on experimental runs with such combinations. The main difficulty run into is the compatibility of the mothproofing agent with the silicon Syl-mer emulsion. Special consideration must be given to the pH of the catalyst. For example, the 72 catalyst being on the alkaline side, will not go together with the strong acid solution necessary for the pickup of the Dieldrin, such as Drewclad, whereas Syl-mer 71 catalyst with a pH between 3 and 6 is excellently suitable.

The recipe for the pad application of Syl-mer 71 is as follows:

The pieces are padded with a solution of 5% Syl-mer 71 and 0.5% Syl-mer 71 catalyst. From the padder the pieces run directly into the tenter dryer and exposed to a curing temperature of approximately 250°F.

United States Patent 3,247,281 was granted to Gagliardi, assigned to Union Carbide Corp. for more effective and nontoxic catalysts for use with silicone water repellent composition.

Chromium Compounds

DuPont introduced in 1951 a Stearato Chromic Chloride under the name of Quilon, a water repellent developed primarily for the treatment of paper products and leather.

Applied by the pad method it imparts also good water repellency to woolen and worsted fabrics. Quilon is supplied as a 30% solution of Stearato chromic chloride in isopropanol pH 2.5. It is applied just before the final drying from a 2% solution on a padder having a pickup of 80% at low roller pressure (40 psi) and at a speed of 15–25 yd/min in order to deposit 0.5% solids of active material on the fabric. The cloth should then be dried at temperatures between 220–240°F.

The usefulness of fatty acid chromium compounds for water repellent treatment of wool fabrics was reported by Engler and Schefer (9).

Fluorochemicals

With the introduction of the Scotchgard brand stain repeller fluorine containing polymers by the Minnesota Mining & Manufacturing in 1957 C° (3M), the textile industry has products available which in addition to good water repellency have the additional quality of oil repellency. Of the various Scotchgard textile chemicals type FC 208 is specially suitable for woolens, worsteds, and wool synthetic blends.

Textile Chemical FC 208 is a non-ionic compound of a fluorochemical resin (fluoroalkyl acrylate) with the following characteristics:

Milky liquid with a minimum solid content of 28%. Typical analysis: Solids 28%, Acetone 18%, Ethylene glycol 9%, and 45% water. The fabric to be treated should be as free as possible of processing agents such as alkalies, soaps and synthetic detergents since these materials affect (*1*) the absorbency of the fabric, (*2*) the penetration of treating bath into the fabric, or (*3*) the water and oil repellency and durability properties of the resultant finish. The use of silicones, and silicone containing antifoams and lubricants must be avoided, since these severely affect oil repellency. A fabric pH between 5.5 and 6.5 is recommended.

The following formulas are suggested by the 3M Company.

1.	Isobutanol	4.00%
	Phobotex f/t/c (Ciba)	0.50%
	Catalyst RB (Ciba)	0.38%
	Sodium acetate	0.10%
	F C 208	3.00%
2.	Isobutanol	4.0%
	Argus DWR Special (Argus Chemical)	2.0%
	F C 208	3.0%

The application of the above aqueous chemical solution to the fabrics involves the following steps:

1. Padding. Padding can be accomplished on regular two roll padders, consistent with usual procedure for applying stabilizing or shrink and crease resistance resins. A wet pickup of 50–70% is desirable. Bath temperatures of 60–100°F are suitable. Experience has shown that on wool fabrics a minimum level of approximately 0.4% fluorochemical resin solids based on the weight of the fabric is normally necessary to obtain Scotchgard brand stain repeller properties. A 50% pickup gives 0.45% solids on the cloth.

2. Drying. Drying temperatures are not critical, may vary between 200 and 300°F (94–150°C).

3. Curing. Curing is accomplished at temperatures between 300–350°F (150–176°C) a minimum cure of 2–3 min is suggested.

4. After washing is generally not required, but is desirable when combinations with crease resistant resins are employed.

In 1967 textile chemical FC 217 was introduced by 3M which gives woolen fabrics an even higher drycleaning durability than FC 208.

DuPont recently introduced a fluorochemical compound under the name of Zepel B for use to impart water and oil repellent properties to textiles. The product, a cationic type, in emulsion form has 13% solids.

For woolens the following formula is suggested: Zepel B 3.5%, Phobotex f/t/c 0.56%, Catalyst RB 0.14%, TLF 1922 0.28%.

The product is most frequently applied by padding at room temperature followed by drying on a frame and curing conditions are about the same as for Scotchgard. A good curing temperature is 340°F (171°C) which will vary from 1 min or less for lightweight fabrics to 3 min for heavy fabrics.

Pittman and Wasley (21,22) found that certain fluoropolymers not only make wool fabrics more stain repellent but also increase their dimensional stability and improve their resistance to bleaches and alkali. They discovered in the 1-(perfluoroacyl) aziridines, a new group of compounds which when applied to wool either in the monomeric state or as a preformed polymer gives a finish which has good oil and water repellency and exhibits some stabilization to felting shrinkage.

Mothproofing

Of the huge number of chemicals claimed as mothproofing agents in the patent and scientific literature, which have already been referred to in *Wool Handbook*, Vol. 1, p. 299, only three or four have been used to a significant extent over a long period in industry, and all of these are organochlorine compounds. The important European work on organochlorine insecticides, which led to the discovery of DDT and of compounds applied specifically for mothproofing (e.g. Mitin FF), has been described in detail

(Läuger, 1944 (16); Läuger et al., 1944 (16)) and makes most interesting historical reading. Other compounds used only for mothproofing are chlorinated, aromatic sulphonamides, known as Eulan U33 and Eulan WA new.

Since the early 1940's, many organochlorine compounds have been manufactured for use as general insecticides and of these Dieldrin has proved to be outstanding as mothproofing agent.

A good summary of the present situation in the mothproofing of wool was given in *Wool Science Review*, Nos. 27 and 28 (38). Extracts from these two articles are given herewith.

These compounds are most easily applied from acidified, aqueous solutions at the boil, when they are sorbed by wool in much the same way as acid dyes. The sulphonic acid groups in the molecules of some of the compounds give water-solubility but also greatly increase the affinity for wool so that, even though the compounds are initially soluble in water, they are removed only very slowly from treated wool and have the fastness properties of good wool dyes.

All of these compounds come very close to being ideal mothproofing agents. Thus, they are all sufficiently toxic to wool-digesting insects to warrant industrial application, they are resistant to repeated washing, drycleaning, light, rubbing, and pressing, etc., are odorless and colorless and produce no undesirable change in the properties of wool. They can be applied easily during some other normal wet process by exhaustion and, in some cases, continuous techniques and can be estimated both on wool and in solution by reasonably simple analytical methods. A disadvantage of Mitin FF and the Eulans is relatively high cost which so far has limited their use, while a disadvantage of Dieldrin is lack of wash-fastness for articles which would be laundered frequently. However, the very low cost, versatility and simplicity of treatment with Dieldrin has allowed its much wider use for general industrial mothproofing.

Over the past two or three years, the use of such compounds has come under close scrutiny, since another requirement of an ideal mothproofing agent is that it should be harmless to mammals at the concentrations employed.

The danger of *acute* poisoning from wool treated with organochlorine compounds is negligible. For example, with Mitin FF at 1% on the weight of wool, a child of 5 kg body weight would have to ingest at once all the Mitin FF from about 60 g wool before any acute toxic effect would be shown (assuming a toxicity for humans equal to one-tenth the LD_{50}, oral, of rats, which is about 1,100 mg/kg body weight); with Dieldrin, at 0.05 per cent on the weight of wool, the figure would be 50 g wool (LD_{50}, oral, rats about 50 mg/kg body weight). In spite of the obvious lack of danger in this regard, Dieldrin has not been recommended for use on baby wear or underwear in order to eliminate even this remote chance.

References

1. Bilbie, H., *Textile. Inst. Industry*, **2**, 10 (1964).
2. Bitterli, W., *Ciba Rev.*, **6**, 32 (1962).
3. Caldwell, J. B., Leach, S. J., Meschers, A., and Milligan, B., *Text. Res. J.*, **34**, 627 (1964).
4. Cednäs, U. M., and Karrholm, E. M., *Cirtel*, **3**, 647 (1965).
5. Cook, J. R., and Delmenico, J., *Cirtel*, **3**, 619 (1965). Cook, J. R., and Delmenico, J., *Text. Mfr.*, **91**, 424 (1966).
6. Davidson, A. N., *Cirtel*, **3**, 565 (1965).
7. Davidson, A. N., and Howitt, F. O., *J. Text. Inst.*, **53**, 62 (1962).
8. Delmenico, J., *Text. J. Australia*, **41**, No. 5, 14 (1966); *Text. Inst. Ind.*, **4**, No. 5, 139 (1966).
9. Engler, A., and Schefer, W., *Textil-Rundschau*, **13**, 643 (1958).
10. Farnworth, A. J., *Text. Res. J.*, **27**, 632 (1957); *Am. Dyestuff Reptr.*, **49**, 996 (1960).
11. Frishman, D., *Text. Industries*, **128**, No. 8, 68 (1964). Frishman, D., *Text. Industries*, **128**, No. 8, 70–80 (1964).
12. Herard, R. A., *Am. Dyestuff Reptr.*, **57**, 488 (1968).
13. Keaton, J., and Stuart-Smith, K., *Text. Inst. Ind.*, **2**, 198 (1964).
14. Kopke, V., and Lindberg, J., *Cirtel*, **3**, 605 (1965).
15. Krantz, H., *I.T.S. Dyeing & Finishing*, **2**, 8 (1965).
16. Lauger, et al., *Helv. Chim. Acta*, **27**, 71 and 892 (1944).
17. Lipson, M., and McPhee, *J. Text. Inst.*, **55**, T512 (1964).
18. Maclaren, J. A., Leach, S. R., and Swan, J. M., *J. Text. Inst.*, **51**, T665 (1960).
19. McPhee, J. R., *Text. Inst. & Ind.*, **1**, 10 (1963).
20. Padro, C. E., and Foster, R. E., U.S. Patent 3, 216, 781, November 1965, U.S.D.A.
21. Pittman, A. G., and Wasley, W. L., *Am. Dyestuff Reptr.*, **53**, 30 (1964).
22. Pittman, A. G., and Wasley, W. L., *Am. Dyestuff Reptr.*, **56**, No. 21, 808 (1967).
23. Scott, K. F., *Text. Industries*, **130**, 301 (1966).
24. Sims, D., and Speakman, T. B., *J. Text. Inst.*, **50**, T473 (1959).
25. Speakman, J. B., *J.S.D.C.*, **52**, 335 (1936).
26. Speakman, J. B., *Proc. Intern. Wool Text. Res. Conf.*, Australia, 1955, E 531.

27. Speakman, J. B., *B.P.*, **775**, 486 (1957) to Marks and Spencer, Ltd.
28. Speakman, J. B., and Wolfram, M. A., *J. Text. Inst.*, **49**, T627 (1958).
29. Sustmann, Cl., *Textilindustrie*, **65**, 372 (1963).
30. Sutcliffe, E., *Text. Industries*, **109**, (1961).
31. Whewell, C. S., *Text. Inst. Ind.*, **1**, 15 (1963).
32. Wolfram, M. A., and Speakman, J. B., *J.S.D.C.*, **76**, 169 (1960).
33. C.S.I.R.O. Division of Textile Industry, Report No. G9, 1959, Siro Set.
34. C.S.I.R.O. Wool Research Laboratories Report, Geelong, 1961.
35. Wool Bureau, *Wool Sci. Technol.*, No. 4, 1961.
36. Wool Bureau, *W.B.*, 5A; *Wool Sci. Technol. Report*, No. 5, 1961.
37. Wool Bureau, *Wool Sci. Technol. Report*, No. 5, 1961.
38. *Wool Science Review*, Nos. 27 and 28, 1965.
39. *Wool Science Review* 31, 1967; *Wool Stretch Fabrics*, 40-51; Hygral Expansion in Wool Fabrics, 12-31.

Part III: Shrinkproofing of Wool

Introduction

The interest in shrinkproofing of wool has been considerably stimulated in the last ten years by the research done at the U.S.D.A. Wool Laboratory culminating in the Wurlan process and the research done in Australia C.S.I.R.O. Wool Research Laboratory in Geelong resulting in the permanganate-salt process. A further stimulus was and still is the market demand of 100% wool garments with permanent creases and pleats which are washable and need a minimum of ironing. Such garments can only be produced by combining a shrinkproofing treatment with a permanent setting. It is estimated that the potential market for shrinkproofing of wool in the United States today is about 30% of the overall production of woven and knitted wool goods, which represents approximately 150 million pounds of clean wool. World-wide the International Wool Secretariat estimates it could be about three times this figure—450 million pounds.

This part deals specifically only with the technical side of processes which have been developed and have become prominent in the last 10 years.

Shrinkproofing of wool top and fabrics has in one way become of age, but still receives considerable attention as a processing problem. As seen from Table 20, pages 292 to 294 of *Wool Handbook*, Vol. 1, over the years many methods have been advocated for producing washable woolens. New methods have been added such as the interfacial polymerization treatment and the use of chloroisocyanuric acid products. In 1963 McPhee (16) stated: Processing to overcome felting shrinkage has advanced greatly over the past 10 to 15 years, although even now perhaps only 10–20% of wool textiles which unquestionably need a shrinkproofing treatment are being treated.

In the United States the following processes commercially applied up to 1965 have been listed and approved by the Army Quartermaster for the treatment of military items (list dated February 1964):

Controlled Oxidation

1. Dylanize SW: Stevensons Inc., New York, N.Y.
2. Dylanize X: Stevensons Inc., New York, N.Y.
3. Kroy: Kroy Inc., Lowell, Mass.
4. Kelpie: Philadelphia Dye Works, Philadelphia, Pa.
5. Schollerize: Scholler Bros., Inc., Philadelphia, Pa.
6. Harriset: Harris Labs., Washington, D.C.
7. Whitaker Process WH-2: Fred Whitaker Co., Philadelphia, Pa.
8. Chatham Continuous Chlorination Method: Chatham Mfg. Co., Elkin, N.C.
9. QM Controlled Chlorination (Knitwear items only)
10. WB-7: Wool Bureau
11. Melafix II: Ciba Co., Fair Lawn, N.J.
12. Chloregal D: Geigy Chemical Corp., Ardsley, N.Y.

Resin Treatments (Active Ingredient—Melamine Formaldehyde Condensate)

1. Resloom M-75: Monsanto Chemical Co., Springfield, Mass.
2. Lanaset: American Cyanamid Co., Bound Brook, N.J.
3. Kymene 557: Hercules Powder Co., Wilmington, Del.

Resin Treatments (Other)

1. Wurlan (interfacial polymerization of hexamethylene diamine and sebacoyl chloride): U.S. Department of Agriculture.

This list, even though it gives methods which meet the U.S. Army specification for maximum felting shrinkages, does not mean they are widely used.

Oxidation Processes

Up-to-date shrinkproofing methods are all based on continuous operation. The two oldest continuous shrinkproofing processes are the Harriset process and the Kroy process. The Harriset process, chlorination in an alkaline solution, was developed during the

Second World War for treatment of top. The treatment provided the U.S. Army with washable cushion sole socks. The Kroy process, chlorination in acid solution, is also a continuous top treatment. The machinery used for the Harriset treatment was a modified five-bowl backwasher, and the Kroy process uses a two-beck machine built by Hunter.

The best known of the chemical oxidative processes which have the least effect on the handle of wool are the permonosulfuric acid process by Stevenson and the permanganate process developed in Australia, and the treatments using dichloroisocyanuric acid compounds (DCCA).

The Permonosulfuric Acid Process

The permonosulfuric acid process also known as the Dylanize process covered by U.S. Patent 2,739,034 (1956) is successfully used on top and pieces as described by Stevenson (23) and Fell (8). The Dylan XC process used for continuous top treatment is performed in a five-bowl backwasher. See Fig. 54.

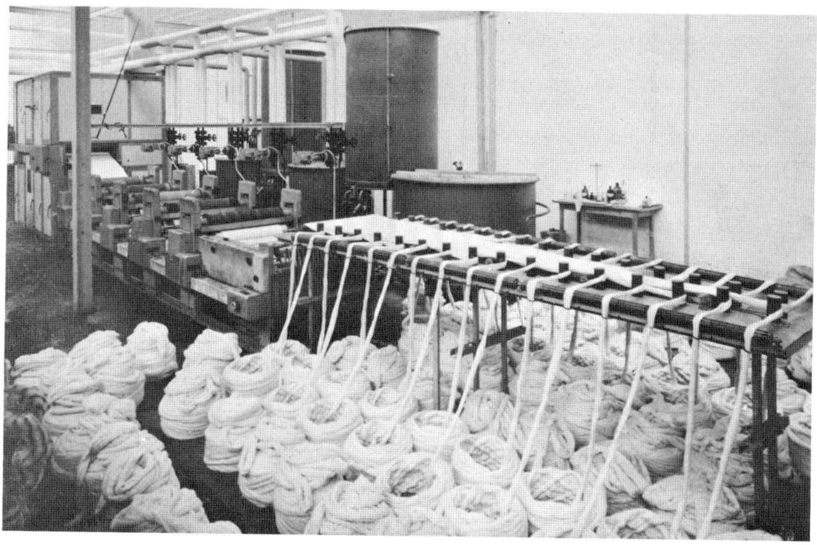

Fig. 54. Dylan XC Top Treatment Range (Courtesy of Precision Processes Ltd.).

Fell (8) gave details on the chemical concentration, contact time and baths temperatures as shown in Table 1. Such equipment is capable of producing 500 lb per hr of treated top.

TABLE 1
Conditions for Permonosulfuric Acid Process

Bath	Chemical	Concentration, %	Contact time, sec	Temperature, °C
1	Permonosulfuric acid	0.3–1.0	15–60	20–50
	Wetting agent	1		
2	Water	Rinse		30
3	Sodium sulfite	5	15–30	30
4	Water	Rinse		
5	Water	Rinse		

A design for machinery for fabrics processing installed in an American mill is seen in Fig. 55. The fabric, mostly 12–13 oz woolen flannels, run at a speed of around 30 yards per minute with a contact time of 20 sec in a bath containing 0.4% permonosulfuric acid and 0.3% Tergitol TMN (wetting agent) with the temperature maintained at 30°C. The permonosulfuric acid is used in salt form, potassium persulfate known under the name Oxone, a DuPont product.

A modification of the Dylan process is covered in the U.S. Patent 3,236,585 (1966) assigned to Precision Processes (Textiles) Ltd., former Stevensons (Dyers) Ltd. In this method permonosulfuric acid is combined with *N*-chloro compounds selected from trichloroisocyanuric acid, dichloroisocyanuric acid and salts of the latter; with a sodium sulfite after treatment. The claim is that the use of these two reagents produces a greater shrink resistance than either used alone.

The type of acid used in dyeing dylanized pieces influences the washing shrinkage. Dyeing with sulfuric acid destroys the effectiveness of the treatment, even with acetic acid a reduction of shrinkage control can result.

Permanganate-salt Process

The treatment of wool with aqueous neutral permanganate to achieve shrink resistance was patented in 1947—Australian Patent

Fig. 55. Dylan Fabric Treatment Range (Courtesy of Precision Processes Ltd.).

123,475. This method was further developed and modified by McPhee (17) and co-workers into the permanganate-salt process. The details on the treatments for loose wool, top, yarns and pieces are compiled in a publication of the CSIRO Division of Textile Industry Wool Research Laboratories—Washable Non-Iron Fabrics (29).

The general directions given are to enter the wet wool into the saturated salt bath (1.16 sp gr), allow about 10 minutes for the wool to equilibrate into the salt, then add the required potassium permanganate solution into the salt liquor to give a 3–5% solution over a period of 5 min, start the treatment at 25–30°C. and run at this temperature for 10 min. The temperature is then raised up to about 50°C so as to exhaust the permanganate in 30 to 40 min. After the permanganate is exhausted the salt solution is returned to the saturator and the goods rinsed in warm water. Then follows the clearing of manganeseoxyd with bisulfite. The preferred method is to add about 10% (on the weight of the wool) of a 60% acetic acid to the machine full of rinse liquor, run for 2–5 min and then add about 7–8% bisulfite on the weight of the wool and run until the wool has returned to the original color. Suitable temperature recommended 30–40°C. The washing off necessary to remove the soluble manganese salts is generally accomplished in two to three rinses of 5–10 min each in warm water.

The Wool Bureau (32) issued in May 1967 a pamphlet WB-7 on the permanganate-salt process specifically for the treatment of lambs wool sweaters. Up to August 1967 the Wool Bureau served 18 processors which signed license agreements. One processor runs the WB-7 process in paddle wheel equipment prior to piece dyeing, about 200 pounds of garment blanks can be treated in a unit every two hours. Another Mill, uses rotary washers and processes per machine about 150 lb of cut and sewn garments every 60 min. In 1967 about 750,000 lb of knit goods were treated in the United States.

So far no other process has been able to match the soft handle obtained in knits with the permanganate-salt process. The continuous permanganate-salt process (known as the G.8, I.W.S. OR) has been developed for card sliver and top as reported by Feldtman (3) and co-workers. The machinery is a modified backwasher. A schematic drawing of the plant is shown in Fig. 56. The two front

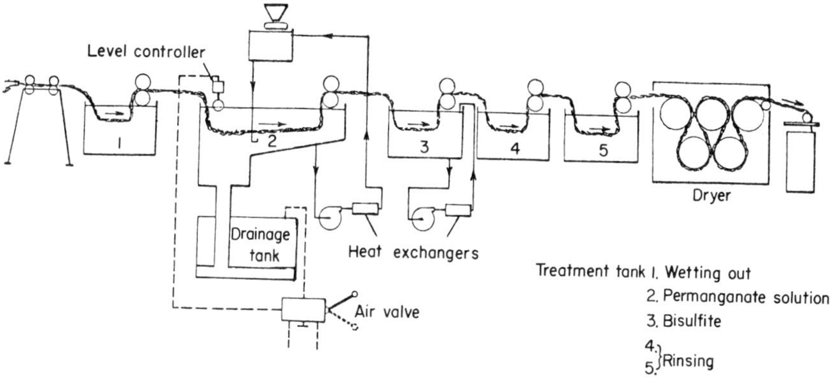

Fig. 56. Schematic Layout of Shrinkproofing Plant. Treatment Tank: 1. Wetting Out, 2. Permanganate Solution, 3. Bisulfite, 4 and 5. Rinsing.

bowls 1 and 2 are made of polyester resin, reinforced with glass fiber, and the three standard steel bowls are lined with the same material. The carrier rollers are also made of polyester resin and are arranged so that card sliver, as well as top sliver, can be processed. The main squeeze rollers are coated with epoxy resin, the top rollers of each set being lapped in the normal way.

Several salts can be used to increase the efficiency of permanganate as a shrink-resistant agent, reported by Williams and McPhee (28), but only sodium chloride, sodium sulfate or sodium nitrate could be economical. For low cost and simplicity generally chloride is used. However, sodium chloride raises corrosion problems which do not occur with sulphate, so that the choice of salt depends on the availability of corrosion-resistant machinery. Most grades of stainless steel are attacked by salt solutions. The most resistant grade is 18/8 Mo stabilized steel, Type 316 in U.S.A. However, this steel is still susceptible to stress corrosion. Hastelloy C (58% Ni, 14% Cr, 5% Fe, 5% W, 17% Mo, 1% Mu, 1% Si, 0.1% C) has been found completely resistant to 5% (wt/vol) permanganate solutions in saturated salt at 70°C; but, since plastic and glass fibers are available, the bowls can be made just as cheaply or more cheaply from them, solving the corrosion problem.

The conditions of treatment in the five-bowl machine as stated by Feldtman and co-workers are as shown in Table 2. The degree of

TABLE 2
Conditions for Permanganate-Salt Process

Bath	Chemical	Concentration, %	Immersion time, sec	Temperature, °C
1	Salt	Saturated[b]	20	40
	Wetting agent[a]	1		
2	Permanganate	2	35	40
	Salt	Saturated[b]		
3	Bisulfite	2	15	38
	Acetic acid	1		
4	Water	Rinse		
	Wetting agent[a]	0.1		
5	Water	Rinse		

[a] Wetting agent: polyoxyethylated nonylphenol.
[b] Specific gravity 1.16.

shrink resistance conferred is greater the higher the temperature of the permanganate bath. In general the useful range is 38–45°C. The rate of production is 250 lb per hr. The consumption of chemicals per 100 lb of top processed is:

potassium permanganate	3–5 lb
sodium chloride	25 lb
sodium bisulfite	10 lb
acetic acid (60%)	5 lb

Good features of this process are that the handle of the wool is unchanged and there is no yellowing.

With the shrink-resistant wool, dyeing should be carried out in solutions with pH values as close to 4 as possible and the temperature kept to the minimum allowing for exhaustion and good penetration of dyes into the fibers. Details of a satisfactory general dyeing procedure have been given by Hine and McPhee (12).

Dichloroisocyanurate Processes

Recent developments have shown that the salts of dichloroisocyanuric acid (DCCA), such as CDB-63 (sodium dichloroisocyanurate) manufactured by the FMC Corporation and Basolan DC from BASF Colors and Chemicals Inc. (potassium salt), are outstanding for providing excellent shrink resistance without affecting the soft and resilient nature of the wool fiber. Monsanto holds U.S. Patent 2,993,747 covering the use of chloroisocyanuric acid compounds for shrinkproofing of wool; formula: $Cl_2Na(NCO)_3$ (see structure). Wool can be made shrink-resistant either by batch or continuous methods.

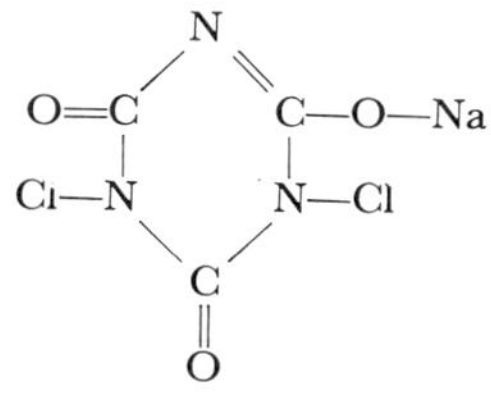

Batch Processing

The advantage of DCC compounds over ordinary chlorination is that exhaustion of the available chlorine is easily controllable.

The bath is prepared at room temperature 10–25°C with the addition of 1% glacial acetic acid and 0.2% nonionic surfactant, then the wool added and the machine run until it is thoroughly wetted. At this point the pH of the bath should be in the 4.5–5.0 range. The predissolved CDB-63 is then added 3–5% (depending on the structure of the fabric) and the machine run until approximately 95% exhaustion of the chlorine has occurred normally within ¾ to 1 hour. For Basolan DC the company suggests to chlorinate in the presence of 10% Glauber salt and 2–3% tetrasodium pyrophosphate anhydrous.

The chlorine is removed by the addition of sodium metabisulfite (1–3 g/liter) pH 3.5–5 and running for 15–20 min at 40–50°C (104–122°F). The liquor is then dropped and the wool given two warm rinses followed by one cold rinse.

Continuous Processing

According to Schiecke (22) three continuous DCC methods have been developed for processing of wool sliver or top; namely, the Basolan method, the Pechiney St. Gobain method (ORCED), and the International Wool Secretariat–IWS method. All three methods work with dichloroisocyanuric acid.

Method I: The continuous top treatment is performed in a five-bowl backwasher similar to the one described for the Dylan XC process (Figs. 54 and 56). The first bath wetting agent, second bath contains the DCC with the necessary acid set at pH 5.5. The third bath bisulfite and acid and the 4 and 5 baths are used for rinsing. The concentration and the temperature of bath two are set according to the contact time (minimum 20 sec).

Method II—Pad-Acid Procedure: This method was devised and patented by Pechiney St. Gobain, who holds the patent rights in many countries.* The process consists of impregnating the wool with a solution of CDB-63 plus a suitable nonionic wetting agent. This is followed by immersion in an acidified bath, a clearing bath and two rinsing baths. Although the pH of the acid bath is low (approximately 2.0), the reaction rate between the wool and the CDB-63 is such that the concentration of the available chlorine is not high enough to cause corrosion. A typical makeup for a five-bowl backwashing range would be as shown in Table 3.

TABLE 3

Conditions for Cyanuric Acid Pad-Acid Process

	Bowl number				
Chemicals (% soln.)[a]	1 CDB-63 bath	2 Acid bath	3 Anti-chlor	4 Rinse	5 Rinse
Nonionic surfactant	0.5	—	—	—	—
CDB-63	8.0	0.8	—	—	—
Hydrochloric acid (38%)	—	0.1	—	—	—
Sodium metabisulfite	—	—	5.0	—	—
Temperature (°C)	25	20	45	45	45
Time (sec)	5	20	15	15	15

[a] These concentrations are for machines which give approximately 50% saturation at the nip.

A slight variation of the method shown in Table 3 involves the use of formic acid instead of hydrochloric acid. With formic acid a

* FMC holds a license under Pechiney St. Gobain's patents. French Patents 1,281,414; 1,281,420; 1,376,756.

pH of 3.5–4.0 is obtained and a reaction time of 22 sec is required to obtain satisfactory reaction between the CDB-63 and the wool. *Method III—Pad-Store Procedure:* This procedure has been developed by the International Wool Secretariat department of product development and its United States branch, the Wool Bureau Inc. Details are contained in Report No. 9 of the Wool Bureau (33), DCCA—A System for Continuous Production of Shrink-Resistant Top, and in the pamphlet of FMC Corporation on CDB-63, both published in June 1967, and in *Wool Science Review*, No. 34 (30).

The treating range consists of a horizontal padder followed by a storage tray which feeds into a three-bowl backwasher. This in turn feeds into a drum dryer (Fig. 57). The appropriate amount of the DCC is padded directly onto the wool along with a suitable nonionic surfactant. The DCCA solution is supplied to the nip of the horizontal rollers at a controlled rate from a reservoir. Any overflow is returned to the reservoir by a pump connected to a drip pan. The storage tray provides a dwell period of approximately two minutes, during which time about 90% exhaustion of the available chlorine will take place. Feed rollers provide positive movement of the top out of the tray and into the first bowl of the backwasher where the wool is cleared with anti-chlor–sodium metabisulfite. In the two remaining bowls the top is rinsed with warm water. In some ranges a fourth bowl is used as a bleach bath with 0.5–1% hydrogen peroxide to improve the color of the top. The dryer can be a drum type or any other suitable unit.

Continuous open width shrink resisting of wool piece goods is also possible with sodium dichloroisocyanurate as has been reported by Easton and Villiers (2). A schematic diagram of the pad-acid process for piece goods in shown in Fig. 58.

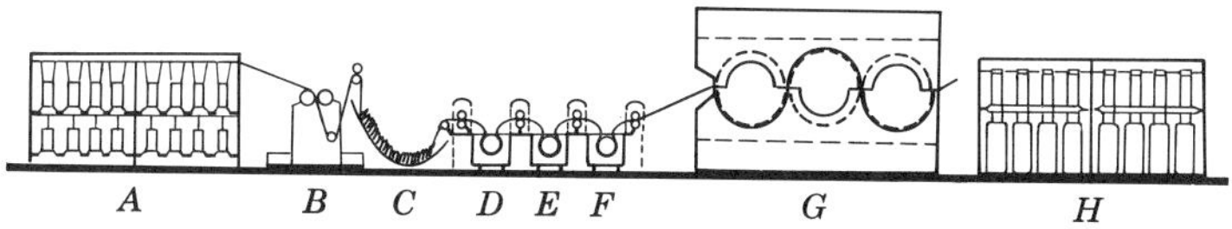

Fig. 57. DCCA—Pad-Store Top Treatment Range.

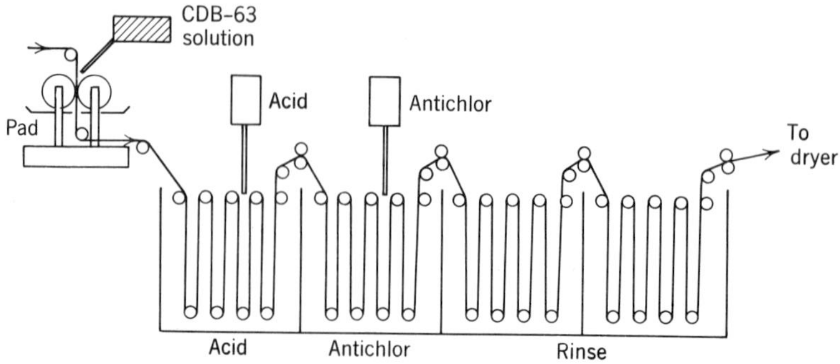

Fig. 58. Schematic Diagram—Pad Acid Process—Piece Treatment Range.

Potassium Bromate Process

The use of bromine as a shrinkproofing agent for wool was investigated by Trotman (37) and the Wool Industry Research Association (38) in 1933. But only recently a suitable and economical bromate process has been worked out by Lewin and Co-workers (35 and 36) for large scale treatment of wool. The batch process is at present used by three mills in Israel.

The instructions (1) for the shrink and felt-resisting of woolen and worsted fabrics (including blankets and felts) and knitted materials with the Lewin-Bromate process are as follows: The batch process is carried out in one main stage in which the wool is treated with potassium bromate, 2.1% on the weight of the wool in the presence of about 0.3 *N* sulphuric acid and 0.5 *N* common salt.

The wool is then antihalogen-treated with 1g/1 sodium bisulphite at 30–50°C, rinsed and neutralized to pH 6–7 with 20 ml of concentrated ammonia per 1 kg of wool. The hand and feel of the wool can be regulated by washing with 2–5 g of soap per liter or with 1% softening agent on the weight of the wool.

The bromate-continuous wool shrinkproofing process is applicable to the treatment of light and heavy woolen and worsted fabrics for obtaining a full shrinkproofing and feltproofing effect. The process consists of impregnating the wool with an aqueous bromate solution at a temperature of 40–50°C and rapidly developing in about 1 *N* sulphuric acid at 70°C. The development stage is followed by mild-reduction and neutralisation stages.

Resin Treatments

The oldest resin treatments which had and some still have wide application for military items such as blankets are the Monsanto Melamine-formaldehyde condensates (U.S. Patent 2,409,906 (1946)) and the American Cyanamid Co. Lanaset process—alkylated methylol–melamine and urea.

The main drawback of these earlier processes is the high amount of resin, 5–10%, necessary to achieve shrink-resistance resulting in a stiff and harsh handle of the wool product in addition to high curing temperatures up to 300°F.

The work of Pardo and O'Connell (19) with polyamide–epoxide resin finishes has shown that prevention of felting shrinkage of the wool fiber is related to surface coating of the fiber. This leads to a masking and with it a smoothing of the scale surface of the wool fiber or an equalizing of the fiber friction so that the difference of both friction coefficients in the fiber tip-root direction and in the root-tip decreases or in the ideal case disappears.

The resin processes developed within the last ten years which have mainly overcome the drawbacks of the earlier treatments are as follows:

Wurlan Process—Interfacial Graft Polymerization

This process was developed at the United States Department of Agriculture Wool Laboratories in Albany, California, and first reported by Whitfield (26) and Fong (9). The process is based on a chemical technique, known as interfacial polymerization, which forms an ultra thin film of a polyamide resin on the surface of wool fibers. Neither drying nor heat curing is necessary. An occasional slightly harsher handle can be overcome by softeners.

The Chemistry of Interfacial Polymerization

Many condensation polymers can be made by interfacial polymerization (IFP), in which reagents in immiscible solvents react at the interface between the solvents. In particular, a water solution of a diamine, such as hexamethylene diamine, reacts almost instantaneously at its boundary of contact with a solution of a diacid chloride, such as sebacoyl chloride, in an inert, water-immiscible solvent, forming a film of polyamide which can be collected continuously. The "polymerization" is based on the Schotten-Baumann reaction of an acid chloride with a compound containing active hydrogen, as shown in Fig. 59.

The practical development of the Wurlan process from a laboratory method to a successful commercial use of the treatment for fabrics has been described by Miller (18). The equipment finally arrived at is shown in Fig. 60.

The clean and neutral to slightly alkaline fabric is delivered to the treating range after fulling, scouring, and scutching. Fabric is fed into the diamine bath; excess liquor is squeezed out by the three-roll mangle shown, and the reaction is completed in the SC bath. Excess solvent with SC is then squeezed out. The fabric is delivered to trucks for subsequent dolly washing to remove residual reagents and solvent (*Wool Science Review*, No. 26 (31)).

Diamine Bath

Referring to the HMDA bath and mangle shown in Fig. 60 the diamine tank is designed to give adequate immersion time at speeds up to 35 ypm. The bottom rolls are driven to reduce fabric tension in the bath. The tank is heated by steam and the tempera-

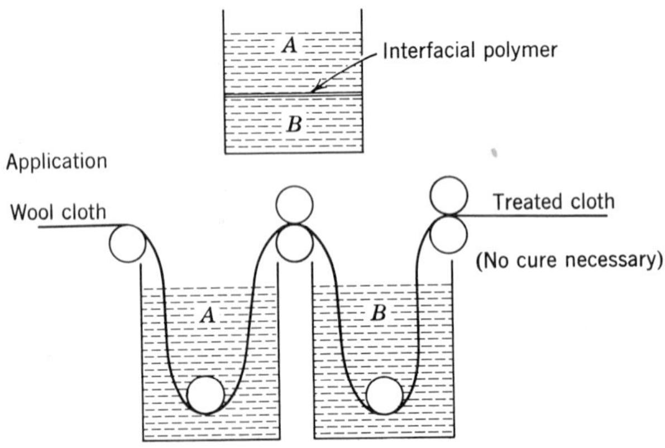

Fig. 59. The Principle of Polyamide Formation at an Interface: "Interfacial Polymerization." *Top*, Chemical Formulas of Reagents and Product; *Center*, Diagram of Arrangement for Forming Polyamide at the Boundary between Immiscible Solutions of the Reagents in a Beaker, *Bottom*, Diagram of Continuous Application to Fabric.

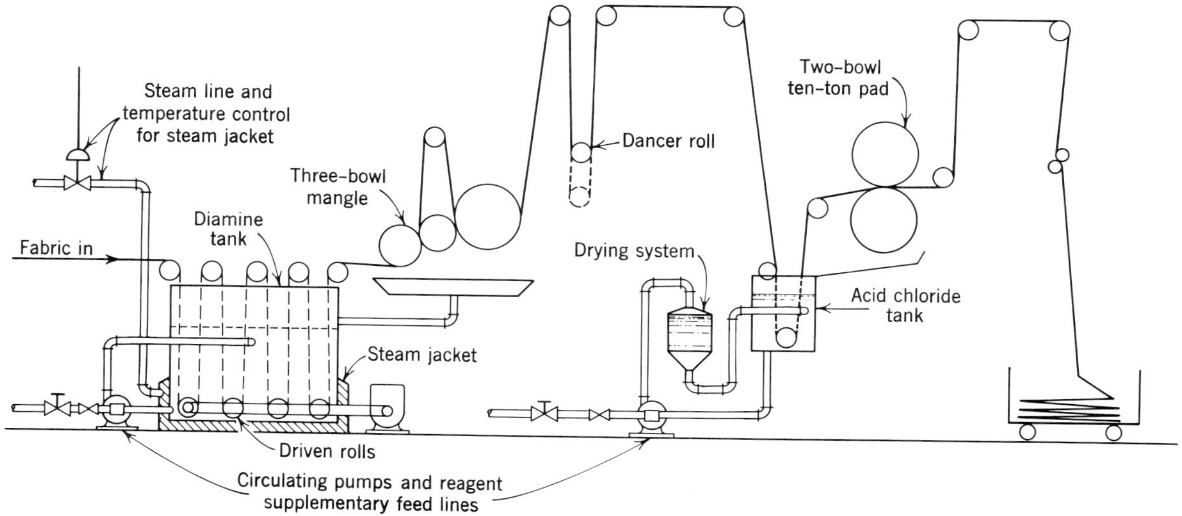

Fig. 60. Diagram of Treating Range Now Used for WURLAN Treatment of Fabric. Efficient, Uniform Expression of Excess Liquid after the First Dip and Continuous Removal of Water and Hydrochloric Acid from the Second Bath are Important Features.

ture maintained by a standard temperature-controlling device. The HMDA solution is continuously circulated by drawing the liquor out at the bottom and pumping it back into the top of the bath through a perforated pipe which runs the full width of the bath. The circulating system also serves to introduce concentrated HMDA solution as necessary to maintain the bath concentration.

In order to achieve the necessary extraction when operating at realistic production speeds, a highly efficient, uniform pad is necessary. For example, the Kuster No. 73 swimming roll pad performs very well in this capacity. Recommended initial composition and specific treating conditions for the HMDA bath for woolen piece goods processing are given in Table 4. The bath composition

TABLE 4

Composition and Processing Data for HMDA Bath

Aqueous Bath	
Composition	1% hexamethylene diamine, 0.5% Na_2SiO_3, 0.05% wetting agent
Volume	200 gal
Temperature	120°F
Immersion (dip-to-nip)	20 sec
Speed	23 yd/min
Add (per piece)	0.6 lb hexamethylene diamine, 0.5 lb Na_2SiO_3, 0.4 oz wetting agent

is checked periodically by acid-base titration. Somewhat different results obtained with sodium carbonate and metasilicate are attributed to pH differences: it appears that equivalent shrinkage control is attainable at higher pH with less HMDA. By using sodium metasilicate in the HMDA bath it was possible to decrease consumption of HMDA from more than 1 lb per piece to 0.6 lb per piece.

Sebacoyl Chloride Bath

The time of immersion in the acid chloride solvent bath is not critical; in fact, the volume of solvent in this tank should be as small as possible. Extraction efficiency is also less important than with the first bath: a standard 10-ton pad works satisfactory.

Since acid chlorides are decomposed by water this second bath must be kept as dry as possible. The preferred method is to circulate the bath through a "molecular sieve" such as Linde AW-500.This removes water and HCl very efficiently. The molecular sieve can be regenerated several times, and though its cost per pound is high, its contribution to the cost of the process is quite low.

The recommended composition and conditions of use of the SC bath are given in Table 5. The amount of molecular sieve needed is

TABLE 5
Composition and Processing Data for Sebacoyl Chloride Bath

Solvent Bath	
Composition	3% sebacoyl chloride in Stoddard solvent
Volume	25 gal
Temperature	ambient (70–75°F)
Immersion (dip-to-nip)	4 sec
Speed	23 yd/min
Add (per piece)	approx. 2 gal 3% sebacoyl chloride in solvent
Molecular sieve	approx. 0.25 lb/piece

a rough approximation because the amounts of water carried over and the rates of hydrolysis are hard to estimate for the range of conditions likely to be encountered.

In summary, the fabric is immersed in a solution containing about 1% hexamethylene diamine and about an equimolar amount of an inorganic base (such as sodium metasilicate) and then is squeezed to a wet-pick-up of 40–50%. The damp cloth is passed through a 3% solution of sebacoyl chloride in hydrocarbon solvent, squeezed to about 15–20% pick-up, and then dolly-washed and dried. Under these conditions with adequate chemical control of the treating baths, it has been possible to treat 100 pieces (of about 80 yd each) per day with little or no variation in the effectiveness of the shrinkproofing.

Wurlan IFP Treatment of Wool Top

The first experiments made for continuous treatment of top were reported by Fong (10) and Ingenthron (13) at the Wool Finishing Conference in Albany in 1964. The commercial adaptation to production on conventional top processing machinery was worked out by the Bancroft people under the name Bancora process as described by Sayle (21). Sayle gives the following details on the continuous top method. The machines used are standard backwashers where only minor changes have been made. The sliver passing through the first bath must be guided that the time from entering the bath until it comes to the squeeze rollers is a minimum of thirty seconds. The three diagrams show the ways this can be accomplished (Fig. 61).

The first way with an airing roller has the advantage that the top is immersed twice. In No. 2 the necessary time was only possible to achieve by a longer airing distance prior to the squeezing. The ideal way is illustrated by No. 3, which is the Fleissner construction. It is important that the pH of the top must be neutral or slightly on the alkali side, its oil content is of secondary importance.

Important too is that each bath has indirect heating pipes to control the temperature. The second bath which contains the acid chloride is followed by two or three baths for neutralization and rinsing. It has been found that a four-bath backwasher is sufficient. In general the speed of the top through the machine is between 3–5 m per min equal to 10–16 ft working with 20–48 slivers. The weight increase through the treatment amounts on the average to 1.5%.

1965/1966 Modifications

The most recent progress in wool chemical finishing based upon interfacially applied resin was covered in a paper by Fong (11).

Three possible modifications of the conventional Wurlan treating process are given. The first experimental modification involves the substitution of a high boiling aromatic petroleum solvent for the

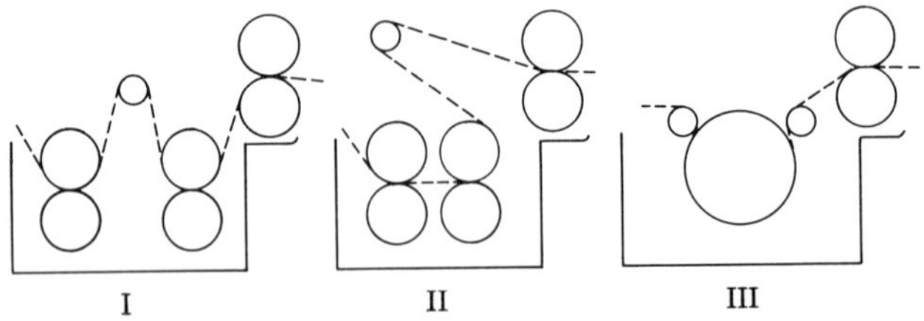

Fig. 61. Wurlan—Bancora—Methods of Sliver Threading in First Bath.

conventional Stoddard solvent. The second experimental system involves use of ethylene diamine, EDA, in combination with SC in an aromatic solvent. The third system involves the use of HMDA in combination with hexanediol bischloroformate. This last system results in the formation of a polyurethane type coating rather than the more conventional polyamide types obtained with a diacid chloride reactant.

Phase Boundary Limited Crosslinking

An alternate interfacial reaction treatment of wool was recently developed in the U.S.D.A. Wool Laboratory which, like the Wurlan treatment, employs two immiscible solvents and the reaction that occurs at the interface. Unlike the Wurlan treatment, the polymer produced is not from combination of two monomers. Instead, a preformed polymer with reactive pendant groups is held in one solvent and a difunctional crosslinking reagent in the other solvent. The interface reaction consists in formation of an insoluble polymer network.

Technically this new treatment is called "Phase Boundary Limited Crosslinking." More briefly it is referred to as the "PBLC" method. As in the case of the Wurland treatment, shrinkage stability is obtainable at relatively low polymer uptake. The hand of the treated materials is in many cases superior to that of Wurlan-treated goods.

The principle, scope and utility of phase boundary limited crosslinking of reactive polymers in wool finishing have been reported in various papers from the U.S. Department of Agriculture Wool Laboratory group as listed under literature. The most recent paper by Whitfield, Remy, and Pittman (27) is on reactive polyesters and polyethers to yield polyurethanes. Urethane elastomers are of special interest as textile finishing agents due to their broad spectrum of properties, availability, and moderate price. Furthermore, the parent polymers contain isocyanate groups which react essentially instantaneously with the amino groups such as are present in wool.

Polyurethane One-bath Process

Reich (20) discussed the use of polyurethanes which are emulsifiable in water. In conjunction with other synthetic polymer dispersions, notably polyacrylates, the polyurethanes are suitable to impart to cotton and rayon better resistance to creasing, abrasion and improves tensile strength and shape stability and to wool materials nonfelting properties.

The nonfelting finish on wool can be produced by a one-bath method. This is possible either in a water or solvent system. Using the water system polyurethane and polyacrylate disperser are combined. Such stabilized emulsions are especially suitable for the treatment of finished wool articles such as pullovers and socks. For other materials the treatment is best done prior to dyeing using the emersion-extraction method. To reach the maximum felt-free effect it is necessary to store the material for 3–5 days and the concentration of the polymers must be such that 1–2% solids are left on the fibers.

Especially simple is the treatment with the solvent system in regular drycleaning units. Polyurethane is easily soluble in organic solvents such as perchlorethylene and is sufficiently stable in the presence of small amounts of water.

C.S.I.R.O. Resin Treatment

Much additional work has been published on shrinkproofing of wool with preformed polymers by the Division of Textile Industry, C.S.I.R.O. Wool Research Laboratories in Geelong, Australia. Some of this work is complementary to the studies by Whitfield et al. The studies were made by Feldtman and McPhee (4) and the latest on the treatment of wool with polyolefins by Feldtman, McPhee, and Pratt (7). This work so far has culminated in the practical application of cheap synthetic polymers from aqueous solutions or dispersions to pretreated wool tops as described by Feldtman, McPhee, and Morgan (5). (Wide patent coverage has been sought on this method.)

Pretreatment of Wool

Earlier work showed clearly that polymer treatments of wool are much more efficient if the wool is given a chemical pretreatment. Suitable reagents include peroxide, ethanolic potassium hydroxide and chlorine solutions. It has been found that the mechanism of pretreatment is to allow better spreading of the polymers over the fiber surfaces.

For continuous processing, the cheapest and simplest technique is to add chlorine from a gas cylinder, at a controlled rate, to a sulphuric acid solution at the required concentration and temperature. The exact conditions depend on the time of immersion of the wool in the solution. Again, however, a modified acid chlorination can be used, if desired.

Machinery and Conditions for Treatment

The machinery used is much the same as that described for the permanganate/salt process, which is similar to that used for most continuous wet processing of wool tops. The essential requirements are a chlorination bowl, in which the wool and liquor can be separated whenever required, a neutralizing bowl, and a bowl containing polymer solution or dispersion, followed by a dryer.

A drawing of the plant is shown in Fig. 62. This uses six bowls in order to obtain complete consistency of results, and to maintain maximum control. The first two bowls are for scouring the tops, which ensures that only thoroughly clean wool enters the chlorination bath.

After scouring, the wool is given a mild chlorination, and it is absolutely essential for optimum results that this chlorination be uniform. This is followed by treatment with sulfite, to remove chlorine, and bicarbonate, to neutralize the acid in the wool. The sliver is then rinsed, impregnated with polymer and dried. After drying, the sliver is very strong (this indicates some adhesion through fiber bonding) and requires processing in a gill box with a

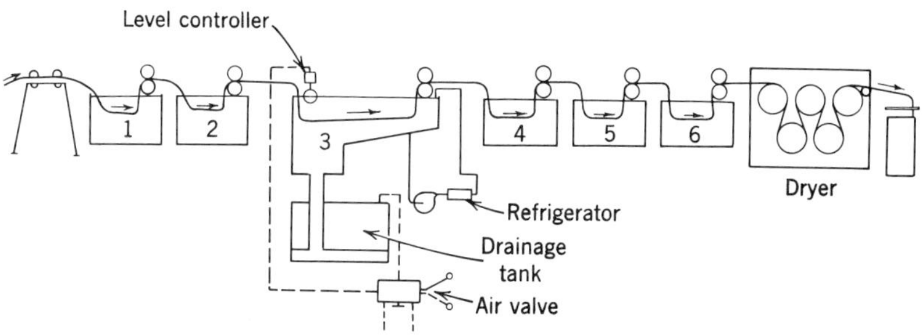

Fig. 62. Schematic Layout of Shrinkprooofing plant, C.S.I.R.O. 1. Scouring. 2. Rinsing. 3. Acid Chlorine Solution. 4. Sulpite/Bicarbonate. 5. Rinsing, 6. Polymeric Solution.

front draft of 1.5–3.5. After this preliminary operation, both the processing back into top and the spinning can be done under normal conditions for the type of wool used.

Choice of Polymer

Almost any type of polymer can be used and the final choice probably will depend on price and simplicity of curing. A polyamide–epichlorhydrin, polyethyleneimine and certain polyacrylates can be cured by drying at normal backwash-dryer temperatures (60–100°C) but some polyacrylates require a higher temperature. However, the polyacrylates are the cheapest of these polymers and give the highest abrasion-resistance, so could prove useful even if some are slightly more difficult to handle. For industrial production, the two major problems requiring careful consideration by prospective users are, (*1*) the design of a chlorination bowl to give uniform treatment, and (*2*) the procedure for dyeing the treated wool with selected dyes of adequate fastness.

J. P. Stevens Process

This process is covered by U.S. Patents 3,165,375 and 3,279,882 and is based on a preformed polymer. The reactive polymer is an amidoaziridine compound. The insolubilization of the coating on the wool occurs through the aziridine group as well as some reaction binding the polymer film to the wool.

The polymer is sold under the name Woolset 36 as a high viscosity milky emulsion with 40% active ingredients, miscible with water in all proportions.

The following directions for the application to fabric are given. Woolset 36 may be applied to wool fabrics before or after dyeing by a pad-dry-cure technique.

The formulation is best applied by a single dip-single nip padding operation. The fabric is then dried and cured. Drying and curing may be carried out in separate steps or combined in a single operation. The following is a typical padding formulation:

Padding Formulation

Woolset 36	60 lb
Buffer 21	3 lb
Made up to 50 gallons	
Dry/cure 275°F	9–15 min

A good scour after curing is necessary.

DuPont Process

The DuPont process is based on the use of Zeset TP, a reactive terpolymer. It is sold as a 15% solution of the polymer in Perclene perchloroethylene. Chemically it is a polymeric acid chloride similar to the types which have been discussed by Remy, Whitfield, and Wasley (27).

Processing directions are given by Bacon and Maloney (1) of DuPont:

Application of Zeset surface modifier to piece goods involves two steps—one of padding on the finish from a solvent bath and one to remove the solvent and cure the finish. The fabric is wet with the polymer solution by padding to leave a known amount of solution on the weight of the fabric. The concentration of polymer in the solution is adjusted to deposit the desired amount (1.5%) on the fabric. The solvent (trichloroethylene) is then flashed off by immersion in hot water for 1–2 min at 180–212°F. This operation also fixes the finish on the fabric by forming hydrolyzed terpolymer which crosslinks with nonhydrolyzed terpolymer to form an insoluble polymer. A schematic drawing of the mill equipment involved is shown in Fig. 63.

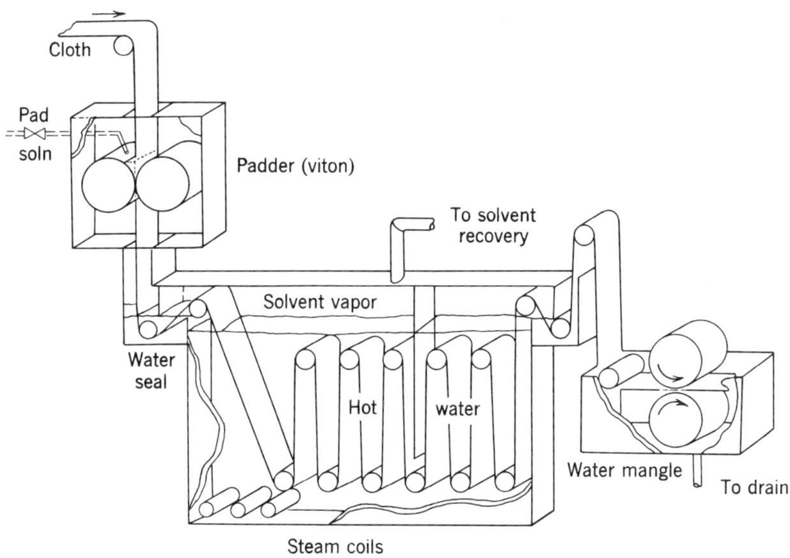

Fig. 63. DuPont Solvent Finishing Range.

Electrical Discharge Processes

The two processes known which were developed in France on the basis of electrical discharge are the Kassenbeck (14) and the Levaux (15). The latter carries the name Disruptex.

Kassenbeck Method

The method is covered by U.S. Patents 2,977,475 and 3,179,482 and is fundamentally a silent electrical discharge process. The effects are produced on wool, specifically wool top, by bombarding it with ions in the glow discharge region of gases. Wool top has been continuously treated by passing it between two electrodes of a gas discharge tube under low pressure (50–100 mm of mercury) and applying high voltage 2–5 kV ac at 500 cps to the electrodes for 1–5 min. In the patent it is stated that approximately 1 kw is sufficient to treat 20–40 kg of top. The treatment improves shrink resistance, breaking strength and dyeing rates.

Disruptex Process

The Disruptex (Disruptomatic) method invented by Levaux is in many aspects similar to the Kassenbeck but it differs widely in the electrode system and is carried out under normal atmospheric pressure. In place of the silent glow discharge the Corona or spark discharge acts on the wool, which has the advantage that no vacuum chamber is required and reaction times are much shorter.

The studies made on wool by Kassenbeck (14), Thorsen (24), and Kodani and von Bergen (25) showed that the effect of electric discharge is confined to the scales. The epicuticle is partly ruptured and the surface structure using the electron microscope shows small cracks, like pockmarks, and longitudinal wrinkles resulting in a rough surface. This physical change in the surface structure seems to be the main reason for the anti-felting properties. The chemical changes, such as increased swelling in $0.1N$ caustic soda and higher dye absorption, seem to be the result of oxidation by the ozone present in the discharge area.

Summary

For each of the eleven methods discussed under the three headings, oxidation, resin and electrical discharge processes, advantages can be claimed as well as disadvantages cited.

In the oxidation processes, specifically in chlorination methods, the pH value plays a decisive part. Working in an alkali solution the hand of the wool stays nearly unchanged and the danger of wool damage is much lower than with acid chlorination. The disadvantage is that the wool yellows more quickly. Working in an acid bath the anti-felt effect is better, also the luster; but the hand is generally more impaired. Tests have shown that in contrast to the older chlorination methods, when working with dichlorocyanuric acid so-called DCCA methods no reduction in breaking strength occurs. Often even increases in strength are found which may be due to the formation of new cross-linkages in the wool structure. The pure oxidation processes, such as the permonosulfuric acid (Dylanize) method, have the advantage of practically no danger of yellowing, rather bleaching, and the effect on the hand is small. *By all resin treatments* the main concern is the change in the hand of the wool, which is altered much easier and stronger than by any other process. This was particularly true of the older method which required a 5–10% resin deposit on the wool surface. With today's polymers, where the additives are 2% or lower and the thickness of the film formed on the fiber surface is ultra thin, the change in the hand is much less. It can be further reduced by adding compatible softeners.

The great advantage is in the fact that no damage to the wool substance is possible, especially with polymers which need no special heat curing or can be cured at low temperatures. The anti-felt effect in general can be classified as excellent. In most cases, fabric treated shows a substantial increase in tensile strength and resistance to flex abrasion. With the Stevens process these amount to 20–30% for the strength and 40–60% in abrasion. However, hand-in-hand with the increases in strength and abrasion goes the increase in stiffness affecting the hand.

Another advantage generally overlooked is the improvement in the colorfastness. This is particularly true for the Wurlan process. Pieces treated before dyeing by this process and then dyed with acid dyes of the milling type have a very good washfastness compared with the low washfastness of pieces dyed with the same dyes but which were shrinkproofed by an oxidation process.

Most resin treatments have been very successful when applied to piece goods such as flannels and knitted fabrics, but considerable difficulties are encountered when treating top and finished garments. In top the main difficulties are caused by fiber bonding, glued together by excessive polymer which makes gilling difficult and wasteful. This problem can be overcome or reduced, as studies at the Albany laboratories have shown, by selecting different chemical agents which form a more suitable type of resin. Unfortunately, some of these chemicals are at present not available at a cost low enough to permit commercial use.

What holds true for the Wurlan type process is also applicable to the one-bath methods using reactive preformed polymers.

In treating finished garments, such as sweaters, with resins, it is hardly possible to get a soft enough hand. Generally for treatment of garments the solvent system is used. In this system high reactive preformed polymers can cause corrosion troubles as it was reported for the Zeset method.

The electrical discharge methods have the great advantage of being applied directly to the dry wool. The Disruptex process seems so far the only commercially used method for the treatment of worsted top. Its main drawback at present is its low production capacity per machine unit—approximately 10 lb per hr against a minimum of 200 lb per hr using the Wurlan-Bancora top treatment. Also the reduction in felting shrinkage is low—around 20%.

References

1. Bacon, O. C., and Maloney, J. E., *Am. Dyestuff Reptr.*, **56**, 319 (1967).
2. Easton, B. K., and Villiers, R. F., *Am. Dyestuff Reptr.*, **57**, p. 523 (1968).
3. Feldtman, H. D., McPhee, J. R., Morgan, W. V., and Parnell, B. G., *Textile Mfr.*, 1965.
4. Feldtman, H. D., and McPhee, J. R., *Cirtel*, **3**, 345 (1965).
5. Feldtman, H. D., McPhee, J. R., and Morgan, W. V., *Textile Mfr.*, 1967.
6. Feldtman, H. D. et al.
7. Feldtman, H. D., McPhee, J. R., and Pratt, *Text. Res. J.*, **37**, 46 (1967).
8. Fell, E. T., *Melliand Textilber.*, **47**, 64 (1966).
9. Fong, W., U.S.D.A. Report ARS 74-22-1961, p. 23.
10. Fong, Willie, Tech. Wool Conference U.S.D.A. ARS, May 1964, pp. 74–29.
11. Fong, W., Personal Communication, 1967.
12. Hine, R. J., and McPhee, J. R., *Text. Res. J.*, **34**, 659 (1964).
13. Ingenthron, W. W., U.S.D.A. ARS 74-29, 1964, Wool Conference.
14. Kassenbeck, P., *Bull. Inst. Textile France*, **18**, 7 (1964).
15. Levaux, R., Disruptomatic Process Masurel, U. S. Patent 3,369, 981, 1968.
16. McPhee, J. R., *Text. Inst. Ind.*, **1**, 10 (1963).

17. McPhee, J. R., *Text. Res. J.*, **30**, 349/358 (1960).
18. Miller, L. A., U.S.D.A. ARS 74–29, 1964, p. 142; Am. *Dyestuff Reptr.*, **54**, 662 (1965).
19. Pardo, C. E., and O'Connell, R. A., *Am. Dyestuff Reptr.*, **47**, 333 (1958).
20. Reich, F., *Textilveredlung*, **2**, 441 (1967).
21. Sayle, J. H., *Melliand Textilber.*, **46**, 1202 (Nov. 1965).
22. Schiecke, N. E., *Z. Textilindustries*, **69**, 82 (186/1967).
23. Stevenson, F. M., and Stevenson, M. C., *J. Text. Inst.*, **53**, 649 (1962).
24. Thorsen, W. J., and Kodani, R. T., *Text. Res. J.*, **36**, 651 (1966).
25. von Bergen, W., *J. P. Stevens Res. Lab. Reports*, (1964-1966).
26. Whitfield, R. E., U.S.D.A. Report ARS 74-22, 1961, p. 15; Winlan ARS 74-29, 1964, p. 25.
27. Whitfield, R. E., Remy, D. E., and Pittman, A. G., *Text. Res. J.*, **37**, 655 (1967).
28. Williams, V. A., *Text. Res. J.*, **32**, 977 (1962); **33**, 444 (1963).
29. C.S.I.R.O., Wool Research Laboratory, Division of Textile Industry, 1961.
30. Wool Science Review, No. 34, 1968.
31. Wool Science Review, No. 26, 1964.
32. The Wool Bureau, Inc., Wool Product Development Report No. 7, 1967.
33. The Wool Bureau, Inc., Wool Product Development Report No. 9, 1967.
34. LaFleur, K. S., U.S. Patent 2, 980557, 1961.
35. Lewin, M., Potassium bromate Process (Lewin Process), U. S. Patent 2923596, 1960; 3106440, 1963.
36. Lewin, M., and Avrakami, M., *Cirtel*, **2**, 351–370 (1965).
37. Trotman, S. R., *J. Soc. Chem. Ind. London*, **52**, 159T (1933).
38. Wool Industry Research Association, B.P., 417 719 (1933).

CHAPTER 14

Carpets

William Ebersold and Werner von Bergen

Everyone connected with the carpet industry must agree with a statement made in 1962 in the Journal of the Textile Institute:

"For practically half a century, the craft of carpet-making remained almost stationary. Then, a few years ago, impelled by the demand created by the rising standard of living, the manufacture of carpets became the focus for a fantastic burst of technological activity. New processes and machines were devised; novel fibers and yarns (novel, that is, for carpets) were engineered to take advantage of an expanding outlet; older textile processes, previously confined to flat cloths, were developed to capture a share of the increased market for pile fabrics; and conventional carpet looms were re-designed to step up production to hitherto unthought-of speeds."

Manufacturers of the traditional Wilton and Axminster carpets all over the world are watching and following the pattern of trade in the United States of America where the tufting process, that offspring of the sewing machine, has since 1951 mushroomed to supply over 85% of the domestic output.

Another process with potentialities is knitting, which has been taught to make loop-pile materials but has not quite overcome some of the technical troubles involved in the production of cut-pile fabrics for floor-coverings.

Both tufting and knitting enjoy the economic advantages of high-speed production, which make them of special interest for the low price-range trade. This has been assisted by the use of inexpensive rayon yarns, which unfortunately, however, lack that resilience which makes wool so desirable a carpet fiber. Moreover, the variety and richness of pattern obtainable in the woven carpet is not as yet as readily available in the newer contenders, which so far have made their major advances in the region of plain, mottled,

TABLE 1
Production of Machine-made Carpets and Floor Rugs*

	Million lb			Million sq yd		
Country	1964	1965	1966	1964	1965	1966
Algeria	0.9	—	—	—	—	—
Australia[a]	—	—	—	9.7	11.2	11.4
Austria	4.0	3.8	4.1	1.2	—	1.1
Belgium	46.7	50.2	59.0	—	—	—
Canada	—	—	—	3.1	—	—
Denmark	13.2	17.8	—	3.5	4.7	—
Finland	—	—	—	—	—	—
France	20.4	21.1	—	6.6	7.1	17.0[f]
German F.R.	73.4	78.3	64.5	20.9	22.3	17.9
Hungary	—	—	—	0.9	1.0	—
Irish Republic	—	—	—	1.2	1.2	—
Iceland	—	—	—	0.1	0.1	—
Israel	0.6	—	—	0.2	—	—
Italy	9.0	9.1	10.4	3.1	3.5	—
Japan	—	—	—	6.4	4.9	4.0
Netherlands[c]	34.7	31.8	32.3	13.3	13.8	16.6
New Zealand[a]	—	—	—	2.9	3.8	—
Norway	0.8	0.8	0.6	—	—	—
Sweden	7.1	7.4	7.8	—	—	—
Switzerland	10.8	11.5	—	—	—	—
United Kingdom[b]	—	—	—	43.5	41.3	39.9
United States[d e]	—	—	—	39.0	—	—
Yugoslavia	—	—	—	2.5	2.7	3.3

[a] Year ended June (Australia) or March (New Zealand) of year shown. [b] Sales in year ended November of year shown, comprising only woven carpets faced with wool or hair. [c] Covers all carpets of soft fibers; weight figures relate to yarn consumed. [d] Deliveries. [e] Excludes tufted carpets (318.9 million sq yd in 1964). [f] Revised series, including production of woven, tufted or other carpets made in any branch of the textile industry; prior to 1966 comprised production by the wool carpet industry, where the pile consisted either of woven or tufted yarn. Figures in weight comprise carpet yarns consumed by the wool industry.

* Of wool and hair, by wool carpet industry. (*World Wool Digest*, **18**, 6 July, 1967).

textured, or simple patterned styles. Printing and sundry mechanical techniques are being developed with the aim of overcoming this limitation.

Table 1 gives the production of machine-made carpets and floor rugs of 22 countries. The United States is by far the largest producer followed by United Kingdom, Germany F.R., Belgium, Netherlands, and France.

United States of America Carpet and Rug Industry

Some of the basic facts about the carpet and rug industry of the United States as reported by the American Carpet Institute Inc. in 1968 are presented in the following paragraphs:

1967 was the sixteenth consecutive year of growth and expansion for the domestic carpet industry. Preliminary data show that manufacturers' square yard shipments of domestic broadloom carpet amounted to an estimated 353.8 million square yards for the year, a 7.8% increase over 1966. Dollar value amounted to an estimated volume of $1,272,000, representing an increase of approximately 5.2% over 1966.

Tufted broadloom carpet reached 318.2 million square yards, 10.2% greater than 1966 shipments. Dollar volume rose to $1,047,000,000, an increase of 9.7%.

Scatter rug volume declined 18% to an estimated 62.3 million square yards valued at approximately $132.0 million. Shipments of automobile and aircraft carpets amounted to 40.2 million square yards or 7% off the 1966 level.

Details of the 1967 shipments and comparisons with 1965 are shown in Table 2.

The U.S. carpet industry has maintained an outstanding growth rate since 1950. Yardage has increased at an annual average rate of approximately 11% per year and dollar value at approximately 7%. Since 1960, however, the rate of growth has been substantially higher by about 14% yardage and 11% dollar value. By way of comparison, since 1960 gross national product and personal consumption expenditures of U.S. consumers increased by approximately 7% per year.

TABLE 2
U.S. Carpet and Rug Supplies
(in thousands of square yards and dollars)

	Square Yards			Dollar Value		
	1965* Final	1966* Final	1967 Preliminary	1965* Final	1966* Final	1967 Preliminary
Broadloom types						
Tufted	260,361	288,710	318,160	$ 884,621	$ 953,952	$1,046,881
Velvet and knitted (1)	28,533	27,104	24,809	172,724	176,387	156,545
Wilton	6,363	6,448	6,124	47,038	50,288	45,195
Axminster	6,621	5,851	4,735	30,618	27,841	22,917
Domestic broadloom total (2)	301,878	328,113	353,828	$1,135,001	$1,208,468	$1,271,538
Wilton and velvet imports	575	560	669	$ 3,521	$ 3,159	$ 3,922
Other machine-made imports (3)	959	831	584	4,860	4,368	3,516
Total all broadloom	303,412	329,504	355,081	$1,143,382	$1,215,995	$1,278,976

TABLE 2 (*continued*)

Miscellaneous categories						
Automobile carpet	48,474	43,298	40,198	$ 89,194	$ 85,435	$ 84,876
Tufted scatter rugs	67,389	75,705	62,310	123,840	144,043	131,953
Domestic braided rugs (4)	13,816	21,006	N.A.	34,329	56,453	N.A.
Other imports (5)	16,203	18,737	16,046	36,848	43,406	36,628
Total miscellaneous	145,882	158,806	N.A.	$ 284,211	N.A.	N.A.
Grand total all types	449,294	488,310	N.A.	$1,427,593	N.A.	N.A.

* Revised

(1) Includes velvet knitted and other woven.

(2) Rugs and carpeting over 4′ × 6′ in size plus a small amount of 27″ commercial carpeting.

(3) Includes tufted carpets.

(4) Includes domestic braided, hooked and miscellaneous types.

(5) Includes hand-made orientals, hooked, braided, tubular and miscellaneous types.

Sources: Bureau of Census, *Rugs, Carpets and Carpeting, Report MA22Q: Tufted Textile Products, Report M22L* and Report IM146 (Imports)

1967 Data on woven carpets are American Carpet Institute, Inc. estimates.

TABLE 3

Domestic Residential and Commercial Broadloom Carpet and Rug Supplies, 1950–1967[a]

	Square yards (In Millions)	Estimated dollar value (In Millions)
1950	85.7	$ 536.7
1951	64.7	449.2
1952	73.1	421.8
1953	79.7	446.1
1954	82.9	434.0
1955	98.3	520.9
1956	110.3	569.3
1957	112.2	567.6
1958	122.6	559.4
1959	144.5	670.1
1960	148.2	667.1
1961	160.9	686.8
1962	194.0	782.7
1963	223.4	882.6
1964	263.8	1,034.5
1965	301.9	1,135.0
1966	328.3	1,208.5
1967 Preliminary	353.8	1,271.5

[a] Source: 1950–61 Woven component estimates from American Carpet Institute, Inc. Tufted data from Bureau of Census, *Tufted Textile Report.* 1962–66 Bureau of Census, *Rugs, Carpets and Carpeting;* 1967 woven component are A.C.I. estimates. Beginning 1962 dollar value F.O.B. mill; previously dollar value of firms selling through wholesalers had been estimated before wholesale discount. Square yards comparable throughout entire period.

Domestic Broadloom Shipments by Type

Tufted broadloom carpet shipments continued their upward trend during 1967 and amounted to 318.2 million sq yd an increase of 10.2 over 1966 representing 90% of total broadloom shipments. Velvet and knitted yardage totaled an estimated 24.8 million sq yd declining about 8.5% and droppping to 7.0% of the total. Shipments of Wilton estimated at 6.1 million sq yd declined to 1.7% of total shipments. Axminster shipments fell to 4.7 million square yards, a drop of 19.1% or 1.3% of the total.

TABLE 4
Broadloom Carpet Shipments by Type[a]
(Percent of Total Yardage)

	Axminster	Wilton	Velvet[b]	Tufted	Total
1951	39.4%	17.1%	34.1%	9.4%	100.0%
1952	36.4	19.8	29.5	14.3	100.0
1953	32.8	18.9	27.9	20.4	100.0
1954	26.3	16.5	24.6	32.6	100.0
1955	24.5	15.0	24.5	36.0	100.0
1956	20.2	14.3	23.4	42.1	100.0
1957	16.4	12.3	21.0	50.3	100.0
1958	12.6	11.1	18.2	58.1	100.0
1959	10.2	11.4	18.9	59.5	100.0
1960	8.5	9.0	15.5	67.0	100.0
1961	6.7	6.5	14.2	72.6	100.0
1962	4.8	5.7	14.9	74.6	100.0
1963	3.9	4.3	13.2	78.6	100.0
1964	3.0	2.5	10.9	83.6	100.0
1965[c]	2.2	2.1	9.5	86.2	100.0
1966	1.8	1.9	8.4	87.9	100.0
1967	1.3	1.7	7.0	90.0	100.0

[a] Source: Bureau of Census and American Carpet Insitute, Inc. Imports amounted to approximately 1¼ percent of comparable domestic shipment.
[b] Includes chenille and knitted carpets. [c] Revised.

United Kingdom Carpet and Rug Industry

In the following paragraphs are presented extracts from the report for 1965 of the Federation of British Carpet Manufacturers.

The Federation, which represents 97% of the British woven carpet industry, states in its report that, as before, the tufted carpet industry is providing increasing strong competition for woven carpet sales. The total British domestic market for pile carpets and rugs in 1965 was 98.2 million sq yd, compared with 96.0 million in 1964.

Of the total, 52.9% was made up of domestic woven carpets and 37.7% of tufted carpets, compared with 54.4 and 33.6%, respectively, in 1964; imports of all kinds made up 6.8% in 1964 and 4.6% in 1965, continuing the downward tendency of the last few years. Tufted carpets amounted to 37.0 million sq yd in 1965, as against 32.3 million in 1964 and 23.9 million in 1963. However, it

TABLE 5

British Domestic Market for Pile Carpets and Rugs[a]
(Million Square Yards)

	1962	1963	1964	1965
Tufted	15	24	32.3	37.0
Woven[b]	57.4	61.5	63.7	61.2
Total	72.4	85.5	96.0	98.2

[a] Source: Federation of British Carpet Manufacturers. [b] Includes imports.

TABLE 6

Production of Carpets by Pile Content in Britain[a]
(Million sq yd and percentage of total)

Category	All wool		50% + wool		Under 50% wool		No wool		Total
	million sq yd	%	million sq yd	%	million sq yd	%	million sq yd	%	million sq yd
Axminster (spool and gripper)									
1964	3.6	9	23.0	57	12.1	30	1.7	4	40.4
1965	2.5	6	20.6	51	14.6	36	2.8	7	40.5
Axminster (chenille)									
1964	0.6	94	—	—	[b]	6	[b]	—	0.7
1965	0.4	54	0.3	39	[b]	7	—	—	0.7
Wilton									
1964	6.1	47	4.7	35	1.2	9	1.2	9	13.1
1965	5.2	36	6.5	46	1.3	9	1.2	9	14.2
Hair/cord									
1964	—	—	0.1	24	0.2	76	—	—	0.3
1965	—	—	0.1	35	0.2	65	—	—	0.3
Total[c]									
1964	10.4	19	27.8	51	13.6	25	2.9	5	54.7
1965	8.2	15	27.4	49	16.1	29	4.1	7	55.8

[a] By members of The Federation of British Carpet Manufacturers (1960). *Note:* Due to rounding, figures may not add to totals shown; percentages rounded up to nearest whole number. [b] Less than 50,000 lb. [c] Including tapestries and sundries not separately specified above.

was not until the quarter ending in November 1963 that official statistics included Northern Ireland's tufted carpet production; the figure for 1963 is therefore an estimate.

Table 6 gives details of woven carpet production by members of the Federation. As to types of carpets produced, Axminsters made up 73% of the total and Wiltons 25%, compared with 74 and 24%, respectively a year earlier. The table illustrates a continuing increase also in Britain in the use of mixture yarns made of wool and synthetic fiber blends—mostly modified rayons and polyamides (nylon).

1963	
Nylon	35%
Acrylics	13%
Wool	47%
Rayon and polypropylene	3%
Cotton	2%
1964	
Nylon	44%
Acrylics	21%
Wool	30%
Rayon and polypropylene	4%
Cotton	(Less than 1%)
1965	
Nylon	45%
Acrylics	26%
Wool	23%
Rayon and polypropylene	5%
Cotton	(Less than 1%)
1966	
Nylon	43%
Acrylics	27%
Wool	20%
Polypropylene	7%
Rayon	2%
Polyester	1%
Cotton	(Less than 1%)

0 20% 40% 60%

Source: AMERICAN CARPET INSTITUTE, INC.

Fig. 1. Estimated surface fiber consumption for broadloom carpets.

Raw Materials

Consumption of surface fibers in broadloom carpet manufacture in 1967 increased to an estimated 563 million pounds, a gain of approximately 8% over the 519 million pounds consumed in 1966.

Together, man-made fibers accounted for around 85% of broadloom surface fiber use in 1967. Consumption of *filament nylon*, the single most important fiber in terms of poundage used, increased 16% during 1967 to a total of 198 million pounds (about 36%) of the total. *Nylon staple* consumption also rose 5% to a total of 58 million pounds. *Acrylic and modacrylic fiber* use totaled 147 million pounds, an increase of 5% over 1966.

Consumption of *polypropylene* amounted to an estimated 33 million pounds, declining about 8% from 1966 use.

Polyester fiber was first introduced in broadloom carpet in 1966, and rose to 35 million pounds in 1967 against 5 million pounds in 1966.

Cotton and Rayon declined.

Figure 1 shows graphically the raw material changes which took place from 1963 to 1966 and Table 7 shows the consumption of the various fibers used in the manufacture of broadloom carpets from 1960 to 1966.

TABLE 7

Estimated Fiber Consumption in Broadloom Carpets[a]
(Millions of Pounds)

	1961	1962	1963	1964	1965	1966	1967
Nylon filament	45	70	90	131	163	170	198
Nylon staple	23	25	30	45	53	55	58
Acrylics[b]	12	22	45	85	125	140	147
Wool	147	149	160	123	112	104	84
Polypropylene	—	—	—	—	18	36	33
Polyester[d]	—	—	—	—	—	5	35
Cotton and Rayon[c]	30	25	16	20	9	9	8
Total	257	291	341	404	480	519	563

[a] Source: Wool U.S. Bureau of Census, *Consumption on the woolen and worsted systems.* Other fibers are trade estimates. [b] Includes modacrylics. [c] Included Polypropylene 1962-1964. [d] Not used in broadloom carpet prior to 1966.

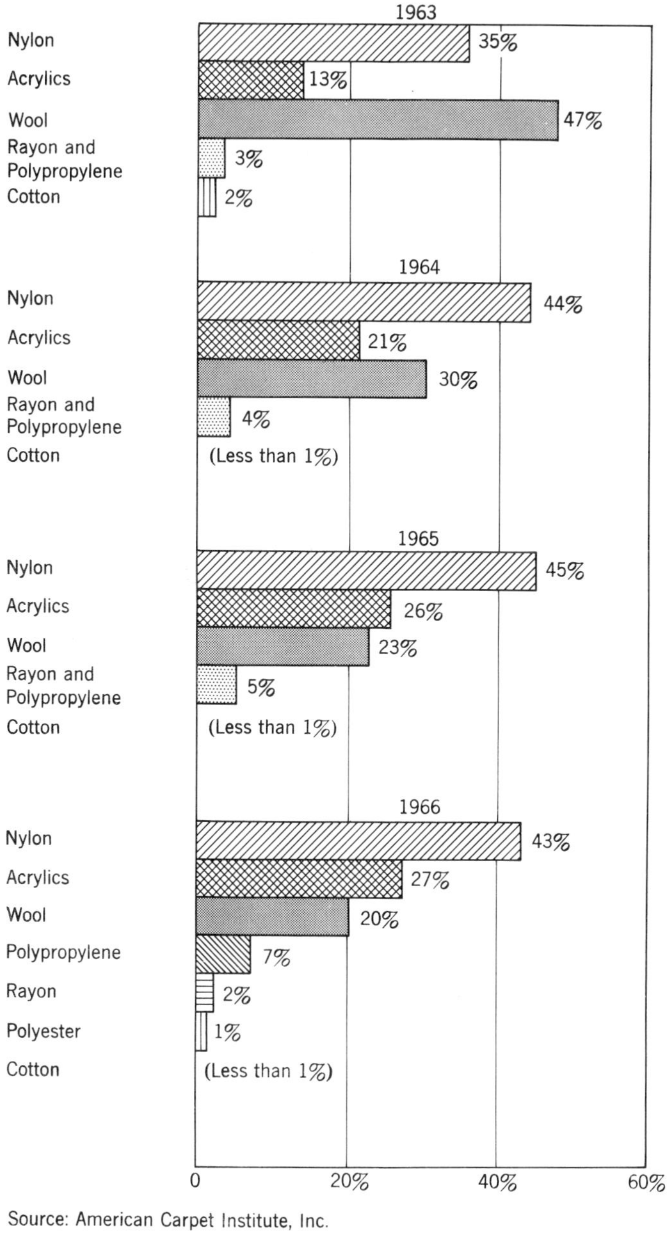

Fig. 1. Estimated surface fiber consumption for broadloom carpets.

Carpet Wool

Carpet wool consumption in 1967 amounted to 84 million clean pounds, a drop of 19% from the 104 million clean pounds consumed in 1966. As a result of this decrease wool continued to lose ground and accounted for 15% of the overall fiber consumption in 1966 as compared with 20% in 1966.

New Zealand and Argentina continued to be the major sources of supply.

The United States Government imposes a duty on all foreign wools to protect the domestic wool grower. The Treasury Tariff Act of 1930 with a few later revisions, permits certain named wools and all wools not finer than 46s to be used duty free in the manufacture of carpets and rugs. In the Tariff Act these carpet wools are referred to as eonomine wools grown on unimproved breeds of sheep and produce a mixture of true wool, heterotypical fibers and Kemp.

The method of grading and marketing carpet wools differs in each geographical area. In general there are two main classifications—shorn and pulled wool. The shorn wools are put into

TABLE 8

U.S. Carpet Wool Imports by Country of Origin
(Millions of Clean Pounds)

	1961	1962	1963	1964	1965	1966	1967
New Zealand	52.2	49.8	66.4	46.2	46.9	52.1	31.0
Argentina	47.6	47.4	41.6	23.2	29.6	33.5	25.9
Pakistan	14.9	12.0	10.7	6.4	6.2	3.6	3.3
Iraq	7.0	5.4	6.6	5.1	3.2	1.8	0.4
Syria	5.2	3.3	5.6	4.7	2.5	4.1	1.9
India	4.1	1.3	1.4	4.3	0.9	1.3	0.3
United Kingdom	6.9	6.4	10.8	6.8	9.2	8.6	6.5
Italy	1.5	1.2	1.9	1.4	1.4	1.0	0.8
Ireland	1.5	1.4	1.1	0.5	1.3	1.6	2.8
Other countries	16.5	15.3	21.8	15.3	7.7	7.0	5.3
Total	157.4	143.5	167.9	113.9	108.9	114.6	78.2

[a] Source: U.S. Department of Commerce.

TABLE 9
Percentages of Each Fiber Type in Carpet Wools by Counting and Weighting

	Types of fibers							
	True wool		Heterotype		Kemp		Colored	
Wool types	By count	By weight	By count	By weight	By count	By weight	By count	By weight
Romney	100	100						
Lanchow	91.03	74.49	5.89	13.18	3.08	12.33	0	0
True Sining	88.65	42.84	9.14	46.19	0.12	0.23	2.09	10.74
Szechwan	87.12	57.65	5.80	27.85	7.08	14.50	0	0
Mixed Sining	84.66	61.51	3.93	16.42	11.35	22.02	0.06	0.05
Woosung	83.72	55.56	0	0	16.27	44.37	0.01	0.07
Aleppo	74.69	43.82	18.87	52.30	2.86	3.15	3.58	0.73
Vicanere	59.22	23.23	40.67	76.67	0.11	0.14	0	0

different classifications, depending on the country involved, such as fleece, second clip, lambs' wool, pieces and matchings, locks, tufts, brokes, crutchings, cots, britch and also into color classifications of white, yellow, fawn, gray, black, and mixed lots. The pulled wools may be called slipes, skin, Haslock or Tanners wool.

The principal physical characteristics taken into account in the selection of wool for carpet are color, strength, and resilience. The demand for pastel colored carpet has added to the importance of color. These wools have a wide range of fiber diameters, and length. Carpet wool fibers range from 1 to 13 in. in length and from 15 to 70 μ in diameter.

Analysis of Carpet Wools

It has been pointed out in volume 1, page 589, that carpet wools consist of a mixture of true wool, hair, heterotypical fibers and kemp. A study of the mixture existing in wools considered as ideal for carpet use was made by Chen. Tables 9 and 10 are from his report.

Carpet wools are not sorted and graded for use. The wool shipments are used in their entirety unless it is contaminated with paint, tar or colored wools. When the best possible white is required, the colored wools are sorted out.

TABLE 10
Fiber Fineness and Length of Carpet Wools

	True wool fiber		Heterotypical fiber		Kemp[a]	
Wool types	Thickness, (μ)	Length stretched (in.)	Thickness, (μ)	Length stretched (in.)	Thickness, (μ)	Length stretched (in.)
Romney	32	8.7	—	—	—	—
Lanchow	19.6	4.3	21.1	4.1	27.5	2.0
True Sining	17.5	4.4	36.3	8.5	—	—
Szechwan	20.6	4.8	33.3	8.5	25.8	2.7
Mixed Sining	17.4	3.2	32.4	7.8	28.4	2.4
Woosung	16.3	2.8	—	—	29.9	1.3
Aleppo	24.0	5.5	41	8.8	39.6	2.0
Vicanere	17.4	2.9	25.7	4.7	—	—

[a] Kemp thickness is arithmetic average of three parts of fiber.

In the United States improved wools not finer than grade 46s may be used in carpets duty free under bond. Present practice of U.S. customs is to make visual examinations of handfuls of wool taken from where the bale covering has been torn or cut. When a lot is doubtful, the bale is broken open and examined. The Tariff Commission seeking an alternate method of examination made a study of laboratory measurement of fiber diameter. In connection with this examination samples from 83 lots of imported coarse wools were tested by the Denver Livestock Division Wool Laboratory, Agriculture Marketing Service, U.S. Dept. of Agriculture during the period November 1, 1958 to June 30, 1959. Table 11 was taken from their report to the Tariff Commission, September 1959, Investigation No. 34.

The shape and size of the medullated portion varies from fragmented medulla spots to continuous and large hollow centers. Medulla appears in fibers of all sizes but is most commonly found in fibers coarser than 30 μ.

Kemp fibers are distinguished from medullated fibers because they are very coarse, brittle and tapered. They always contain large medulla and appear as flattened tubes in cross section. The distinction between Kemp and medullated fiber is in some instances arbitrary. The Denver Wool Laboratory followed the rule that if the medulla accounts for more than ¾ of the diameter the fiber is recorded as Kemp. Because of the ribbon shape the average diameter was measured by the bi-diameter cross section method. Most Kemp fibers measure over 50 μ by this method.

Named, similar and other unimproved wools can be distinguished from improved wools by their wide dispersion of fiber diameter. None of the unimproved wools had a coefficient of variation less than 33% and none of the improved wools had a coefficient of variation as high as 33%.

Unimproved wools, with a relatively predominant fine fiber and a wide range of coarser fibers, invariably have many more fibers under 30 μ, fewer fibers between 30 and 55 μ and many more over 60 μ than do the improved wools. With the exception of the Cordova samples, none of the unimproved wools had less than 3.5% of fibers over 60 μ and none of the improved had as many as 3.5% over 60 μ.

Distribution of number of fibers by diameter does not reveal the extent of the physical difference between these wools. Distribution of the volume of fibers by diameter more clearly depicts the nature of these physical differences. Because of their greater bulk the proportional effect of coarse fibers in total is much greater than in number. In the unimproved wool, which contains the greater proportion of large fibers, the difference between the distribution in number and volume of fibers is pronounced.

Fiber diameter, the percent of medullated or the percent of Kemp cannot be used to determine whether a wool is unimproved or improved. In the samples reported the unimproved wools ranged between 25 and 40 μ and the improved between 32 and 40 μ. The percent of medullated varies widely in both wools. Kemp was present in all the unimproved and in half of the improved wools but the Kemp in the improved wool was generally limited to trace amounts.

Figure 2 shows a cross-section of the middle of a lock of Scotch Blackface wool. It has been analyzed by the Denver Wool Laboratory. The sample had an average width in microns, of 38.8 and a coefficiency variation of 50.9%. The cross section shows the three types of fibers present, namely: the fine wool-hair forming the undercoat, the medullated coarse hair forming the outer coat, and the very coarse, ribbonlike Kemp fibers.

TABLE 11
Analyses of 83 Lots of Imported Coarse Wools for Carpets

Type, visual grade, and condition	Country of origin	Average width, μ	Standard deviation, μ	Coefficient of variation, %	Medullated fibers, %	Kemp fibers, %	Percentage of distribution over 60 μ
		Named Wools					
Cordova	Argentina	31.3	11.8	37.8	7.3	0.2	2.12
Cordova	Argentina	30.9	12.0	38.8	9.8	0.8	2.82
Scotch Blackface	United Kingdom	38.8	19.7	50.9	34.9	11.4	14.13
Scotch Blackface	United Kingdom	35.3	15.1	42.9	30.2	8.7	7.34
Haslock	United Kingdom	36.9	15.1	41.0	34.0	9.9	9.31
Haslock	United Kingdom	36.7	14.8	40.3	32.1	7.6	11.37
Welsh Mountain	United Kingdom	32.1	12.5	38.9	12.8	3.1	3.43
Welsh Mountain	United Kingdom	33.7	12.5	37.2	16.6	2.6	3.79
Cyprus	Malta	34.0	17.4	51.3	18.2	3.7	8.31
Cyprus	Malta	34.5	18.4	53.4	17.4	5.9	9.01
Sardinian	Italy	36.2	18.2	50.4	20.5	4.9	11.74
Sardinian	Italy	32.5	18.8	57.8	11.1	3.2	11.69
Aleppo	Syria	36.0	16.9	46.9	12.4	2.3	11.19
Aleppo	Syria	33.7	15.6	46.2	10.5	1.2	6.74
Aleppo	Syria	34.4	17.0	49.3	9.8	1.6	9.48
Aleppo	Syria	36.5	17.8	48.8	9.3	1.9	8.14
Aleppo	Syria	31.3	13.9	44.3	7.5	2.4	3.94
Bagdad(Awassi)	Iraq	30.0	13.1	43.8	10.1	1.4	3.57
Bagdad(Awassi)	Iraq	32.0	13.6	42.7	7.1	0.4	4.20
Iranian	Iran	35.0	15.8	45.1	8.4	1.4	8.21
Iranian	Iran	36.5	16.5	45.3	12.8	0.9	8.87
East Indian							
Joria, white	India	39.5	14.4	36.4	43.6	9.8	9.68

Vicanere, white	India	36.3	15.1	41.6	40.9	5.2	8.87
Bicanere, superior white	India	33.3	13.0	38.9	34.3	4.7	4.00
White	India	31.7	15.4	48.7	11.3	8.0	6.81
White	India	31.5	15.4	48.7	11.2	7.9	6.67
Montgomery, pale yellow	Pakistan	31.4	13.8	44.0	21.2	5.3	4.99
White	Pakistan	28.1	13.7	48.6	10.8	2.3	3.79
Khorrassan superior white	Pakinstan	31.6	15.5	49.0	16.6	7.1	6.50
Mongolian white	Outer Mongolia	25.2	15.5	61.8	2.9	6.0	6.18
Mongolian white	Outer Mongolia	26.4	18.0	68.1	2.4	7.9	8.94
Irish Blackface	Ireland	32.3	15.3	47.4	18.1	4.4	6.80
Karakul, steel grey	South Africa	30.8	15.0	48.6	7.8	0.2	5.30
Karakul, steel grey	South Africa	30.7	14.5	47.3	7.9	1.3	4.25
		Improved Wools					
BA November	Argentina	36.9	10.4	28.1	6.6	—	2.87
BA Burry	Argentina	33.9	9.1	27.0	7.9	—	0.50
BA pulled	Argentina	34.7	9.4	27.1	6.2	—	0.68
BA pulled	Argentina	35.7	9.8	27.5	4.3	0.2	0.69
BA, 5-6s	Argentina	35.8	9.9	27.7	3.3	—	1.68
BA Super Lambs	Argentina	31.7	10.0	31.4	18.7	0.6	0.87
North German fleece	Germany	36.4	11.0	30.2	18.0	0.2	2.61
North German fleece	Germany	35.8	10.0	27.9	2.4	0.1	1.00
North German fleece	Germany	36.3	10.9	29.9	21.4	0.4	2.12
North German fleece	Germany	35.9	10.7	29.9	19.9	0.4	1.98
Deep Irish Wethers, fleece	Ireland	37.7	9.7	25.8	20.6	0.2	1.93
Deep Irish Wethers, fleece	Ireland	37.8	8.7	23.1	23.0	0.1	0.69
Super Select Wethers, fleece	Ireland	37.9	10.1	26.6	27.3	0.1	2.13

TABLE 11—(Continued)
Analyses of 83 Lots of Imported Coarse Wools for Carpets

Type, visual grade, and condition	Country of origin	Average width, μ	Standard deviation, μ	Coefficient of variation, %	Medullated fibers, %	Kemp fibers, %	Percentage of distribution over 60 μ
Super Select Wethers fleece	Ireland	36.7	10.3	27.9	28.0	0.2	2.04
Irish Blackface	Ireland	39.5	9.5	24.2	26.1	—	2.06
Fleece	New Zealand	37.2	9.0	24.3	2.4	—	1.13
Fleece	New Zealand	38.1	9.3	24.5	6.8	—	1.06
Fleece	New Zealand	38.2	9.6	25.2	4.2	—	1.49
Fleece	New Zealand	37.9	9.2	24.2	4.2	—	0.80
Fleece	New Zealand	39.4	9.3	23.6	10.9	—	2.19
Fleece	New Zealand	38.0	8.9	23.5	4.0	—	0.68
Fleece	New Zealand	38.4	8.5	22.2	1.8	—	0.31
Fleece	New Zealand	37.3	8.6	23.1	4.9	—	0.68
Fleece	New Zealand	37.4	9.0	24.2	6.4	—	0.93
Fleece	New Zealand	37.7	8.6	22.9	4.2	—	0.75
Fleece	New Zealand	37.7	8.4	22.3	3.8	—	0.37
Fleece	New Zealand	37.1	9.4	25.2	4.8	—	0.93
Fleece	New Zealand	37.0	9.1	24.7	5.2	—	0.36
Fleece	New Zealand	37.8	9.8	25.9	9.8	0.1	1.37
Fleece (B style)	New Zealand	37.6	9.5	25.3	6.6	0.1	1.11
B Fleece	New Zealand	36.2	9.2	25.4	5.2	—	0.81
B Fleece	New Zealand	34.0	9.2	27.2	5.2	—	0.75
2nd Shear	New Zealand	34.0	9.4	27.5	7.9	—	0.37
2nd Shear	New Zealand	34.1	9.3	27.1	6.1	—	0.38
2nd Shear	New Zealand	34.5	9.1	26.4	6.5	—	0.37

Crutchings	New Zealand	37.9	10.6	28.0	17.2	0.3	2.01
Crutchings	New Zealand	38.4	10.3	26.8	14.9	0.1	2.06
Crutchings	New Zealand	36.9	8.7	23.6	14.1	0.1	0.56
Crutchings	New Zealand	36.6	11.1	30.3	18.6	0.1	2.11
Crutchings	New Zealand	37.4	11.1	29.7	15.4	0.2	2.67
Crutchings	New Zealand	39.5	10.4	26.4	14.4	0.1	2.24
Crutchings blend	New Zealand	35.8	10.1	28.2	13.9	—	1.69
Crutchings blend	New Zealand	35.3	10.4	29.5	12.4	0.1	2.06
Pulled	New Zealand	33.2	9.4	28.3	14.2	—	0.50
Pulled	New Zealand	33.0	8.8	26.6	13.6	—	0.44
Pulled	New Zealand	33.2	8.8	26.5	11.6	—	0.31
Pulled	New Zealand	32.9	9.4	28.4	11.6	—	0.07
Pulled	New Zealand	33.2	8.7	26.1	15.9	—	0.12
Pulled	New Zealand	33.2	9.0	26.9	16.0	—	0.12
Dag, shorn	New Zealand	34.5	9.7	28.1	10.7	—	0.86
X Bred fleece	U.K.–Scotland	35.6	12.7	35.7	35.8	1.8	4.43
Devon fleece	United Kingdom	37.7	11.4	30.3	49.3	2.9	3.04
Devon fleece	United Kingdom	37.6	11.5	30.5	47.9	3.5	3.24
Combing fleece	Uruguay	36.9	9.1	24.7	4.7	—	0.62
Combing fleece	Uruguay	36.3	9.1	25.2	4.8	—	0.94
Combing fleece	Uruguay	33.7	8.2	24.3	4.3	—	0.24
Heather	Yugoslavia	33.8	15.5	40.0	9.4	1.1	10.94
B Fleece	Australia	36.0	8.9	24.6	7.4	—	0.56
Monte Criolla	Uruguay	32.4	10.3	31.9	19.4	0.2	1.31
Blends of Named or Similar and/or Improved Wools							
Blend Cordova	Argentina	31.8	10.8	34.8	5.1	0.9	1.92
Blend Cordova	Argentina	30.8	10.9	35.5	7.1	0.4	1.83
Blend Londonshire (shorn)	United Kingdom	33.7	15.2	45.2	13.9	2.0	5.37
White carpet blend	Holland	31.0	15.1	48.8	7.0	2.9	5.18

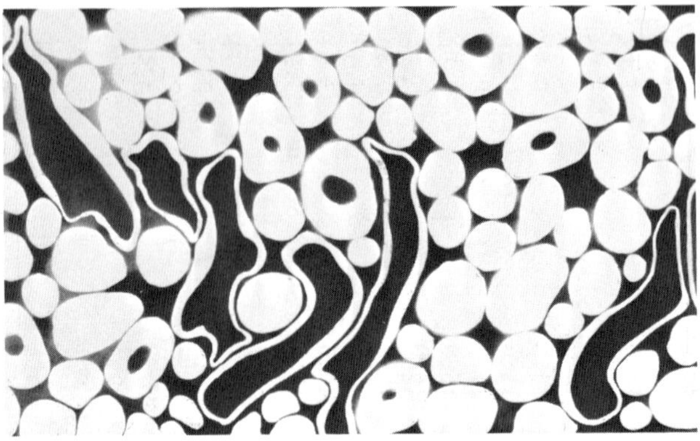

Fig. 2. Cross-section of the mid-section of a lock of Scotch Blackface Wool X250.

Other Carpet Fibers

The complexities of international affairs since the late 1930's made the availability of the common and desirable carpet wools uncertain and unpredictable. The conditions coincident with availability, international and national economic and political circumstances have frequently caused wide raw material cost fluctuations that have incurred extreme variations of profit and loss to the industry. These situations pointed out the need for a restraining influence that could assist in stabilizing the cost of the industry's principle raw material. Obviously, domestic produced synthetic materials not only offered a degree of protection against spiralling and uncertain prices but also insured a more stable and predictable supply. The potentials of this situation was brought to the attention of the fiber producing chemical companies and with this interest their research and development activities were intensified. The introduction and growth of synthetic fibers in a field that appeared to belong exclusively to wool has been phenomenal. Rayon and acetate introduced in carpets in the 1930's were the first synthetic fibers used by industry. Nylon staple was introduced in the 1940's and the acrylics and modacrylics became available in the 1950's. The extent of this growth in the United States of America is illustrated in Table 7.

The American Carpet Institute, Inc. made the following statement in the 1958 edition of "Basic Facts About the Carpet and Rug Industry":

"Viscose, acetate, nylon and acrylic types represent the results of years of intensive research and the industry's answer to dwindling wool production and unstable wool market. They are not substitutes but a permanent addition to the raw material supply. Because they can be engineered by the chemical industry into controlled shapes, contours, lengths, and colors, floor covering with high style, values and bright colors can be produced which have great appeal and service."

The ability of the chemical industry to manufacture textile fiber to detailed specifications enabled the carpet producer to work with raw material having uniform physical properties at a predictable cost and availability. It also enables smaller raw material inventory with a quicker capital turnover. This type of control demonstrates the real plus values of the man-made fibers as compared to nature grown wool. All of the carpet type synthetic fibers can be obtained with general properties summarized as follows:

Staple length: 3 and 4 in. are most common. Other lengths available if desired

Denier: 15–18 are most common. 8, 12, 22, and 25 denier modifications available with some synthetics

Color: All available as white or natural. Solution or dope dyed available in most synthetic categories

Luster: Bright, semidull, and full dull available

In addition to the staple fibers, many of the synthetic carpet types are available in continuous filament form with modifications that impart bulking properties. Although synthetic fibers are experiencing rapid acceptance for floor coverings, wool is the time proven fiber and is used as the reference standard. The following paragraphs summarize some of the values imparted to carpet by several commonly used synthetic fibers or fiber categories.

Rayon

The first synthetic fiber to gain ready acceptance in the carpet industry was viscose type rayon, and the reason was directly associated with the economic factors of price and availability. Rayon also had an additional value—superior white color. As a result, many well styled, low priced fabrics were made available to the consumer. Rayon, however, has elements of deficiency that limit usage in pile floor coverings. The important negative values are:

Ease of crushing and lack of recovery
Rapid soiling and difficult maintenance of appearance
Ready staining and difficult stain removal
Poor flex and abrasion resistance
Water browning or ring staining with water spillage

Nylon

The major values of nylon are in the areas of color, durability, and stain removal. Nylon is among the whiter fibers and makes possible pastel colorings and shades that cannot be made on wool.

The flex life and durability of nylon is outstanding and these particular properties assure the fiber a position in floor coverings. Nylon can be used as a blending fiber with other components to increase the durability of marginal constructions to an acceptable level. In loop pile constructions, staple nylon has a tendency to fuzz. The high strength of the fiber prevents ready breakage and removal in service with the result that the pills formed may create a serious appearance problem. This pilling tendency can be overcome by complete impregnation of the base of the tufts with backsize, stiff brushing and shearing in the finishing. It can be further reduced by using parallel spun yarns with properly applied backsize. The bearding and pilling can be completely eliminated with continuous filament yarns and backsize. With the proper selection of staple or filament and backsize, most types of carpet constructions can be prepared without having severe pilling problem.

Nylon is a nonconductor with low dielectric loss. It is a generator of static electricity that can be an annoyance to carpet users under certain atmospheric conditions. The biggest seller in 1967 in the United States was continuous filament nylon which captured 36% of the fiber market while nylon staple captured 10%.

Acrylics

The acrylic fibers became an important factor for carpet use in the mid-1950's. This class of fibers was intended to fulfill the long sought requirement of being a practical and acceptable substitute for wool. The acrylic fibers in appearance are similar to wool but there are some basic and important differences. The fibers are hydrophobic in character, and this results in a natural resistance to staining and also accelerates the release of stains and soils upon wet cleaning. The felting action of wool is not present in the acrylic fibers, and this makes possible the maintenance of texture of acrylic pile fabrics both during service and wet cleaning. This appearance maintenance property is a major advantage when compared to wool carpeting. In the important properties of crush, crush recovery, and durability, the acrylic fibers are compared to wool. Closest to wool comes the Monsanto Type 71 bicomponent acrylic carpet fiber with its helical, three-dimensional non-reversible permanent crimp.

As a fiber class, processing difficulties had to be overcome in the general areas of skein and piece dyeing, twist setting, and carpet finishing. Acrylic and modacrylic staple fibers accounted for 26% of the United States carpet fiber market in 1967.

Continuous filament high bulk acrylic yarn in carpet deniers is available in experimental quantities. Compared to nylon, the difference in durability and present costs will limit the use of this yarn form.

Modacrylics

The general properties of the modacrylics are similar to acrylics with one important exception. The acrylic fibers are flammable, and the modacrylics are nonflammable. Continuous high bulk modacrylic yarn in carpet deniers is available in experimental quantities. The negative factors of the acrylics also apply to this yarn.

Olefins

In the category polypropylene staple fiber in carpet deniers has created interest for use in floor coverings. Because of low specific gravity, the fiber offers greater bulk and coverage than any other carpet fiber. The flex life and durability are sufficiently similar to nylon to enable the industry to utilize these factors and make serviceable fabrics using less face fiber weights. However, full utilization of these factors is limited by the fiber's lack of resistance to and recovery from crush. Some compensation of these negative factors can be made by increasing the face fiber density in the construction. In addition, the fiber is slippery and will come out of its anchorage to beard. This is particularly evident in loop pile fabrics. It can be used in cut pile fabrics provided the base of the tuft is securely anchored.

Polyester

Polyester fiber, first introduced in broadloom carpet in 1966, has made significant inroads in the carpet market. In 1967 it rose to an estimated usage of 35 million pounds, about 7 times the 1966 consumption. Producers expect this figure to double by the end of 1968.

Staple Fibers

The synthetic staple fibers as a complete class have certain processing similarities to wool. They can be blended, carded, spun, and twisted by conventional equipment, and yarn constructions within the normal range of wool yarns can be produced. In general, it can be stated that woolen system spun yarns produced from synthetic fibers are smoother and more uniform than wool yarns. These processing similarities combined with the specific properties discussed in the preceding paragraphs insure continuing growth and interest in these fibers for pile floor coverings.

Backing Materials

In addition to the various surface fibers, the carpet industry also consumes quantities of backing yarns in woven carpets and pre-woven backing for tufted carpets. Although no exact figures are available, it is estimated that during 1967 the industry consumed approximately 30–32 million pounds of jute and kraftcord or paper yarns and approximately 5–7 million pounds of cotton and rayon backing yarns for woven carpets.

Import figures of pre-woven jute backing for use in tufted carpet manufacture are given in Table 12. India, the major source of jute backing, accounted for 80% of imports. Kraftcord, cotton and rayon were used for backing materials and in 1965 backings of polypropylene fiber were introduced into the market to replace jute and kraftcord, which rot and shrink outdoors. While no reliable data are available on the latter materials their use is increasing. The newest innovation is flowed-on foam backing, mostly high density polyurethane.

Grosicki and Thangiani (6) discussed the improvements in carpet characteristics as the result of back sizing processes. Rhodes (10) gives an outline of the evolution which has taken place since 1948 in the field of carpet backing fabrics.

TABLE 12
U.S. Imports of Prewoven Jute Backing over 100 inch in Width
(In thousands of linear yards)[a]

	India	Other countries	Total
1963	68,141	15,946	84,087
1964	83,118	19,898	103,016
1965	88,096	20,167	108,263
1966	97,858	24,476	122,334
1967	110,660	22,092	132,752

[a] Source: U.S. Department of Commerce, Report IM-145. Substantially all of the yardage imported is in the form of 9 ft, 12 ft and 15 ft widths with 12 ft predominating. A rough estimate of square yards may be obtained by multiplying by 4.

Blending of Carpet Fibers

Color, diameter, length, strength, resilience, and loftiness are the main characteristics to be considered for carpet pile yarns. No one kind of wool contains all of these characteristics to the degree desired by the carpet manufacturer so that it is common practice to blend a number of kinds together to satisfy the needed requirements. The different characteristics must be blended together to make it possible to produce yarns to a definite specification. Blending minimizes the variation between lots obtained at different times from different places. The blend components are proportioned to retain the desirable attributes of a particular wool and to minimize the undesirable. Thereby, the manufacturer is less dependent on a single source of supply and it is possible to vary the wools to maintain product quality and cost standards as related to procurement, availability and price. The use of a wide range of wool components in a blend makes it easier to maintain the finished products important characteristics regardless of availability and the seasonal variations occurring in wools. Manufacturers consider the composition of their blends a trade secret.

A common method of blending is to distribute the different wools directly from the bales onto a long table feeding the bale breaker of the wool scouring train. The wools are tumbled and mixed as they go through the scouring and drying units. For greige yarn manufacture the scoured wool is piled on the floor and then torn down vertically as it is fed into a mixing picker. After the mixing picker, it is again put into a pile and cut down vertically as it is fed into the card feed box. For stock dyed yarns the scoured wool is sent directly to the dyehouse and after dyeing the numerous dye kettle loads are successively piled on top of one another. The pile is cut vertically and fed into a mixing picker. This is usually turned over twice before carding to insure thorough mixing of the various dye kettles making up the order.

Recently, because of a style trend to pastel shade that has created a need for maximum utilization of white wools a method identified as type scouring has come into use. The type wools are scoured as received and not blended with any other type on the feed table. After scouring, the wool is sampled and baled. The sample is used to rate the whiteness and then use the wool most advantageously in a blend. The bales of the various components making up a blend are proportionately stationed around a number of feed boxes that then deposits the wool in small bits onto a fast moving belt feeding a mixing picker. After the mixing picker, the wool is baled and the bales are numbered. The numbers are randomized and the bales are placed around a feed table and put through the system again. This is repeated until it is acceptable for carding.

Synthetic fibers can be blended with each other or with wool using either of these methods. The yield of finished yarn from a given amount of fiber is greater for synthetics than for wool. The starting proportions of the fiber components in the original blend put up will be different from that in the finished yarn. The addition of synthetic fibers to wool will also increase the percent yield of finished yarn over that normally obtained from 100% wool. In order to make up a blend for a predetermined ratio of fibers in the finished yarn it is necessary to adjust for normal water content and to have experience with a particular card and spinning set up. This is important because of the Fibers Labelling Act put into effect in 1960.

It is accepted as good practice to limit a single blend component to a minimum of 10% of the total blend put up. It is very difficult to obtain uniform fiber dispersion of smaller amounts. To insure uniform blending of wool and synthetics in combinations of a ratio greater than 2 to 1 it is good practice to preblend. This consists of taking the smallest portion and putting it with an equal amount of the largest component. One turnover of the preblend is sufficient. The preblend is then added to the balance of the predominant component. Two turnovers are usually sufficient but this could depend on the requirements of the finished product.

Carpet Yarns

Woolen, worsted and semiworsted yarns used in carpet and rug manufacture are much heavier in count than those used in the apparel industry. Worsted yarns are generally used in Axminster oriental styles and Wilton patterns for commercial and specialty installations. Woolen yarns are predominantly used but in the last few years the half worsted yarns have gained considerable ground as base yarns for tufted carpets. Mixtures of wool with one or more of the following synthetic staple materials are common in carpet yarns today and may be spun on either system:

Acetate, Acrylic, Mod Acrylic, Nylon, Polyester, Polypropylene, Rayon.

Recently continuous filament yarns of acetate, rayon and nylon have been introduced to the trade. They are generally referred to as *textured*, *lofted*, or *bulk* yarns that require less face yarn content in fabrics to obtain a coverage equal to staple yarns. Examples of continuous Nylon 6-6 filament yarns are: Antron Type 826, 1500 denier 80; BCF Nylon, 1300 denier 68, 2600 denier 136 and 3700 denier 204.

Several different stylings from wool or wool–synthetic blends are obtained exclusively from the actual yarn construction. These textures or stylings are referred to as: *Plush*, *Saxony*, *Twist* or *Frieze*, *Back-twist* or *Pull-down*, and *Novelty* or *Textured* yarns. Except for novelty yarns, these effects are generally achieved by modifications of yarn count in combination with the singles and ply twists.

Systems used to express yarn count or weight include the *woolen* count, or actual number of yards per ounce; *run*, or number of 1600-yd units per pound; *worsted*, or number of 560-yd units per pound; *metric* or number of 1000 meter units per kilogram, or *tex* or number of grams per 1000 meters.

TABLE 13

Carpet Yarn Number Conversion Table

Woolen, yd/oz	Run, 1600 yd/lb	Worsted, 560 yd/lb	Metric, 1000 m/kg	Tex, grs/1000 m
20	0.20	0.57	0.64	1550
31	0.31	0.89	1.00	1000
40	0.40	1.14	1.29	778
50	0.50	1.43	1.61	622
100	1.00	2.86	3.27	310
150	1.5	4.29	4.84	207
200	2.0	5.71	6.45	155

Most carpet face yarns range in singles count from 40 to 200 yd/oz with spinning twists from 3 to 6 tpi. Generally, not more than four singles are used in a ply yarn construction. Twist ranges of two-ply yarns are 0.5–12.0 tpi; three-ply from 0.5 to 9.0 tpi; and four-ply from 0.5 to 6.0 tpi. The plied yarn count in all weaves and tufting ranges from 20 to 80 yd/oz. Combinations of these twists and counts create the end yarn types described in the following paragraphs for wool or wool-synthetic blends on either the woolen or worsted system. Some of the synthetics, when used by themselves, do not retain sufficient "memory" for heat or twist setting and are not applicable.

Plush Yarns

These are low twist in single or multi-ply. Visually, they have a smooth or velvet effect and create good carpet face coverage. An example is a two-ply, 60 yards to the ounce yarn with three turns per inch spinning twist in the Z direction and a ply twist of 2 turns, S direction.

Saxony Yarns

These are referred to as balanced twist set yarns and produce an individual tuft appearance in cut pile constructions. A three-ply, 30 yards to the ounce yarn with 3.0 Z spinning twist and 3.0 S twist in the ply followed by a heat set process would produce this particular style.

Hard Twist Yarns

These are more textured saxony yarns obtained by increasing the ply twist in the above example from 3.0 S to 5.0 S.

Frieze Yarns

Twist or frieze carpets have a unique texture as the side and tip of the cut tufts are randomly seen on the carpet surface. The tips are darker in shade than the sides of the tufts. This texture is obtained from an unbalance in the twist and a ply twist sufficient to render a kinky yarn in the relaxed state prior to twist setting. The saxony yarn above with 8.0 turns ply twist would produce this effect.

Backtwist or Pull-down Yarns

Axminster fabric styles were broadened after World War II with this type yarn. The drawn down effect creates a lower level pile height when used with standard Axminster "plush" or "saxony" type cut pile yarns. This patented process involves a multi-ply, twist-on-twist yarn which is heat set and then back- or untwisted beyond zero. The yarn, when cut in the weaving operation, tends to return to the first ply twist and thus contracts in length.

Novelty Yarns

Many unusual variations of yarn manufacturing not described before are grouped in this category. These include slub, flake, and nub yarns which are created by special mechanical attachments to the card or spinning frame. Special novelty effects made on the twisting frame are described in Chapter 6.

Carpet Backing Yarns

Many types of materials are used to form the woven carpet back, which is composed of filling or weft yarns, and stuffer and chain, called warp yarns. Those most generally employed for this purpose are given in Table 14.

TABLE 14
Carpet Backing Yarn Data

Type material	Yarn Count[a]	Ply	Number ends used
Filling yarns: (weft)			Per Pick
Cotton	11s–2s	1–6	1–4
Jute	8–20	1–2	1–4
Kraftcord—Paper	10–16 (Jute count)	1	1–4
Stuffer yarns: (warp)			Per Dent
Cotton	11s–2s	2–6	1–4
Jute	10–20	1–2	1–4
Kraftcord—Paper	12–16 (Jute count)	1	1–4
Chain yarns: (warp)			Per Dent
Cotton	11s–4s	2–6	1–4
Jute	10–20	1–2	1–4
Rayon	1100–2200 d.	1–2	1–4

[a] The count used in each case is the count for that individual material, namely: cotton: number of 840-yd hanks of yarn per pound; jute and kraftcord: The weight of 14,400 yards of yarn; rayon: denier, grams per 9,000 meters.

General Description and Preparation for Weaving

Weaving is the operation of forming a fabric by interlacing yarns or filaments to make a continuous web. The filling or weft yarns are those yarns running across the width of the fabric, while warp yarns run lengthwise in the carpet. In carpet weaving reference to "warp" includes the chain, stuffer and the pile face yarns. The stuffer yarns are used to give the fabric "body" and "hand" and not to bind in the filling yarns as do the chain warp yarns. The face and stuffer yarns are held together by the filling and the chain yarns. The sturdiness of the back contributes to the stability and general service performance of a carpet.

The service quality of a carpet is controlled by the density of the face materials. Density is achieved by the combination of (*1*) "pitch" or number of tufts across the width, (*2*) the number of tuft rows per inch in the warp direction, and (*3*) the face yarn count. Pitch is often expressed as the number of dents per 27 in., for example: 216 pitch is 8 dents per in. The face pitch for velvet and Wilton carpets ranges from 90 to 256, the predominant pitch for Axminster is 189. Pile height contributes to the style and luxury attributes of a carpet but it does not make the same contribution to serviceability and utility as the weight obtained by density.

Many of the yarns used in carpet must be specially prepared for use in a loom. The filling yarns are made into cops and these are dense, tightly wound coreless packages that will fit into the loom shuttle. Frequently, but not always, the cops are sized in a pressure vessel with water soluble gums that are selected for reasons of good penetration, adhesive and rewettable properties. The amount of size pickup is about 15%. The sized cops are used in a moist condition to assist in the yarn release and feed smoothly from the shuttle and also to improve the fabric stability and the pile bind. Weaves that use wavy, serrated or undulating wires use dry filling to reduce the friction and facilitate the withdrawal of the wires. Unsized Kraftcord or paper yarns also is used for filling.

Most stuffer yarns for velvet and Wilton carpets are sized on a slasher with rewettable starch and gums and wound onto a beam. The amount of size pickup is up to 15%. Kraftcord or paper stuffer yarns and the stuffer yarns for Axminster are not sized but are wound onto a beam for the loom. The function of stuffer yarns is to give the carpet hand and body and the addition of size to the yarn further enhances the contribution of stuffer yarns to these attributes. It also increases the strength and reduces chafing and fiber shedding in the loom.

Usually, chain yarns for velvet and Wilton are sized with rewettable gums and a small amount of starch on a slasher and then wound onto a beam. The size pickup ranges from 5 to 8%. Rayon chain for Axminster is sized with latex or resin to reduce slipperiness and avoid filament breakage but cotton yarn for the same purpose is not sized. Many of today's weaves do not require sized chain yarns but the yarn has to be converted from a supply package onto a beam that will fit the loom with the number of yarn ends dictated by the fabric pitch.

The face yarn for the conventional velvet loom is put onto a magazine creel fitted with a sufficient number of spindles to accommodate all the different pitches to wind 18 or 36 in. sections of carpet width onto a "*mill*." After the complete number of sections for a prescribed carpet width are put upon this *mill* the yarn from all the sections is simultaneously transferred to a loom beam in one operation. The reasons for using a mill are for efficiency, reduce yarn waste, and conserve floor space. Some velvet looms will take the yarn directly from a yarn supplier's package placed in a creel behind the loom. The Wilton loom draws face yarn from creel frames, each made up of spools or packages of the properly selected colored yarns.

The carpet designer and the colorist are using a square or check paper to draw the pattern or figure. Each square represents a tuft in the carpet.

A complete unit of the decorative figure or a pattern is generally made 9, 12, 27, or 36 in. in width and length for symmetry and the general purpose of adaptation and application to various shapes and dimensions of floor space that might use carpet. The design may be constructed alternately across the width or length to make what is called a "drop" match or a "set" match and this is shown in Fig. 3.

Carpets can be grouped in the following style categories:

Type	Description
Plush	Level cut pile surface using balanced low twist yarns
Twist	Irregular cut pile surface using frieze or unbalanced twist yarns
Plain texture	Multilevel loop pile surface in a pattern
Tweed	Moresque colored yarns in level and multilevel pile heights
Embossed	Level pile surface with tuft omission or with selected coverage of filling with face yarn
Figured	Level and multilevel pile surface using different color and twist yarns controlled by a pattern

The designer and the colorist have the greatest latitude when working with spool Axminster. Theoretically, the number of colors is limited only by the number of tufts in the carpet but in practice it is controlled by the economics of small dye lots and the ensuing yarn waste. The Axminster designer may use different twist and

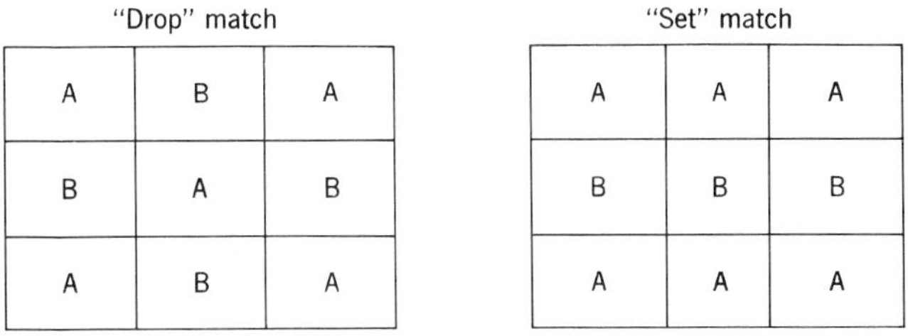

Fig. 3. Method of design construction.

yarn counts for plush, tweed and backtwist yarns for pull-down, multilevel, or embossed effects.

The Wilton designer is usually limited to the use of six colors but he can introduce more by splitting a frame. Additional style effects may be obtained by:

1. Combination of pile wire heights and the use of serrated wires.
2. Embossed effects can be obtained by holding the pile warp under the pile wire for 1 to 3 wires.
3. High level or floats can be made by holding the pile warp over the pile wire for 1 to 3 wires.
4. Cut and loop pile by using wires with and without cutting blades.
5. Multilevel loops and special effects by shearing.

Velvet carpets are designed primarily for use of plain colors in plush, twist, and plain textures by using balanced and unbalanced twist yarns, serrated wires and wires with and without cutting blades. Tweed colorings are obtained by using moresque yarns.

Chenille carpet designing is very similar to spool Axminster. The designer is unlimited in use of yarns and colors.

Knitted carpets are designed for plain colors, textures and tweed loop pile effects by use of yarn. Geometric designs in color and texture are made by planting colors in respective guide bars and controlling the tension of the yarn feed.

Machine tufted carpets are designed for plain plush and loop pile surfaces. Plain and patterned textures are obtained by use of yarns, pattern devices and oscillation of the backing cloth. Moresque yarns are used for multi-colored tweeds and colored stripes are made by planting in the creel. Patterned cut and loop pile and patterned two color loop make use of pattern devices and two needle bars. Figure 4 shows cutting cards for a Wilton pattern, at Lee's Glasgow plant.

Velvet Carpet

The simplest of all carpet weaves and is used mostly for solid colors. When closely woven, velvet carpet is long wearing and rich looking. The range of colors and texture variations is almost limitless. A plush effect results when the yarn is straight and the pile cut. Tightly twisted yarns in cut pile provide a frieze surface. Uncut looped pile gives a pebbly texture. For sculptured effects, the pile is woven in different heights to form a pattern or hit-and-miss effect. Combination of cut and uncut pile produces attractive surface texture.

Fig. 4. Cutting cards for a Wilton pattern. (Courtesy of James Lees and Sons, Division of Burlington, Industries, Inc.)

The velvet loom is primarily a single shuttle, plain and flat loom with cam operated harnesses. Warp yarns for stuffer, chain and pile are fed into and threaded through heddles in each respective warp harness from beams mounted on the back of the loom. The chain warp ends on a beam are divided into two sections and the ends from each section are threaded through its own respective harness frame, thus with one harness frame in a high position and the other in low, a V opening or shed is formed by the chain warp ends.

The filling yarn enclosed in a shuttle is inserted into and through the shed and this insertion is called a pick. The filling yarn is projected across the loom on top of the bottom chain warp, under the stuffer warp and then is returned over the top of the stuffer warp under the top chain warp. The chain yarn harnesses then reverse their position of high and low and bind together the filling, stuffer and pile warp yarns. The harness action for the face yarn is so timed to allow the introduction of a pile wire under the face yarns coincident with the shuttle pass for the bottom "pick" or "shot." The face yarn harness is dropped to loop the yarn over the pile wire and then the shuttle passes over the face yarn and under the top chain warp as previously described for the top shot of filling. A cross section of a conventional two shot cut pile velvet weave is shown in Fig. 5 and a schematic of velvet weaving is shown in Fig. 6.

A "through the back" two shot velvet carpet is made by putting the face pile yarn on the bottom shot. Loop pile is made by omitting the cutting blade from the pile wire. Nonraveling type of carpet is made by applying sufficient latex backsize to this weave.

Another conventional but special purpose velvet construction is the 3 shot velvet shown in Fig. 7. This is used in commercial carpets and for worsted or other slippery face yarns to retard any tendency to sprout by improving the pile bind.

Certain special style effects have been obtained with a 4 shot velvet weave.

A variation from conventional velvet weaves is the insertion of the wire with the top filling shot. A cross section of a "wire on shot" weave is shown in Fig. 8. This loom uses 5 harnesses, 2 for pile, 2 for chain and 1 for stuffer. The cut shows the 2 pile yarns in the same dent shedding over alternate noncutting wires. Some variations of this weave use pile yarn draws ranging from ½ to double pitch.

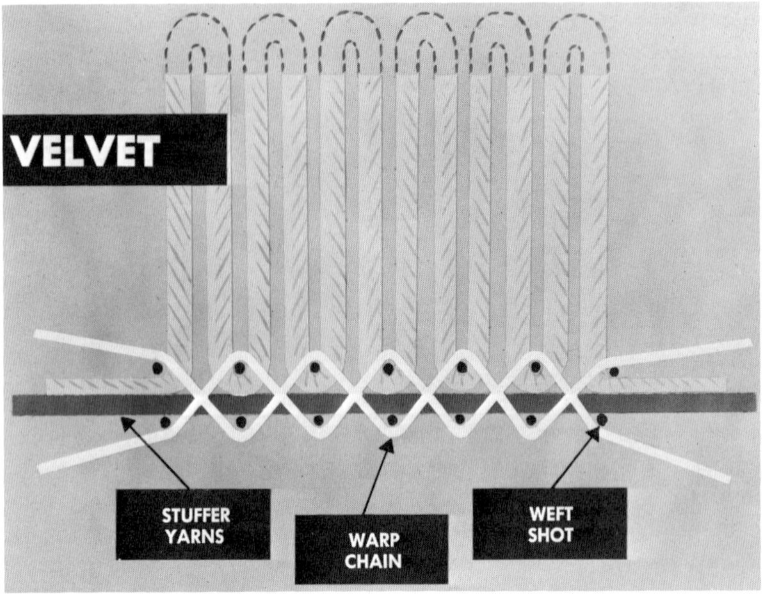

Fig. 5. Conventional two shot cut pile velvet weave.

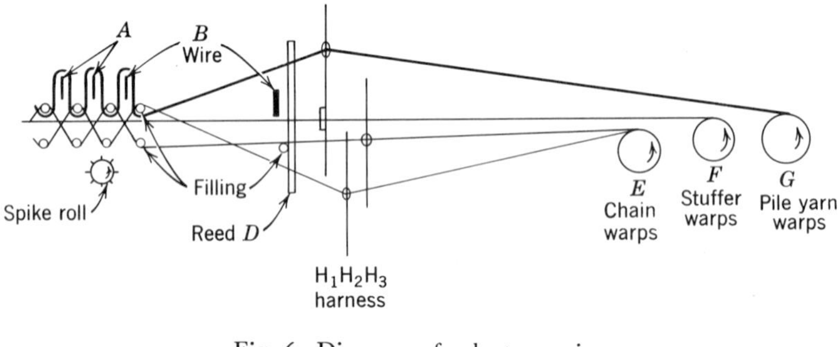

Fig. 6. Diagram of velvet weaving.

A cut or loop pile velvet weave with texture can be made by putting the pile warp ends over 2 wires and then binding under the next filling shot. This is made on a loom equipped to weave with wire inserted with either the bottom or top shot. This is called a Double Wire Velvet. A cross section of a 5 harness weave equipped

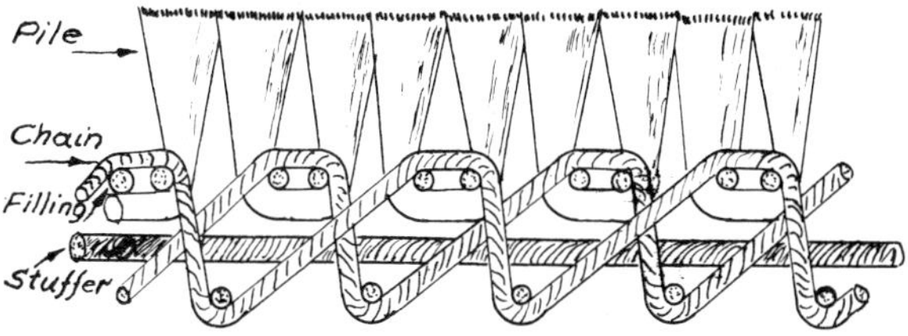

Fig. 7. Three shot cut pile velvet.

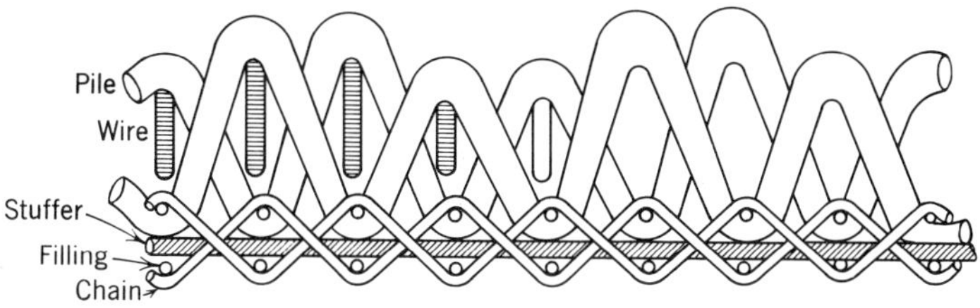

Fig. 8. Wire on shot velvet.

Fig. 9. Double wire velvet.

with special cams to put the pile warp ends over 2 wires is shown in Fig. 9. This shows one of the many possible variations in which 2 noncutting wires are planted alternately in the wire box to produce pile texture.

Velvet carpets with an irregular and designed surface employ serrated, undulating or profile wires. A sketch of this type of wire appears in Fig. 10. The standard carpet wires are ground to varying depths as designed to create a pattern. The pattern is made by "back robbing" or drawing back pile warp yarn from the previous row by the highs in the wire when it is withdrawn. The pattern

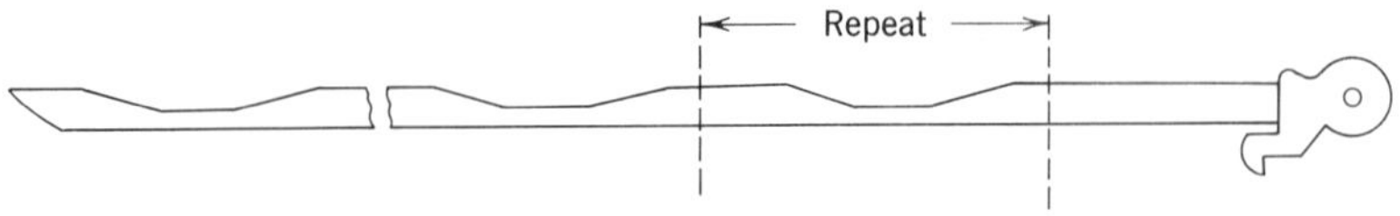

Fig. 10. Patterned profile wires.

repeat across the wire may be 9 or 18 in. Generally 40 wires are used to make a complete pattern.

New velvet weaves have been developed that are characterized as single insertion or single shot weaves. These are made on a modified loom that uses four harnesses, two for chain and two for pile. It does not use any stuffer yarns. The filling may be inserted with either a needle or a shuttle and it can operate with the pile yarn on a beam or yarn suppliers' packages from a creel. The loom weaves the face yarn through the back and permits many weave formations by varying the draw and the pile wires. The carpet is heavily backsized with latex or resin to obtain "hand" and a pile bind that exceeds conventional velvet weaves. A cross section of the weave is shown in Fig. 11.

Another variation of single shot velvet weaving employs wire motions operating on each side of the loom. The pile wire is inserted alternately from each side for each pick and this results in some increased productivity. The filling insertion is with a shuttle.

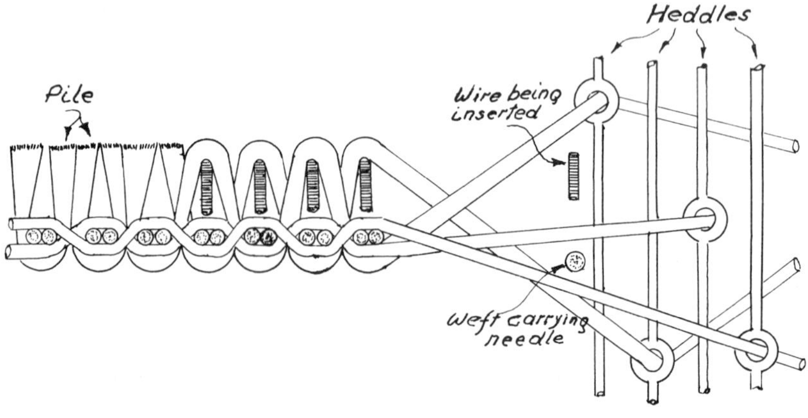

Fig. 11. Single shot or modified velvet.

The loom is equipped with 4 harnesses, two for chain and two for pile warp. The heddle uses an eye and slot; pile yarn is threaded through the eye and chain yarn through the slot. This loom puts the face yarn through the back and makes possible many weave variations by manipulating the draw and the pile wires.

Velvet type weaves are used primarily for the production of plain colors or the multicolor effects obtainable by using moresque yarns. This weave is efficient in the use of the more expensive face yarns and the manufacture is relatively more economical than other weaves. The pile yarns may be soft or hard twist types with the textures ranging from smooth plush to frieze and high-low patterned surfaces. The color patterned limitations of velvet carpet frequently determines the need for those weaves with greater color selectivity.

Wilton Carpet

Wilton carpet takes its name from a town in England where it was first made. It is woven on a loom with a specialized combination of a Jacquard system. A distinguishing feature of the loom is a series of pattern cards perforated like player-piano rolls. These cards automatically regulate the feeding of different colors of yarn into the loom as shown in Fig. 12.

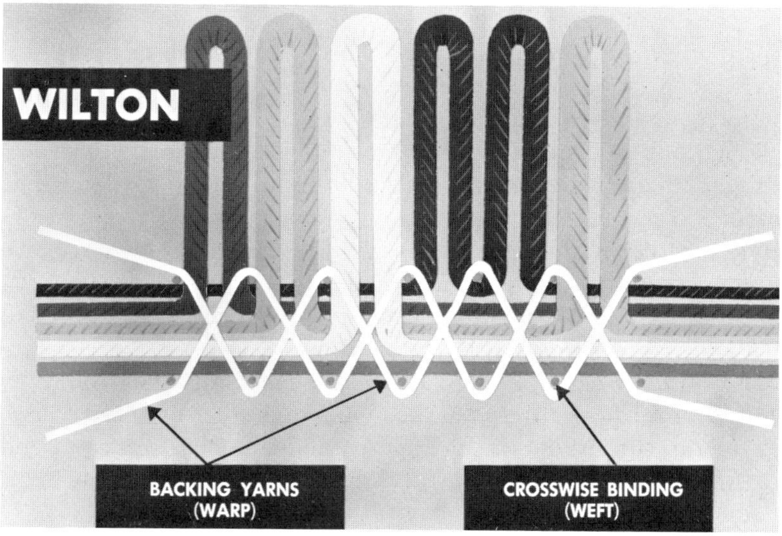

Fig. 12. Conventional two shot Wilton weave.

Though the number of colors possible to use on this loom is not unlimited, the Jacquard mechanism assures accurate reproduction of intricate patterns with great clarity. One color at a time is drawn up as pile, the others are buried beneath the surface. This feature gives additional body, resilience and strength to Wilton carpet, and is responsible for its "hidden-value" reputation.

Good quality Wilton is a luxurious and long-wearing carpet with deep, dense pile. It is available in a wide range of multicolor designs and also solid colors. Many new texture variations, including modern carved effects, still further widen the choice of Wilton carpets and rugs.

Wilton carpets are made on a Jacquard equipped loom that can make either loop, cut pile, combination of loop and cut, and level or multilevel pile surfaces. Basically, this is a velvet loom with the face yarn banked in a creel instead of on a beam and fitted with a mechanism that uses punched cards to select the colored yarns that appear on the face of the carpet. The system was invented by Joseph Marie Jacquard of Lyons, France, in 1800. Each card controls the tuft selection for one row across 9 or 18 in. of carpet and the cards are laced together in consecutive order to complete a pattern. The older system referred to as Brussels, which at one time meant a 2 shot low loop pile Jacquard carpet, uses a card with large punched holes for 9 in. of carpet width. A more modern system referred to as fine index uses smaller holes and controls 18 inches of carpet width. Skilled operators working from a design pattern transfer that design check by check, each check representing a tuft, to a plate equipped for hole punching. When the plate is completely set a full card is punched out or "cut." The number of holes in a card will depend on the pitch and whether there is a hole or a blank space determines which end is selected. The Brussels system selects by means of a blank, and the fine index system uses the punched holes. A newer system, referred to as modified or 3 position Jacquard employs the combination of both blank and hole to select and position the face yarn ends.

The number of colors used in a pattern usually determines the number of frames or banks of face yarn in a creel required to hold the yarn to supply the loom. Each frame will carry sufficient ends of a single color to put an end in each dent for the entire width of the carpet. The number of frames used also signifies the number of face yarn ends to be found in a carpet dent and serves as identification of the wilton weave. A Wilton loom equipped with either the Brussels or fine index Jacquard may use from one to six colors while the modified or three position Jacquard is limited to three. More colors may be introduced by planting or splitting a frame.

The colors that do not appear on the face are buried in the back. Because of this, claims are made for hidden value but this is difficult to explain, except as it contributes to costs. The yarns that are buried in the back help to make a stiffer and more compact back; however, this is not the position of most usefulness for a face yarn. Multiple framage alone does not insure serviceability but well constructed Wilton carpets are highly regarded for their ability to withstand the heavy service demands of commercial installations.

Jacquards for Carpet

The face of a card for a five frame Brussels Jacquard is shown in Fig. 13. The dimension of the card is $17\frac{1}{2} \times 2\frac{5}{8}$ in., the punched holes are $\frac{1}{4}$ in. in diameter. A card controls the yarn selection for 9 in. of carpet width. The left half of the card controls the yarn selected for the face of the odd numbered carpet dents and the right half selects the yarn for the even numbered dents. The left side of the card shown becomes the leading edge as the cards rotate around the Jacquard cylinder. Starting at the top left of the card in Fig. 13 and counting across the first horizontal row, holes 1, 2, 3, and 4 are punched out and this would cause the yarn in the fifth frame to be lifted over the carpet wire and appear on the carpet surface in the first dent. In the Brussels Jacquard the fifth frame is selected when the holes for all the other frames are punched. Continuing to read the same row in the right half of the card and starting the count over again, holes 1, 2, 3, and 4 are punched the same as the left side of the row causing the same action to be repeated for the second dent. Moving down to the second horizontal row the space where the number 1 hole would appear is blank and numbers 2, 3, and 4 are punched out. This will cause the color in number 1 frame to be lifted over the wire and appear in the surface of the carpet in the third carpet dent.

A complete reading of the holes and blanks in the card of Fig. 13 and the frames selected for the pile tufts follows:

TABLE 15
Reading Charts for Five Frame Brussel Jacquard

No. yarn/dent	No. frame/color	No. yarn/dent	No. frame/color
1	5	16	4
2	5	17	1
3	1	18	3
4	1	19	3
5	1	20	3
6	5	21	3
7	3	22	2
8	3	23	2
9	4	24	2
10	1	25	2
11	4	26	1
12	4	27	5
13	4	28	4
14	4	29	1
15	4	30	1

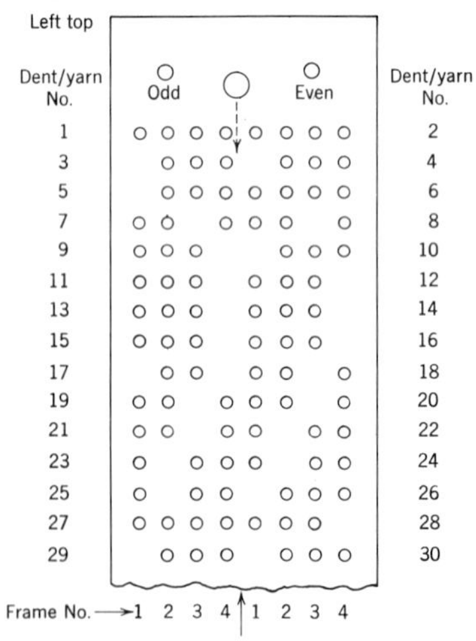

Fig. 13. Section of card for Brussels Jacquard.

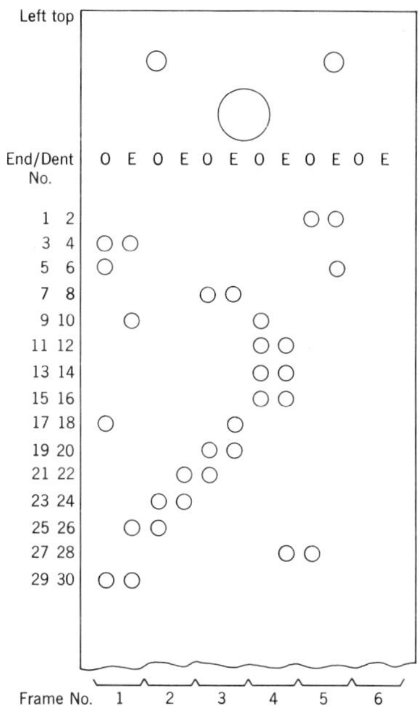

Fig. 14. Section of fine index card for Jacquard.

The face of a card cut for the same design as shown in Fig. 13 using the fine index is shown in Fig. 14. This system groups together 2 dents per row. The left punch in a frame space selects the ends for the odd numbered dents and right punch selects for the even numbered dents. The left edge of the card shown in Fig. 14 is the leading edge as the cards rotate around the jacquard cylinder.

A schematic of the Brussels card Jacquard head mechanism is shown in Fig. 15. The perforated cards for the complete design are laced together to make a continuous belt that revolves around a perforated "cylinder" (B) which may be square, pentagonal or hexagonal in shape. The perforations in the cylinder are indexed to match all possible holes that could be punched into a card. The cylinder presses the face of a card against a needle board holding the needles (C) that protrude about 3/8 in. beyond the board's surface. The needles (C) are maintained in this position by the weight (G) hung on the cord which exerts an effect on the needles

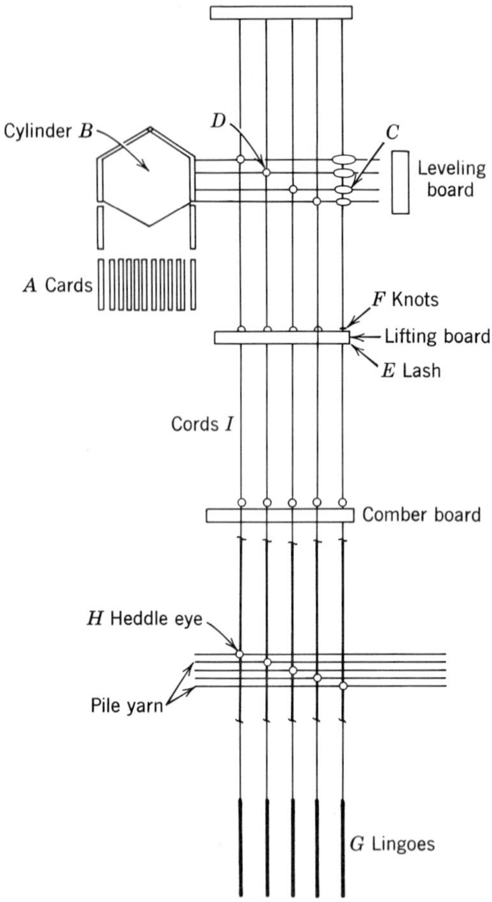

Fig. 15. Schematic of the Brussels card Jacquard head.

similar to that given by a spring. Wherever a hole occurs in the card the needle enters the cylinder and leaves the position of the vertical hung harness cord unchanged. When a needle encounters a blank space in the card, the needle moves back and into the needle board changing the position of the eye in needle (*D*) and that in turn pushes the harness cord (*I*) so that the knot (*F*) in the cord (*I*) will be caught in the narrow section of the keyhole notched lash board (*E*) and picked up when the lash board is raised. The cords that are raised carry the respective face yarns that they control, while all the other face yarns stay in the bottom of the shed. When the selected face yarns are raised, a "pile wire" is inserted into the shed and then the lash board is lowered to form a loop over the wire. A shot of filling follows immediately to secure the loops. The needles that were displaced in the needle board are assisted back into position by a "padder" or "levelling board."

After sufficient number of rows are woven to insure a bind, the wire is withdrawn leaving the yarn that was looped over it as pile. If a cutting blade is placed in the end of the wire the loops are cut as the wire is withdrawn. At one time the name Wilton was applied exclusively to cut pile Jacquard carpet and Brussels meant a loop pile Jacquard. Today, Wilton means simply a Jacquard weave and it may be either cut or loop pile.

The function of the comber board, also called comper board, is to raise all face yarns simultaneously for the insertion of the bottom filling shot.

A schematic of the Jacquard head mechanism for fine index is shown in Fig. 16. The performance of this mechanism is faster, more dependable and durable than the cord Jacquard. The operation is similar to the Brussels. A perforated cylinder (*B*), shaped square, pentagonal or hexagonal, carrying cards (*A*) punctuated with the complete design laced together to form an endless belt presses the face of the card (*A*) against a needle board (*D*) with needles (*C*) protruding ⅜ in. beyond the board surface. The blank space in the card pushes the needle back into the needle board (*D*) and this in turn shoves against stiff vertical wires (*E*) with a hook on each end called "hooks." The lower hooks rest in a grate (*G*) and the upper hooks are placed in a griff (*F*) just over but not resting on the griff knives (*K*). The griff (*F*) carries the knives (*K*) that engages the hook (*E5*) that is left in position to be picked up by knife (*K5*) when the griff is raised. The hooks, heddles and lingoes are all tied together with linen cord and threaded through a comber board (*T*). The lingoe weights keep everything taut and in a vertical position. The pile face yarn is threaded through the eyes (*N*) of the heddles. When needles (*C*) *2, 3, 4, 5* are pushed in and hook (*E*) is left in place, the face pile yarn in frame 1 will be lifted by the griff (*F*) and looped over the pile wire. As the griff is lowered, the face yarn loops over the pile wire and a shot of filling

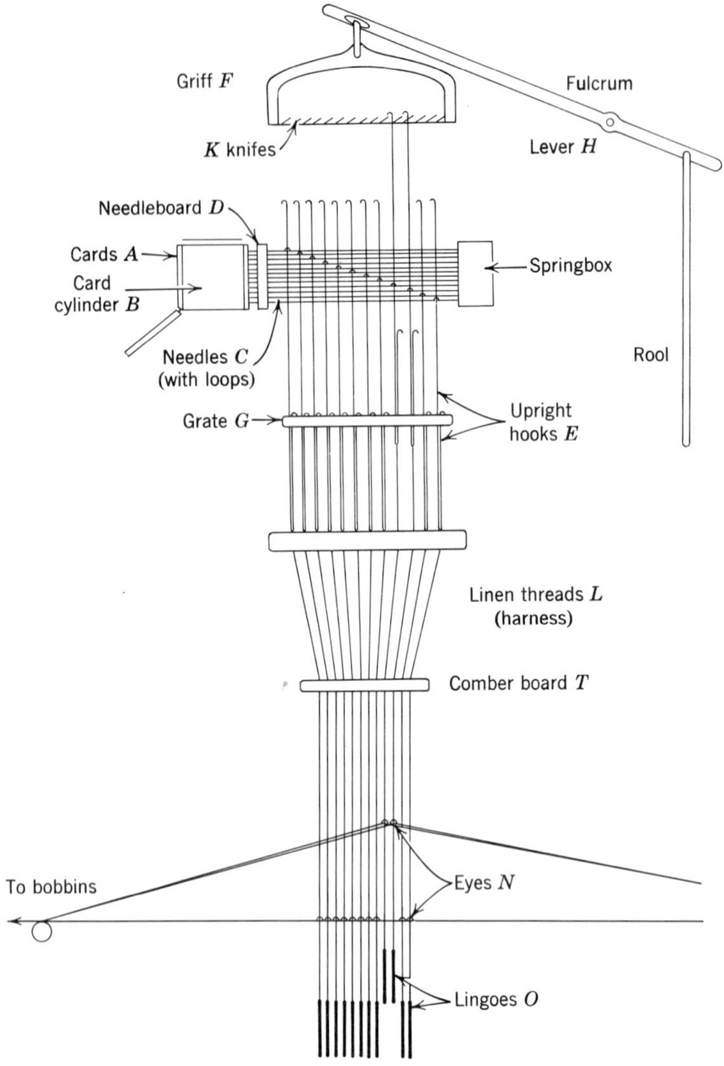

Fig. 16. Schematic of the Jacquard head mechanism for fine index card.

is inserted to hold the looped pile. After a sufficient number of rows are woven to insure the bind, the wire is withdrawn leaving either loop or cut pile.

As the griff moves 9 in. vertically to lift the selected ends of face yarn from the bottom of shed and put it in a top position for the insertion of the pile wire, the grate raises 4½ in. and lifts all the remaining ends of face from the bottom of the shed and places them in a neutral position for the introduction of the bottom filling shot.

The needles are returned to the original position of protruding beyond the needle board surface by springs contained in spring-box at the back end of the needles.

Brussels and fine index carpet Jacquards select a single end of face pile yarn to appear over the pile wire in a dent and positions all the other remaining ends of face yarn in the neutral position of the shed for placement with the dead warp or stuffer. The need for new and different style effects has brought into use the three position and modified Jacquard. Each of these systems is limited to three frames of color because a frame requires two hooks fastened together by a tail cord for each heddle. Both use a modified fine index Jacquard mechanism equipped with fixed and mobile knives. The fixed knives are placed at a point midway between the griff in the highest position and the grate in its lowest position.

The modified Jacquard can select 2 ends of face yarn each for placement in one of four different positions within one carpet dent. This information is punched into a card by using combinations of two punches, one punch and one blank, one blank and one punch and two blanks to identify the selection. It is possible to place the end in the neutral position with the dead stuffer, to put the end over the top filling shot or float over the filling, to put the end over the wire, to put the end over the wire and hold to float over two or more wires. One card contains all the information required to select the face yarns for one row of carpet 18 in. wide—bottom shot, pile wire and top shot.

The three position Jacquard can select three ends of face yarn, each for placement in one of four different positions within a shed or carpet dent for each shot of filling. It uses the same code of punch and blank as the modified Jacquard but it requires one card for each shot or two cards to complete one row of carpet 18 in. wide.

Card cutting plan and the resulting surface effects are shown in Fig. 17.

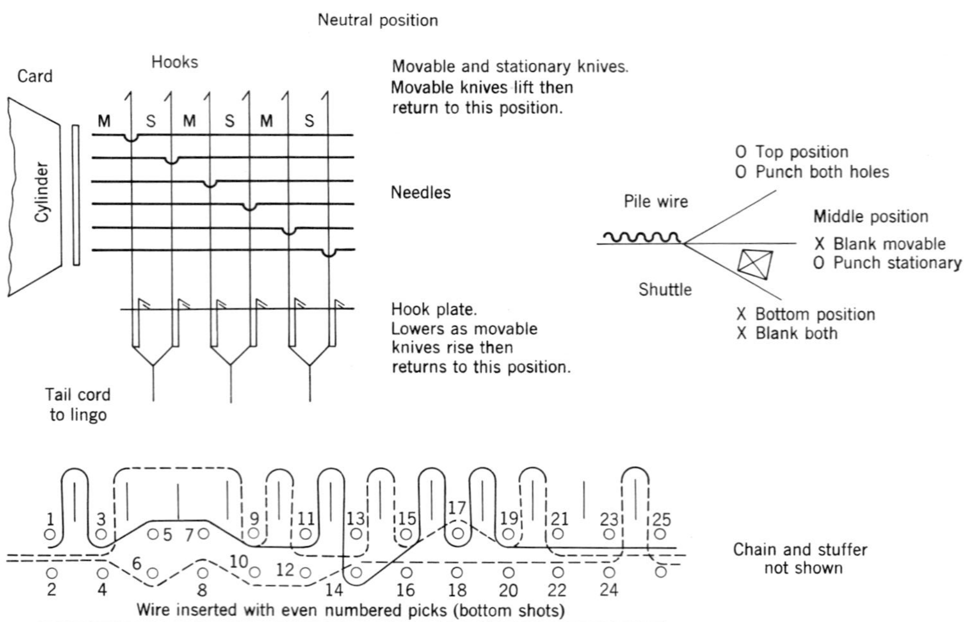

1	2	3	4	5	6	7	8	9	10	11	12	13	14	15	16	17	18	19	20	21	22	23	24	25	Pick No.	
X	O	X	X	X	X	X	X	X	X	X	O	X	X	X	O	X	O	X	X	X	X	X	X	X	Movable	1st frame
X	O	X	O	O	O	O	O	X	O	X	O	X	X	X	O	X	O	X	O	X	O	X	O	X	Stationary	
X	X	X	O	X	O	X	O	X	O	X	X	X	O	X	X	X	X	X	O	X	X	X	X	X	Movable	2nd frame
X	O	X	O	O	O	O	O	X	O	X	O	X	O	X	O	O	O	X	O	X	O	X	O	X	Stationary	
X	X	X	X	X	X	X	X	X	X	X	X	X	X	X	X	X	X	X	X	X	X	X	O	X	Movable	3rd frame
X	O	X	O	X	X	X	O	X	X	X	X	X	O	X	O	X	O	X	O	X	O	X	O	X	Stationary	

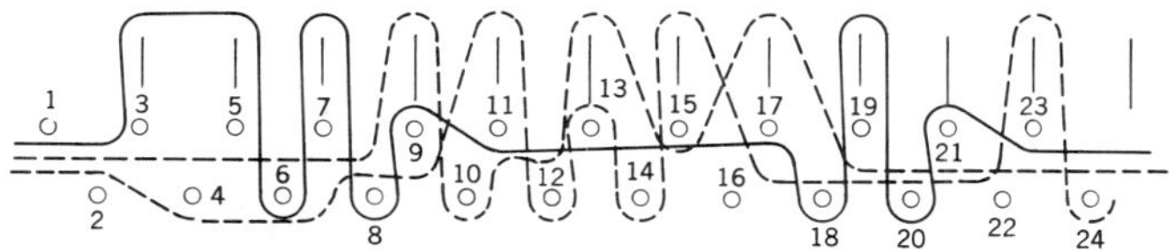

1	2	3	4	5	6	7	8	9	10	11	12	13	14	15	16	17	18	19	20	21	22	23	24	Pick No.	
X	X	O	X	O	X	O	X	X	X	X	X	X	X	X	X	X	X	O	X	X	X	X	X	Movable	1st frame
X	O	O	O	O	X	O	X	O	O	X	O	X	O	X	O	X	X	O	X	O	O	X	O	Stationary	
X	X	X	X	X	X	X	X	O	X	X	X	O	X	X	X	O	X	X	X	X	X	X	X	Movable	2nd frame
X	O	X	O	X	O	X	O	O	X	X	O	O	O	X	O	O	O	X	O	X	O	X	O	Stationary	
X	X	X	X	X	X	X	X	X	X	O	X	X	X	O	X	X	X	X	X	X	X	O	X	Movable	3rd frame
X	O	X	X	X	X	X	O	X	O	O	X	O	X	O	O	X	O	X	O	X	O	O	X	Stationary	

Fig. 17. Card cutting plan and surface effects of the three position Jacquard.

Wilton Weaving

The loom is set up for weaving by placing the pile yarn that has been wound onto spools in their respective creel frames behind the loom. The chain and stuffer yarns are slashed and put on beams. Figure 18 shows a cross section weaving of a three frame 2 shot Wilton carpet. The pile yarn spools A, B, C are put into their respective frames, then threaded through their specified harness *G1*, *G2*, *G3*, and then through the reed and onto spike roll. K. The warp and stuffer yarns are threaded through harnesses *F1*, *F2*, *F3* and then through the reed onto the spike roll K. The number of stuffer ends or dead warp can be varied for hand and compactness. The three frame fabric as shown will have six strands of yarn in each reed space or carpet dent—three pile yarns, two chain warp yarns and one stuffer yarn.

A three-shot weave is used to obtain better pile bind to avoid sprouting of slippery yarns and to improve the sturdiness of the carpet. A cross section of a 3 shot 3 frame Wilton is shown in Fig. 19.

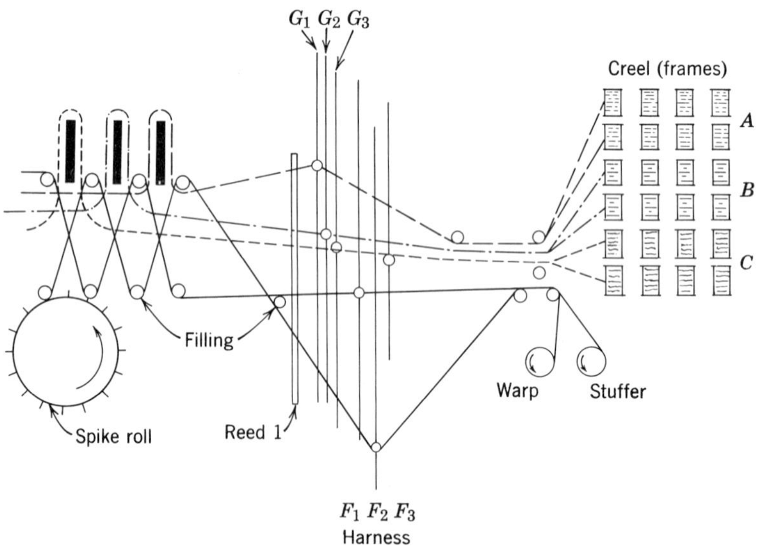

Fig. 18. Diagram of Wilton carpet weaving.

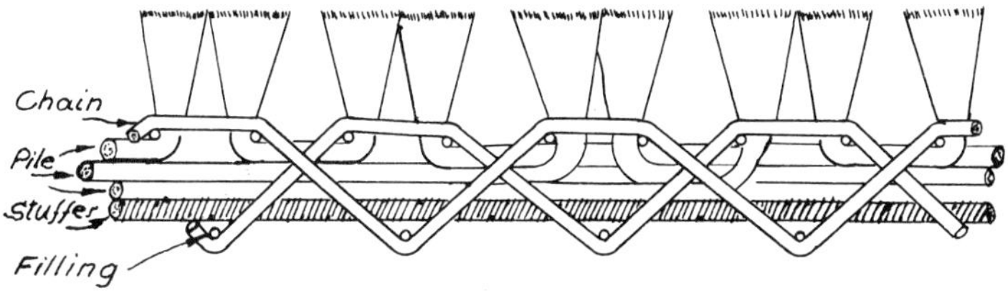

Fig. 19. Cross-section of three shot three frame Wilton weave.

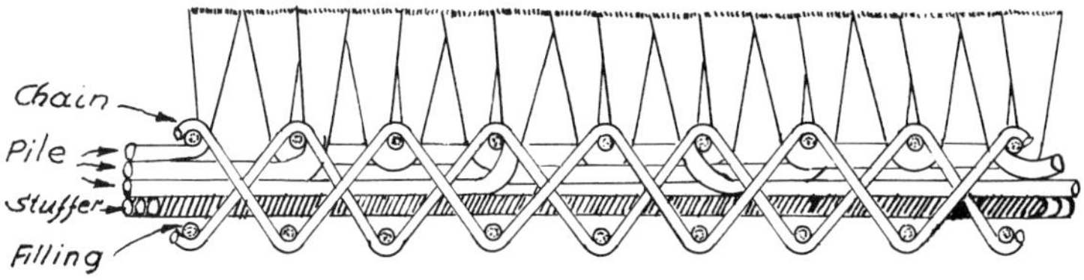

Fig. 20. Cross-section of two shot three frame Wilton weave.

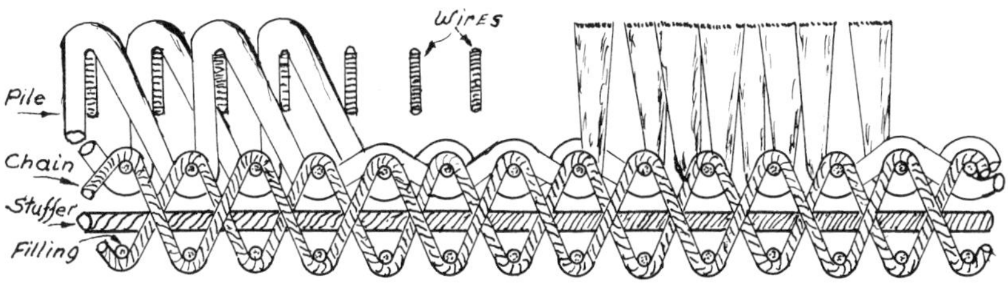

Fig. 21. Cross section of two shot two frame Wilton weave.

Other modifications used are:

Two shot 3 frame Wilton weave a pile end working over a wire, floating over two wires, working alternately over the top filling shot, floating over two filling shots, floating over three filling shots and buried between top and bottom shots. See Fig. 20.

Two shot one frame Wilton weave the pile yarn working top filling shot and over a single wire and floating over subsequent wires, and a two shot two frame Wilton weave (Fig. 21) showing a pile end over a wire and top shot alternately and then working over and under 3 top shot alternately.

Axminster

Also named for a town in England but the loom was invented by an American. A specialized carpet loom is used that allows an almost unlimited combination of design and color. Changing of patterns on this loom is a slow, expensive process. Separate lengths of "ends" of colored yarn must be set with absolute accuracy for a faithful reproduction of the pattern.

Patterns in Axminster carpet are offered in endless variety—stylized, geometric, classic, modern and floral. Thus, Axminster approximates the versatility of hand-woven rugs. The pile is cut except in a few special weaves. Almost all the yarn appears on the surface. Goċd quality Axminster has a compact pile with excellent crush resistance that gives long service. One distinguishing feature of this weave is that the back is heavily ribbed; it can be rolled lengthwise but not crosswise.

The power loom for the weaving of spool type Axminster was developed in the United States by Halcyon Skinner, around 1870.

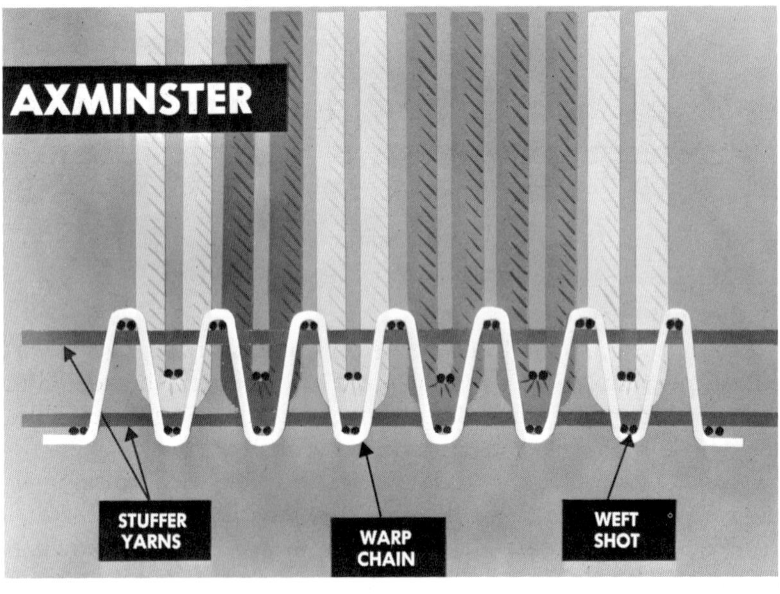

Fig. 22. Conventional Axminster weave.

Since the spool type Axminster loom is the most widely used in the United States, its operation will be described first.

As in any figured fabric, the first step in the production of Axminster carpet is the designing. The design is drawn and colored on paper which is ruled in both directions. Each space (check) on the paper represents one tuft in the finished fabric. The designer and colorist have the greatest latitude when working with spool Axminster. The number of colors used in a design is theoretically limited only by the number of tufts in a pattern and the number of colors that can be dyed. From a practical standpoint the number of colors is limited by cost of dyeing small lots of yarn and the type of design desired.

After the required yarns are dyed to the desired colors they are "set" on spools. The setting operation consists of winding parallel ends of the correct color yarns onto metal spools. The setting operators, working from the colored design, place packages of the dyed yarn on a creel or bank in the sequence in which they appear in one row across the fabric. The yarns pass through a reed to keep them running parallel thence into the setting machine where they are wound onto the spool. The amount of yarn wound on each spool is controlled by a measuring device. After the yarns are wound on the spool a comb-like clamp is applied to the yarn to hold the ends in their proper sequence and the spool is numbered to correspond to its place in the design.

After setting, the spools are threaded by pulling each yarn end through tubes on a tube frame. On wide looms multiple spools are threaded through the tubes of a single tube frame so that all of the pile ends required for one weftwise row are supplied from a single tube frame. The spools for broadlooms are usually 36 in. or 54 in. in width while the tube frames for a wide loom are 9 ft, 12 ft, 15 ft to correspond to the width of the loom. The threading is done by a machine consisting of a row of needles which are pushed through the tubes of the tube frame, engage the ends of yarn on the spool then retract through the tubes, drawing the yarn ends with them. The tubes of the tube frame guide the pile ends through the backing material warps during the weaving operation.

When the tube frames have been threaded they are taken to the loom and mounted on two endless chains (left and right), which have been arranged so that there is one link in each chain for each tube frame to be used in the loom.

The backing of an Axminster carpet consists of one or more chain warps, stuffer warps and the filling (or weft) yarn. The chain and stuffer yarns are supplied from large spool-like beams arranged at the back of the looms and their motion is controlled by drawing the ends through heddles which in turn are raised or lowered by cams. The filling is inserted by a long metal rod, known as the needle, having an eye through which the filling is fed. The needle travels through the warps, across the loom, carrying a double end of yarn. When the needle is completely inserted a small shuttle traps the yarn by passing through the bight of the yarn. The needle retracts through the warps leaving the double yarn in the fabric. There are normally three double filling insertions per row of pile yarn; however, one, two, or four may also be used.

While the backing of the carpet is being woven the pile is inserted by the tube frames. The tube frames are removed from the chains, which carry them, by the clutch arms and carried downward so that the tuft tubes pass through the backing warps. The loom reed swings forward, carrying with it a filling yarn, to trap the pile ends projecting from the tuft tubes, then the tube frame is raised to draw off a measured amount of the pile yarn. After the tube frame is raised a pair of knives operate shear-wise across the loom to cut all of the pile ends, freeing the tube frame so it may be replaced in the endless chains. The free end of pile yarn projecting through the bottom of warps is folded upward around the filling yarn to form the second end of the tuft. This folding is done either by a comb working from the underside of the loom or else by pushing the filling and tuft over metal projections called the "nose" board, depending upon the type of loom.

The positions of the warp ends are changed and the additional fillings are inserted. The chain carrying the tube frames is advanced to present the next tube frame to the clutch arms and the cycle repeats for the next row of pile.

A schematic cross section of a Spool Axminster Loom "Nose Board" type is shown in Fig. 23.

There are many weave variations which may be used on the spool Axminster loom by changing the position of the warps with respect to the filling, the number of fillings per row of pile, and

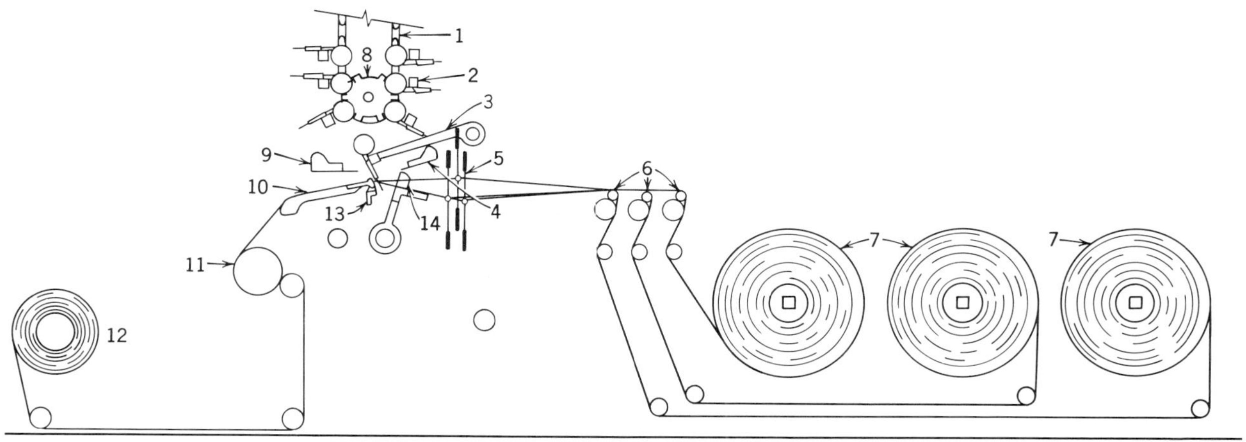

Fig. 23. Schematic cross-section of a Spool Axminster Loom. "Nose Board" type. (1) Endless chain which carries tube frames; (2) tube frame; (3) clutch arm; (4) Back knife stock and knife; (5) heddles; (6) warp tension devices ("jumbos"); (7) warp (chain and stuffer) beams; (8) pull-over sprocket; (9) front knife stock and knife; (10) breast plate and nose board; (11) take-up (spike) roll; (12) carpet roll-up; (13) cleaner; (14) reed-mounted on lay.

selecting the filling around which the pile is bound. The two most widely used variations are shown below in Figs. 24 and 25.

The second most widely used type of Axminster loom is the Jacquard Gripper. This type loom is reported to have been invented about 1890 by Brinton and Greenwood of England.

The weaving of the backing fabric on the Jacquard Gripper loom is the same as that described for the Spool loom, but the selection and insertion of the pile yarn is entirely different.

The main disadvantage of this type of loom is the fact that the colors which can be used in a reed space are limited to twelve.

The pile yarn is wound on packages and placed on a creel behind the loom, as in Wilton weaving. One end of each color yarn, to be used in a reed space, is drawn through one of the eyes in a guide or carrier. As the name implies, a Jacquard is controlled by the design which has been punched into cards. Depending upon which holes have been punched in the cards the Jacquard lifts all of the ends selected to be used, into a horizontal row across the loom. A series of grippers grasp the pile ends, draw off a given length of yarn, hold the ends until they are cut, then inserts the cut length of yarn into the backing where they are folded to become the pile tufts.

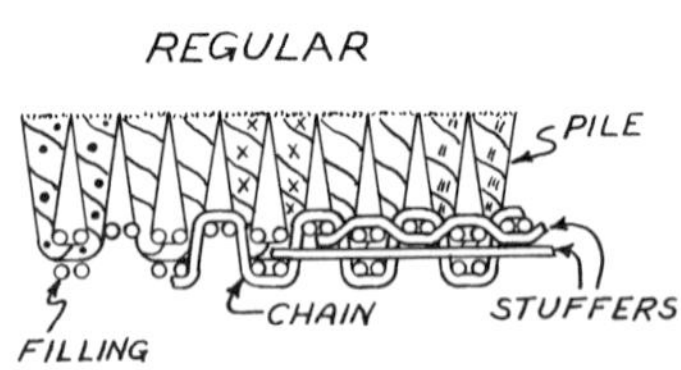

Fig. 24. Cross-section of Regular Axminster weave.

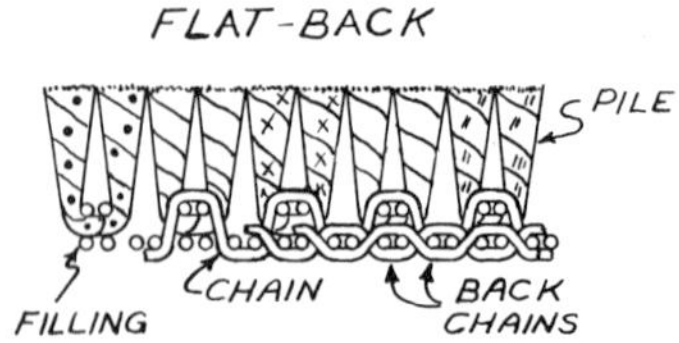

Fig. 25. Cross-section of Flat back Axminster weave.

The weaves possible on a Jacquard Gripper loom are approximately the same as those produced on a Spool Axminster loom.

A third type of Axminster loom is known as a Spool Gripper, and is a combination of the Spool loom and the Jacquard Gripper loom. The designing, setting and threading are the same as for a Spool loom. The threaded tube frames are placed in endless chains on the loom and are advanced one by one to a position where the projecting pile ends are engaged by grippers, the desired length of yarn is cut off and the grippers insert these lengths of yarns into the backing fabric.

The Axminster process has a decided advantage in that it can produce a design in color without having the unused pile ends running in the back of the fabric, as in Wilton carpet weaving.

Chenille

This is a thick soft carpet that requires two looms for weaving. On one loom, the pile surface is woven in a blanket that is cut into furry strips. These are woven into the base of the carpet on another loom. Chenille can be woven in any pattern, color, shape or size up to 30 ft wide. It is often custom-woven.

Chenille carpet, also called Chenille Axminster, was first made by James Templeton of Scotland in 1839.

The manufacture of chenille carpet is costly and it is considered a luxury carpet. It is the only carpet in which the design is formed by the weft for in all the others the function of the weft is to combine with the chain to form the woven fabric. The method of weaving involves many hand operations and this makes it possible to create chenille rugs in any size, shape, pattern and color. Today, this weave represents a relatively small portion of the carpet yardage manufactured and there are only a few makers in the United States.

The chenille fur used as the weft yarn of the chenille carpet is cut from a leno woven blanket made on a flat loom using fine cotton yarn in groups of 6–8 ends for the warp and using woolen or worsted yarns for the filling. The distance between the groupings of warp yarns governs the length of the tufts as the woven blanket is split at the mid point between the grouped warp ends. The filling yarns of this loom are wound onto small bobbins in as many colors as the design requires. One filling shot corresponds to one check on the design paper and the colored bobbins are changed in the shuttle by hand as frequently as the design necessitates. The number of shots per inch will vary from 12 to 20 or more depending on the yarn count and the quality to be made. The total number of filling shots in the blanket will be equal to the number of tufts required to make the width of carpet specified.

A blanket is woven for each row in the lengthwise repeat of the pattern. After the blankets are woven they are passed through a machine with revolving knives cutting the blanket into strips that are then numbered to identify the part of the pattern it represents. The cut strips are moistened and passed over a heated roll with V-shaped grooves that fold the fur strips so that the cut ends are brought together and face in the same direction. The fur strips are sorted, tied together in proper sequence, put onto bobbins or into containers in preparation for the second loom.

The second loom has a fine "catcher" warp, chain warp and stuffer warp. The catcher thread holds the chenille fur in place and must not be visible from the carpet surface. This necessitates a fine count yarn of high quality cotton. The chain and stuffer yarns are usually cotton of coarser count and lower grade. Each warp is put onto a beam, set up behind the loom and threaded through the eyelets of the harnesses similar to a velvet loom. The filling yarn of the back or basic weave may be jute but in the better qualities coarse woolen yarns mixed with cattle-hair are used. Usually there are two to four shots of filling yarn in the back to each row of chenille fur held in by the catcher warp.

The chenille fur is entered into the shed while catcher warp is raised and the loom is stopped. The fur may be carried back and forth by a finger-guide or passed as a bobbin by hand from weaver to weaver who are spaced 4–5 ft apart across the width of the loom. Each row of fur is hand combed and dressed to match the preceding row. The loom is started again and the next two or four shots of basic back are inserted.

Chenille carpets are brushed, steamed and finished in a manner similar to other carpets.

Miscellaneous Carpets

Tapestry carpet manufacture had as its purpose a process of making colored and designed carpet more economically than that used for Wilton carpets. Drum printed pile warp yarn carried the design and the carpet was woven on a velvet loom. The preparatory processes are expensive and the appearance of the finished product inferior to Wilton. This type of carpet succumbed to competitive attrition and is no longer being made in the United States.

Ingrain carpets, also known as Kidderminster and Scotch, was a double faced pile-less carpet that was reversible. The design was made by using colored warps and the design on one face was usually different from the other. This carpet is no longer being manufactured in the United States.

Smyrna carpet uses flat chenille fur instead of the V-shaped type fur that is used for Chenille carpet. This weave uses a double chain warp and is double faced. This type also has given way to other types of carpets in today's competitive market.

Knitted Carpets

Knitted carpets are fabricated in one operation as are woven carpets. Unlike weaving however, the knitting process loops together the backing yarn, stitching yarn and pile yarn with three sets of needles—much the same way hand knitting is done. To give additional body to the fabric, a coat of latex is applied to the back. As in tufted carpets, a second backing is frequently added as shown in Fig. 26.

Since a single pile yarn is used in the knitting operation, as in velvet, knitted carpets are usually solid colors or tweeds, although in recent years patterns have become possible.

Knitted carpets usually come with uncut loops, both single level and multi-level. A cut pile can also be achieved with modifications in the knitting machine.

The most widely used type of machine is Raschel warp knitting machine equipped with either plush points or hooks for forming the pile.

The backing and pile yarns may be supplied from either beams or creels and usually consist of the knitting yarn which binds the fabric together, an inlay or filling yarn plus the pile yarn.

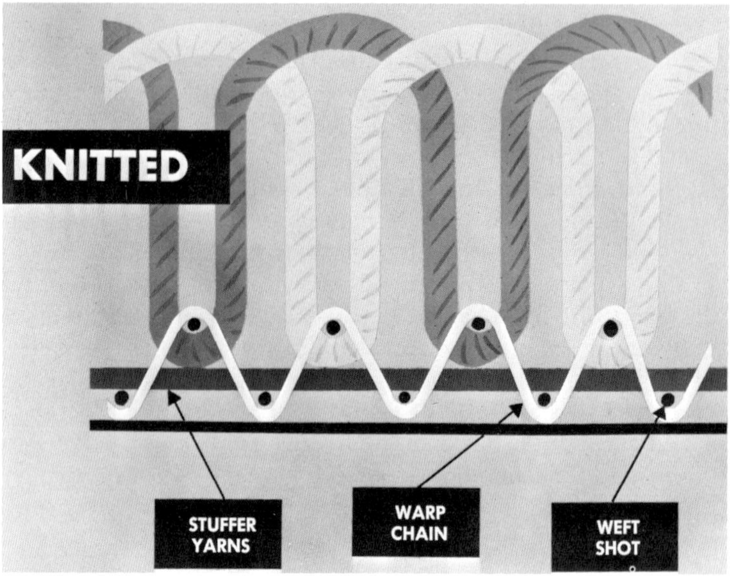

Fig. 26. Knitted carpet cross-section.

The majority of the knitted carpet produced is loop pile, however, by equipping the machine with plush points fitted with blades it is possible to produce cut pile fabric. It is also possible to make cut and loop designs in the fabric by alternating plain and cutting plush points and selecting the yarns which pass around the different points.

The guides swing to the front of the machine to complete the cycle. No shifting of the guides occur at this time.

Another recent entry in the floor covering field is the furlike fabrics produced on a modified circular knitting machine. The standard circular knitting machine is modified by placing several miniature cards around the circumference of the machine.

The backing of the fabric is a conventional circular weft knit cloth and the pile is composed of fibers which are knitted into the backing stitch.

The cards on the machine are supplied with a pre-carded sliver usually composed of a blend of synthetic fibers. Through the action of its main and worker cylinders the sliver is again broken down into its fibers which are removed from the main cylinder by a doffer cylinder.

As the needle cylinder revolves in the knitting machine the needles are raised so that the needle hooks pass through the wires of the doffer cylinder. In passing through the doffer wire the needle hooks remove the fibers from the doffer wire. After the needle has passed out of the doffer a weft yarn is engaged by the needle hook and the needle is forced down to knit both the weft yarn and the fibers into the backing.

For each revolution of the needle cylinder a row of pile is produced for each card supplying the pile fibers. For a machine equipped with six cards and running at 25 rpm the production rate is 150 rows of pile per minute.

The fabric as knitted on the machine is tubular and must be slit and opened to a flat surface for finishing. The pile, as knitted, is very rough and requires extensive shearing and finishing to produce a smooth furlike surface.

There have been some limited types of designs produced in these fabrics by selectively supplying different color fibers to the needle hooks.

The Tufted Process

Tufted carpet is manufactured in two separate operations. First the primary backing either of woven or of nonwoven material is produced. In the second operation the tufts are needled through the primary backing. The needling-through of the tufts is done by a machine that operates somewhat as an oversized sewing machine, with thousands of needles working simultaneously. To hold the tufts permanently in place, the backing is coated with a heavy layer of latex. In actual service the wearability of tufted carpets that were backsized only has been insufficient; therefore on a high percentage of the yardage a second or double backing (see fig. 27) together with a backing compound is added to lend additional "hand" and dimensional stability. This double backing has helped to improve the characteristic that makes for one of the major differences between tufted and woven floor coverings and has aided its competitive position.

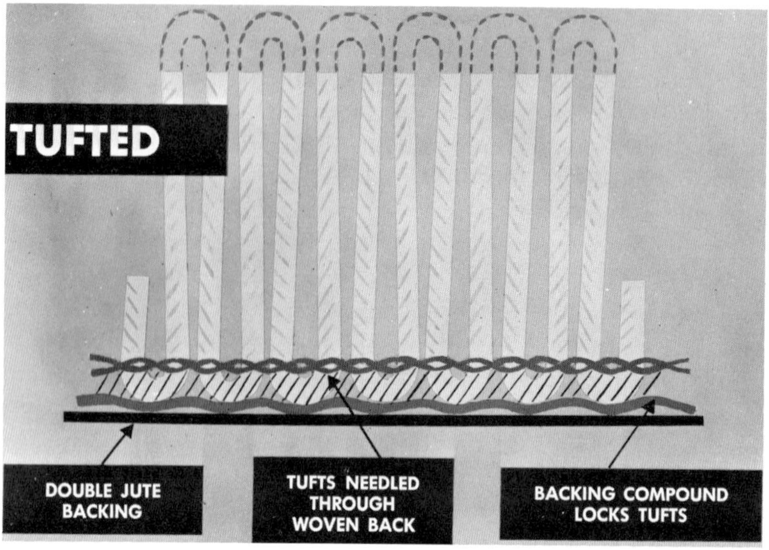

Fig. 27. Tufted carpet cross-section.

The variety of textures possible in tufted carpet is increasing, as a result of recent innovations in tufting equipment. The tufted pile can be multi-level, cut or uncut. Striated and carved effects can be achieved. The texture can be looped or plush, to mention a few of the possibilities.

Although tufted carpets were originally made only of single color or moresque yarns, they are now being made in multicolor patterns, with an increasing variety of textural effects. Because of continual refinements in the tufting process, tufted carpets will probably continue to become more versatile in design and appearance.

The handicraft known as tufting is very old. Historical records tell us that hand-tufting existed during the Egyptian Empire and it has continued in one form or another throughout the centuries. The primitive craftsman wove loose mats of reeds and with a crude needle inserted flax yarns or strips of leather through the reed mat to form soft pile.

This primitive and simple handicraft provided the means whereby American Colonist women made useful and decorative articles for their homes from the meager things they had available. However, handicrafts also go through cycles of popularity and evolutionary changes. The ordinary sewing needle was replaced by a special kind of a needle for inserting or hooking the yarn into a piece of cloth. Enterprising suppliers to the handicrafter invented a variety of manually operated needling devices to accelerate the insertion of tufts and hasten the completion of the article. The designs were stenciled onto the cloth and the cloth was suspended in a frame for tufting. The loops were formed on the side opposite the point of insertion and if cut pile was wanted the loops were cut and trimmed with hand scissors after the tufting was completed. The back was treated with size and adhesive and then finished with cloth or burlap that was sewn onto the back.

The next step in the order of progression was to increase productivity and this led into the use of power and machinery.

Today the custom tufter uses both manual and power driven single needle hand tools and also multi-needled sewing machines. The loops made with single needle hand tools are hand cut and trimmed. Carved designs are made with power driven hand shears. The pile is held in place by backsizing with a liberal quantity of latex.

Machine tufting

Even though the handicraft of tufting rugs has been in existence a long time the present machine carpet tufting industry stems from the bedspread handicraft of Miss Catherine Evans of Dalton, Georgia, in 1895. In 1922, as the demand grew and the commercial possibilities became apparent, men entered the field and a very considerable homecraft industry was developed in sections of Georgia, Tennessee, Alabama, and South Carolina.

The patent records reveal tufting machinery inventions dating back to 1860 but the invention of a commercial power tufting machine in 1930 by Mr. Glen Looper is thought by many to have been the major contribution in the start of this new industry. The single needle machine was perfected and the multineedle machines were built starting with 2 and then progressing to 4, 8, 12, and 24. Today the industry is using as many as 2,434 ($\frac{5}{64}$ gauge) needles in a machine 15 ft wide.

This industry is relatively new but the tufting of the first rug is obscure. The rug industry started by making bath mats and scatter rugs using a single needle machine that looks like and performs similar to a domestic sewing machine. It had a stitch rate as fast as 3500 per minute and the backing cloth—cotton muslin or duct, or jute—was manipulated by hand. The increasing demand for this product and the need for higher production efficiency led to the development of 2 and then 4 needle machines running 2500 stitches per minute and spaced $\frac{3}{8}$ in. apart. In 1947, a modified sewing machine equipped with 24 needles spaced $\frac{3}{16}$ in. apart was used to tuft the entire surface by hand-guiding the backing cloth repeatedly through the machine until the entire surface was covered. As the number of needles was increased it was necessary to find a method of controlling and feeding the backing cloth mechanically. Such a machine was developed for the making of bedspreads about 1940 and it was called a "yardage machine." Figure 28 shows first yardage machine ever built. This machine was 120 inches wide, the needles were spaced $1\frac{1}{8}$ in. apart and it stitched at the rate of 550 per min. In 1949, the design of the yardage machine was reinforced, adapted to carpet manufacture and made commercially available to the industry by Cobble Brothers of

Fig. 28. Yardage tufting machine—first machine built. (Courtesy The Singer Co.) By Joseph Cobble at Chattanooga, Tenn.

Chattanooga, Tennessee. This machine was 120 inches wide, the needles were spaced ⅜ in. apart and the stitch rate was 550 stitches per minute.

The cross section of a tufting machine is shown in Fig. 29.

In tufting wide seamless carpeting, the pile is formed when a multiple of needles acting in unison are punched or projected through a previously woven backing fabric of jute or cotton, carry-

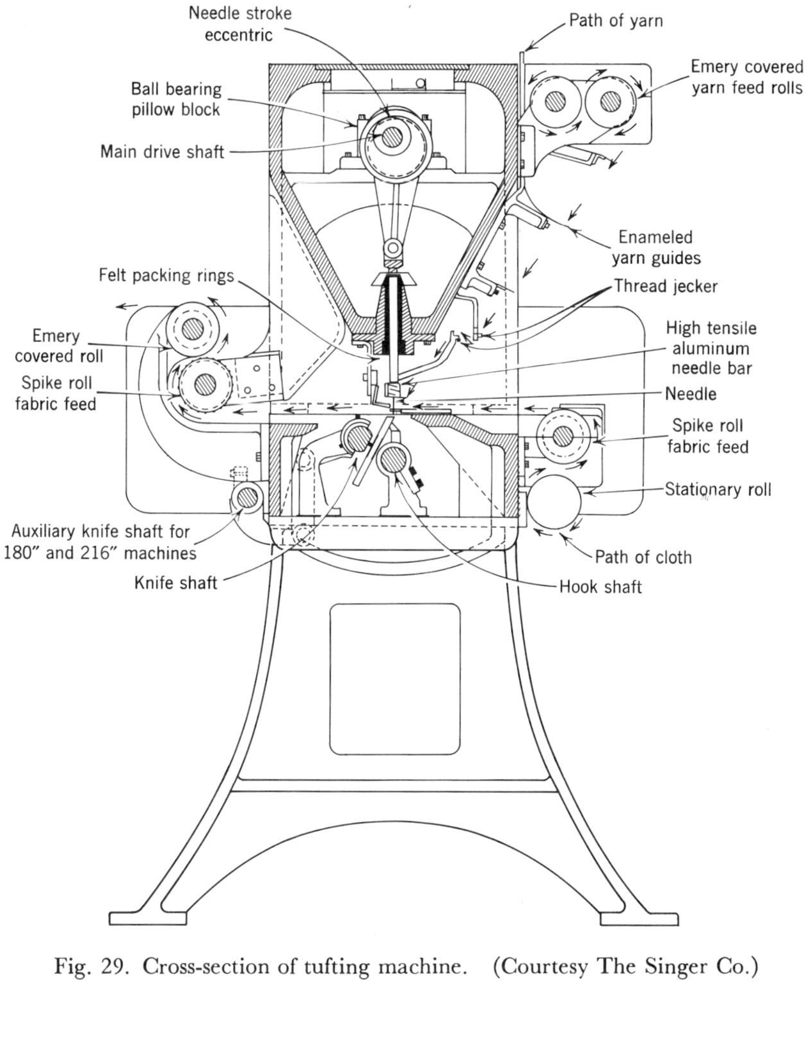

Fig. 29. Cross-section of tufting machine. (Courtesy The Singer Co.)

ing ends which are engaged over metal loopers (Fig. 30), hooking L2 to form loops which produce the ultimate pile. The loops suitably remain uncut on the machine producing loop pile carpet (*LL*) or can be cut with the addition of cutting blade (knife Fig. 31) to act in unison with the pile looper, to cut the top center of the loop to form a cut tuft (Fig. 31). The needle in forming the pile tufts goes through the backing fabric to a point (*L2* Fig. 30) at

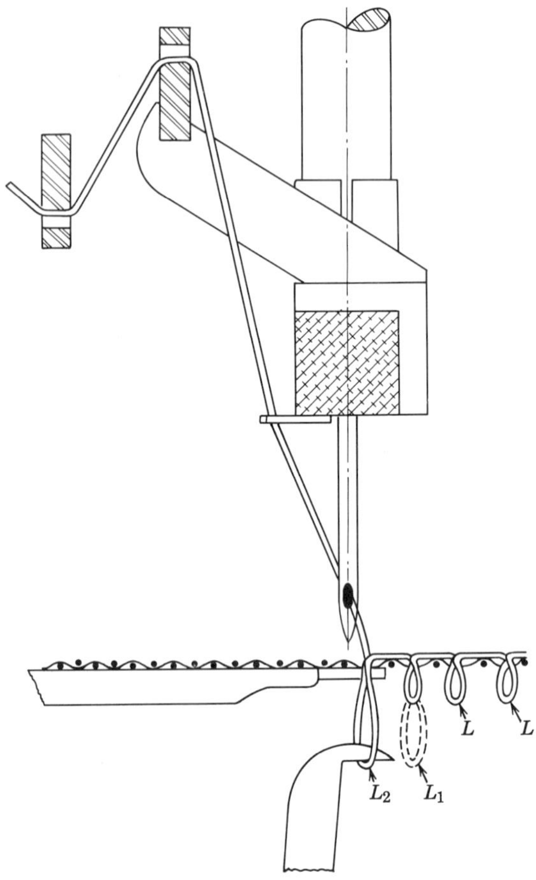

Fig. 30. Diagram of pile formation in a tufting machine—production of uncut or loop pile carpet.

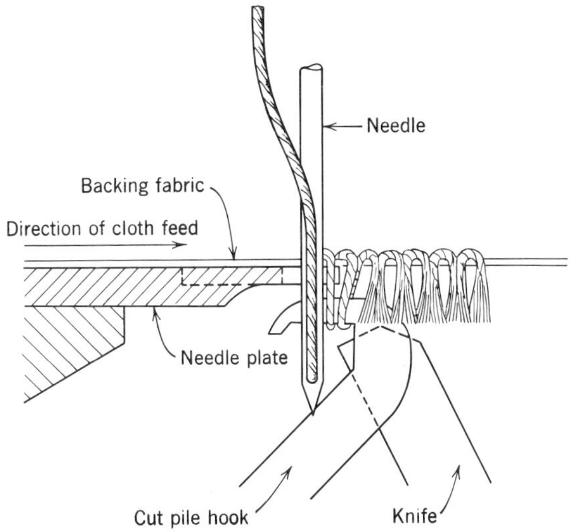

Fig. 31. Diagram of cut pile formation in tufting. (Cobble Brothers Manual.)

which the eye of the needle is at or beyond a position that corresponds to the height of pile, and then the looper engages and the needle reverses, bringing the sides of the pile yarn through the same punched hole in the backing fabric. The pile tuft, unlike that produced by weaving, has an extremely narrow base, the sides of pile yarn being substantially together at the backing fabric. The needle momentarily remains out of the backing fabric, while a horizontal feed roll pulls the fabric through the machine in a forward warp-wise direction, whereupon the needles again puncture the backing fabric to form another horizontal (weft) row of pile (stitches) tufts. The yarn to be fed to the respective needles in the multiple needle bar, extending across the width of the machine, is drawn from packages (cones) in a creel or from a beam, by continuous yarn feed rolls or bars, through metal or plastic tubes.

Starting with the initial year of 1949 the manufacture of the tufted type of floor covering has grown at a much faster rate than any of the other types of soft floor covering materials.

The many machine and product improvements made during this period contributed to increasing the demand for the product. The first carpets were loop or cut pile, piece dyed and made of cotton yarn. The use of spun rayon staple yarns followed and then either stock dyed or solution colored fiber was used to overcome the inherent piece dyeing defects of side-to-center, and end-to-end shade variations. Multicolored twist yarns were added for tweed style effects. Today, all available fibers that will produce a yarn of sufficient strength to run in a tufting machine are being used. The yarn count is limited to the opening in the eye of the needle and the spinning methods producing large knotless packages of yarn are preferred.

The new style innovations made possible by the many rapid machine developments has held the buyer's interest and increased the demand for tufted carpets. A surface pattern effect is obtained by the "loop robbing" principle similar to that used in weaving velvet carpets with serrated pile wires. The previous tufted loop is pulled down to a lower level by limiting the amount of yarn to make the following loop. Simple geometric designs made on machines equipped with two pairs of rolls with two speed control for yarn feed appeared in 1952. Within the years following through to 1954, more intricate geometric designs appeared as Super Tufting machines using 3, 4, 6, 7, and 9 pairs of rolls were developed (Fig. 32). Also, during this period the service quality of tufted carpets was improved with the introduction of the 1/8 inch gauge machine to make carpet with greater pile density.

To overcome the limitations of geometric designs and obtain the more flowing style effects made possible by curved lines or scroll patterns, a 5/32 in. gauge machine using 20 pairs of rolls to control the yarn feed for every two ends was introduced. Within a short time the 2 end control became a limiting design factor and a single needle control using 120 pairs of rolls was introduced in 1957. The cable 120-roll photoelectric scroll pattern attachment, which is the most versatile and widely used pattern attachment for tufted carpets is shown in Fig. 33.

Another development of this period was the introduction of a product containing a combination of cut and loop pile scroll patterned textures made on the Whitfield machine by Cabin Crafts of Dalton, Georgia. This is a 3/16 in. gauge machine equipped with 2 sets of loopers, one for making loops and the other equipped to make cut pile. The needles in normal operating position will make

Fig. 32. Tufting machine with six roll pattern attachment. (Courtesy James Lees & Sons, Division of Burlington Industries, Inc.)

loop pile and any pair of needles can be deflected by an air motivated cylinder to engage the cut pile looper.

Since the introduction of loop-cut textures of the Whitfield machine, numerous other methods to reproduce similar textures have appeared in patent literature. A machine producing loop or cut pile using the same needle and only one looper with cut pile hooks is made by The Singer Co., Cobble Division.

This basic loop/cut machine is much the same as an ordinary tufting machine. However, the inner parts of the tufting mechanism have been changed and redesigned to allow cut pile or loop pile to be produced from the same needle and using only one special hook.

Fig. 33. Tufting machine with 120 roll photoelectric scroll pattern attachment. (Courtesy The Singer Co.)

Designs are made in relief or sculptured pattern by the use of either the Universal Type Pattern Attachment, the Single End Control Scroll Type Attachment or the 7 Roll Pattern Attachment.

Any design can be made on the loop/cut machine . . . repeat patterns . . . medallion patterns . . . floral or geometric.

With the Universal Type Pattern Attachment or Single End Scroll it is never necessary to rethread the machine to change patterns. Replacing the pattern drum accomplishes a complete pattern change.

Only one pattern drum and its set of controls are used and through a unique hook arrangement the height of pile determines whether it is left in a loop form or is cut with the highest pile being cut and a slightly shorter pile being left in a loop form.

The pattern mechanism of the early controlled two speed yarn feed rolls was directed by a pattern chain using high and low links to select the fast or slow speed to drive the yarn feed rolls. A following development used a revolving metal cylinder upon which was placed a cut out of the design in paper or linoleum. Electrically charged scanning fingers, one for each set of yarn feed rolls follows the revolving pattern and when the finger makes contact with the metal cylinder a solenoid is activated and changes the yarn feed speed for that particular roll to slow speed and makes low loops. When the finger rides the uncut portions of the revolving pattern, the circuit is broken and the feed rolls operate at the fast speed and make high loops. This now has been simplified, by using a

Photoelectric System—same as the Scroll attachment. Using this method, patterns are drawn in the form of light and dark areas. The light areas represent the low pile height, and the dark areas the high. The patterns are placed over a clear drum which has a light mounted inside. As the drum revolves, the light is either blocked from or shines through to the photocells (these take place of fingers) mounted approx. $\frac{1}{16}$ of an inch above the pattern. These photocells are wired to the clutches controlling the speed of the yarn feed rolls, and this is how the pile height is regulated.

The product of another pattern control device known as "bar pattern attachment" was introduced by Mohasco in 1957. A bar (Fig. 34) containing a cut for each end of yarn to make a definite height loop, similar to serrated wires used in velvet weaving, carries the design for one row of stitches (Fig. 35). The entire pattern is fastened together to form an endless chain (Fig. 36) and used to control the feed of pile yarn to the needles. Heretofore only two pile height controls were in use and with this rather simple mechanical device any surface pattern effect can be made in numberless variations of pile heights.

Another attachment made the shifting of needle bars possible which eliminates the danger of stripes when using different colored yarns in alternate needles.

The equipment and methods used for finishing tufted materials is similar to that used for woven carpets.

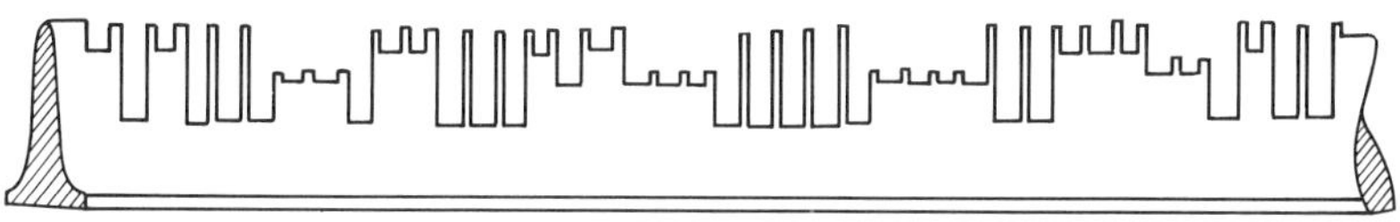

Fig. 34. Bar pattern attachment—Mohasco.

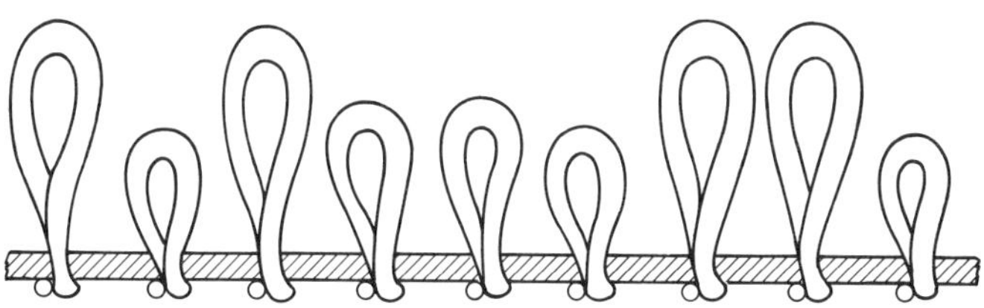

Fig. 35. Loop height variations produced with bar patent attachment.

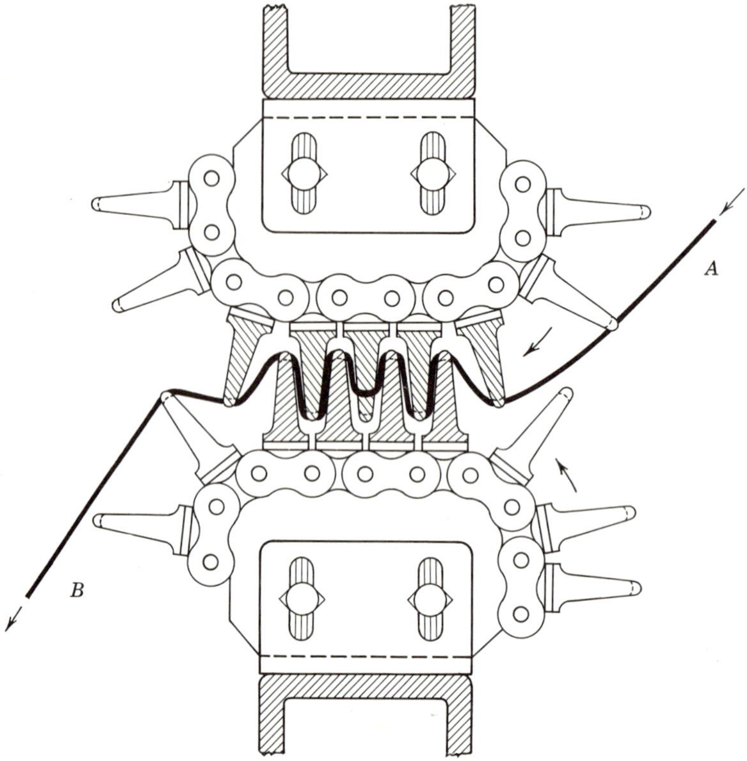

Fig. 36. Endless chain used to control the feed of the pile yarns to the tufting needles.

Carpet Finishing

Facial appearance is greatly enhanced by the degree and workmanship of carpet finishing. Processes usually involved in the finishing operations are:

Burling or mending
Face and/or back steaming
Brushing, back beating, napping, shearing
Backing cloth or scrim application
Backsizing and drying
Inspection

In the burling or mending operation, the carpet roll, 25–50 yd in length is given a close inspection on the face and back. Missing tufts, broken warp or filling threads are hand repaired as are minor weave imperfections.

Steam application to the back and face make the yarns more open or fuller and "blooms" the cut pile yarn thus giving better surface coverage. This operation is quite necessary for "pull-down" or backtwist yarns mentioned in the "Carpet Yarn Section" as it is through the steam that these yarns are drawn down creating the multilevel surface.

Brushing is done with a rotary cylinder covered with nylon or bristle and cleans the surface of cut fibers and other foreign material. Back brushing or beating is also common in the trade.

Napping is an operation that lays the surface pile in the correct direction and also gives better coverage to the fabric. This is done by the use of stainless steel fancy or card wire which may be straight or "knee-hooked" in shape. This wire is mounted to a cylinder which revolves against the carpet.

A large degree of human element is present in the shearing operation as it is here that the operator must use his judgment in the amount of cutting or shearing. Loop pile fabrics may be sheared very lightly to remove the fuzzy or protruding hairs on top of the loop or surface pile. Random cut and uncut effects are obtained by shearing the high loops of a multilevel loop-pile carpet.

A backing cloth may be added to the fabric for increased dimensional stability and additional "hand" or body. This is a loose flat weave or knitted fabric and is referred to as burlap-jute or scrim which may be kraftcord or man-made materials. These cloths may have a weight range from 4 to 12 oz a sq yd plus the additional weight of the adhesive used to bind it to the carpet.

Most fabrics made today are backsized with a latex coating material using single or double trough with a revolving applicator roll. This operation gives weight, improves pile bind and body or "hand" as well as improved back appearance. The size used may be synthetic latex or pure gum compounds and titanium dioxide is frequently used for whiteness. Starch or dextrine sizes are about outmoded. Amounts used or applied may range from 2 to 30 oz a sq yd.

Before packing for shipment or storage, most mills have an intensely lighted area for exposing the entire roll for a final inspection, grading and registering. A more detailed description of the methods used in commercial testing of carpets and their evaluation and quality control is given by Cohen (3). The American Society of Testing Materials D 13 on textiles issued the following test methods for pile floor coverings–carpets:

D 2646–68T	Backing Fabrics
D 1116–65T	Black Carpet Beetle Larvae Attack, Resistance of Pile Floor Coverings to
D 2404–67	Jute Backing Fabrics
D 2401–67	Service Change of Appearance of Pile Floor Coverings
D 1335–67	Tuft Bind of Pile Floor Coverings
D 1486	(*Tufted Pile Floor Coverings*) *Discontinued 1968—Incorporated in Methods D 418*
*D 418–68	Woven and Tufted Pile Floor Coverings

Dyeing

In the past the classical broadloom carpets were made from stock-or skein-dyed yarns. After the introduction of the tufted carpets it soon became apparent that piece-dyeing would give maximum flexibility with minimum inventory. These inventory and process savings, according to O'Mahony, (7) have made the piece-dying route in the United States the overwhelming favorite for the production of broadloom carpets either in becks or by continuous methods.

In 1960, piece-dying comprised only 17% of the 150 million sq yd broadloom carpet market. By 1966, piece-dyed nylon had an estimated 48% of the entire 328 million sq yd broadloom market. It is estimated that possibly, the quantity of piece-dyed nylon carpet produced during 1968 will exceed half of the entire broadloom carpet market—a remarkable change in less than ten years.

This tremendous growth was started by the advent of filament nylon carpets and pushed ahead by the introduction of the resist, deep-dying, ultra-deep-dying, and now, cationic-dyeable nylon yarns which permit the stylists exceptional freedom in the production of colorful designs at modest cost. These developments mean that multicolor piece-dyed styles made with the differential-dyeing nylon yarns cost far less to produce, design for design, than do predyed yarn styles.

Considerable efforts have been made by the technicians of the International Wool Secretariat in England and the Wool Bureau with differential-dyeing woolen yarns to produce multicolor piece-dyed wool carpets. The Wool Bureau has introduced two and three color techniques under the tradename "MULTY KROM." The bases of these techniques are the principles and methods already discussed in Chapter 11, pages 694-703.

Batch-dyeing of carpets in a beck follows the same principles as ordinary piece-dyeing of woolen and worsted fabrics discussed in Chapter 11, pages 748-754, taking in account, however, the heavier weight, wider width, and texture retention.

Continuous Dyeing of Tufted Carpets

For solid shades and multitone or multicolor effects which can be attained with modified synthetic tufting yarns, variations of the pad-steam continuous dyeing technique have been employed.

Systems developed for the continuous dyeing have been based on a variety of machines, such as those of the Bradford Dyers Association, Beloit-Kleinewefers Textile Machinery Corporation, Fleissner GMGH & Co., Edward Kuester Maschinenfabrik, Machinenfabrik Zimmer, and the Stalwart Dyeing Co., Ltd.

In discussing the various machines Zimmerli (13) states the following: The Fleissner continuous dye range for tufted carpet employs a 2-roll dye padder, followed by initial steaming on a perforated drum, and finally loop steaming. Screen drums are used for washing and drying (Fig. 37). The Beloit-Kleinewefers range features a back filling technique of applying dye liquor to the carpet, with its tufts downward, from which point it is absorbed into the tufts (Fig. 38). Loop steaming and washing by jet-type washers follow. The Stalwart continuous dye range for carpet consists of a special dyestuff application unit which effects a greater percentage pick-up than the competing units, followed by a horizontal steamer and washing section.

The Kusters system separates the dye addition and fixing operation into a series of separate steps, each of which is closely interrelated with the others, but by itself is precisely controlled (Fig. 39). First, the carpet passes through an immersion pan containing

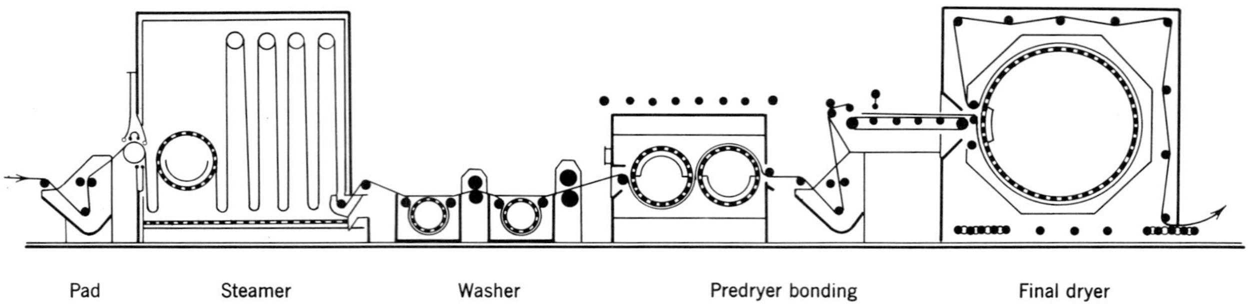

Fig. 37. Fleissner continuous dye range for tufted carpets.

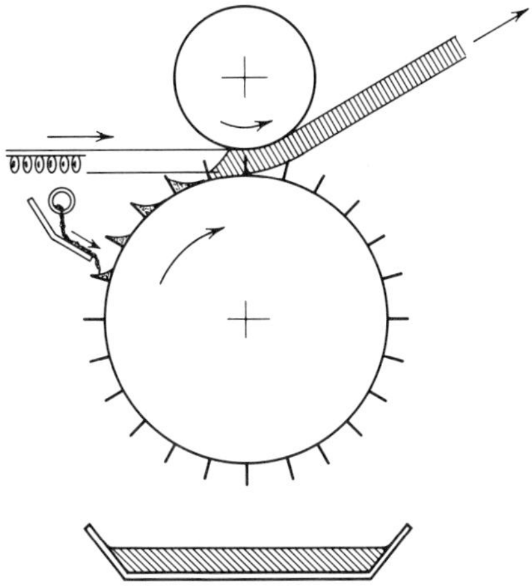

Fig. 38. Carpet dye system. (Beloit-Kleinwefers.)

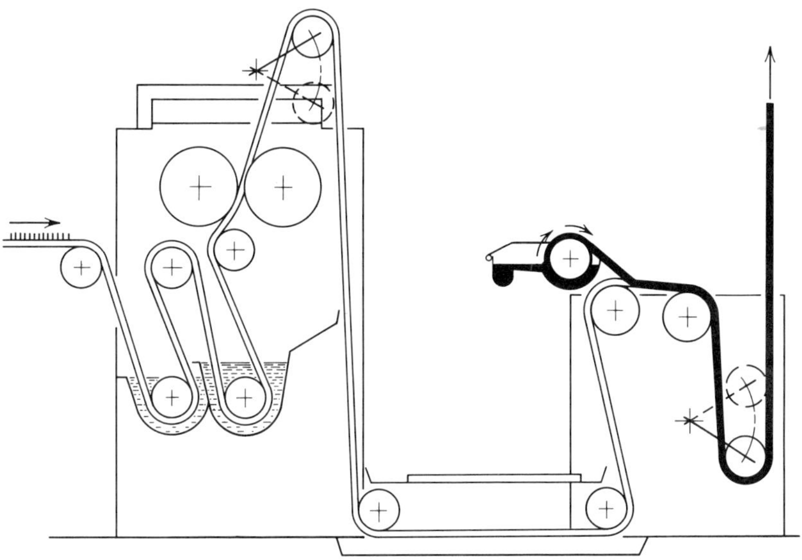

Fig. 39. Carpet dye system. (Kusters.)

wetting agent, and any chemical assistants required by the dyestuff to be used. Next the carpet passes through a nip effected by a swimming roll and mating roll, where the pick-up is established, preconditioning the carpet to receive color. The dyestuff solution is metered continually into a small capacity pan, into which a pick-up roller dips. A novel system of doctor blades is used to transfer an even coating of dyestuff solution from this pick-up roller to the surface of the carpet.

Having absorbed the dyestuff, the carpeting passes into the steamer. After passing over the first needle roller, the carpet starts a 20-ft descent in the first of 7 festoon loops, and during this passage fixation takes place.

After leaving the steamer, the carpet passes into a first cooling bath, then through a light-duty squeezer into two washers equipped with cog belt agitators, which compensate for the slow linear speed of the carpet by creating a dividend, high-turbulence circulation laterally across the face of the carpet from center line to selvages. After washing, the carpet receives its final extraction prior to drying on a subsequent tenter frame.

This continuous dyeing system (Fig. 40) is successfully operating commercially at speeds up to 15 ypm.

Carpet printing

One of the most dramatic developments is the recent adaptation of the automatic screen printing process to provide the color and patterns desired in tufted and nonwoven carpets.

Bradford Dyers Association, Ltd., Machinenfabrik Zimmer, Stalwart Dyeing Co., Ltd., and others have developed machines for printing carpets.

The Zimmer screen printer for moderate height of pile utilizes magnet-actuated roller squeegee (Fig. 41). The Bradford Dyers Association system uses conventional squeegees, plus a vacuum to assist penetration (Fig. 42).

Screen printing techniques allow the carpet designer to use patterns beyond the present scope of tufting equipment, and also to provide colors and patterns for homogeneous nonwoven carpeting such as needlepunch.

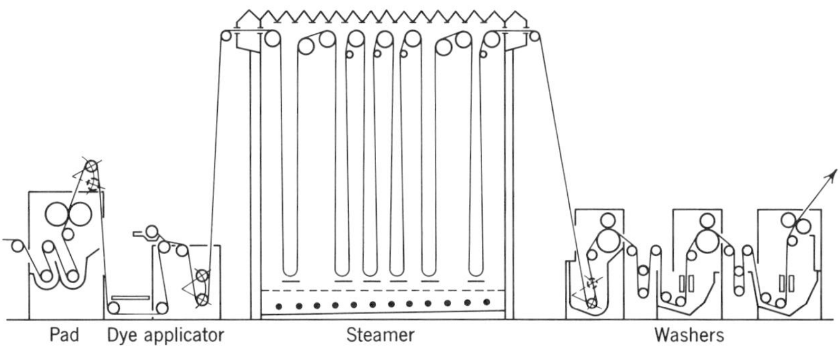

Fig. 40. Carpet dyeing range. (Kusters.)

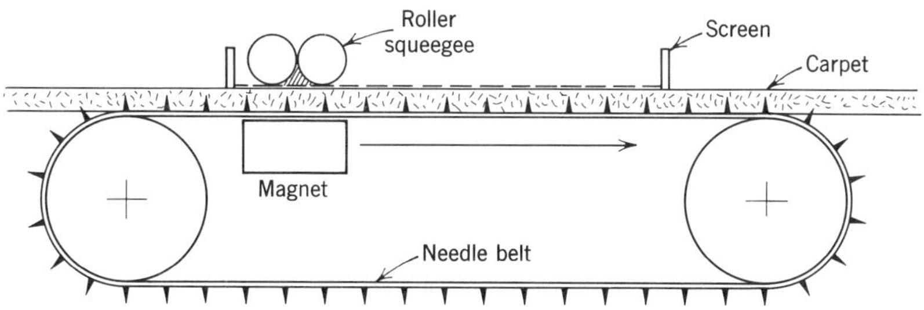

Fig. 41. Screenprinting for carpets. (Zimmer.)

United States Carpet Industry Organization and Equipment

Concurrent with the growth in carpet shipments which took place over the last decade and a half was the change in the market structure of the industry. The Census Bureau's data, summarized in Table 16, shows that there were 159 individual companies with 245 plants manufacturing carpet and scatter rugs in 1963. The number of companies manufacturing tufted carpets had increased by 23%, compared with 1958, while woven carpet and rug manufacturers had declined by 15% over the same period.

TABLE 16

Number of Companies Manufacturing Woven and Tufted Carpets and Rugs, 1958–63[a]

	1958	1963
Woven	60	51
Tufted	88	108
	148	159

[a] Source: U.S. Bureau of Census, *1963 Census of Manufactures*, Floor Covering Mills, Table 1, p. 4.

Historically, production facilities were centered in the Middle Atlantic states and New England but today most of the major producers have spinning, weaving and tufting facilities in Virginia, North Carolina, South Carolina, Georgia, Mississippi and Tennessee. Plant location statistics for 1963, summarized in Table 20, indicate that tufted carpet production is heavily concentrated in Georgia. California has also become an important tufted carpet production area.

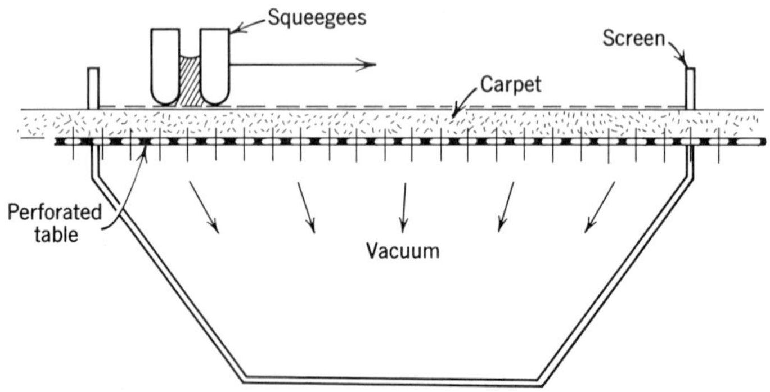

Fig. 42. Screenprinting for carpets. (Bradford Dyers Association.)

Weaving and tufting equipment in operation

Census Bureau data for 1966 and 1967 provides the most recent information on carpet tufting and weaving equipment in place.

Between 1964 and 1965 the total number of looms in place had declined from 2,050 to 1,687 or 18%. The number of tufting machines in place through 1966 is shown in Table 18.

TABLE 17

Carpet Looms in Place, 1966[a]

	Loom type				
Loom size	Axminster	Wilton	Velvet	Other	Total
27 in.	113	242	120	7	482
3 ft under 9 ft	10	43	23	66	142
9 ft	119	69	71	11	270
9fi to 12 ft	110	164	230	26	530
15 ft	33	108	110	12	263
Over 15 ft	—	1	3	4	8
Total	385	627	557	126	1,695

[a] Source: U.S. Bureau of Census, Current Industrial Report, *Rugs, Carpets and Carpeting M 22 Q (66)*.

TABLE 18

Tufted Carpet Machines in Place, 1962–67[a]

	Number				
Tufted machine size	1962	1963	1964	1966	1967
Under 9 ft	1,453	1,492	1,510	1,183	1,217
9 ft	76	76	72	131	148
12 ft	75	81	76	121	96
15 ft	286	335	356	494	532
Over 15 ft	42	46	64	40	64
Total	1,932	2,030	2,078	1,969	2,057
Custom tufting machines: all sizes	244	232	303	198	161

[a] Source: Same as Table 17 for 1962–1964 data. 1966–67 data, U.S. Bureau of Census, Current Industrial Report. *Tufted Textile Products, M22L(67)-2.* 1965 data not reported.

As is fairly evident from the machinery data shown in Tables 17 and 18 broadloom carpet production in the United States today is concentrated in wider widths. No exact figures are available, but it is probable that 85% or more of the total output was in 12 foot or wider widths in 1966. These widths can be sold as wall-to-wall carpet or cut into rugs of all sizes and dimensions. Statistics on custom tufting and area rugs are not available but it is believed that their volume is relatively small.

Employment

Employment trends in the U.S. carpet industry in recent years has been as follows:

TABLE 19

Number of Employees in the U.S. Carpet Industry[a]

	Woven	Tufted	Total
1958	18,151	11,528	29,679
1959	19,746	12,487	32,233
1960	18,404	13,272	31,676
1961	16,410	14,787	31,197
1962	14,476	17,528	32,004
1963	13,398	19,589	32,987

[a] Source: 1963 Census of Manufacturers, *Woven Carpets and Rugs; Tufted Carpets and Rugs, Report MC 63 (p)-22D 1 and 2,* Table 1, page 1 both reports.

References

1. Angus, G. B., Basic Structures and Fibers used in Carpet Manufacture, *Text. Inst. Ind.*, **3**, 313 (1965).
2. Beal, W., Modern Dyeing Techniques in the Carpet Trade, *JSDC*, **83**, 3 (1967).
3. Cohen, I., Carpets: Commercial Testing, Evaluation and Control of Delivered Quality, *Text. Inst. Ind.*, **6**, 147, 181 (1968).
4. Ellis, P., *Ciba Rev.*, **4** (1961); page 2, A Comparison of Traditional and Tufted Carpets; page 10, Technical Development in Machine Made Carpet Production; page 24, Tufted Carpets.
5. Freeland, V. D., *Rev. Textile Progress*, **16**, 179 (1964).

6. Grosicki, Z. D., and Thangiani, S. B., Improvement in Carpet Characteristics Resulting from Backsizing Processes, *Text. Inst. Ind.*, **3**, 59 (1965).
7. O'Mahony, G. M., Impact of Differential-Dyeing Carpets, *Am. Dyestuff Rep.*, **57**, 36 (1968).
8. Phillips, W. E., Artistic Design Trends in Machine-Made Carpets, *Ciba Rev.*, **4**, 6 (1961).
9. Porter, I. S., Modern Developments in Wilton Carpet Weaving, *Text. Inst. Ind.*, **2**, 4 (1964).
10. Rhodes, T., Carpet Backing Fabrics, *Canadian Text. J.*, **48** (1968).
11. Stewart, R. G., Scouring Techniques and Fadable Yellow in Carpet Wools, *Am. Dyestuff Rep.*, **55**, 361 (1966).
12. Sellers & Co., Carpet Printing and Anglo-Austrian Venture, *Man-Made Textiles*, 1964.
13. Zimmerli, K., Finishing Machinery, *Modern Textile*, **44** (1968).

CHAPTER 15

The Manufacture of Felts Felts–Nonwoven

THOMAS J. GILLICK, JR.

Historical Background

Wool felt may be defined as a fabriclike or fibrous structure obtained by the interlocking of wool fibers by a suitable combination of mechanical work, chemical action, moisture and heat without spinning, weaving or knitting. It may consist of one or more classes of fibers; wool, reprocessed wool and/or reused wool with or without admixture with other fibers either natural or synthetic.

In all cases the materials discussed in this section are obtained as a result of fiber to fiber friction or entangling and can, and do exist without nonfibrous additives or bonding agents.

Wool felt is one of the oldest fibrous structures known. Evidence of the early use of felt has been found in the nomadic tribes of central Asia, who used it for tents as well as clothing. It furnished protection against wind and rain and yet was easily rolled up and transported. Later it was used as a floor covering to replace rushes and coarse straw mats. Felts of this type also made early religious and ceremonial headdresses. It appears that felt was known in India as early as 325 B.C. Marco Polo, the Venetian, reported that the Mongols of the thirteenth century made circular felt-covered tents and covered their carts with black felt tarpaulins to keep out the rain. The Chinese considered white felt to be peculiarly sacred. The Greeks used felt for a head covering and later, as lining for helmets, as well as for heavy rain-proof garments. Julius Caesar employed felt as a shield against arrows and as a covering for wooden raiding towers and other military equipment. History mentions a light brigade of Caesar's which was equipped with breastplates of felt. A complete felt-making plant was discovered in the ruins of Pompeii.

To this day, primitive tribes of the desert—Mongols and Berbers—still use felt for clothing and tents and make it by almost the same crude process used by their ancestors. They often add horse hair or camel's hair to the wool. Exhibits of the early pieces of felt and the methods by which they were made are found in museums both in this country and abroad.

The wool felt industry in the United States consists of approximately ten manufacturers with more than ninety percent of the total annual production confined to six companies. The larger of these companies represent in most cases a gradual consolidation of many of the smaller specialty felt companies which came into being in the second half of the nineteenth century and early decades of the twentieth century. The estimated total annual consumption of fibers of all types is thirty million pounds.

"For many years, the Felt Association, Inc. served as the industry source for marketing and manufacturing standards. This function is now performed by The Felt Manufacturers Council, a division of the Northern Textile Association."

The manufacture of wool felt is, of course, by no means limited to the United States. Canada, England, the countries of Western Europe, South Africa and Australia are all represented by well-established manufacturers. Prior to World War II many felt plants were known to exist in the countries of Central Europe and in the Soviet Union. Little is known of the industry there today, but there is every reason to believe that it continues with equal or increased importance.

Manufacture of Felt (1) (11) (12) (24)

Fibers

The felt industry utilizes a greater variety of raw stock types than any other branch of the textile industry. The choice of fibers for a given felt is generally dependent upon the end use properties required for the finished felt and in some cases the wool content may be as little as ten percent of the blend (5) (6).

Wool grades employed run from US 48s to 70s generally in one inch to two inches staple for the production of top quality felts. These wools are scoured, usually carbonized, and frequently depitched.

Noils, card wastes, tops and thread wastes are also used to a large extent, while cotton, silk, hair, kapok, rayon and other synthetic fibers are used in blends for specialty end use purposes.

The manufacture of felt is conducted by a separate division of the wool textile industry, using a general method that has been handed down from antiquity and which probably antedates weaving.

The felting process is distinct and separate from other textile construction methods. It is done on specially designed machinery which cannot be used for other textile purposes. The first and last stages of felt manufacture are however, the same as those used in the production of woven woolens. In both felting and weaving the raw fibers must be selected, graded, scoured, mixed, and carded. In each case, moreover, the process ends with washing, dyeing, drying, pressing, cutting, inspecting, and other finishing operations.

The essence of the felting process is the hardening and fulling. Hardening is peculiar to felting alone; fulling may appear as one of the finishing operations in other woolen and worsted manufacturing but in felting it is a basic operation.

Blending or picking

Designated proportions of specified wools, or wool with other fibers, are weighted out according to batch requirements. The mix, or blend, is spread on the floor in layers until the batch is complete. It is then fed into the mixing picker, a machine which blends, opens, and redistributes the fibers. The stock is run separately through the blending and mixing picker and then into a bin, where it is blown and air blended until the fibers are thoroughly opened up and uniformly mixed. See page 122, Volume II, Part 1.

Carding and fabricating

After picking, the blended raw stock is put through a breaker card equipped with a Bramwell, or other automatic weighting feed.

Fig. 1. Six-cylinder white goods carding set for wool felt. It consists of four breaker cylinders, one finisher straight card, and a finisher crosser card to the left. (Courtesy The Felt Association, Inc.)

From the breaker, an Apperly or Broadband automatic feeding system delivers the roving or web to a standard type wool finisher card further to blend and comb the fibers. The finishing card, or "former," as it is often termed, delivers the stock in web formation to an endless forming apron. The forming apron delivers batts ranging from $\frac{3}{4}$ to 1 in. in thickness and 36–80 in. in width.

The customary method of fabricating "roll" felt is to cross alternate webs from multiple-card sets to provide transverse, as well as longitudinal strength. "Sheet" felts, which are usually the thicker, high-density types, are fabricated in smaller pieces to satisfy practical milling requirements and are frequently formed on a Blamire lapper card, with successive batts crossed by hand laying.

This 90° layup provides a maximum of dimensional stability. These batts are laid up in considerable thickness, sometimes to as much as 3 ft. (See Fig. 1).

Hardening

The batt has now arrived at the operation which is peculiar to felting and which has no direct parallel in the textile, or any other, industry. The object of this hardening process is to give the batt enough consistency to permit easy handling in the subsequent fulling process.

The batt is put onto the lower platen of the hardener (see Fig. 2). This platen is of varying dimensions to accommodate the different sizes of batt. An apron of burlap, properly moistened, is laid over the batt and other batts are layed on top of the first until the desired weight and thickness is attained. Steam is now turned on and the pieces saturated. The top platen of the press is then let down onto the hot, moist blanket and the hardening process is started. The top platen oscillates rapidly and transmits its motion to the mass in combination with moisture and heat. This motion induces felting of the wool fibers between the batt layers which make up the felt. The hardening process will reach its limit when the interlocking of the fibers can proceed no further under interfacial pressure and vibration. In the hardening process, roll felts may be reduced in thickness from as much as 12 in. in the batt form to 1 in. in the hardened state. Sheet felts are generally formed into thicker batts, ranging up to 3 ft for a sheet that is reduced in hardening to 3 in.

Fulling

Most of the fulling action is carried out by means of direct controlled impact in the conventional one-, two-, three-, and four-hammer mills (see Fig. 3). These mills are bin-like receptacles with plano-concave breasts, opposed to which are impact hammers having stepped faces which conform roughly to the contour of the breast. Driven by crank or eccentric mechanism, the hammers roll,

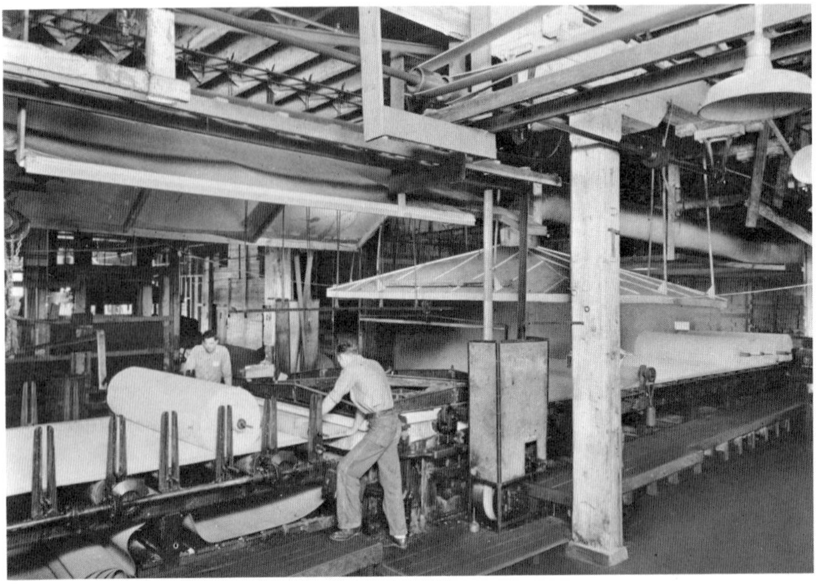

Fig. 2. Hardener on which carded batts are laid is at right of machine. After being passed over the steam chest and under the plate, the batts are rolled up for the next process of fulling. (Courtesy The Felt Association, Inc.)

lift, pound, and squeeze the felt batt against the breast, then allow it to fall back, and turn it so that a different face is presented for the succeeding power stroke. (For the action of fulling, see Vol. 1 Chapter 3, p. 199.) Prior to fulling, the hardened batt is treated with a fulling agent which serves as a lubricant during the operation and is subsequently removed. For the lighter felts, warm water is sufficient. For soft-finished pad and millinery felts, soap is preferred. Very dense, heavy-duty mechanical, glass-polishing hair felts, and sheet felts are most readily fulled with the assistance of a sulfuric acid solution, 2° to 6° Twaddle, equal to 1.5 to 4.5%. This treatment is known as acid fulling.

After treatment with the appropriate fulling agent, the felt is folded and rolled in a definite pattern to obtain the desired width and lengthwise shrinkage, and loaded in the fulling mill. Fulling

Fig. 3. Hammer fulling of wool felt. The hardened felt is being laid in the door (the heavy wood is shown immediately in front of the operator's leg on the left), which is raised and the hammers are put into operation. (Courtesy The Felt Association, Inc.)

continues for periods of three minutes, for lightweight felts, to twelve hours or more for hard felts, sheet and wheel felts. Periodically during the process the felt is removed from the mill, straightened, pulled to remove wrinkles and prevent it from felting upon itself, and then refolded and replaced for further milling. Dimensional shrinkages attained by fulling approximate 10–20% for pad weights, 40% for backchecks (the densest grades of roll felts) and 75% for hard sheet felts.

While the felt industry has been able over the years to avail itself of somewhat conventional processing equipment for many of its operations, the steps of hardening and fulling have required special equipment. Traditionally, England and Western Europe have been sources for these machines.

Many improved and enlarged hardening machines have been brought into the United States from England during the last decade and improvements in conventional hammer fulling machines have provided machines as shown in Fig. 4, which have been produced in the United States.

Strong impetus has been put into development of so-called "continuous" hardening and felting equipment.

The greatest revolutionary change in the wool felt industry took place after World War II in the fulling section with the introduction of new semicontinuous and continuous machines. According to Dietrich, (4) the Pilgrim-Step Roller Fulling Machines, built by Krafft-Göbel in Germany, are increasingly replacing the conventional Hammer Fulling Mills. These machines are suitable for fulling light and heavy felt out of wool or hairs; depending on the purpose for which they are used, they are equipped with special accessories.

The working principle is mainly one of a two-direction felting action. The Pilgrim motion, which consists of a rhythmical forward and backward movement of the upper and lower rollers,

Fig. 4. Roll felt about to be unloaded from an improved type 4 hammer fulling mill. The "breast" is laying upon the bench and the hammers may be seen behind the felt roll. (Courtesy American Felt Co.)

favors this action. The felting of the core of the felt is assisted by the continuous pressure cycling. In addition, goods can be exposed to sideways impulses caused by an axial vibration of the upper rollers.

The Pilgrim-Step Roller Mill works with 19 pairs of rollers made of strong steel pipes. The speed can be regulated from 0.36 to 3.6 meters per minute (½ to 4 yards per minute), and the feed and takeoff can be controlled independently. The distance between the rollers can be set from 2 to 70 millimeters (0.08 to 2.7 inches). The upper vibrating rollers produce a specially good compacting crosswise to the movement of the goods and can be set for the different qualities of felt from 250 to 550 double motions.

The diagrams in Figure 5 show two models of the Pilgrim-Step Fulling Machine. Approximately 60 such machines were in use in

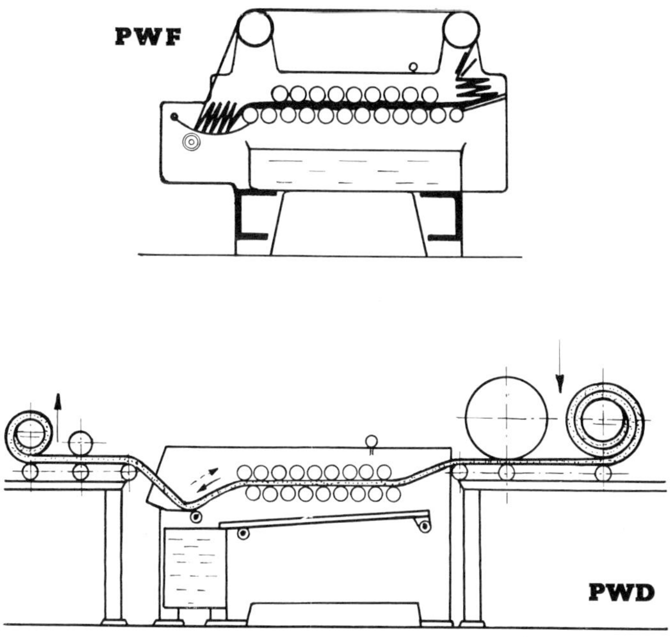

Fig. 5. PWF Pilger step roller fulling machine for thin felts up to 2.5 mm thick; PWD Pilger step roller fulling machine for felts over 2.5 mm in thickness. (Courtesy Krafft Göbel KA Textil machinenbau, Germany.)

1964 in most of the felt factories around the world. Other constructions are covered by German Patent 1,139,808 and U. S. Patent 2,951, 276 (1960). A similar equipment is the Cassé machine built in France by Marcel Cassé. The Hammer Fulling Machine still keeps its place in processing high-density felts.

Wet finishing and dyeing

The wet finishing and dyeing of roll felts is done in conventional dolly washer or reel type textile wet processing equipment. Scouring, neutralizing, bleaching, dyeing, and after treatment methods closely parallel those of the woolen and worsted cloth trades. Dyeing is usually done only on felts for the apparel and decorative trades and the bulk of roll felt is sold in the natural undyed state.

Sheet felts are washed and excess fulling acid is removed by the use of centrifugal extractors fitted with spray nozzles. After water spraying and rinsing, the sheets are finally extracted and dried in forced hot air ovens. For convenience, the sheets are hung on rack trucks which may be easily moved in and out of the drier.

The roll felts are rinsed and extracted by the use of roller extractors as shown in Figure 6.

Light weight roll felts are dried on a continuous basis in endless pin clip forced convection hot air dryers. The denser or thicker grades are usually dried on tentering racks or frames in heated rooms or ovens in a batch process operation.

Finishing

The dried roll felts are finished by shearing, pouncing (Fig. 7), napping, and pressing, or any suitable combination of these. The finished thickness of almost all roll felt is set by either roller or platen pressing, and the product of the industry is sometime erroneously referred to as "pressed felt." Each sheet is inspected for weight, thickness and then trimmed or framed to the desired dimension.

Sheet felts are generally produced 36 sq in., although there are some exceptions. Commercially available thicknesses range from $\frac{1}{8}$ in. to 3 in., with a weight range available from 2 lb/sq yd to 8 lb/sq yd in the $\frac{1}{4}$ in. thicknesses, up to 80 lb/sq yd in the 3 in.

Fig. 6. Rubber covered roller extractor used for rinsing and extracting roll felts. (Courtesy American Felt Co.)

thickness. On specialty orders, sheet felts can be produced as thin as 1/16 of an inch in certain densities.

Roll felts represent the predominant product of the industry, and are manufactured in continuous lengths from 10 to 60 yards in widths from 54 to 80 in. Weights may range from 3 oz/sq yd to 18 lb/sq yd in a thickness range from 1/32 of an inch to 1 1/4 inch.

Properties and modification

The wide range of fiber blends, types, weights and thicknesses of wool felts produced today is perhaps the most significant testimonial to the use, properties and versatility of this class of nonwoven materials. Commercial Standard CS-185-52, "Wool Felt," lists sixteen separate classifications under sheet felts. In addition, eleven separate items are listed under the general heading of roll felts for the apparel and decorative trades.

Fig. 7. Firm density mechanical roll and sheet felts are finished to precise thickness through use of a pouncing machine. (Courtesy American Felt Co.)

A listing of major properties which makes the aforementioned breakdown necessary would include the esthetic and warmth factors which contribute to apparel and decorative usage, but also such mechanical factors as wear resistance, oil absorption, capillary flow, thermal conductivity, coefficients of friction, porosity, compressional resistance, resistance to ageing, and energy absorption (13).

In recent years, due to increased demands for ever widening applications, it has been found desirable to modify the properties of felt by combining, impregnating or surfacing with other materials. Natural and synthetic elastomers and polymeric resins are generally used for impregnating, while sheet and film materials such as rubber are used for combining. In many cases, solvent activated or pressure sensitive adhesive coatings are also used to widen end use properties. Table 1 summarizes some treatments of wool felt and their special application.

The many mechanical properties of wool felt have led to engineered industrial applications which include filtration, sound

TABLE 1
Wool Felt Treatments and Related Design Applications

Treatment	Added property	Applications
Chromate impregnation	Corrosion resistance, flame resistance	Gaskets for refrigerants and heat exchangers; seal for dissimilar metal insulation
Graphite impregnation	Lubrication	Anti-friction lubricating pads, washers and seals
Natural and synthetic rubber impregnation	Flexible stiffening and strengthening	Backing and cushioning for plastics and rigid materials
Phosphates, ammonium sulfamate	Flame resistance	Clothes dryer seals
Dihydroxydichlorodiphenylmethane	Fungus resistance and impregnations	Funginert; acoustical and thermal insulation
Natural and synthetic rubber laminations	Waterproofing and rigidity	Seals and washers
Aluminum or lead foil laminations	Properties of the metal	Optical, acoustical and thermal insulation
Latex coatings	Reinforcement and antiskid	Electric motor and fan bases

absorption, thermal insulation, shock absorbing, cushioning, padding, packing, surfacing, spacing and frictional. Since felt possesses all of these properties and functions at all times, it is not unusual in many applications for the use of felt to be related to two or more of these properties (7)(8)(9)(10).

Standards and specification

The development of the wool felt industry has led to a mutual recognition between manufacturers and consumers for the need for standards and specifications for use in governing selection, acceptance and use of the many individual types and grades of wool felt. In addition, standard methods of testing have been developed to strengthen the use of specifications (2).

TABLE 2
Supplementary Engineering Data on SAE Grade Felts

SAE spec. no.	Abrasion wear felt to felt 10,000 rubs/8 in. stroke; % decrease in thickness at stated loads in oz/sq in.			Coefficient of static friction; Felt against				Oil absorption and capillary flow SAE No. 20 oil at 70 F								Thermal conductivity K-factor (1 in. felt) Btu/hr/sq ft/in./F
	3.8	7.0	10.2	Wood	Glass	Metal	Oiled metal	24 hr absorption; % weight increase	Wicking rate in./min lapsed time: 5	10	15	30	60	90	120	
F-1	1.0	2.3	2.1	.37	.26	.22	.18	185	0.5	0.7	0.8	1.1	1.6	2.0	2.2	0.36
F-3	2.1	2.1	3.0	.37	.26	.22	.18	265	0.5	0.7	1.0	1.3	1.7	1.8	2.1	0.36
F-5	3.0	3.0	6.0	.37	.26	.22	.18	320	0.6	0.8	1.0	1.3	1.7	1.9	2.0	0.30
F-15	4.1	7.6	18.5	—	—	—	—	570	0.6	0.9	1.1	1.4	1.6	1.8	1.9	0.24

* Thermal and acoustical felts blended with kapok fiber have tested values of 0.21 K factor and 0.80 sound-absorption coefficient at 512 cps for 1-in. thicknesses of 3.2 lb per sq yd weight.

The Society of Automotive Engineers in 1923 adopted as a standard "Felts," covering the types of felt generally required for automotive usage. This mechanical roll felt standard has been the basis for several hundred individual company felt specifications in use today. Federal Specification CF-206a, the general military procurement felt specification in current use represents a consolidation in 1954 of a total of seven individual but overlapping service branch and Federal felt specifications. Under Committee D-13 on Textile Materials of American Society for Testing Materials,(15) there is an active Subcommittee A–10 on Felts which has prepared and is responsible for two Standards. These are ASTM D-461 "Standard Methods of Testing Felts," and ASTM D-2475 "Standard Specification for Wool Felt." The latter covers the chemical and physical requirements for roll and sheet felts. In Table 2 are given some engineering data on Society of Automotive Engineers Felts.

TABLE 3
Production of Pressed Felts[a]
(Thousand lb)

Country	1962	1963	1964	1965	1966
Australia[b]	—	—	4,009	4,514	—
Austria	1,155	1,155	979	1,219	1,486
France	4,678	5,154	4,822	4,224	4,623
Japan	19,100	20,448	23,274	25,219	27,533
Portugal	33	67	82	116	237
Sweden	807	540	437	506	492
United Kingdom[c]	8,801	7,668	7,360	7,170	6,810
United States	—	18,378	16,000	20,414	—

[a] Excludes needle loom felts.
[b] Thousand square yards. Year ending June of year shown.
[c] Total consumption of materials in pressed felts and hat felt manufacture.

References

1. Anon., American Felt Company, Glenville, Conn., 1951. *Felt, A Mechanical and Engineering Material.*
2. Campbell, J. B., *Materials and Methods*, August 1950. Engineering Properties of Felt.
3. Dietrich, H., *Textilpraxis* **11**, 1 (1956). Wool Felt Manufacturing and Its Important Machines.

4. Dietrich, H., *Melliand Textilberichte* **45**, 618 (1964). Technical Development Stage of Wool Felt Industry.
5. Gillick, T. J., *Electrical Manufacturing*, April 1958. Synthetic Fibre and Wool Felts, Their Design, Characteristics and Uses.
6. Gillick, T. J., *Industrial and Engineering Chemistry*, August 1959. Natural and Synthetic Fibre Felts.
7. Gruberg, L. D., *Steel*, February 1952. Felt for Finishing.
8. Lehmberg, W. H., *Mechanical Engineering*, February 1945. Mechanical Properties and Uses of Wool Felt.
9. Lehmberg, W. H., and Johnson, E. L., *Product Engineering*, June 1946. Felt Wick Lubricating Systems for Bearings and Machine Parts.
10. Lehmberg, W. H., *Kent's Mechanical Engineers Handbook*, Chapter 5, Wiley, New York, 1950. Felt.
11. Pope, N. B., and Marshall, J. F., American Felt Company, Glenville, Conn., 1946. *The Story of Felt.*
12. Pope, N. B., The Felt Association, New York, 1950. *Everybody Uses Felt.*
13. Riley, M. W., *Materials and Methods*, December, 1956. Treated Felts.
14. Sustmann, C. C., *Ciba Review*, November 1958. *Ciba Rundschau* No. 139, 1958. Felt.
15. American Society for Testing Materials. *Standard Methods of Testing Felt*, ASTM: D461–67. *ASTM Standards for Textile Materials*, Part 24, 1968.

The Manufacture of Woven and Needled Felts

CARL GORDON

Felts which are made by the woven or woven-and-needled processes consume about three percent of the total wool used in the U.S., i.e. about 6.3 million pounds per year.* On a world basis, the quantity would be more than doubled (Est. factor 2.5). In spite of this large usage, woven felts are not as commonly known as pressed or non-woven felts. Approximately 80% of the amounts mentioned are used in papermachine felts which carry the newly formed wet sheet of paper from the wire on which it is formed through squeeze rolls to remove excess water. It might well be asked why woven felts are called felts. To answer this we turn to the beginning of papermaking when only hand sheets were made. Each sheet was removed from the hand dipped screen and placed on a square of pressed felt with another felt on top of it. Further sheets and felts were built up alternately until a stack was ready to be placed in a screw type press where the excess water was squeezed out.

In the early 19th century the Fourdrinier machine was invented to make paper continuously by flowing dilute paper stock onto a moving wire screen. The sheet was transferred from the wire to a woven blanket which then carried it through squeeze rolls to remove more water. Since the squeeze roll action was replacing the sheet-felt-sheet-felt . . . sandwiches which were squeezed in the screw press, it was natural that the woven fabric should be called

* U.S.D.A., "The Wool Situation," October 1964.

by the same name,—felt. This is undoubtedly the principal reason for its name. A secondary reason, however, is that many of the woven felts look much like a pressed felt, so much so that one would almost need to dissect them to be sure which is which. Compactness of weave and heavy fulling (frequently more than a 2:1 width difference) make the woven goods lose their yarn structure appearance. This is aided sometimes by added "cover" from napping (Fig. 8).

It may be helpful to describe a paper machine felt, its purpose and the job that is required of it. It may be thought of as an endless belt which is as wide as a paper machine. This may be as much as 350 in. In lengths, felts vary from as little as 30–40 ft to nearly 300 ft. Felt sizes may be better appreciated when it is realized that they average approximately 3.5 oz per sq ft (range 2–6) so that a single felt will often weigh 600 lb or more. It is common for the felt to be

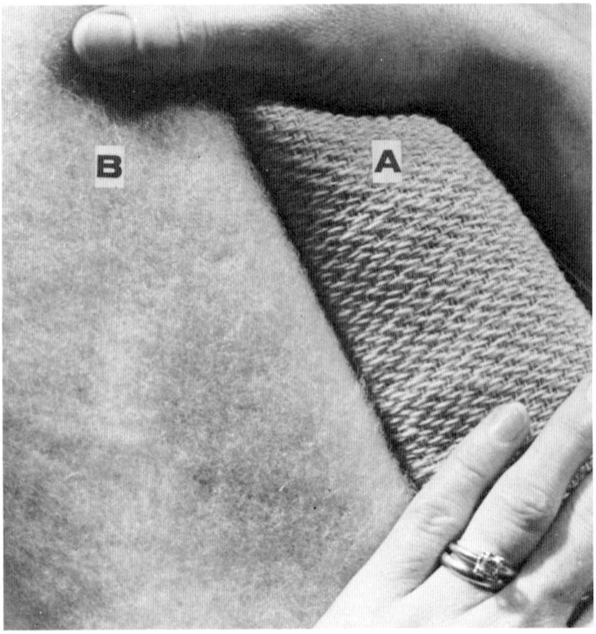

Fig. 8. Sections *A* and *B* are of the same fabric before and after fulling, respectively. Note how fulling has transformed the yarn structure to a uniform surface.

driven only by the press rolls with the felt having to drive all the other rolls. Thus it becomes obvious why strength and stability are prime factors in the success of a felt.

Strength alone is not enough however, since another factor, that of removing water from the sheet at the press rolls is also a paramount requirement. This is commonly called "openness" and becomes more and more important as the speed of a machine increases. With present day conditions, production rates often make the difference between success and failure and much emphasis is therefore placed on machine speeds. Machines which 10 years ago were running 600–1000 ft/min are now running 1200 to 1800, and machines which made tissue grades at 1800–2500 ft/min are now running 3000–4000.

Still a third fundamental requirement in the making of woven felts is that of giving the sheet of paper a "finish" which is needed for its intended use. Naturally some sheets are much more critical from this standpoint than others. Designing felts to meet the needs of the paper industry is both a science and art.

The relationship of these three basic characteristics have frequently been shown by means of a very simple illustration. Imagine an equilateral triangle with each corner representing one of the three characteristics, "Strength," "Openness" and "Finish" respectively and pivoted at its center. If the "Strength" corner is elevated, the "Openness" and "Finish" corners are depressed. A similar condition takes place with each of the other factors in turn i.e. as one is elevated the other two are lowered. Thus, making a felt to fill the requirements of a particular machine position is a continual triangular dilemma. A felt might call for (*1*) using large-strong yarns for greater strength, (*2*) using fewer yarns for better openness, or (*3*) many small yarns for better finish. It is obvious that all of these things cannot be done, in fact not even two of them. However, sometimes other means of fulfilling some of these requirements simultaneously can be found, and will be discussed below under "Synthetics," "Treatments" and "Needled Felts."

Use of Synthetics

Originally, all woven felts were made of wool. However, with the advent of synthetic fibers, it was found that a percentage of these

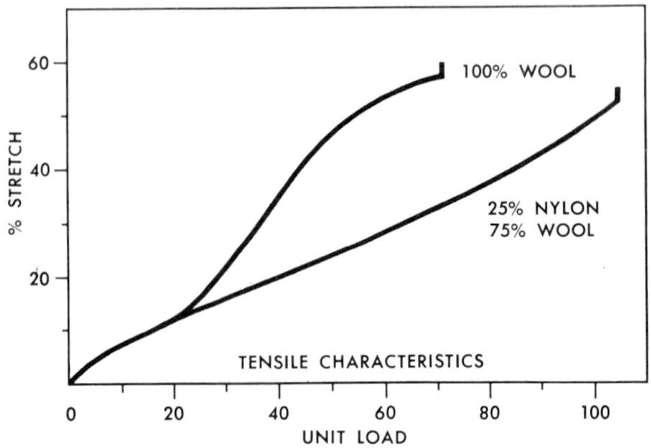

Fig. 9. Adding 25% nylon to the wool blend increases the strength by approximately 50%. After a unit load of 20 the stretch is reduced. Both increase stability.

could be incorporated with the wool to given enhanced properties of greater wear and strength and thus longer life. This was such a radical change from previous thought and practice that it was patented in 1947.*

Fulling has always been an important step in the making of felt for it is a leveler and compacter which improves the "Finish" of the felt and so also the sheet of paper made from it. Synthetics do not "full" in themselves although it is possible to obtain adequate fulling when reasonable amounts are blended with wool. Experience has indicated that synthetic fibers may be used up to 75% where wear resistance is important. This takes advantage of much of their beneficial characteristics without significantly affecting the other desirable felt properties of resilience and openness.

Figure 9 shows typical curves of stretch vs. load for an all-wool felt compared with one having a 25–75 nylon–wool blend. It may be noted that after the initial portion of the curve the synthetic–wool blend stretches less, thus making it more stable. Following the ideas mentioned above, this makes it possible to obtain added strength (and wear) without too severely influencing the "openness" or "finish" requirements. Although nylon–wool blends are mentioned above, polyester–wool blends act similarly.

* U.S. Patent No. 2,423,828 (July 15, 1947) by Z. J. Chagnon, assigned to Albany Felt Company, Albany, New York.

Synthetic blends are particularly beneficial when paper mills use higher stock temperatures. Some mills are doing this for different reasons but primarily to take advantage of the greater fluidity of water at the elevated temperatures, with the result that the presses can express more water from the sheet of paper and the felt can "handle" the extra water by acting more open. However, as the temperature increases, the elongation of the felt is greater so that at the same load it might "stretch off the machine." Figure 10 shows that an all-wool felt is 20% weaker at 130°F, and 29% weaker at 170°F than at 70°F. Figure 11 shows how polyester content gives felts greater stability and strength. For example a 25% polyester felt has 27% more strength at 130°F than an all-wool felt at 70°F. Even at 170°F this same felt is 18% stronger than the all-wool at 70°F. (Other similar comparisons are apparent from the chart.)

Ultimate strength, however, is not usually the most important factor in a felt. More important is its stability under a load of approximately 20% of its strength. Thus, elongation at 20% of ultimate strength will determine whether it is sufficiently stable to do its required job for 250,000 to 1,000,000 revolutions without stretching and narrowing significantly.

It may be noted from the data in Fig. 12 that if a paper mill increases the stock temperature from 70 to 170°F, an all-wool felt would elongate 15%—an amount that would probably cause the felt to be removed. However, if the felt had 25 or 45% polyester in the lengthwise yarn under the same conditions, the felt would have elongated 11 or 7% respectively, a condition which might be tolerated.

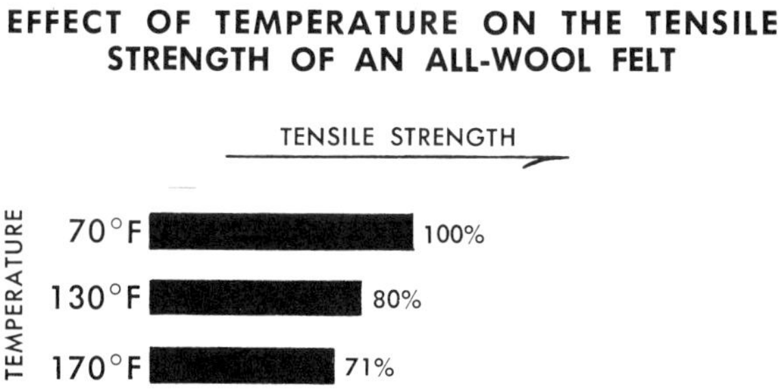

Fig. 10. Elevated temperatures lower the strength of wool felts.

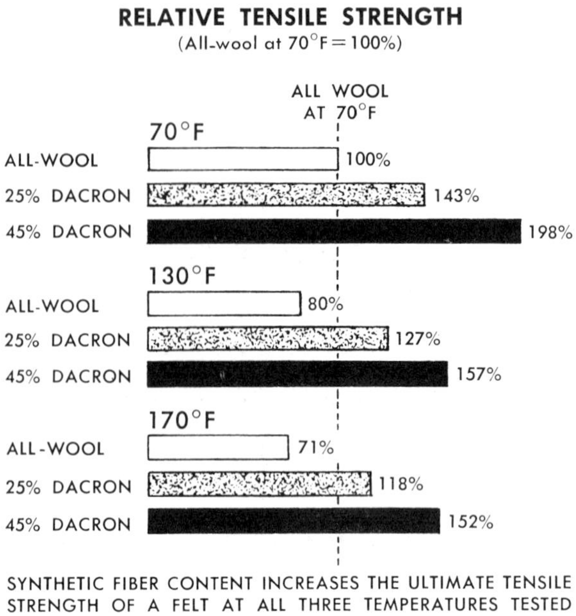

Fig. 11.

Synthetics therefore again help to impart more "strength" or stability to a felt without significantly impairing its other two properties of "openness" and "finish."

Synthetics are here to stay in the woven felt industry as may be realized by the fact that approximately eighty percent of all papermakers' felts now contain some synthetic fiber. While several types of synthetics can be included in the blends for paper machine felts, those in greatest use are the nylons and polyesters. It is realized, however, that other fibers may be added to this group. It is estimated that 35% of all fiber for paper machine felts is synthetic, and this proportion is tending to increase.

Comparisons

Basically, many of the steps in the manufacture of woven felts are the same as those in the manufacture of normal woolen fabrics. However, there are certain differences mentioned below. Paper machine felts will be the product referred to for most of this dis-

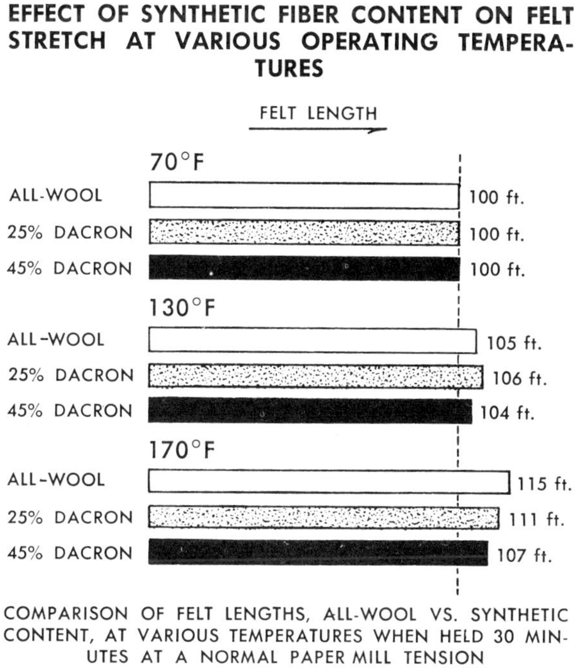

Fig. 12. Adding 25 or 45% of Dacron polyester fiber to the wool blend improves felt stability (decreases stretch) at elevated temperatures.

cussion, but differences in other important industrial fabrics may also be included.

Wools: Paper machine felts are a quality product and require the best wools obtainable. The word "best" however means different things for different types of felts. Sometimes it refers primarily to fineness, to softness, or fullability, to strength or to staple length, but in every case only sound wools are selected. Wools from most of the wool producing countries of the world are used. Many times a certain grade of wool is only available at one time during the year, so it is necessary to estimate one's requirements for an extended period. This means that a large and expensive inventory is needed to maintain uniform quality through the year.

Sorting

Sorting is not too different from that practiced for normal woolen manufacturing, though it is important to have every grade as uniform as possible and relatively free from burrs as well as tender or britchy grades. It is interesting to note that in the last few years artificial light ("daylight" fluorescent tubes) has been usable for wool sorting, a step that for many years was considered impossible. This has extended possible working hours, particularly on short and dark winter days.

Scouring

For many years the most commonly used scouring materials were soda ash and soap. The alkaline soda ash solution had a deleterious effect on wool, particularly if there was any stoppage during the processing. Current practice is to use synthetic detergents with no alkali present except that which comes from the salts on the wool itself. These have a pH similar to bicarbonates and are therefore much less harmful to the wool than carbonates such as soda ash. In order to secure effective scouring with these less alkaline solutions, the temperature in the first two bowls is raised to 135 or 140°F. The third and fourth bowls (and fifth if available) are progressively lower in temperature by about 5 degrees each. In spite of the higher temperatures, the more neutral solutions leave the wool in a far better and sounder state. See page 36, Part 1.

Blending

Although blending is done to a certain extent in the grease wool itself, most of the blending is done at the picker. Frequently a preblend is made prior to the completed blend in order to work in the different grades more uniformly. It is a real science to use the right wools for the right jobs. Long strong wools would be chosen for the lengthwise yarns in the felts while shorter and finer wools would be chosen for the crosswise yarns in order to develop bulk and better fulling properties.

Picking

Oil and water are added, usually as an emulsion, at this point, the oil for lubrication during carding and the moisture to bring the wool above the equilibrium level for the humidity maintained in the card room. It is preferable that during carding the wool give up rather than pick up moisture. All carding is done at a relatively higher moisture content so the fiber is more flexible and breaks less.

Cards

Cards generally have three main cylinders with a metallic breast, Peralta rolls, intermediate broad-band feed and tape condensers. Card widths range from 60 to 84 in. with capacity for 96–120 ends.

Even though the cards are capable of making the above mentioned number of ends, much of the time the ends are doubled or even quadrupled because of the larger sizes of yarns required for coarse felts. Thus, in a card which could produce 96 ends, only 48 or even 24 are taken off. Some cards are run continuously on the doubled basis and then the tapes are so arranged that the doubling takes place ahead of the rub aprons (U.S. Patent No. 2,101,370 by H. H. Knight, Davis and Furber Machine Company) thus making a more homogeneous end with less tendency to cause trouble in spinning.

Spinning

From a woven felt standpoint the unusual part is in the large range of yarn sizes, from 70 to 2100 tex (Table 3). The degree of twist or the "twist constants" used also have a wide range, namely from 3 to 32 on the basis of turns per centimeter and tex units (Table 4).

TABLE 4

	Usual ranges of yarn sizes for felts of the following types		
	Fine	Medium	Coarse
Grain/100 yards	100–300	400–600	600–2,000
Tex	71–213	284–424	424–1,420
Woolen runs	4.37–1.46	1.09–0.73	0.73–0.22
Worsted	12.50–4.18	3.13–2.08	2.08–0.62
Cotton	8.33–2.78	2.08–1.39	1.39–0.417

Twist constants have a real value not only as a means of indicating the angular degree of twist for any size of yarn but also in showing the relationship between single and ply yarns during the plying process. It is well known that if Z twist, for example, is applied to the singles and S twist is used in the ply, the twist of the singles is removed by the same number of turns as that inserted in the ply. Figure 13 shows the twist constant of the single yarn on the ordinate and the curve shows how this changes as more and more ply twist is inserted (shown on abscissa). This particular chart applies to only 2-ply yarns but similar ones could be made for any other numbers of plies. In using the chart, if we start with a singles yarn having a twist constant 22Z and insert a ply twist equal to T.C. 12S, the singles yarn is reduced from 22Z to 5.3Z twist constant. As a second example, if we start with a low degree of twist in the singles in the Z direction, say twist constant 15Z, and

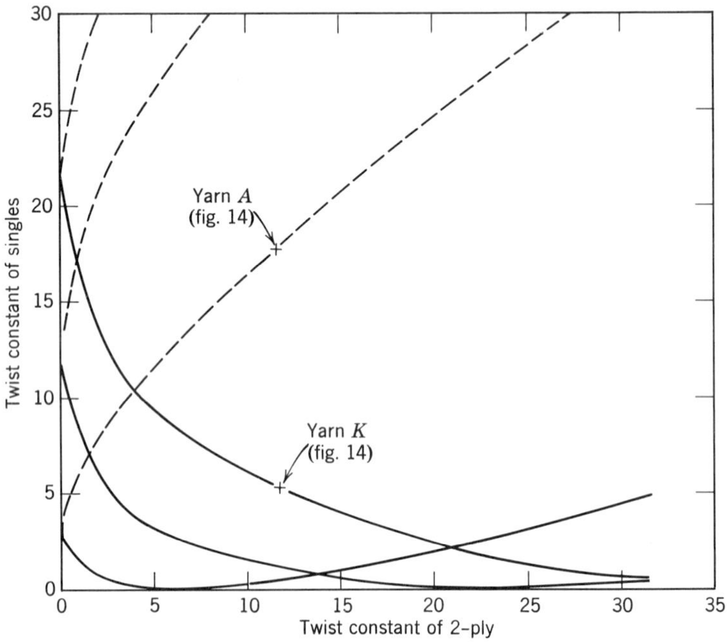

Fig. 13. Twist constants of singles yarns are followed in their degree of twist while they are made into 2-ply yarns of opposite twist direction (solid lines) or the same twist direction (dotted lines).

we insert a large amount of ply twist in the S direction then all of the singles twist is removed and some extra is added in the S direction, so that in the final yarn the twist is S direction in both single and ply. As a third example, if for some special reason the ply twist is inserted in the same direction as the singles, for example both Z twist, then the twist constant of the singles goes up very rapidly as the ply twist is inserted. These are shown in the dotted lines of the chart. (Fig. 13).

For anyone involved in designing or using yarns, it would be well for him to redraw the above chart using units of weight and twist with which he is most familiar. The fundamental value of these twist relationships should not be obscured because of unaccustomed units. From Fig. 13 and Table 5 one can visualize how the singles twist changes when 2-ply twist is inserted. They also indicate when twists are unrealistic or impossible.

A study of yarns which were made in different combinations of single and ply twists indicates they are quite different in appearance. Those yarns which end up with single and ply twists in the same direction look much more like single yarns than like ply, whereas those yarns which end up with a significant single twist constant in the opposite direction from that of the ply, stand out prominently as 2-ply yarns (Figure 14). Table 5 gives the data for the two yarns "K" and "A" shown in Fig. 14. The positions of these two yarns are also shown on the curves of Fig. 13.

TABLE 5

Yarn Letter	Yarn Size (Tex)	Twist in Turns per Cm		Twist Constants**		
				Singles		
		Singles	2-Ply	Put-in	After Plying	2-Ply
"K"	178	3.54–Z	1.82–S	22.4–Z	*5.3–Z	*11.8–S
"A"	178	1.34–Z	1.82–Z	3.2–Z	*17.8–Z	*11.8–Z
"T"	178	1.34–Z	2.65–S	3.2–Z	*3.1–S	*25.0–S

* Condition in completed yarn.

$$** \text{ Twist constant } = \frac{(\text{Turns/Cm})^2 \text{ x Tex})}{100}$$

K

A

Fig. 14. Twist of the "A" yarn is in the "Z" direction in both singles and ply. (It looks like a singles.) In the "K" yarn, same size, the singles is in the "Z" direction and the 2-ply in the "S" direction. (It looks like a ply.) *Note:* The important point is the direction of twist of the singles and ply in the *final yarn.* A yarn such as "T" in the table (not photographed) would also look like a singles, since during plying all of the original "Z" twist of the singles would be removed and some "S" twist would be added.

Weaving

Weaving of paper machine felts with the accompanying dressing and drawing-in processes, differs radically from that in the average cloth mill. This is not because of the weaves, for these consist mostly of standard types such as a plain, straight or broken twill, sateen, four harness filling back and face, etc. The unusual feature is in the size of the looms, which may extend to approximately 700 in. in width (Fig. 15). You might well ask why these must be so large. It may be realized from what was said earlier that these felts must run on the paper machine full width without any imperfection or difference in tension. Since many paper mills are now 300 in. or more in width, and since there is frequently a 2:1 difference between the woven and the final width of the felt, it becomes apparent that to make felts for these machines the looms must be 600

Fig. 15. Felt loom, 600-in.

or more inches wide. This is true even when they are made with the loom warp as the lengthwise yarns in the final felt. In all felts made this way the ends must be joined by hand as indicated below, a long tedious process. To eliminate this step, some felts that are relatively short, i.e. up to 100 or 110 ft long may be woven endless, that is, tubular. The loom warp then becomes the crosswise yarn in the final felt. It may be realized that as the looms become wider longer felts can be made endless.

Joining of felts is a very interesting process and one which is characteristic of felt making. It is a necessary step for all very fine felts and for all felts which are too long to be woven endless in the available looms. The first step is to bring the two ends of the felt together on a table (Fig. 16). Each of the felt ends has a fringe possibly 2 ft long. The joining consists in "weaving" each yarn, one at a time, into the body of the felt in the following steps:

1. Yarn end No. 1 on the lower fringe is tied to the corresponding yarn on the upper fringe.

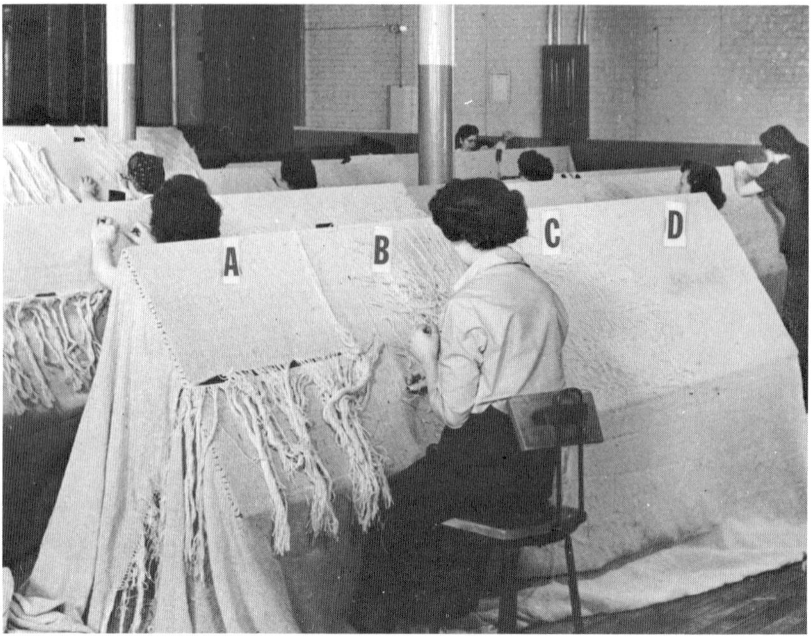

Fig. 16. Joining felts: (*A*) Not joined; both cloth ends with free warp yarns. (*B*) Individual yarns drawn from lower to upper cloth, or vice versa. Both warp yarns emerge at the same crossover yarn. (*C*) Long "tails" have been shortened by cutting and one yarn has been "hooked" by the other. (*D*) "Tails" have been cut very short (ready for fulling).

2. This end is followed by eye into the upper fabric and it is pulled so as to draw the lower end snugly into the upper cloth.

3. The next corresponding pair of ends is handled similarly but this time the upper end is drawn into the lower fabric. This cycle is repeated for each end across the entire felt width which may comprise up to 8000 separate yarns.

4. The tails are then cut off a few inches above the fabric.

5. At this point, both upper and lower threads emerge from the fabric at the same two crosswise threads and a "pin-point" imperfection would result if the ends were cut off at the fabric surface. Therefore, one end is hooked by the intersection opening so that there is a short overlap instead of a slight gap. It should be mentioned that this joining area is staggered over several feet of felt length.

6. Finally, the ends are cut off close to the fabric. In doing this, care must be taken to hold the scissors parallel to the fabric so as not to inadvertently cut or damage any threads in the body of the felt (Figure 16).

Burling includes removing all knots and, as in the joining operation, the two untied ends are made to lie flat and to overlap slightly. The overlap is drawn in with the same weave pattern as the rest of the felt. Burling also includes an inspection for any other weaving imperfections, since minor corrections can be made at this point which would be impossible to correct after fulling.

Finishing is another process which has special methods for the woven felt industry. By this we mean that the extent of fulling, and the requirements of fulling to precise length, width and tension are unique. A typical example may show this better. A felt for paper mill use may be ordered 100 ft long and 100 in. wide since it is expected to run on the paper machine at that size. However, while running it is required to drive numerous felt carrying rolls and be pulled over suction boxes or other cleaning equipment so that it is under considerable tension—often 20–30% of its breaking load. In order to furnish a felt for this job, it might be woven 100 ft long and approximately 200 in. wide. During the fulling, the length might be fulled in to 85 ft and the width to 107 in. Then when the felt is pulled out to its expected running length of 100 ft, the width would be drawn in to the required 100 in.

Let us consider now the fulling step which caused the size differences. Many theories and explanations as to the mechanism of fulling have been presented. (See *Wool Handbook*, Vol. 1, page 199.) Without going into these in detail, a few points may be apropos. Fulling is always done with the material in a wet condition and mechanical working of the fabric is necessary. As the material is worked, some of the fibers are stretched while others are compressed. Those which are compressed either move forward or into a loop while those which are stretched retract as they are released from the tension which caused the stretching. This small disarrangement of fibers would not be too significant except for the fact that the wool fibers have a surface scale structure which allows them to move in the root direction much more easily than in the other. Because of this DFE (directional frictional effect) and the fact that the fibers are able to recover from stretch, the disarrangement mentioned above results in an interlocking of the loops, or entanglement of the mass of fibers. Thus, a compacting of the fabric with a decrease in size takes place. In all of this operation the fibers themselves do not shorten nor do they change in diameter, indicating again that entanglement, which incidentally is not a reversible process, is the basic feature of fulling.

Wool has another property which is most helpful for felts and certain other industrial fabrics, namely, its ability to be stretched while wet and after drying in this stretched state, to retain this size until it is subsequently wet up. Felts for example are dried under high tensions, sometimes as high as 40 pounds per inch of width. This makes them long enough so they can be installed around the different rolls of the paper machine. When they are wet up in preparation for making paper they contract, become flat and smooth, also develop the necessary tension to drive the rolls around which they are installed.

The entire feasibility of using woolen jackets as covers for rolls is made possible by this property of wool to remain in its stretched condition until it is wet up. A heavy fabric, often weighing 8 ounces per square foot, is woven endless, that is, tubular. It is then fulled to a size perhaps 20% less than the circumference of the roll it is to cover. While it is still wet it is stretched to a size several inches larger than the roll circumference and dried. The jacket maintains this size (except for a minor original pull back) and so can be pulled over the roll for installation. When it is in place, it is wet up and it tries to shrink back to its fulled size, but is prevented from doing so by the roll. Thus a tight, snug and resilient covering is possible.

The above instances may be thought of as extreme cases of what the average woolen fabric manufacturers call "relaxation shrinkage." To them, it may be a problem—a thing to be avoided. In fact the "sponging" of woolen and worsted fabrics, and the compressive shrinking process are designed to remove relaxation shrinkage which may have been dried into the fabric. So, the same property which is a problem to one phase of the industry is a boon to another, and even permits a usage which would not otherwise be feasible.

All finishing machinery for papermakers' felts: washers (Fig. 17), fulling mills (Figs. 18 and 19), nappers, etc., are specially made to handle endless goods. In some cases, rolls are supported for loading and unloading by air operated jacks while the end bearing support is swung out of the way. In others, the entire roll is built with two bearings on one side only. Naturally, this requires very heavy construction on the cantilever principle.

Washing is done after fulling with the felts passing continuously between two rolls above a large tub (dolly washer type). Sometimes as much as 1200 lb of felt are washed at a time, in fact some single felts weigh over 800 lb, although the average is approximately 150 lb. Washing removes not only the lubricating oil which was added at the picker, but fulling soaps, floor dirt and residual wool grease not removed in the original scouring.

Napping is done as required, by the use of "gigs" in which natural teasels (a kind of burr) are used. These are mounted at a small angle from the horizontal on rotatable spindles. The "teeth" or "hooks" of the teasel grab into the fabric, pulling gently but persistently on the available fibers to bring up a nap. Each alternate spindle is "angled" in the opposite direction so that a slight sidewise

Fig. 17. Large stainless steel dolly-type washer where 1000 lb of felts can be washed at one time.

Fig. 18. Rotary fulling mill, capable of handling large felts. Rolls are cantilevered for easy loading of endless goods. The side of the trap box has been raised for better viewing.

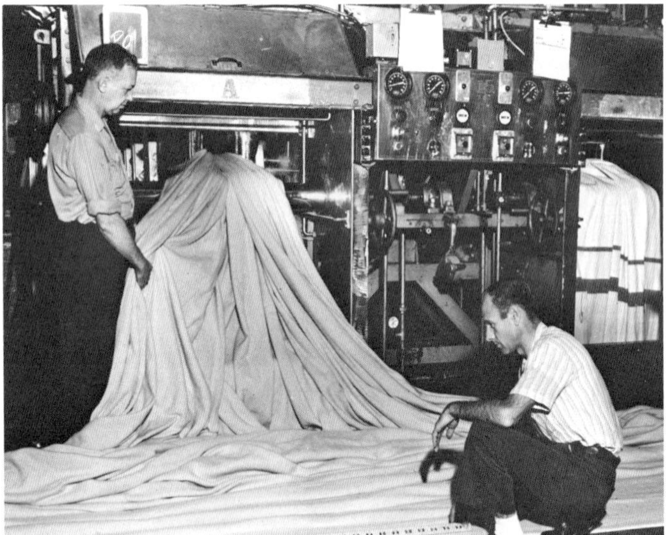

Fig. 19. Measuring width of felt after a fulling run.

pull of the fabric to the right is immediately countered with an equal pull to the left (Fig. 20).

Within the last five years, synthetic teasels have been made and are currently in use. (U.S. Patent No. 3,080,637 (March 12, 1963), by William H. Dutt, assigned to Albany Felt Company, Albany 1, New York.) These are more uniform than the natural ones, and have much longer life. See page 822, Chapter 12.

Singeing is also done on certain felts in which difficulties from protruding fibers might be anticipated. It is done by a controlled flame playing on the felt while it is in motion. The degree of singe might vary from a mere burning of the longest protruding fibers to a hard singe half way down to the warp yarns.

Use of Treatments

Chemical treatments can be chosen which will improve the felts in the following ways:

Improved:

1. Stability
2. Strength

Fig. 20. Synthetic teasles mounted on napper.

3. Start-up
4. Water handling

Increased resistance to:

5. Abrasion
6. Chemical damage
    (*a*) Chlorine
    (*b*) Alkali
7. High temperatures
8. Bacteria
9. Fiber shedding
10. Filling up with foreign matter

Sometimes felts are improved by adding properties to them which they never had and sometimes by decreasing their limitations. For example, they can be made more rugged and stable so that they will hold their size while running under the tensions demanded of them, also so they will wear longer; or they can be made more resistant to degradation from chlorine which is commonly used in paper mills to retard the growth of slime or other microorganisms. Note the significant difference in the stretch curves in Fig. 21. The original untreated felt is strongest and if it is left untreated but exposed to chlorine it stretches considerably more and strength drops off. However, both treatments shown (Z5 and H12) stretch less (after exposure to chlorine) than the original felt. It may be noted that the ultimate strength has decreased in this case but the important point to note, as mentioned earlier, is the stretch at the normal running load (about 20% of strength) and in this area the treated felts, even after exposure, are even lower than the original.

Treatments can also be used which will retard bacterial growths which so often weaken felts. Even after prolonged use on the paper machine, they can be shown to be effective. Figure 22 illustrates how a treatment (W6) has retarded quite dramatically loss of strength due to bacteria. Consider the two solid lines which represent strengths of felts which were incubated for 12 days with bacteria "A" found in a felt which had been used on a paper machine. During this time the untreated (blank felt) had lost more than 60% of its original strength while the treated (W6) one lost 10%. A comparison of the dotted lines shows even more difference

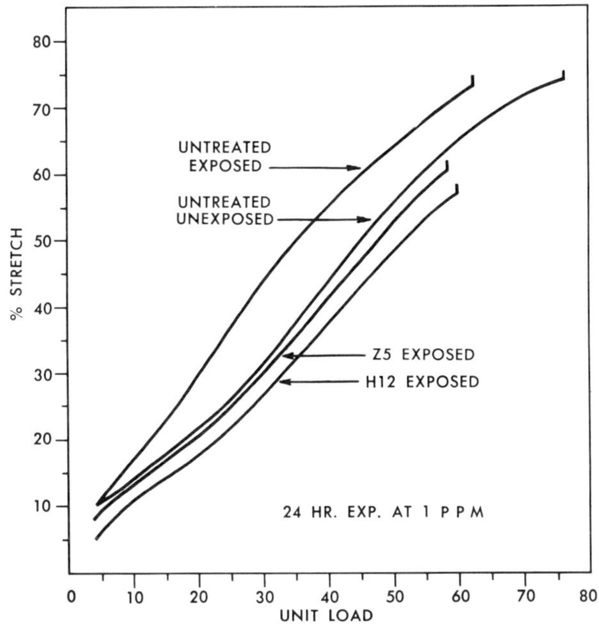

Fig. 21. Felts with Z5 and H12 treatments show lower stretch after 24-hour exposure to 1 ppm of chlorine than the original unexposed felt. Without any treatment the same chlorine exposure caused excessive stretch even at moderate loads.

for felts incubated 20 days with bacteria "B," also found in a used felt.

In Fig. 23 is shown how another treatment, H12 has improved a felt's resistance to wet abrasion. After 1000 cycles it lost only 13% of strength in contrast to 43% for a duplicate specimen which was not treated.

It was noted earlier that felts exposed to high temperature paper mill stocks can be made more stable by using synthetic fibers in the blend. In a similar manner they can be made stable by certain chemical treatments. Figure 24 shows how treatment Z5 causes a felt to run at a shorter length (more resistance to stretch) than untreated felts. At 130°F the difference caused by a specific treatment was 2 feet in its running length and at 170°F it was 10 ft. These differences are highly significant in the successful operation of a felt.

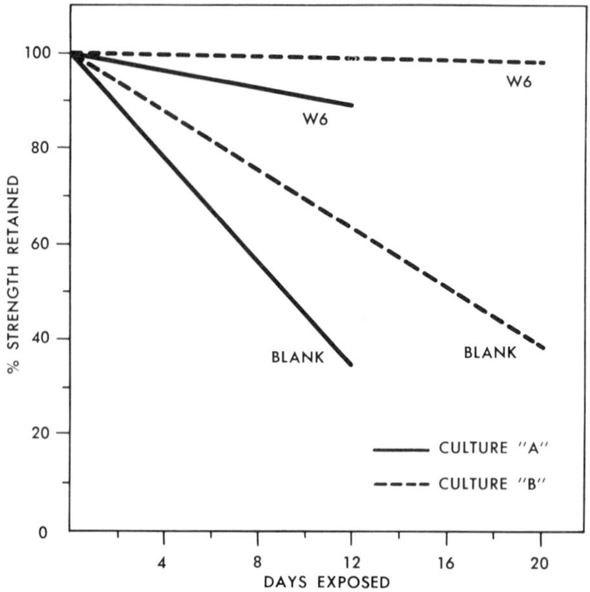

Fig. 22. Treatments can significantly retard bacterial degradation of felts. Untreated ones lost 60% strength, whereas treated ones only 10%. (Solid lines.)

Let us now consider the nature of treatments. Some treatments produce chemical modifications of the wool, a crosslinking of its molecular ladder-like structure, while others are an addition of a resin to accomplish the required change. Often a combination of the above can be used, namely the application of a resin that also bonds chemically to the fiber. As examples of the chemical modification type are (*1*) the use of benzoquinone as described in a French patent about 1913, (*2*) that of treating with chromium salts followed by a reduction with sodium sulfite as in U.S. Patent No. 2,284,387, and (*3*) another chemical modification is by the use of metal thiocyanates as in U.S. Patent No. 2,515,181. An example of the resin type is the use of a melamine resin such as that indicated in U.S. Patent No. 2,503,629.

In addition to the above, there are treatments which change only the surface characteristics of the felt. A wetting agent would be the simplest example of this but there are others that affect the felt similarly, or oppositely, for a prolonged period and thus alter

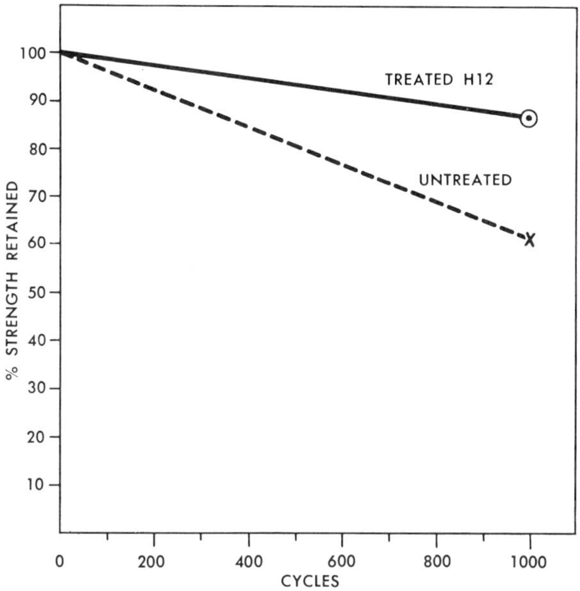

Fig. 23. Abrasion resistance is improved by treatment H12.

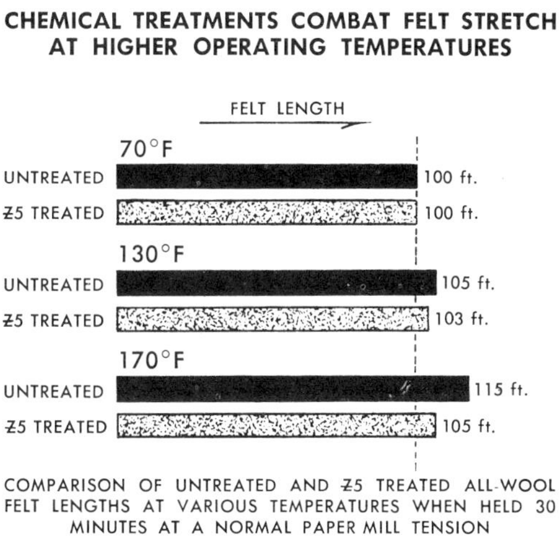

Fig. 24.

its water-handling properties. References on these materials, both wetting agents and water repellents are too numerous to mention here.

Thus, to a degree, chemical treatments have again raised the level of the "dilemma triangle" since "strength" and "openness," or "strength" and "finish" can be improved simultaneously.

An idea of the importance of treatments can be attained by realizing that the majority of all felts receive some type of chemical treatment. A greater range of treatments is possible for felts than, for example, for clothing because in clothing, color, permanence-of-color and any impairment of "hand" are of serious consequence. With felts, where the chief function is mechanical, these factors are usually not very significant.

Drying, as was referred to earlier, is done by putting the felt around two rolls. One of these rolls is large in diameter (4–6 ft) and heated with steam, while the other is unheated and is mounted on a carriage which moves along a track so the felt can be pulled out to the desired length and tension (Fig. 25). It is run around these rolls until dry. Auxiliary heat may be applied with infra-red or heated air to speed the drying which may take from 20 min on small felts up to 3 hr on large ones.

Needled Felt

Some papermakers' felts are made by taking a woven supporting fabric and needling into it a carded web or batting (Figs. 26*a*, *b*, *c* and 27). In these needled felts, the base fabric is specially designed for the purpose and is made of yarns of suitable blends of wool and synthetic fibers. The web is formed by blending and carding wool and synthetic fibers and is laid on the base just prior to passing through the needle loom (Figs. 28 and 29). Several layers of web may be used, the felt passing through the needle loom after each layer is added. The needling is accomplished by a large number of needles on which are barbs which carry the fibers of the web into the base as the needles are pushed repeatedly into the base (Fig. 27). Following needling, the felt is processed through appropriate finishing operations, then washed and dried in a manner similar to conventional felts. This type of product has more fibers which are perpendicular to the surface than the conventional all woven material.

Fig. 25. Drying long felt on steam-heated roll, with carriage and stretch roll holding it out to proper length. Infrared auxiliary units are frequently used.

The advent of the needled felt has added new dimensions to felt characteristics and is making possible a greater level of production and performance in the paper industry.

The earliest patent on needled felts was applied for in 1939 by P. T. Gates.* Felts made by this method were used in the asbestos-cement field with considerable success. In mid-1959 needled felts were introduced to the paper industry. This was a new application of the earlier method of manufacturing.

Today, less than ten years later, the needled felt has been proved and used successfully on almost every type and kind of paper machine. Every day improvements are being made as additional

* U.S. Patent No. 2,271,295 (Jan. 27, 1942)(Percival T. Gates)(Filed 1/10/39); 2,581,790 (Jan. 8, 1952)(Percival T. Gates); 2,588,228 (Mar. 4, 1952)(Percival T. Gates).

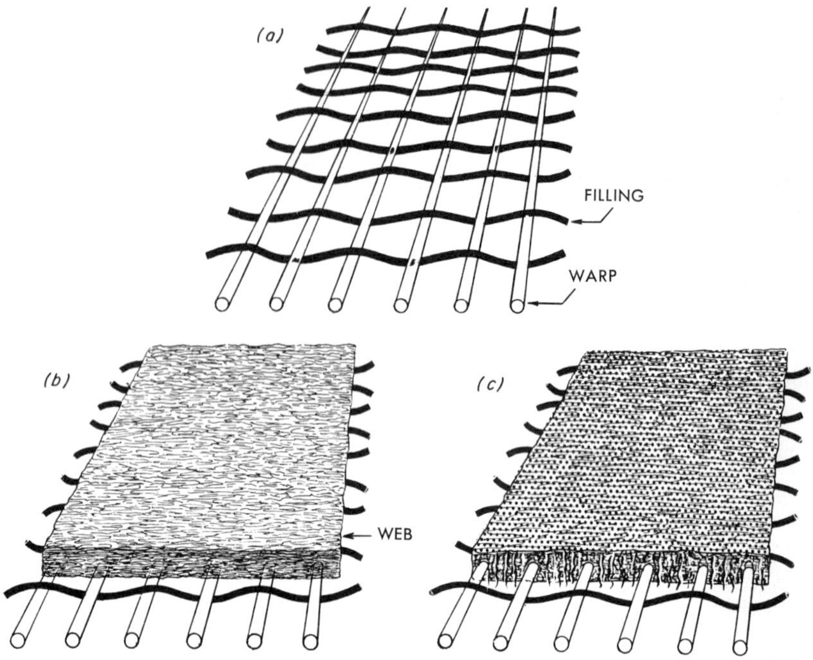

Fig. 26. Building of needled felt, by diagrams. (*A*) Woven base fabric. (*B*) Web of blend fibers laid on base fabric before needling. (*C*) Final fabric with web needled into base.

research and field experiences are adding to techniques and knowledge of this new product.

One of the recent advances in this area is a warp oriented base which after needling is knuckle-free. This maximizes the use of lengthwise yarns and needled web while eliminating the effect of the cross-wise yarns.

Referring again to the triangle of felt requirements, the needling process has given new dimensions to what is possible within the same felt. With more fibers in a vertical direction, drainage is better and since many of the web fibers fill in the areas between yarns, felts are smoother and the paper sheet finish is improved. Also it has been found that with needled felts a higher percentage of synthetic can be used in the web than was possible in conven-

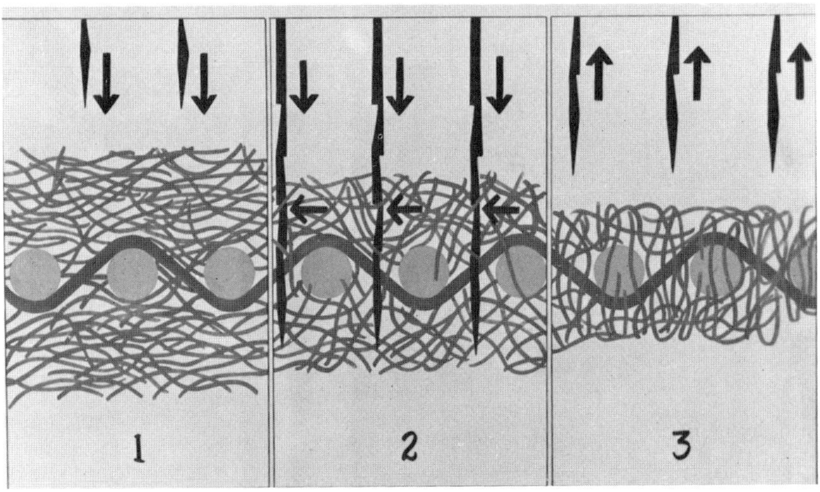

Fig. 27. Compacting the needled felt. (1) The loose batt of fibers is shown over the base fabric. Needles have not yet contacted them. (2) The needles have forced the batt fibers into the base as each needle barb catches more fibers. (3) The needles are removed and the composite needled felt remains. *Note:* The drawing shows the web on both sides. In practice the needling on the reverse side would be done separately and after the felt was turned over.

tional felts. This has thus improved the strength and wear resistance of the felts for longer life. Needling therefore has in effect raised the entire level of the triangle since "strength," "openness" and "finish" have all been improved—setting new standards for the life and performance of felts. So much success has been obtained from needled felts that approximately 55% of all felts are of this type. This is indeed a radical change for a brief ten year period.

Prospects for the Future

Technological advances in the manufacturing of felts and in the paper industry have been almost phenomenal in the last fifteen years. Paper machines are using new methods of forming the sheet, new suction pick-up methods for transferring the sheet to the felt, new grooved rolls for greater efficiency in the press sections, new felt cleaning methods and new drying techniques. All of these have

Fig. 28. Felt on needle loom.

contributed to the higher and higher speeds of paper machines. Concurrently with these developments, improvements have been made in the felt industry, (*1*) in the use of synthetic fibers, (*2*) development of chemical treatments, and (*3*) in the needled felt process.

Fig. 29. Needle loom.

These combined advances in the paper and felt industries have resulted in a substantial increase in the amount of paper which is produced per pound of felt used. As a result, although the paper and paper board produced has increased from about 32 million tons in 1950 to 50 million in 1967 the poundage of felts used has actually decreased. Worldwide use of felts will be keyed to the increasing worldwide consumption of paper.

Bibliography

Felt Facts, Albany Felt Co., Albany, N. Y.

Unpublished material from the research laboratory of the Albany Felt Co., Albany, N. Y.

Wool—Its Chemistry and Physics, P. Alexander and R. F. Hudson, Chapman & Hall, London, 1963.

Wool—A Study of the Fibre, S. G. Barker, H. M. Stationery Office, London, 1931.

Bulletins, Huyck Felt Co., Albany, N. Y.

Bulletins, Nordiska Maskinfilt AB, Halmstad, Sweden.

Magazines and procedings of the paper industry: *Paper Trade Journal*, *Southern Pulp and Paper Manufacturer*, *Technical Proceedings of TAPPI and CPPA* (Technical Association of the Pulp and Paper Industry and Canadian Pulp and Paper Association).

CHAPTER 16

Physical and Chemical Testing

FRITZ F. KOBAYASHI AND WERNER VON BERGEN

In *Wool Handbook*, Vol. 1, Golub has presented the sampling and testing of loose wool fiber, as well as considered the various aspects of sampling as they relate to the significance of the test results. This chapter is devoted to the testing of general intermediate and finished products such as top, roving, yarn and fabric as well as such final products as carpets and socks.

An appreciable period of time has elapsed since the publication of Volume 1; for this reason, recent progress and the development of new testing instruments are included in this chapter. Before considering various methods of testing, the philosophical question, "What is the purpose of testing?" might be asked. The usual answer to this question may be summarized in a statement which indicates that the test result is useful in assessing a characteristic of a material which is of scientific or commercial importance. Answers of this type suggest that background knowledge of what is an acceptable value for the test result is available, and that the real criteria is frequently a comparison of the test result with some established value. This comparison is, in fact, testing by inference or evaluation. For example, the numerical values for the breaking strength and breaking elongation of a yarn may not in themselves be of prime importance. The real importance of these numerical values may lie in the fact that past experience indicates that yarns having breaking strengths or breaking elongations below a certain level will result in lower weaving or knitting efficiency.

Fig. 3. Forte Electronic Moisture Analyzer for Bales.

A distinction should also be made between an objective test and a subjective test. Objective properties are measurable and the measurement can be expressed numerically. Subjective properties are not measurable in the exact sense of the word. In many cases subjective properties are assessed visually or tactically and depend to some extent on personal preference. Ideally, an objective test is to be preferred since the test result is expressed numerically, and these numbers can be referred to many days or months later. However, in a significant number of situations subjective testing is the accepted procedure. This is usually because either the expense of objective testing or the elapsed time to obtain a test result is not consistent with the need. The reliability of any test result is dependent on the adequacy of sampling. A discussion of sampling is given in Volume 1. The industrial or commercial significance of a test result is dependent on the selection of the appropriate property tested, the known accuracy and precision of the test method, and the adequate interpretation of the test result in terms of the material tested.

The testing of woolen and worsted intermediate and finished material is similar in many respects to the testing of other industrial materials. Many test methods exist. Some of these are standard methods established by national or international bodies. A list of ASTM and IWTO test methods pertaining directly to wool and wool products are shown in Tables 1 and 2. These listings indicate the magnitude of the available test methods and serve as a reference to methods which are well established in both national and international circles.

TABLE 1

ASTM Specifications, Methods, and Recommended Practices

	Wool Fibers
Specifications for:	
D 1381-61	Fineness of mohair top and assignment of grade
D 2252-66	Fineness of types of alpaca
D 419-66	Fineness of wool and assignment of grade
D 472-63	Fineness of wool top and assignment of grade
Methods of test for:	
D 1283-57	Alkali-solubility of wool
D 1294-63 T	Breaking strength of wool fiber bundles—1-in. gage length
D 2524-66 T	Breaking tenacity of wool fibers, flat bundle method—1/8-in. gage length
D 1909-65	Commercial moisture regains for textile fibers, table of
D 1776-67	Conditioning textiles and textile products for testing
D 1060-65	Core sampling of raw wool in packages for determination of percentage of clean wool fiber present
D 1282-66 T	Diameter, average, fiber in wool top, card sliver, and scoured wool, resistance to air flow as an indication of
D 2130-61	Diameter of wool and other animal fibers by microprojection
D 1906-62 T	Effective gage length by evaluation of clamp error in single fiber testing, estimation of
D 1574-66	Extractable matter in oven-dried wool
D 1575-61	Fiber length of wool
D 519-66	Fiber length in wool top
D 2462-66 T	Moisture in wool by distillation with toluene
D 1576-64	Moisture in wool by oven-drying
D 2525-66 T	Moisture content, devising a sampling plan for various forms of wool for the determination of
D 1770-66 T	Neps, vegetable matter, and colored fiber in wool top
D 2165-61 T	pH of aqueous extracts of wool and similar animal fibers
D 2142	*(Rapid Estimation of Staple Length of Wool Tops)—Discontinued 1967*
D 1282-66	Resistance to air flow as an indication of average fiber diameter of wool top, card sliver, and scoured wool
D 1771-62	Sample reweighing of lots of packaged raw wool

TABLE 1 (*continued*)

D 2525-66 T	Sampling plan for various forms of wool for the determination of moisture content
D 1060-65	Sampling, core, of raw wool in packages for determination of percentage of clean wool fiber present
D 2491-66 T	Staple crimp in grease wool locks
D 1234-54	Staple length of grease wool
D 1113-65	Vegetable matter and other alkali-insoluble impurities in scoured wool
D 1334-67	Wool content of raw wool—commercial scale
D 584-66	Wool content of raw wool—laboratory scale
Recommended practices for:	
D 2264-64 T	*Calculating number of tests to be specified in determining average quality of a textile material (see Part 24)*
D 2118-66	Establishment of standard moisture content for wool and its products
D 861-64	*Tex system, use of, to designate linear density of fibers, yarn intermediates, yarns, and other textile materials (see Part 24)*

Other test methods have been developed by individuals or small groups. Many of the latter, though arbitrary or empirical, provide useful information for the purpose intended. The user of any method must recognize the advantages and limitations of the method. A particular test method may provide the user with a result which is quantitative, semi-quantitative, or qualitative in nature. Some test methods provide numerical test results which are related to a true zero value, while other methods provide only a relative ranking or a deviation from an assumed starting point. Test methods may be classified as direct or indirect. The direct methods actually measure the desired characteristics while the indirect methods measure a property associated with the desired characteristics. The latter methods are used because direct methods are either not available or the elapsed time or applied time required to perform a test by a direct method is prohibitive. The procedure used in any indirect method must be followed implicitly since deviation from it may introduce serious errors.

Many volumes have been written on the physical and chemical testing of textile materials. In selecting material for inclusion in this chapter, the basis used was to include those materials which were judged to be desirable for one of the following reasons. The method is useful for mill control purposes: the method is of particular

TABLE 2

I.W.T.O. International Wool Textile Organization Specifications of Test Methods prepared by the Sub-Committee for Test Methods and Values and adopted by the I.W.T.O. Technical Committee

Note: This list replaces the list of Specifications circulated in 1966.

Title	Code number
Method of determining "Barbe" and "Hauteur" for wool fibers using a comb sorter	IWTO-1-66 (E)
Method for the determination of the pH value of a water extract for wool	IWTO-2-66 (E)
Method of test for the acid content of wool	IWTO-3-65 (E)
Method of test for solubility of wool in alkali	IWTO-4-60 (E)
Method of determining wool fiber length distribution using a single fiber length measuring machine	IWTO-5-65 (E)
Method of test for determining the mean diameter of wool fibers by the Air-flow apparatus	IWTO-6-66 (E)
Recommendations on the precautions to be taken in commercial transactions when using micron values of wool fiber diameters as determined on Air-Flow instruments (*on yellow paper*)	uncoded
Method for the determination of the diethyl ether-soluble matter in combed wool sliver	IWTO-7-60 (E)
Method of determining wool fiber diameter by the projection microscope	IWTO-8-66 (E)
Method of test and assessment for proofness of wool fabrics against the common (webbing) clothes moth	IWTO-9-62 (E)
Method for the determination of the dichloromethane-soluble matter in combed wool sliver	IWTO-10-66 (E)
Method of test for the solubility of wool in urea-bisulphite solution	IWTO-11-65 (E)
Method of test for determining the percentage of medullated fibers by the projection microscope	IWTO-12-64 (E)
Method of sampling raw wool in bulk for fiber diameter and/or length measurement	IWTO-13-64 (E)
Method of test and assessment for proofness of wool fabrics against the Black Carpet Beetle, *Attagenus piceus* (Oliv.)	IWTO-14-64 (E)
Method for the colorimetric determination of cystine plus cysteine in wool hydrolysates	IWTO-15-66 (E)
Method of test for wool fiber length using a WIRA Fiber Diagram Machine	IWTO-16-67 (E)
Glossary of terms relating to Tops (*partly on yellow paper*)	uncoded
Method of test for determining wool fiber length distribution parameters by means of an Almeter	IWTO-17-67 (E)
Method of test for determining short-term irregularity of linear density of wool slivers, rovings and yarns by means of a Zellweger-Uster Evenness Tester	IWTO-18-67 (E)
Method for the determination of the I.W.T.O. clean wool content in greasy wool	IWTO-19-68 (E)

importance to the testing of wool containing textile materials; or the method is used for testing a characteristic which is frequently the cause of consumer complaints.

The Relationship and Function of a Laboratory

The laboratory is primarily a service department much like the accounting department in that it exists to provide useful information to the producing units as well as to purchasing and sales. The principal service rendered by a laboratory in a woolen or worsted manufacturing establishment includes the following:

1. Checking of incoming raw materials for quality and uniformity.

2. Routine testing during the course of manufacturing operations to indicate when these operations are proceeding in control according to the established procedure and when they are not.

3. Assisting in overcoming certain unusual difficulties which are encountered during manufacturing operations.

4. Recommending (after careful screening) new products and methods either for the improvement of products or to reduce manufacturing costs.

5. Providing photographic or other evidence for the impartial settlement of claims.

6. Finally, making tests on the finished fabric to insure that the product meets the specifications which have been established.

In planning and equipping a laboratory, careful consideration should be given to the amount and type of testing anticipated, and the number of people involved so that efficient use is made of both personnel and floor space.

Jankowski (61) shows a typical floor plan for a dyeing laboratory. This is illustrated in Fig. 1. While this is primarily designed for a dyeing laboratory, it serves to illustrate the utilization of space for efficient operation.

Physical Testing

Before considering test methods for specific items or quality characteristics, attention will be given to some general methods which apply to all wool products. These are: Atmospheric Conditions, Moisture, and Weight-Length Relationships.

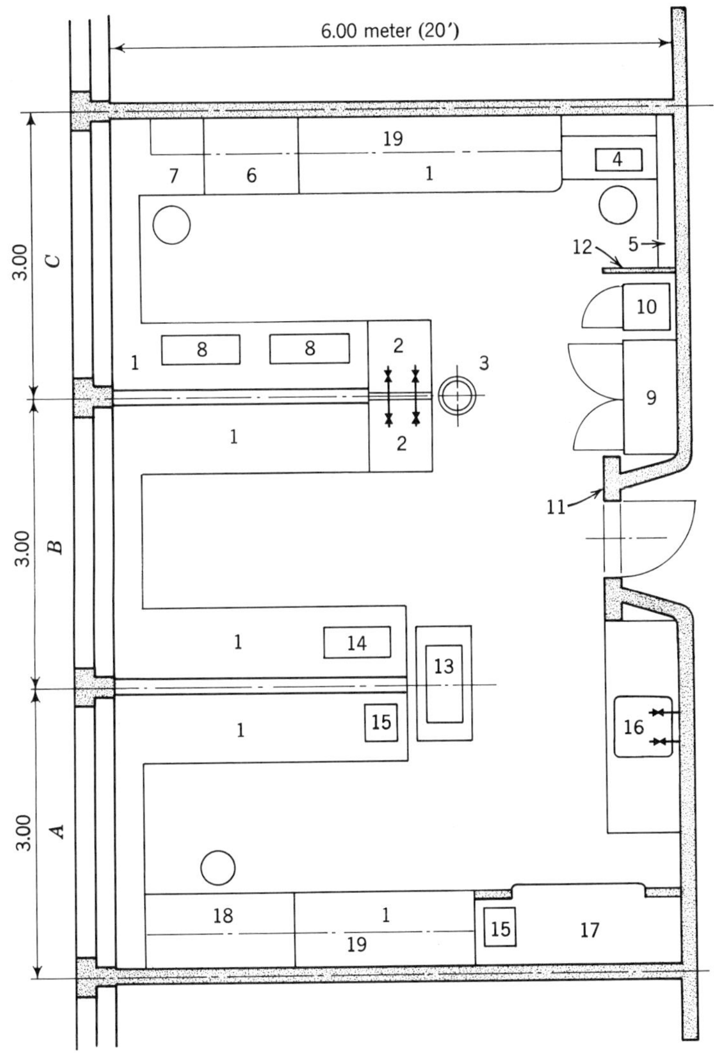

Fig. 1. Layout for a Five-Man Laboratory.

Atmospheric conditions

The physical properties of wool and many of the fibers used in blends with wool are markedly affected by the moisture content of the material undergoing test. It is quite difficult to measure and adjust the moisture content of a material undergoing test. It is, however, a relatively simple matter to define the atmospheric conditions with which the material is in equilibrium and under which the test is performed. The International Standards Organization has established the Standard atmosphere for testing as, "An atmosphere which has a relative humidity of 65 $\pm$ 2 percent and a temperature of 20 $\pm$ 2°C (68 $\pm$ 4°F)." Since the amount of moisture contained by a material in equilibrium with the standard atmosphere will depend on whether the material approaches equilibrium with the atmosphere from the "high" or "low" side, a preconditioning is included in the ISO recommendations. This consists of a partial drying to insure that equilibrium is approached from the "dry side." The atmospheric conditions recommended provide a useful means of standardizing on the conditions of test.

Moisture

The area of testing for moisture is covered in Volume 1. However, several new instruments have become available since publication of that volume. Also included here are certain instruments particularly useful for in plant control purposes.

Golub has described the common drying techniques in Chapter 10, Vol. 1. In addition to those described, an instrument using this principle has been developed recently by Mackay (79) at C.S.I.R.O. This device consists of a "drying unit" and an auxiliary weighing unit. The former consists of a means of forcing heated air through a specially designed canister containing the test specimen. The weighing unit is calibrated directly in terms of percent regain. In use, the canister is placed on the balance, sufficient wool is added to the canister to result in the balance reading 0% regain. The canister is then placed on the drying unit for a predetermined period of time, after which it is removed and placed on the balance. The latter indicates the weight loss directly in percent regain. This unit is illustrated in Fig. 2*a* and *b*.

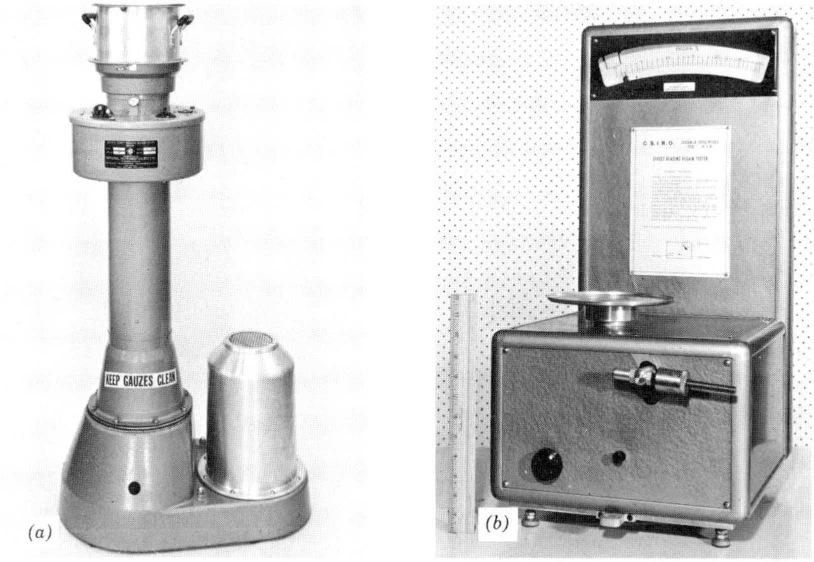

Fig. 2. (*a*) C.S.I.R.O. Direct Reading Moisture Tester; (*b*) Balance for C.S.I.R.O. Regain Tester.

In any discussion of the determination of moisture by drying methods, it is important to recognize the types of errors that may be introduced. Two of the common errors frequently not recognized are the presence of volatile materials other than moisture, and the influence of the relative humidity of the air entering the system. If volatile materials are present, the loss in weight due to this material is included in the total weight loss which is attributed to moisture.

It has long been recognized that variations in the temperature used for the drying procedure will seriously affect the results obtained and for this reason various specifications include tolerance limits with respect to this variable. The recognition of the influence of the relative humidity of the air entering the drying system on the equilibrium moisture content of wool so dried is relatively recent. Basic information related to this phenomenon was developed by Weigerink (120) in 1940 and recalculated by Toner, Bowen, and Whitwell (107) in 1947. A graphical illustration of the magnitude of the error to be expected as a function of the humidity of the air entering the system is shown in Wool Science Review, No. 19, 1961. (5) A discussion of the effect of environmental conditions on commercial weight is given by Sustmann, (104) Sustmann and Henning (105).

Several rapid instrumental methods are available for the assessment of moisture content of wool. These depend on the measurement of some characteristic of the material which varies with moisture content. These may be classified as dc resistance, ac impedance, or AC capacitance type instruments. Some of these instruments if properly calibrated can be very useful tools. They may be used for estimating the moisture content using relatively small samples and the test results are obtained essentially immediately. Several of these instruments have been evaluated by Toner (108, 109, 110). The use of this type instrument, while rapid and convenient, must be done with the recognition of the limitations of an indirect method.

Relatively recently, an instrument has been developed by the Forte Engineering Company. Von Bergen (114) and Breen (18) discuss the use of this instrument. It is basically an AC capacitance type meter. However, the design of the instrument and particularly the design of the "sample" holder is much improved over earlier instruments. With this instrument, the material to be tested is placed in a chamber, the chamber is closed, and an indicating meter is read. The value obtained from the indicating meter is converted to moisture content by reference to a conversion table. The entire operation requires about 30 seconds. In addition to the "sample holders" available for loose wool and top as described in Chapter 10, Volume 1, specific models of the instrument are designed to accommodate an entire cone of yarn, an entire bolt of cloth, or an entire bale of wool. Figures 3, 4, and 5 illustrate Forte model 900 used for entire bales of wool, model 700 for yarn packages; and model 800 for bolts or rolls of cloth.

Because the method used is an indirect one, a method of verification of the operating characteristics is required. The manufacturer provides a number of metal bars for this purpose. When one of these is inserted in the instrument in place of the test specimen, the instrument response is equivalent to a stated moisture.

content. It is usually sufficient to verify the consistency of operation of the instrument at two levels of equivalent moisture content to insure that no instrumental errors have developed. This verification of the instrument can be completed in a few minutes and does not require special skills.

The following two examples serve to illustrate the savings that prevail as well as the differences in approach to sampling, when the Forte testing device is used.

Example 1. For the determination of moisture content of yarn on cones using the conventional techniques, a portion of yarn is re-

Fig. 4. Forte Electronic Moisture Analyzer for Yarn.

Fig. 5. Forte Electronic Moisture Analyzer for Cloth.

moved from the outside of a number of cones. This procedure is based on the assumption that the moisture content of the yarn is uniform throughout the cone. Following the testing by conventional oven drying techniques, the yarn is regarded as sampling waste. Using the model designed for yarn on cones, entire cones are tested. This eliminates both the question of consistency of moisture content throughout the cone, and the attendant sampling waste.

Example 2. A manufacturer indicates that the moisture content of individual pieces of wool cloth varies from 4% to 11% depending on the prevailing weather conditions at the time of finishing. If these pieces are shipped on a "weight at the time of shipment" basis, the amount of loss to the company is appreciable. If, however, the weight at the time of shipment is clerically adjusted to a standard moisture content by testing each piece, this loss can be eliminated.

Weight–Length Relationships

Measurement and control of the weight–length relationship throughout manufacturing is a prime consideration for two reasons which are not mutually exclusive. For accounting purposes recognition must be made of the fact that the raw material is purchased in weight units and the final product is sold in length units with the expressed or implied requirement that the material weigh not less than a prescribed amount per unit length. For technical reasons, uniformity of the final product both within and between pieces is contingent on adequate control of this relationship at all stages of manufacturing.

Yarn Number and Linear Density

Yarn number is defined as the linear density of a textile strand. This is frequently referred to in trade practice as size or count.

ISO recommends the universal yarn count system. The recommended system is the Tex system. It is a direct one and the basic units are grams per kilometer. Using this system, if 1000 m weigh 1 g the yarn is said to be 1 tex. If 1000 m of yarn weigh 3 g the yarn is 3 tex. See Chapter 3, "Woolen Spinning." Control methods used within a plant consist of measuring a length of material and weighing this length. Most difficulties and interdepartmental misunderstandings concerning these values may be traced to one or more of the following causes. The moisture content of the material has changed between successive tests on the material; the tension used in establishing the length is not constant; or the weighing mechanisms are not in calibration. Table 3 shows the typical length and weight units used in the United States for measuring materials at various stages of processing. Bornet has established a rating system for the coefficient of variation of yarn number of yarns spun on the worsted system. This rating is based on 120 yd skeins. It is equally applicable to combed cotton yarns, and yarns made of rayon, acetate, and synthetic fibers made using the cotton system. This is shown in Table 4.

The expected variation in *average* yarn number is of importance to both the spinner and the yarn consumer. This cannot be expressed on a single percentage for all yarn since the method of

TABLE 3
Typical Measuring Units and Methods of Expression of Linear Density

	Length measured, yd	Weight units	Results expressed
Worsted system			
Card sliver	5	oz	oz/5 yd
Comb sliver	5	oz	oz/5 yd
Top	5	gr	gr/yd
Pindrafter sliver	4	gr	gr/yd
Roving	40	gr	gr/yd or gr/40 yd
Yarn	120	gr	Yarn no.
Woolen system			
Card roping	—[a]	gr	gr/yd
Yarn	120	gr	Yarn no.

[a] 2 yd from a roping spool. For a spool containing 60 ends, the total length taken is $2 \times 60 = 120$ yd.

TABLE 4
Rating of Yarn Number Variation (Coefficient of Variation) Based on 120 Yard Skeins Developed by Bornet

Rating	Coefficient of variation in %
Excellent	Below 2.0
Very good	2.0–2.5
Good	2.5–3.0
Fair	3.0–3.5
Poor	Above 3.5

manufacture (woolen or worsted) and the actual yarn number affect the limits to which the spinner can control the yarn number.

ASTM D 2644-67T and D 2645-67T list the maximum permissible deviation of the average yarn number from the specified yarn number.

These are:

For woolen spun yarn: ±7.5%
For worsted spun yarn: ±4.5%

"Sample" Preparation

For the preparation of small quantities of material, a sample card is a useful tool.

Since the diameters of the workers and strippers are quite small, the sample card has certain limitations from both the standpoint of blending efficiency and the ability to process relatively long fibers. Manufacturers of these devices provide accessories such as lap drum, or other means of collecting the product. An auxiliary device used in conjunction with the "sample card" by the insulated wire industry is essentially a rotating trumpet. This device acts as a side delivery condenser. A sample blender consisting of a breast and three cylinders is available. The entire unit occupies a space 2 × 2 × 3 ft. A device of this type is shown in Fig. 6.

A device for preparing "yarn" is a useful adjunct to the sample card. This device consists of a single spindle usually mounted horizontally and driven by a small vari-speed motor.

An experienced operator using a sample card and a spinning device can prepare a sample of yarn in about 10 min.

Unevenness

The term unevenness is defined by ASTM as: The nonuniformity of the linear density of a continuous strand or portion of a strand. At this time commercially available testers for the measurement of this property are of two basic types. Neither of these types measures linear density directly. Instead, they measure a property associated with linear density. One type uses as a detection element a pair of condenser plates. A change in the mass of the material drawn through these plates results in a change in capacitance of the system. This is amplified and transmitted electronically to a recorder. The other type uses as a sensing system a loaded tongue and grooved roll. Variations in the linear density of the material passing through these rolls results in a deflection of one of these rolls. This mechanical deflection is converted to an electrical signal by means of an impedance coil. The electrical signal is amplified electronically and transmitted to a recorder. Both types of instruments are equipped with sensing devices of various sizes to accommodate strands of widely differing linear densities. Typical recorder charts for a 5 run woolen yarn and 1/36s worsted yarn are shown

Fig. 6. Sample Card Unit with Auxiliary Blender.

in Fig. 7. Unevenness testers utilizing other sensing systems have been proposed. Typical of the sensing systems suggested are a vibrating string method, a photoelectric profile method, an air flow method, and a method using a Beta radiation emitter and a suitable detection device. For either technical or economic reasons testers based on these principles have not become commercially available.

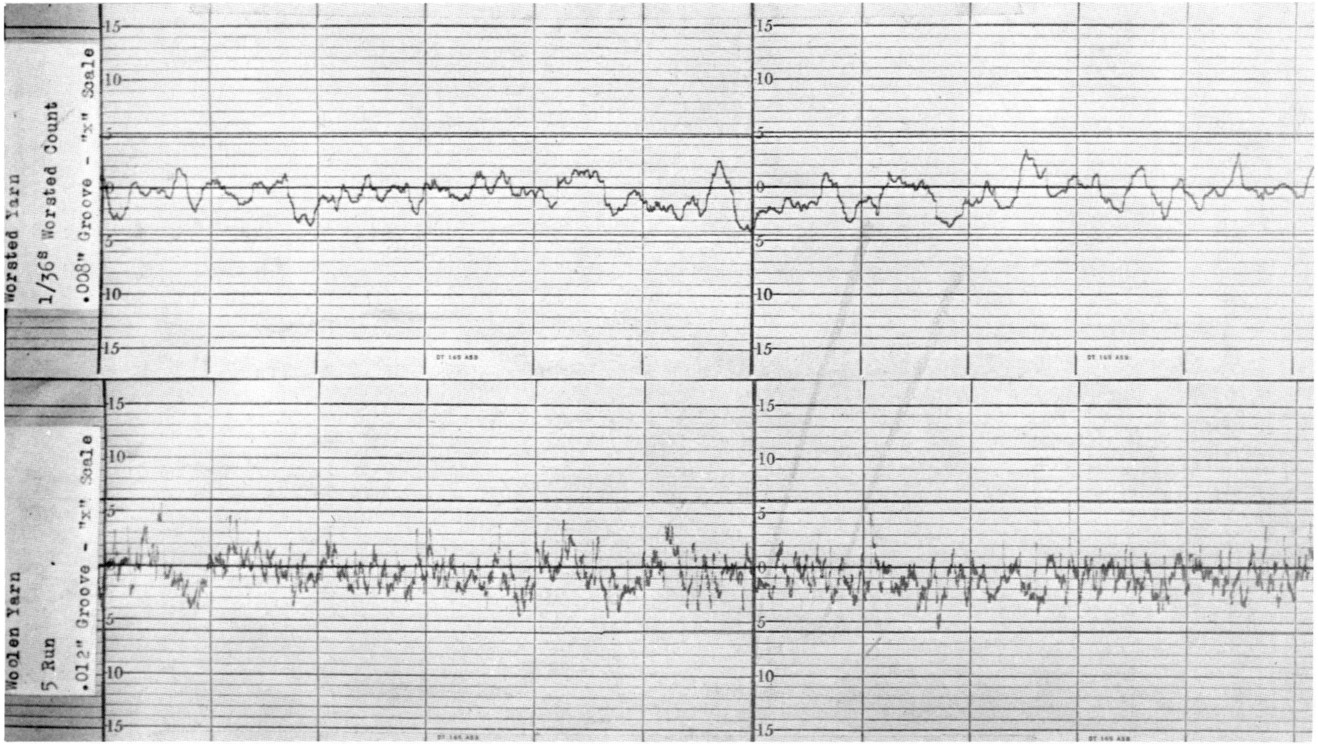

Fig. 7. Recorder Charts for Woolen and Worsted Yarn.

Unevenness testers utilizing condenser plates as a sensing element have been criticized because of the errors of measurement which can occur during the use of this type instrument. Foster (43) shows that for cotton and rayon the moisture content of the material contributes to errors. Hearle (53) shows that the degree of change of dielectric properties with changing moisture content is dependent on the frequency used. Hearle and Walker (54) present data to show that differences in dielectric properties of various fibers can contribute to errors. Mack (78) indicates that density, twist, dielectric constant, and geometrical configuration can all contribute to errors. The manufacturers of capacitance type unevenness testers claim that the instruments incorporate certain features which minimize chances of errors of measurement.

All of the capacitance type instruments measure deviations from an assumed mean value. None of these devices, however, are capable of measuring from a "zero base line." The mechanical sensing (tongue and grooved roll) type tester, when properly calibrated, measures deviations from a "zero base line." These deviations can be used to measure linear density as well as unevenness. The results obtained by the uses of this type instrument are not influenced by variations in moisture content of the material, differences in the dielectric properties, differences in twist, or differences in form factor. Because of the inertia of the loading system, the rate of testing using this type of device is appreciably slower than that possible with a capacitance type tester. The maximum testing rate using a properly loaded tongue and grooved roll type device is about 12 yd/min. Capacitance type testing devices are equipped to test at rates up to 100 yd/min.

The evaluation of the trace plotted by the recording device of any tester depends on the nature of the information desired. The manual methods require the measuring of deviations of the trace from a reference line. These deviations are measured for a number of successive length increments. A computation utilizing these deviations results in a number which expresses unevenness. The details of this type of computation as well as description of sampling and other methods of interpretation are given in ASTM-D 1425-67. Manual methods of chart evaluation are time consuming. For this reason, instrument manufacturers have developed auxiliary devices. The relationship between deviation of linear density and the mean linear density of the strand as determined by manual evaluation of the test chart or by automatic evaluators performing a similar operation is illustrated in Fig. 8. Other automatic evaluators either indicate the coefficient of variation directly or provide a pair of numbers from which the coefficient of variation may be computed.

The integrators available for use with capacitance type testers are of two types. These are the "point to point" and "fading memory." The "point to point" integrator starts and stops integrating at precisely known time intervals. Using this type device as recommended, the value obtained is not biased with respect to time. The "fading memory" type integrator operates continuously and the integrated value obtained is biased in that the value obtained is effected to a greater extent by that portion of the sample tested last. However, both types will provide a close approximation to the true integration provided the manufacturer's instructions regarding the speed of the strand and time of integration are followed. An accurate and reliable integrating device for unevenness measurements made using a mechanical sensing type tester was described by Vose and Plummer (116). This device used electronic amplification and a pair of watt hour meters as the accumulating device. The General Electric Integrator supplied with the Pacific Evenness Tester is to the best of the author's knowledge the only automatic evaluator commercially available as an accessory to an unevenness tester which performs a true integration under all conditions. The inherent advantages and limitation of the various testers and evaluating devices, *if recognized*, will not preclude the use of these devices as useful tools for the evaluation of material in process and finished products.

In addition to providing a means of expressing unevenness as a numerical value, the testing devices provide valuable information for detecting malfunctions of operating equipment.

Most instrument manufacturers publish unevenness values obtained with their instrument for various stages of production. Tables of this type serve as a useful guide in the evaluation of material in process. Because the various instruments and methods of use are not the same with respect to such things as sensing element, sensing length, method of evaluation, length of material tested, etc., comparison of unevenness test data using different test-

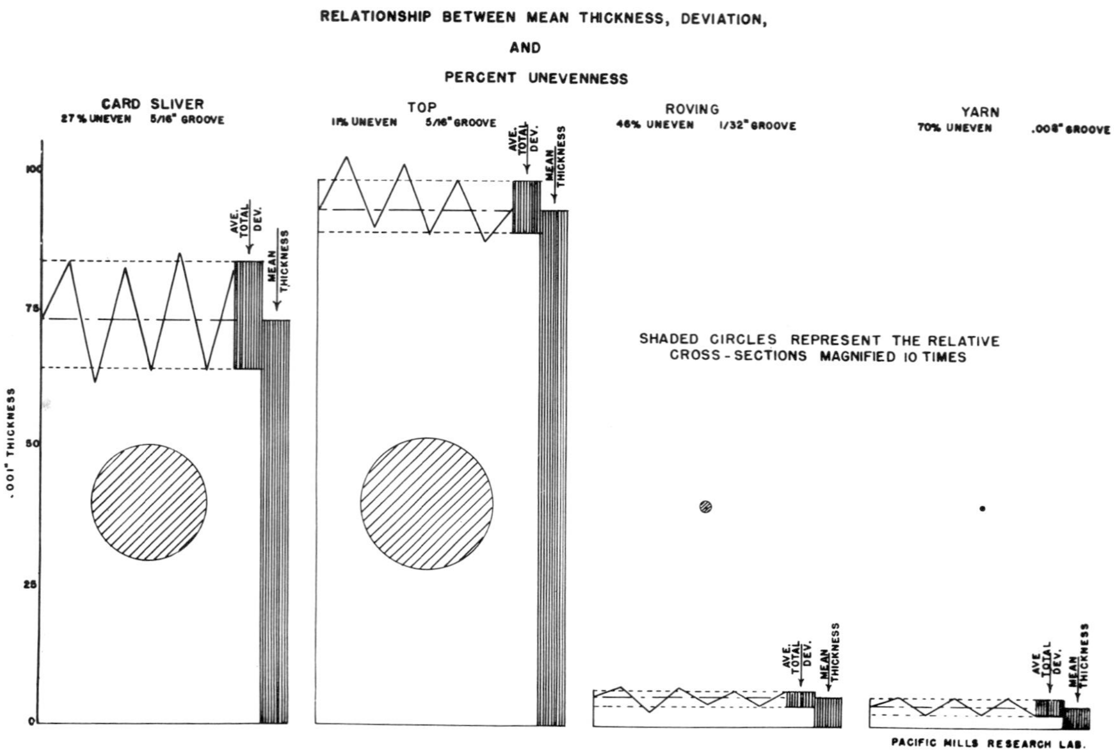

Fig. 8. Pacific Evenness Tester.

ing devices leads to considerable confusion. Waters (115) published a series of equation for translating values obtained by one instrument system to those obtained by another. These data were obtained from tests made using cotton yarn. It would seem quite likely that the form of the equations would be generally the same for other fibers. For both research and testing purposes it is frequently desirable to compare observations made on a particular material with a computed theoretical minimum. Huberty (58) was one of the first to propose such a system. Bornet (16) has proposed a relatively simple computational system for rating the unevenness of yarns spun on both worsted and cotton systems. This system is independent of yarn number and fiber fineness. Figure 9*a*, *b*, *c*, illustrates capacitance sensing type instruments of Uster used for testing top, roving and yarn covered by IWTO-18-67, method of test for determining short term irregularity of linear density of wool sliver, rovings, and yarns by means of a Zellweger-Uster Evenness Tester.

The Pacific instrument as indicated earlier is designed to measure from a "zero base line." As a result of this feature the instrument can be used to measure the mean linear density of the strand. The latter may be computed from the test information obtained for evenness measurement. The values of linear density obtained using this method are independent of the moisture content of the material being tested.

In many cases it is desirable to know the frequency of "abnormally" thick spots in a yarn. At least one manufacturer of capaci-

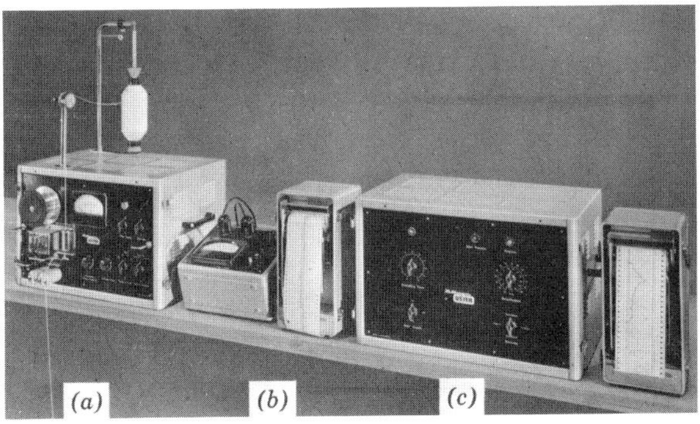

Fig. 9. Uster Evenness Tester.

tance type testing equipment (Uster) has marketed a device which counts the number of such defects in a length of yarn. (See neps.) An instrument based on an optical detection is also available for measuring the frequency of thick or thin spots.

The same manufacturer has also utilized the capacitance measuring principle in the design of a device suitable for the measurement of yarn during the winding operation. This device may be considered as a production tool rather than a testing tool. It replaces the conventional mechanical slub traps commonly used in the winding operation. Mechanical slub traps are relatively insensitive to so-called torpedo slubs and, of course, are worthless for detecting "thin spots" in the yarn. The capacitance type detectors are said to be useful in eliminating both of these type imperfections.

The capacitance measuring principle has also been applied to the rapid assessment of fiber length distribution of top. Early work on the application of the principle was done by WIRA. (See *Wool Handbook*, Vol. 1, page 756). An instrument developed by the Belgian Technical Institute is commercially available. The time required for the testing of a sample is said to be 5 min. (See length.) IWTO-17-67.

Wool Fibers and Top

The principles of core boring as a means of sampling wool in bales for the determination of clean wool fibers present has been discussed by Golub in Vo. 1. Since publication of that volume, the IWTO has published a method for the estimation of top and noil yield and scouring yield. This method differs in some respects from that of ASTM. Various countries or trade regions have different definitions regarding wool so that both the analytical methods as well as the methods of computation can be somewhat different. Descriptions of four of these methods may be found in Bulletin of the Textile Center for Scientific Research and Control, Roubaix, France. The promulgation of methods of this type indicate the growing universal acceptance of scientific methods for the evaluation of industrial raw materials. Further evidence of the acceptance of this type sampling and testing is the USDA Field Laboratories. These laboratories (Fig. 10) operate in the field during

Figure 10. Wool Laboratory Field Testing Equipment. Left Front: Coring Tools, Right Front: Staple Length Sorting Equipment, Left Back: Scouring Tub and Drying Over, Right Back and Center: Projection Microscope Port-Air Airflow Instrument.

shearing time and provide a laboratory assessment of various wool quality characteristics.

Pointing (86) has presented several excellent review articles concerned with the measurement of wool quality. An abbreviated discussion of the scientific measurement of wool quality is given in *World Wool Digest*. These papers indicate the broad extent of acceptance of scientific methods for the evaluation of various properties of wool which have long been evaluated subjectively. Many national governments and regional standards associations such as the Panamerican Standards Association are actively engaged in the promulgation of scientific methods for the evaluation of wool. The British Wool Federation standard contract for wools purchased from Commonwealth countries now includes an option of laboratory evaluation rather than the traditional combing trial.

In January 1966 the new standards for wool grades became effective in the United States. These include limits for the mean fineness of four grades not previously listed. These grades are:

"finer than 80s, 62s, 54s," and "coarser than grade 36s." These standards also include limits for the dispersion of fiber fineness in each grade and an explicit method of classifying a material whose mean value of fineness falls in one grade but whose distribution of fineness exceeds the limit for that grade. Official samples of all grades are available from United States Department of Agriculture. These may be purchased either singly or in complete sets.

McMahon and Whitely at the University of New South Wales have conducted extensive work relating objective assessment of wool fineness in tops to the subjective evaluation in current use. The growing availability and use of air flow measurements for the measurement of fiber fineness has been described by Golub in Vol. 1. Since publication of that volume another instrument has been developed by Wasastjerna (118). Also, the Technical Committee of the I.W.T.O. (International Wool Trade Organization) has issued a supplement to IWTO-5-66:

> "Recommendations on the precautions to be taken in commercial transactions when using micron values of wool fiber diameters as determined on Air-Flow instruments."

It is frequently desirable to estimate the fineness of wool in terms of denier or estimate the fineness of a fiber expressed in denier in terms of mean width. Table 5 shows the manner in which this estimation can be made. In using this type of conversion it should be recognized that while wool is essentially circular in cross-section, many of the other fibers are not. This difference in form will not usually introduce a serious error for industrial purposes.

Vegetable Matter and Neps in Top

The cotton industry has been concerned with the amount of nonfibrous vegetable matter in the raw cotton it receives. An instrument for evaluation samples of cotton for the amount of these materials present has been developed by the Shirley Institute, and is known in the trade as the Shirley analyzer. The use of this instrument for the estimation of the amount of vegetable matter in scoured wool is described by Talpay (106). An English language description of this method is available in Platts Bulletin. An extract of the latter may be found in Textile Industries (4). However, since

TABLE 5
Mean Diameter in Microns of Fiber of Circular Cross Section

	Fiber density[a]				
Denier	1.14	1.18	1.32	1.38	1.50
2.0	15.7	15.5	14.6	14.3	13.7
2.5	17.6	17.2	16.3	16.0	15.3
3.0	19.2	18.9	17.9	17.5	16.8
3.5	20.8	20.4	19.3	18.9	18.1
4.0	22.2	21.8	20.7	20.2	19.4
4.5	23.5	23.2	21.9	21.4	20.5
5.0	24.8	24.4	23.1	22.6	21.7
5.5	26.0	25.6	24.2	23.7	22.7
6.0	27.2	26.8	25.3	24.7	23.7
7.0	29.4	28.9	27.3	26.7	25.6
8.0	31.4	30.9	29.2	28.6	27.4
9.0	33.3	32.7	31.0	30.3	29.1
10.0	35.1	34.5	32.7	31.9	30.6
15.0	43.0	42.3	40.0	39.1	37.5
20.0	49.7	48.9	46.2	45.2	43.4

[a] Typical fiber densities: nylon, 1.14; most acrylic fibers, 1.18; wool, 1.32; rayon, 1.50.

this method is a mechanical separation, the results obtained may differ somewhat from the ASTM methods which are based on a chemical separation. The method mentioned earlier for the determination of vegetable matter yields only a single value for the amount of vegetable matter present. Frequently it is desirable to know the amount of vegetable matter present by type or size. This is particularly true for wool top. A method for the evaluation of wool top for the amount and size of vegetable matter particles is given in ASTM D 1770-66. Figure 11 taken from this method illustrates the classes of vegetable matter. Since the presence of a few large pieces of vegetable matter are usually more serious than a larger number of very small pieces, at least one top manufacturer has established a point weighting system. In this system the various size pieces of vegetable matter as illustrated in the ASTM classification are assigned point values. The larger size pieces are assigned a higher point value. Specifications are then written to establish a

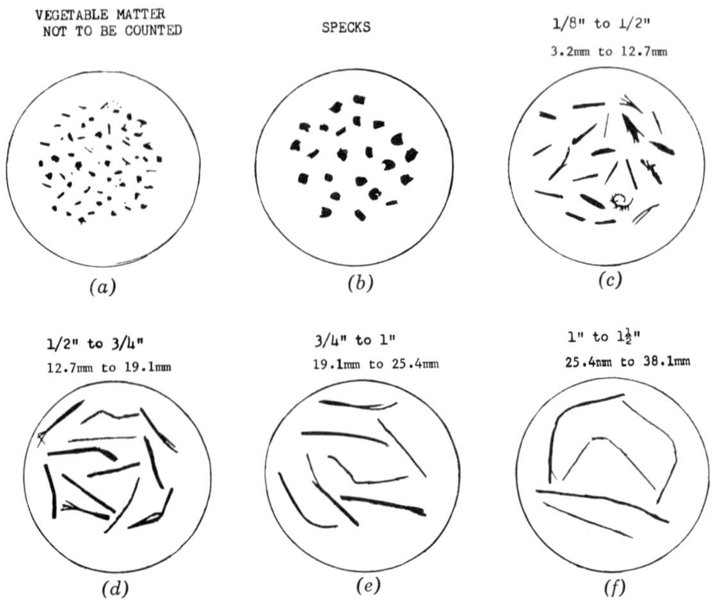

Fig. 11. Vegetable Matter Classes—ASTM-D-1770.

limit of the total number of points associated with vegetable matter which are permissible in a given weight of top. The method cited also contains procedures for the evaluation of wool top for the presence of neps and colored fibers. Figure 12 is a reproduction of the nep scale used to categorize the neps found. The classification of neps by size according to this scale provides a reproducible numerical method of evaluation.

Linnert (77) presents a method of evaluating the frequency of neps is a card web and Basu (12) has developed methods of assessing the frequency of neps in the product of the comb. Both of those papers deal with cotton system equipment. It is quite likely that the techniques would be applicable to wool. Reference is also made to the Glossary of Terms Relating to Tops by I.W.T.O. Subcommittee.

Opening out and examination of Top for vegetable matter, neps, colored fibers and faults, such as slubs, rubber waste and fly waste

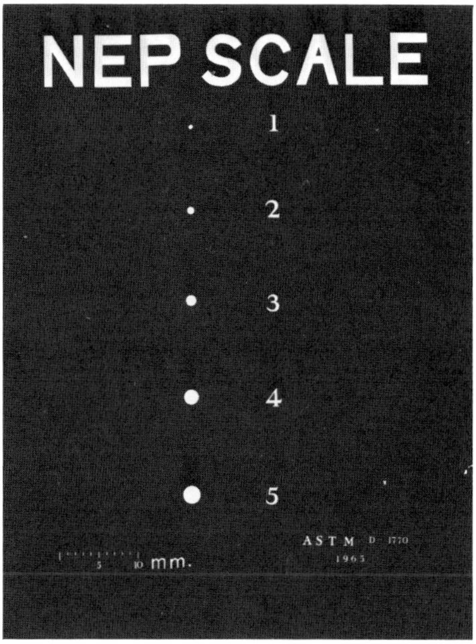

Fig. 12. Nep Scale—ASTM-D-1770.

can, with advantage, be carried out with a so-called 'Schleier-strecke" (Veil Draw Box).

This comprises a draw-box modified so that the sliver is drawn into a thin web, which is then conveyed over an illuminated screen. A commercial version of this apparatus is the Toenniessen Top Tester seen in Figure 13. The counter in front of the screen facilitates the recording of the various faults. A glossary of terms relating to faults in Tops has been published by International Wool Secretariat, Research Department.

Douglas (37) describes the claimed advantages for the capacitance type nep counting device for yarn.

Strength of Wool Fiber Bundles

The alteration of fiber strength in processing can account to a large extent for the change in strength of the finished fabric. A method for defining this property is given in ASTM 1294-58. This method consists of two parts: the preparation of fiber bundles

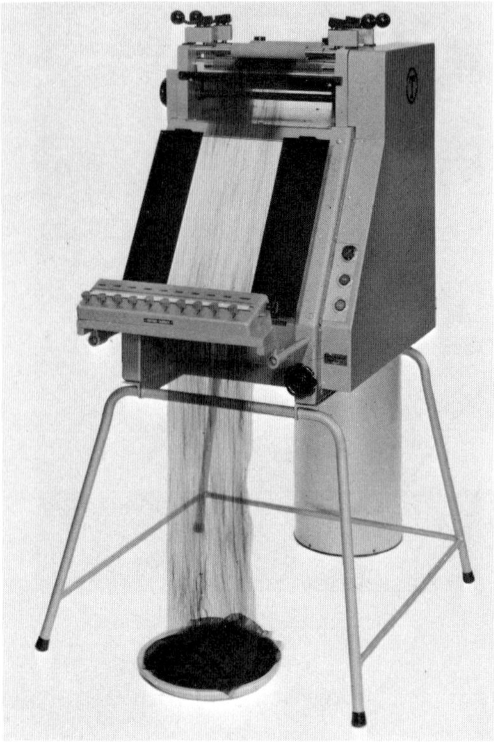

Fig. 13. Toenniessen Top Tester.

under carefully controlled conditions and the breaking of those bundles using a pendulum type tester. The method of computation permits the expression of the results in either psi or g/tex. Barella and Sust (10) have investigated the use of the Stelometer for the measurement of both strength and elongation of wool fiber bundles and found that the instrument appears to provide a rapid solution to the industrial evaluation of these properties (Fig. 14). Ratcliffe and Wood (89) indicate that the instrument may have limitations. Barella et al. (11) further indicate that the instrumental method is adequate for measuring changes in mechanical properties as a result of chemical treatment.

Lie and Sumer (74) developed a pair of adapters to facilitate the use of the jaws of the Pressley tester to the Instron. They used this

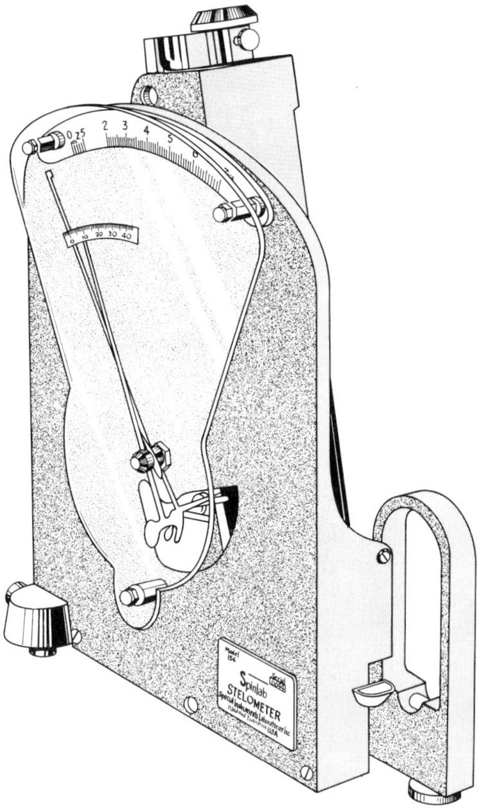

Fig. 14. Stelometer.

combination to study some elastic performance characteristic of cotton and man made fibers and concluded that the equipment combination used was satisfactory for evaluating fiber bundle characteristics.

Fiber Length

A relatively new instrument for evaluating this property is the AL-Meter (Fig. 15). This operates on a capacitance sensing principle. The method of sample preparation (Fig. 16) and the computational methods are extremely simple. For these reasons the instrument provides a very rapid method for assessing the fiber length distribution of wool top. Details of the method are given in IWTO-17-67.

The WIRA tester is a capacitance type tester. Calculations based on the data obtained using this instrument provide a result in terms of a cross sectional area biased mean length. This is the French "Hauteur" IWTO-16-67 method.

The AL-Meter with its associated automatic computer and nomograph provide the cross sectional area biased mean length, the weight biased mean length, and the coefficient of variation of the cross sectional area biased mean length, but no measure of the distribution of the weight biased mean length. Interlaboratory tests using the WIRA instrument and the AL-Meter have been conducted by the IWTO. Data from these studies indicate that the two instruments yield slightly different results. The fact that the various instruments and methods of computation yield numerical values which differ does not negate these as useful tools. An excellent comparison of the Schlumberger and AL-Meter is provided by Dillie et al. (36). In addition to the average length and a measure of dispersion, the consumers of top are frequently interested in the amount of short fiber present. A common value used in the United States is the amount of fiber under 2 inches in length. Various spinners have different requirements for this value but a

Fig. 15. AL-Meter.

Fig. 16. AL-Meter Sample Preparation.

typical desired value is: not more than 20% under 2 in. These numerical values refer to measurements made using the Suter or similar instrument.

Noilage

In the worsted system an important process variable is the amount of noil produced as a function of the amount of material in process. The method of testing consists of collecting and weighing the comb sliver and noil produced in a given length of time. Both of these are weighed separately, and suitable computations are performed. The computation most widely used in the United States is:

$$N/(T + N) \times 100 = \%\ \text{noil}$$

The computation preferred in Britain is:

$$T/N = \text{tear}$$

In either case:

N = weight of noil
T = weight of comb sliver delivered

In order to eliminate computation a special weighing device may be used. This balance system includes two pans or platforms. The sliver is placed in one and the noil in the other. At equilibrium the pointer indicates the amount of noil directly. A typical device of this type is shown in Fig. 17.

The amount of noil obtained is dependent on a variety of factors. Among these are the combing system used, the amount of short fiber present in the raw material, the manner in which the material

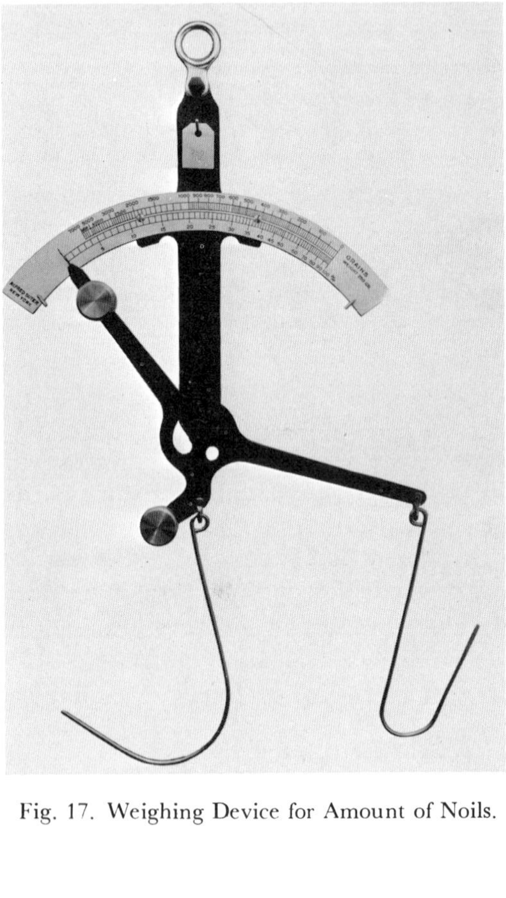

Fig. 17. Weighing Device for Amount of Noils.

was prepared for combing, and the quality characteristics desired in the combed material. Typical noilages range from about 8% to about 20%. The lower values are usually associated with long coarse wools and the higher values with short fine wools.

Roving

The same general methods described for top apply to roving.

Yarn

As indicated earlier, one of the primary considerations is uniformity of the weight and length ratio. This factor strongly influences both the weaving or knitting efficiency as well as some performance characteristic of the end product. The methods of testing and tolerances are given under general methods. Also included at that point are discussions of evenness measurement and the significance of these measurements. Bornet (16, 17) has developed a method of rating yarns for both short-term and long-term unevenness. In addition to the quantitative measure of evenness, it is frequently desirable to have standards for visual comparison. This comparison is made by winding the yarn to be examined on a set of panels and comparing the panels prepared in this way with a set of reference panels. Figure 18 illustrates the use of this type of examining system.

Yarn strength, elongation, and recovery

Yarns may be tested for strength in one of two forms. These are skein or single strand. The method using skeins is generally much less sensitive than the method using single strands. In addition to the lack of sensitivity in the measurement of strength, the skein method leaves much to be desired in the measurement of elongation. In spite of these limitations, the skein method is widely used for mill control and acceptance testing purposes. In addition to the differences in the form of the test specimen, three fundamentally different types of testing equipment are available for the testing of the breaking load of yarn. These are: the constant rate of traverse (pendulum), the constant rate of load (incline plane) and the constant rate of specimen extension. The first two types are

Fig. 18. Seriplane Tester.

relatively insensitive because of the limitations of inertia. However, for mill control purposes the constant rate of traverse is the most widely used instrument. A graphic presentation and an analysis of the relationship between motion of the component of both the constant rate of load and constant rate of traverse type instrument is given by Skinkle (135). For research purposes two instruments of the constant rate of specimen extension type are available. One of these is designed to use a resistance strain gauge as force measuring system. This is typified by the Instron machine. The other type uses a spring and impedance coil as a measuring system. Typical of this type is the instrument made by Zwick and illustrated in Fig. 19.

The skein method utilizes the constant rate of traverse type machine. The details of this method of test are given in ASTM D 1578-67. Meaningful elongation measures cannot be obtained by the skein method. In brief, a number of skeins are prepared using a yarn reel. Each of these skeins is mounted in the appropriate accessory spools attached to the testing machine and broken.

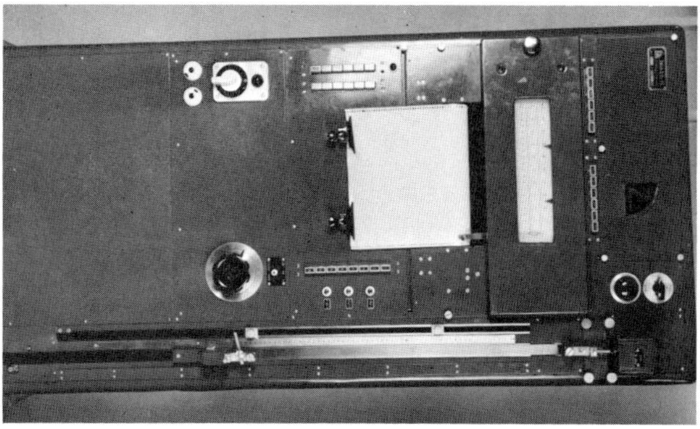

Fig. 19. Zwick Electronic Tensile Testing Machine.

The breaking strength of each skein is measured. The reporting of tests made using skeins must be accompanied by the circumference of the reel used, and the number of wraps per skein. The skeins used for determination of yarn number may be used for breaking strength. If this is done, the average breaking strength may be corrected to the breaking strength at some specified yarn number. If an indirect yarn numbering system is used, the following method of computation applies:

$$S_2 = (C_1 S_1)/C_2$$

where

S_1 = observed breaking load
S_2 = adjusted breaking load at the specified yarn number
C_1 = observed yarn number
C_2 = specified yarn number

The coefficient of variation of breaking strength by the skein method using 80 yard skeins should not exceed 10. for worsted spun yarns and 6.5 for woolen spun yarns.

Single strand tests may be made using any one of the various types of testing devices described above. This test is made by breaking individual strands of yarn mounted in the clamps of the testing machine. Any of the various types of machines equipped with autographic recording equipment will plot a trace of the load vs elongation properties of the test specimen. The breaking load and ultimate elongation can be read from the chart. For research purposes elongation at various loads may be determined. If recovery characteristics are desired, the testing devices such as the Instron or Zwick instruments can be adjusted to perform cyclic tests to some load less than the breaking load.

Automatic single strand testers are available. These testers will make a predetermined number of breaking strength tests on yarn from a single bobbin or a group of bobbins. This type of equipment automatically records the breaking strength and ultimate elongation obtained on a single test and then proceeds to continue the testing, so that at the end of the time period, a record of the individual values of breaking strength and ultimate elongation are available. A typical unit of this type is built by Zellweger-Uster.

All of the above tests are essentially static tests. They provide a measure of the strength of a given length of yarn under some specified test conditions. A dynamic test which the author has found useful consists of a tensioning device and a pair of power driven draw off rolls. The yarn from a cone or other package is fed through the tensioning device and draw off rolls. The tensioning device is adjusted to some desired point and the draw off rolls started. The yarn breaks whenever the tension is in excess of the breaking load under the conditions of test. A predetermined yardage is tested and the number of breaks occurring in the yardage is recorded. This test in effect gives the number of weak spots in the length of yarn. A device of this type is described by Crabtree and Rae (34).

Twist of yarn

The term twist is used to indicate axial or helical rotation of the components of yarn. In the United States twist is expressed in turns per inch. In Europe the twist is expressed in turns per meter. In addition to the magnitude of twist the direction of twist must be indicated.

Basically the method for the determination of twist in yarns made using the woolen and worsted systems can be classified as mechanical or optical methods. The mechanical methods are used whenever sufficient yarn is available for test. The optical method is used only when the amount of yarn available for testing is very small, or the yarn for some reason cannot be removed from a fabric, or if it is desired to know the twist at a single point for research purposes. The mechanical methods are classified as the direct counting method and the twist-untwist method. These are described in detail in ASTM D 1422-65 T and D 1423-66 T. The twist-untwist method is subject to certain errors particularly if the twist has been set either by steaming or by prolonged storage. For this reason ASTM recommends that this method not be used as a research or acceptance test. Detailed analyses of the differences between the two methods have been made by Worner (127) and WIRA (126).

The basic apparatus, known as a twist tester, is illustrated in Fig. 20. This device is designed with one rotatable clamp which cannot move axially and one clamp which may be moved axially but is not rotatable. The rotatable clamp is attached to a resetable readout device which indicates the total number of revolutions of the clamp. The nonrotatable clamp is designed to permit the pretensioning of the test specimen when the direct method is used. For the twist-untwist method an auxiliary device is used. It is designed to apply the same predetermined tension to the test specimen at the start and finish of a test but permits the removal of this tension during a portion of the test.

For the direct counting method, a different pretension device is required as a part of the apparatus.

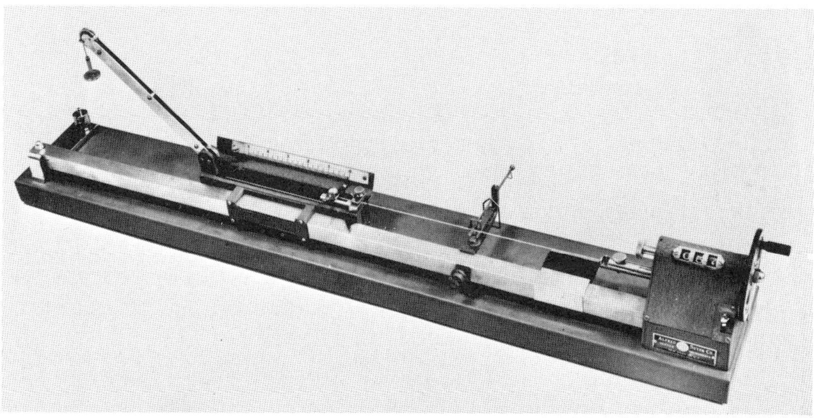

Fig. 20. Twist Tester.

The optical method consists of measuring the angle between the axis of the yarn and the component (either fiber or single yarn) and the diameter of the singles and plied yarn at this point. Using these values the twist in the singles and in the plied yarn can be calculated. Schwarz (101) gives a detailed description of both the method of measuring and the method of computation.

The amount of twist inserted is, of course, a function of the end use of the yarn. The commercial limit is usually about 7.5% of the stated twist. Depending on the manufacturing system and the yarn number it is frequently possible to control the twist within somewhat closer limits than these.

Blend Distribution of Yarns

This term is used to describe the homogeneity of a blend.

The methods of measuring and evaluating blend distribution have been discussed by Burtt and Woo (27), Coplan and Bloch (33), Hamilton and Cooper (52), and Fuchino and Veno (47). Long term uniformity of axial distribution is a desirable characteristic. However, ideal uniformity either radial or rotational may not necessarily be desired for either aesthetic or utilitarian reasons. Cooper (32) presents a critical review of the statistics of blend irregularity.

Fabric Testing

In general, the performance characteristics of a fabric are dependent on the interrelationship between fiber, yarn and fabric properties. For woven fabric the nature of the weave and the number of picks and ends are of primary importance. For knit fabric the knitting construction and the number of coarses and wales per inch are the principal factors. The number of picks and ends per inch for woven fabric may be determined by counting, using the method described in ASTM D 1910-64.

Interpretation of these data must be made with the recognition that the fabric construction may undergo appreciable changes during the various operations following removal of the fabric from the loom or knitting machine and during subsequent processing.

Strength and elongation and recovery

As in the case of yarn strength, several types of equipment and methods of testing are available. The equipment used for fabric strength is either of the constant rate of traverse (pendulum) type or the constant rate of elongation type. Fabric clamps are available for use with these machines. Two different types of breaking strength tests are made. These are the grab and ravel strip tests and are discussed in detail in ASTM D 1682-64. These differ primarily in the method of specimen preparation. When using the grab test the test specimen size is 4 $\times$ 6 in., the larger dimension being the direction of the test. The fabric is clamped in the testing machine equipped with jaws having a 1-inch face width mounted 3 in. apart. The machine is started and the fabric broken. The ravel strip test differs in that the test specimen size is $1\frac{1}{2} \times 6$ in., with the long dimension in the direction of test. Yarns are removed from the sides of each test specimen until the remaining fabric is one inch wide. The test specimens are then broken in the testing machine equipped with appropriate size jaws. If the testing machine is equipped with an autographic recorder the elongation at the rupture may be obtained. In general, the test results obtained using the grab method may be the same or greater than those obtained using the ravel strip method. In most woolen and worsted fabrics the grab method will yield higher results. This is primarily because in the grab method the yarns not gripped by the jaws contribute to the strength of the material. The minimum breaking strength required by cutters varies with the garment made from the fabric. The warp and filling breaking strengths in pounds that have been found acceptable for various apparel items are shown in the table on page 1160.

Ferguson (42) discusses and suggests certain danger levels of cloth specification with respect to consumer requirements. Cutters of high quality apparel wear usually require a strength substantially greater than this. A typical high grade men's wearcutter requires that the fabrics have a minimum breaking strength by the grab method of 40 lb.

The International Wool Secretariat (I.W.S.) in conjunction with International Wool Textile Organization (I.W.T.O.) tested and analyzed approximately 1000 wool woven apparel fabrics to establish their breaking strength level.

TABLE 6

Woolen and Worsted Fabrics Breaking Strength in Pounds[a]

	Minimum		Normal	
	W	F	W	F
Men's wear suits and jackets; woolen	25	20	30	25
Men's wear tropicals and worsted	30	25	35	30
Men's wear regular worsted suitings	35	30	40	35
Men's top coats and overcoats	30	25	35	30
Woolen shirting	23	20	28	25
Worsted shirting	23	25	32	30
Women's wear coats and suits	23	20	28	25
Wool skirts	25	20	30	25
Dress goods	20	18	25	23
Uniform fabrics				

[a] ASTM Grab Test.

The specifications for the Woolmark label operative as from March 1, 1969 require tensile strength figures (kg/5 in. strip) on the following minimum basis:

Men's and boys' major garments	20 kg
Women's and girls' major garments	10 kg
Accessories	0

The specification makes no distinction between woolen and worsted fabrics. To broaden it, I.W.S. suggests the end-product-specific specification shown in Table 7.

For knit goods, particularly if the material is not appreciably felted, conventional breaking strength testing equipment is not applicable. For evaluating fabrics of this type, a measure of bursting strength is used. Two types of bursting strength are available. These are referred to as the ball bursting and diaphragm bursting types. The former is designed to be used for the testing of fabrics having a relatively high elongation. The latter is designed to be used for the testing of fabrics having a relatively low elongation. The requirements of these testing machines are described in ASTM D 76-67 and the use thereof in ASTM D 231-62.

TABLE 7

Categories	Worsted		Woollen	
	Kg[b]	Lb	Kg	Lb
Men's and Boys'				
Coats	20	44	15	33
Suits and trousers	25	55	20	44
Jackets	20	44	15	33
Shirts, dressing gowns	15	33	15	33
Schoolwear	Special		Special	
Women's and Girls'				
Coats	15	33	15	33
Costumes and skirts	20	44	15	33
Slacks	25	55	20	44
Dresses[a]	15	33	10	22
Blouses, underwear, Nun's veiling, nightdresses	10	22	10	22
Schoolwear	Special		Special	
Accessories	0	0	0	0

[a] To accommodate very lightweight dress materials, a possible modification here could be a special category for fabric less than, say, 150 g/m^2, the minimum standard being, say, 10 kg.

[b] British Standard Method for the determination of breaking load and extension of strips of woven textile fabrics. It was found that the British 5-cm (2 in.) strip method yields approximately 20 percent higher results than the ASTM Grab Test.

Stretch Fabrics

The recent introduction of so called stretch fabrics requires the use of test methods suitable for measuring the extent to which a fabric will distort under relatively low loads, and also the extent to which the material will recover from this distortion. Those parameters are of importance in the designing and marketing of this type fabric. Several methods have been proposed to measure these characteristics.

A method currently under consideration by subcommittee B9 ASTM D-13 may be used to measure both the extent of deformation under load, and the unrecovered deformation when the load is removed.

Burton (26) has developed both electrically operated and pneumatically operated testing devices for the assessment of stretch fabrics.

Von Bergen and Clutz (115) developed a test method for stretch fabrics which is a modification of an elongation test developed by E. I. du Pont de Nemours for the evaluation of fabrics containing "filament stretch yarns." Test specimens 8 × 24 inches with the long dimension in the direction of the test are prepared for testing in both the warp and filling. The test specimens are marked so that a pair of marks are 2½–3 in. apart near the center of one end. A similar set of marks are placed 18 in. away. One end of the fabric is suspended from a horizontal clamp at least 6 in. long mounted so that the fabric hangs free. The opposite end of the fabric is placed in a similar clamp and a weight attached to the clamp. The total weight of clamp and weight is 5 lb. One minute after the application of the load, the length between the marks is read using an appropriately calibrated scale such as a shrinkage rule. If the difference between the two sets of marks is greater than 0.5% the clamp should be realigned. The weight and clip are removed after 5 min and the length measured immediately. The method of computation is:

$$\frac{\text{extended length} - \text{original length}}{\text{original length}} \times 100 = \%\ \text{elongation}$$

$$\frac{\text{recovered length} - \text{original length}}{\text{original length}} \times 100 = \%\ \text{unrecovered elongation}$$

A test method suitable for quality control purposes has been developed by E. I. du Pont de Nemours. This method consists of six steps. These steps taken directly from Technical Information on Processing Stretch Woven Fabrics, are:

1. Cut fabric into 22-in. × 3-in. specimens for testing. The stretch yarn must be in the 22 in. direction.

2. Put a 2-in. bench mark on the test specimens.

3. Attach metal clips or other suitable clamps at least 3 in. wide to each end of the specimen.

4. Hand the specimen by one clip and load the sample to 5 lb.

The 5 lb weight is not just an arbitrary or convenient weight; it has been chosen in an effort to correlate with the stretch measurement obtained by the "New York Hand Stretch Test."

5. After 1 hr, measure the distance between the original 20 in. bench marks and record as "extended length."

6. Remove the weight and clip and measure the fabric. Allow sample to recover for 6 hr and again measure the distance between the original 20 in. bench marks. Record these lengths as "initial recovered length" and "recovered length." The percent stretch and percent unrecovered stretch are then calculated using the following formulas:

$$\text{percent stretch} = \frac{\text{extended length} - \text{original length}}{\text{original length}} \times 100$$

$$\text{percent unrecovered stretch} = \frac{\text{recovered length} - \text{original length}}{\text{original length}} \times 100$$

The measurement of fabric stretch is of significance in fabrics other than the so called stretch fabrics. Von Bergen (113) has indicated that the percent elongation in the filling direction using the latter method, is indicative of the amount of relaxation shrinkage to be expected in London shrinking. Baird (8), using a much more complicated apparatus, found that the elongation in the filling direction under a load of 4 kg yielded results which were essentially equal to the relaxation shrinkage of the material under test.

Thickness, compression, resilience

Since woolen and worsted type fabrics are deformed by load, the measurement of thickness must be made under specified conditions. ASTM D 1777-64 defines thickness of a textile material as " . . . the distance between the upper and lower surfaces of the material, measured under a specified pressure." The same method of test provides a guide to the pressures used for various fabrics. These are soft fabrics, 0.005–0.5 psi; moderate fabrics 0.02–2.0 psi; and firm fabrics 0.1–1.0 psi.

This method is a general one and for this reason does not specify the area to be measured. ASTM D 418-42, and D 1486-57 T apply to woven and tufted pile floor coverings, respectively. These methods of test require the use of a presser foot 3 in. in diameter and a pressure of 0.1 psi. Similarly ASTM D 461-67; a method of test for felt requires a presser foot having an area of 1 sq in. and a load of 10 oz. Because of the ease with which some fabrics are compressed, and in some cases the difficulty of precisely measuring the small forces involved, a presser foot having an area of at least 5 sq in. is desirable.

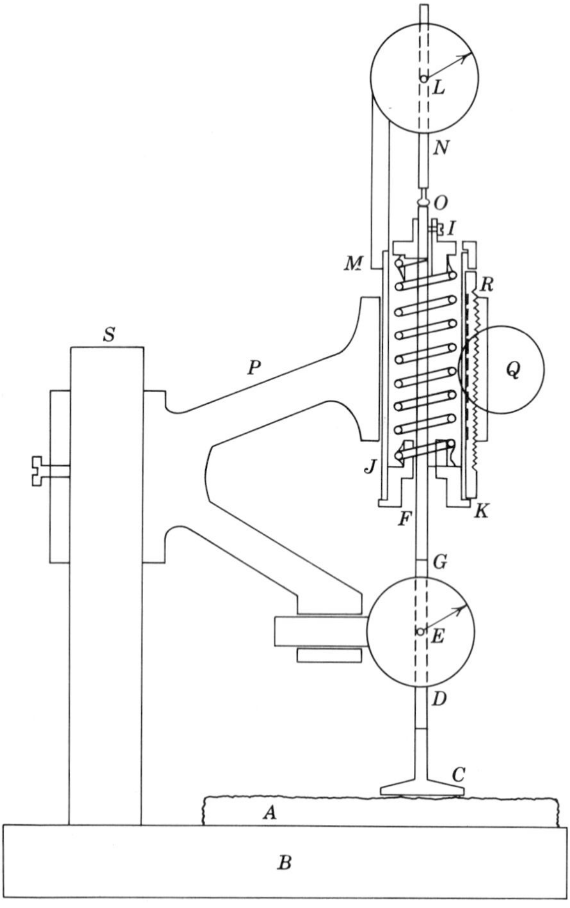

Fig. 21. Schematic Diagram of Compressometer.

Henno and Joubet (55) have developed a method for measuring the thickness and compression of bulked yarn and fabric.

Schiefer (98) has developed a Compressometer which permits the measurement of thickness under a series of loads. This device can be used to measure both compression and recovery.

A schematic diagram of the instrument is shown in Fig. 21. The specimen, *A*, is placed upon the anvil, *B*. The foot, *C*, is circular in shape and varies from 1 to 5 in. in diameter. It is fastened to the bottom of the spindle, *D*, of the lower dial micrometer, *E*. The lower surface of the foot is plane and parallel to the upper surface of the anvil. The rod, *F*, is fastened to the top of the spindle at *G* and to the top of a helical spring, *H*, at *I*. The bottom of the spring is fastened to the tube *J*, at *K*. The upper dial micrometer, *L*, is fastened to the top of the tube at *M*. The spindle, *N*, of the upper dial micrometer rests on the top of the rod at *O*. The tube may be moved up or down relative to the frame, *P*, by turning the knob, *Q*, of the rack and pinion, *R*. By turning the knob the foot may be lowered upon the specimen. The pressure which is applied to the specimen by the foot may be ascertained from the upper dial reading and a calibration curve of the spring. The upper dial indicates the elongation of the spring. The distance between the face of the foot and the anvil, i.e. the thickness of the specimen, is indicated on the lower dial. Each dial is graduated to read directly to 0.001 in.

With this instrument, thickness can be readily determined to .001 in. at pressures ranging from .005 to 3.0 psi. It was found that the best correlation between the thickness of apparel fabrics and warmth was obtained when the thickness was determined at a pressure of .005 psi using a 5-in. diameter presser foot.

For research purposes an Instron type testing machine is available for making thickness, compression, and recovery measurements. An accessory for compression measurement is also available for the tension model. Solmaz (102) using this device developed a series of prediction equations for knit deep pile fabrics made using man made fiber. These prediction equations relate the construction factors, fiber fineness, fiber length, and weight per unit area to energy to compress and resilience. It is quite likely that similar relationships can be developed for various fabrics made using the woolen or worsted system. Apparel fabrics are universally sold on a basis of weight per unit length. For this reason the criteria usually used to classify fabrics with respect to end use is weight per linear yard. Typical of the range of weight and range of thickness of apparel fabrics are the data shown in Table 7. In view of the discussion concerning thermal conductivity which follows a more reasonable classification would use thickness as a criteria for end use. A classification of this type is shown in Table 8.

TABLE 7

Weight and Thickness Range of Wool Apparel Fabrics

Use	oz/linear yard—56 in.	Weight oz/sq yd	Thickness .005 lb/in.
Ladieswear			
Dress material	6– 9	4 – 6	.020–.035
Suiting	8.5–12	5.5– 8	.030–.070
Coating	12–18	8 –12	.050–.250
Menswear			
Tropical worsted	8–10	5 – 6.5	.025–.035
Suiting and panting (summer)	8–12	5 – 7.5	.030–.070
Suiting, winter	12–16	7.5–10	.050–.090
Top coating	16–20	10 –12.5	.070–.150
Overcoating	20–32	12.5–21	.130–.250

TABLE 8

Thickness Range of Wool Apparel Fabrics

	Thickness in inches
Ladieswear	
Dress material	.020–.035
Suiting	.025–.040
Coating	.040–.250
Menswear	
Summer suiting and panting	.025–.040
Winter suiting	.040–.090
Top coating	.070–.125
Overcoating	.125–.250

Thermal properties, comfort, and warmth

Historically wool has held an unrivaled position in certain apparel fields because of its accepted superior insulating properties. Schiefer has shown that the insulating value of textile materials is due essentially to the airspace in the fabric and is essentially independent of the fibers comprising the fabric. Thus fabrics of equal porosity* and equal thickness will have essentially the same thermal conductivity.

This may be taken as an implication that the advantages in this respect historically claimed for wool are mythical. This is not true. A principal advantage of wool in apparel fabrics is its superior ability to recover from deformation during use and in this way maintain a high "insulating value" throughout its useful life.

Since the term warmth is used to denote that combination of properties which contribute to the desirability of an item for protective apparel, it includes a combination of properties such as thermal transmission, vapor permeability, wicking and heat of wetting.

Thermal transmission

By thermal transmission is meant the rate of energy flow through a fabric in a direction perpendicular to the plane of the fabric. ASTM D 1518-64 T describes a method for measuring this characteristic. For use as outer wear, in cold climates a low thermal transmission is desirable. However, individuals performing relatively mild exertion will exude an appreciable amount of perspiration. This may exist largely in the form of vapor if the exertion is below some level and will be in the form of both liquid water and vapor if the exertion is above this level. For this reason the ability of a fabric to transfer both vapor and liquid water are important in the evaluation of fabric for comfort purposes. Two relatively simple tests designed to measure the diffusion of water vapor through a textile material have been used by Fourt and Harris (44).

A method for measuring the transfer of liquid water by wicking was developed by Rees (99). In this method the fabric is placed in

* The ratio of the volume of air or void contained within the boundaries of a material to the total volume (solid matter plus air or void) expressed as a percentage.

contact with a fritted glass disc. The amount of water transferred is measured by a manometer. Fourt and Monego (46) have described a sophisticated system for measuring the total amount of energy (both heat and moisture) flowing through a fabric. The latter method is an ideal research tool.

The diffusion of water vapor through textile has been studied by Fourt and Harris (45); and Whelan, Goodings, and Turl (119). The psychological effects of clothing are discussed by Fourt and Harris in Newburgh's book.

Abrasion resistance

The term abrasion is used to denote the wearing away of any part of a material by rubbing against another surface. The measurement of abrasion resistance may be made using any one of a

Fig. 22. Stoll—CSI Abrasion Tester.

number of abrasion testing machines. Six of these and the methods for operating them are described in ASTM 1175-64 T. The recommended procedure lists the methods as:

A. Inflated diaphragm
B. Flexing and abrasion
C. Oscillatory cylinder
D. Rotary platform
E. Uniform abrasion
F. Impeller tumble

Methods A and B use the same instrument. This is illustrated in Fig. 22. In Method A the fabric is distended by means of a diaphragm and may be abraded either unidirectionally or multidirectionally (Diagram, Fig. 23). In the Method B the fabric is subjected to simultaneous unidirectional flexing and abrasion. Method C is designed to subject the fabric to unidirectional abrasion while under defined tension, while Method D is intended

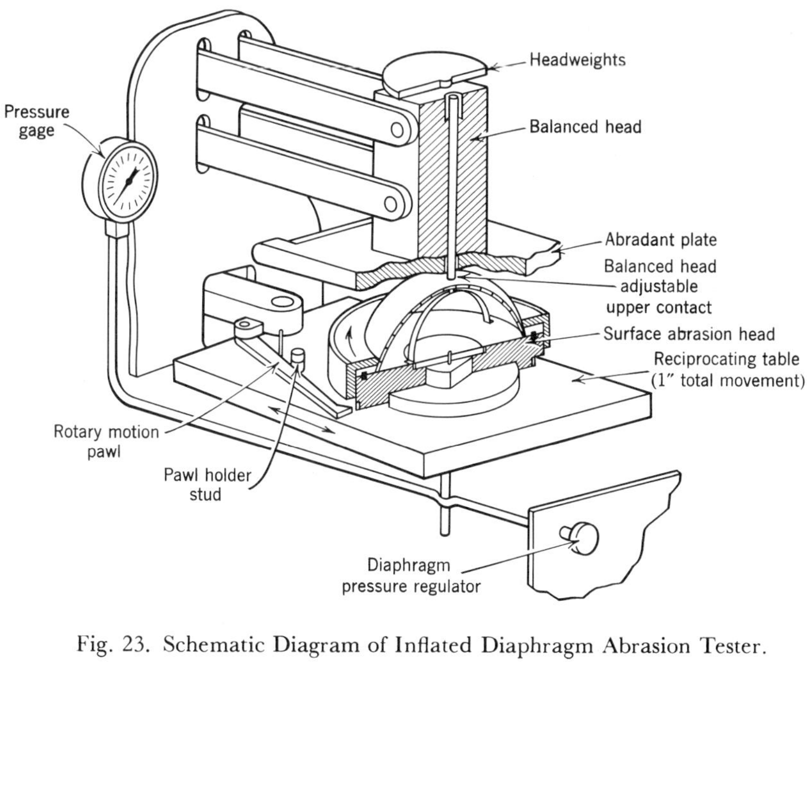

Fig. 23. Schematic Diagram of Inflated Diaphragm Abrasion Tester.

TABLE 9

Particulars of Abrasion Machines

Name	Motion	Abradant	Load in specimen	Assessment of end point	Effective dimensions of specimen (approx.)	Ref.	Methods
Schopper	Circular, reverses every 500 cycles	Abrasive paper	1000 g	Loss in bursting strength or cycles to hole	50 cm^2 (circular)	Koch *et al.* (1942)	E
Martindale	Lissajous figure	Worsted fabric	200 g or 790 g	Weight-loss or cycles to thread breakage	2.5 cm diam	Martindale (1942)	E
Accelerotor	Rotating impeller 1500–3500 cycles/min	Abrasive paper	—	Loss in wt. or visual	Square to weigh 3 g	A.S.T.M. D 1175 AATCC 93-66 T	F
Wyzenbeek	Osc. cylindrical abradand 90 cycles/min	No. 0 emery paper	900 g and 900 g tension	Loss in breaking load	4.7 cm × 23 cm	A.S.T.M. D 1175	C
Schiefer	Circular, uniform relative velocity	Steel blades, abrasive or fabric	4,500 g	Thickness or weight-loss	1.25 cm to 5 cm diam	D 1175	E

TABLE 9 (*continued*)

Taber	Rotary 70 cycles/min	Abrasive wheel	250 g to 1000 g	Loss in breaking load	10 cm diam	D 1175	D
Stoll-Quartermaster	(*1*) Linear or multi-directional 125 cycles/min 2.5 cm stroke	Abrasive paper	2250 g or less	Cycles to hole or visual assessment	9.4 cm diam inflated diaphragm	D 1175	A
	(*2*) Flexing over blade or bar 2.5 cm stroke	—	—	Cycles to rupture, loss in breaking load or visual assessment	2.5 cm × 20 cm	D 1175	B

Selected Bibliography

Beck, L., *et al.*, *Melliand Text.*, 1965, **46**, 1095, 1221 and 1346.
Committee of Dir. of Textile Research Associations, *J. Text. Inst.*, 1964, **55**, p. 1.
U.S. National Bureau of Standards, Report No. 2319, March 1953.
Tannenhaus, S. J. and Darby, M. E. Rep. No. 36 of U.S.A. Quartermaster, 1950.
Dowlen, R. P., *Text Res. J.*, 1966, **36**, 80.
Winks, J. A., *Text. Inst. & Ind.*, 1965, **3**, 26.
Weiner, L. I., and Pope, C. J., *Text. Res. J.*, 1962, **32**, 350.
Martindale, J. G., *J. Text. Inst.*, 1942, **33**, T151.
Lee, J. S., *et. al.*, *Text. Res. J.*, 1961, **31**, 540.
American Society for Testing Materials. D1175.

to subject the fabric to rotary rubbing action. Method E subjects the fabric to a uniform multidirectional abrasion.

All of the instruments used for the methods mentioned above are provided with a means of adjusting the load on the fabric abradant interface and have interchangeable abrasive surfaces so that one may select the combination of these which are known or judged to be the most satisfactory for testing a particular fabric. The measurement of abrasion resistance may be made either by running the test specimen to the first visible hole, complete destruction, or to some fixed number of cycles. If the latter procedure is used the test specimen may be either examined visually or it may be further evaluated for breaking strength or air permeability.

A list of the important operating characteristics of eight different abrasion machines in widespread use is printed in Table 9; further details of these may be found from the Bibliography. This list is not, however, exhaustive and at least 100 different instruments have been described in the literature. Of the machines in use, perhaps the Schiefer tester and the Accelerotor may be said to possess unique features.

This list was published in the *Wool Science Review* No. 32, July 1967. The Martindale instrument is used extensively in wool laboratories in the British Commonwealth (Fig. 24).

Fig. 24. The Martindale Abrasion Machine.

Schiefer (99) has developed a novel method for evaluating fabrics either abraded instrumentally or in use. This method indirectly measures the mass of the material remaining after abrasion or wear by means of a "capacitance gage."

Burton (25) has developed a modification of the Martindale Abrasion tester which is said to overcome some limitations of the original device.

Beck et al. (13) have presented an evaluation and summary of international interlaboratory tests for fabric abrasion.

Pilling

By pilling is meant the development of balls of fiber on the surface of the fabric. These balls are attached to the fabric by fibers. ASTM D 1375–67 lists three methods for evaluating pilling. These are:

1. The appearance retention method
2. The brush and sponge method
3. The random tumble pilling tester

This method of test indicates that Methods *1* and 3 are applicable to both woven and knit fabrics. In evaluating a fabric, test specimens are exposed to the conditions defined by the method after which the incidents of pills may be evaluated either by counting or subjective examination. Fig. 25 illustrates the apparatus used in Method 2.

Shedding

By shedding is meant the loss of fiber from the fabric due to the mechanical action suffered in use. No standard method is recognized for the evaluation of this phenomenon. Von Bergen (113) has indicated that the Accelerotor is a useful testing device for the evaluation of this characteristic. The device is operated at 2500 rpm without a liner for this test. The fiber separated from the fabric after running 5, 10, and 15 min is collected and weighed. The total amount of fiber separated from the fabric and the rate at which it separates is indicative of the amount of shedding. A 5–6% loss during the first five minute period and a total of

Fig. 25. Brush and Sponge Pilling Tester.

9–10% loss for the 15 min period are typical values obtained from fabrics known to be acceptable for women's coatings.

Dimensional Stability

The term dimensional stability as used herein may be defined as the change in the dimension of the fabric due to a given set of conditions. The changes may be classified as hygral, relaxation, or felting. The term hygral is suggested by Baird (8) and refers to the change in dimension due to changes in the environmental humidity. Increasing humidity results in an increase in dimension. This phenomenon is reversible.

The movements of three wool fabrics, a gabardine, a flannel, and a tropical worsted with relative humidity changes are shown in Fig. 26.

By relaxation is meant the recovery of strains set up in the fabric during processing. This temporary set is usually associated with tensions applied to the fabric particularly during drying. Release of these strains results in a decrease in the linear dimensions of the fabric, hence the dimensional change is one of shrinkage. By felting is meant the migration of fibers in the material when the latter is subjected to mechanical action. This fiber migration results in a consolidation of the fabric. Again the dimensional

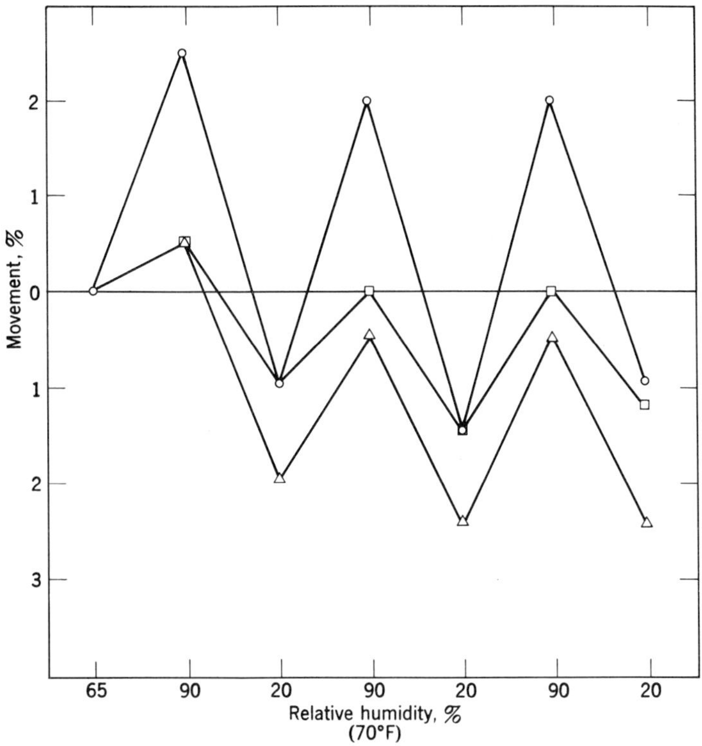

Fig. 26. Movement of Fabrics with Relative Humidity.

change is one of shrinkage. (In a fabric on which the shrink resistant treatment has been excessive, the fabric may stretch rather than shrink.) The term consolidation as used here is not to be confused with the term consolidation as used by some workers to indicate compensational changes of dimension of knit structures.

The occurrence of the three types of dimensional changes are dependent on the fibers from which the fabrics are made. Hygral changes are to be expected in any fabric containing hydrophylic fibers. The magnitude of change expected being dependent on the amount of hydrophylic fibers present and the extent of fiber response to changes in humidity. Relaxation shrinkage may be expected in all fabrics, but the magnitude of the changes would be small for those fabrics made wholly or largely of hydrophobic fibers. Felting shrinkage, of course, only occurs in those fabrics containing animal hair fibers.

The term sponging shrinkage is used to denote that shrinkage occurring during the sponging operation. This operation is designed to remove residual relaxation shrinkage from the fabric prior to cutting and thereby minimize fabric shrinkage after the garment is made. The latter could result in either a shrinkage of the garment and/or undesirable appearance where the fabric is stitched to a component which does not shrink to the same extent.

The term press shrinkage is used to denote the shrinkage of a woolen or worsted fabric during pressing. The most common press in use is the Hoffmann press. This device is designed to exert pressure on the fabric while steam is blown against it. In commercial practice this is followed by a vacuum cycle in which a vacuum is exerted on the fabric through the bottom portion of the press. Press shrinkage has been investigated by Davidson (35) and Sustmann (104). The latter reached the following conclusions concerning the shrinkage of selected worsted fabrics shrunk on the Hoffmann Press.

> "The release of latent tensions in the fabric on the Hoffmann Press is mainly brought about by the natural moisture in the fabric . . . and only to a minimal extent by additional moisture present in the steam. Shrinkage on the Hoffmann Press decreases as the vacuum time is prolonged, but it increases with increasing duration of the steam-treatment (up to a certain point). There are undoubtedly variations in the shrinkage effect in different types of fabrics, depending on the combination of steaming and vacuum conditions. The cause of such variations may be readily traced in the case of the worsted fabrics tested in this investigation. For equal vacuum times the change in the fabric dimensions increases with longer steaming times.
>
> Detailed further investigations of the dimensional changes carried out on the same fabrics, using a hot iron press of the Spuhr-type made by Frank Weinheim and Schopper, showed that all the shrinkage results obtained by this method under the various experimental conditions were higher than those obtained on the Hoffmann Press."

Baird (7) has described two methods of test for relaxation shrinkage. These are designated as "Immersion in Water" and "Use of Steam (Hoffmann) Press." The manipulations involving the press, taken directly from the publication, are:

> *Hoffmann Pressing*. Place the samples on the bed of the press and steam for 5 seconds without the head locked down. If there is contact between the head of the press and the sample it should be sufficiently light to permit the sample to move. Then allow 10 seconds baking with the head locked, using only a light pressure. Finally, give 5 seconds vacuum again with the head unlocked as for steaming so that contraction of the sample may take place freely. This procedure should be carried out at least three times with 10 minute intervals between pressings to allow cooling and partial reconditioning of the fabric.

Tailorability

Hygral shrinkage has been investigated by Baird (9), Cednas (29) and von Bergen and Clutz (115). The latter have demonstrated that these phenomena are largely responsible for the unattractiveness of certain garments at high humidities, the unattractiveness being primarily the puckering of the fabric.

The reason for the puckering lies in the expansion and contraction of the fabric with changing humidity. Figure 26 shows the movement of three fabrics: tropical worsted (8½ oz); woolen flannel (12 oz); and worsted gabardine (13 oz)—during 3 humidity cycles covering a range of from 20 to 90% humidity.

The graph shows that the possible maximum movement in the case of gabardine is 3½%. Using a pair of 50 in. long pants as an example shows a leeway in which they can shrink or expand up to 2 inches.

Figure 27*a* and *b* show dramatically the effect that a change in relative humidity from 20 to 90% can have on a man's gabardine jacket. This change in appearance is not necessarily due solely to residual relaxation shrinkage in the fabric as delivered to the garment cutter or to hygral shrinkage. In some cases, due to errors in cutting or stitching garments, garments may be produced which are not completely satisfactory with respect to pucker. In these cases the press room supervisor can instruct the press operator to tension the garment during pressing. This removes the pucker temporarily but, in fact, induces strains in the fabric, i.e., more relaxation shrinkage.

Felting and relaxation shrinkage

From the production standpoint the two types of shrinkage of primary importance are relaxation shrinkage and felting shrinkage. Since about 1940 there has been increasing interest in the development of "minimum care" fabrics including washable woolen and worsted fabrics. In testing for dimensional stability the factors relaxation shrinkage and felting shrinkage must be measured successively if meaningful results are to be obtained. AATCC method 99-1960 T describes a method for the evaluation of both relaxation and felting shrinkage for both knit and woven goods. This method is restricted to use in the testing of materials containing at least 50% wool and that the wool contained therein has been treated to render it shrink resistant. The relaxation portion of this test method consists of conditioning, marking and measuring certain dimensions on each test specimen, soaking the test specimen

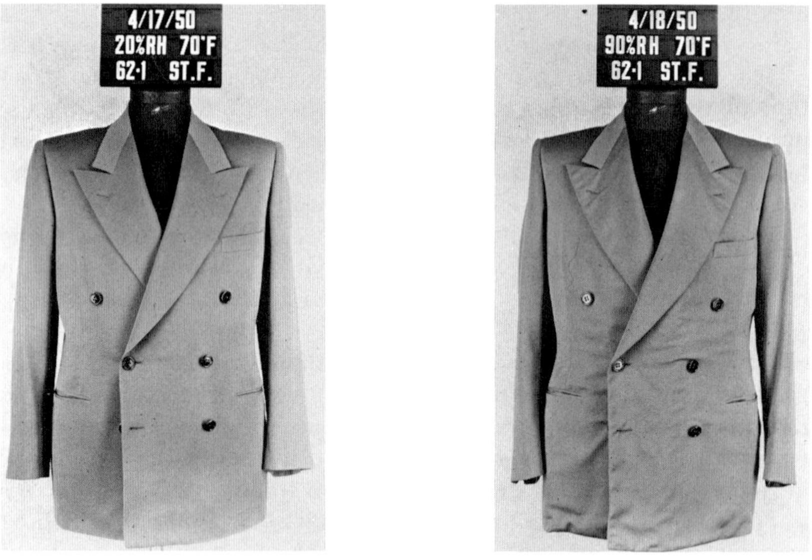

Fig. 27. Man's Jacket (*a*) at 20% RH and (*b*) 90% RH

in an alkaline solution for a stated minimum time followed by extracting, drying, reconditioning and remeasuring the dimensions measured earlier.

$$\%\ \text{relaxation shrinkage} = \frac{\text{original measurement} - \text{relaxed measurement}}{\text{original measurement}} \times 100$$

The laundering portion of this test method includes a washing cycle followed by two rinsing cycles.

A very rigorous test for shrinkage in laundering is that contained in CCCT 191b method 5554. This method is an accelerated test designed to evaluate military service fabrics. Methods using strongly alkaline solutions suppress felting shrinkage of fabrics treated by certain methods and in this way are not realistic when compared to domestic washing in this country. A test method using an anionic detergent at a pH not exceeding 8 and a temperature in the range of 100–105°F is consistent with the type of laundering these fabrics would receive in domestic washing. Since the trend in home washing in the United States is the incorporation of automatic dryers, it is also desirable to incorporate a tumble drying in the method of test.

Feldman and McPhee (41) have studied the relationship between the construction of fabric, type of shrink-resistant treatment, and the shrinkage during machine washing and tumble drying.

Krasney et al. (68, 69, 70, 71) has made a comprehensive study of minimum care fabrics, and points out that in addition to the criteria of shrinkage, the fabrics marketed as a minimum care should also be evaluated for two other characteristics. These are both related to appearance. A garment made from a given fabric may show excellent resistance to shrinkage and from this standpoint be acceptable as a "wash and wear" item. However, the appearance of the garment with respect to surface modification may be such that the garment owner would not wear it unless it was pressed. This surface modification may take two forms. These are "mussiness" and "fuzziness." For evaluating "mussiness" Krasny has extended the ranking system appearing in AATCC Tentative Test Method 88-1958 by including an additional category for materials better than classification No. 5. This extended ranking system is used to evaluate the fabric's propensity to develop wrinkles *during laundering*. The latter should not be confused with the testing of fabric propensity to wrinkle during wear.

For evaluating "fuzziness," Krasny used both a compressometer and a visual ranking. The correlation between these two methods is relatively poor over a broad range of fabrics. However, these workers found that for any one fabric pretreated in several ways, a good correlation between the two methods of evaluation was obtained.

The evaluation of a fabric for dimensional stability, mussiness or fuzziness should be a multistage test. Depending on the type and intensity of treatment, the fabric may show one of the following two types of performances or, more likely, some intermediate type of performance. The two extremes are:

1. A significant change in dimensions or surface characteristics during the first stage and essentially no further changes during successive stages, or

2. A very slight change in either dimensions or surface characteristics during the initial stage but a cumulative increase in shrinkage or cumulative decrease in surface appeal as a result of successive stages.

Illustrative of the latter are some of the studies made by the National Institute of Dry Cleaning (21, 22, 23). One such study describes the effect of incorporating water in the solvent or the introduction of water in the solvent as a result of high relative humidity on the formation of surface fuzz on the back of wool gabardine cleansed with this solvent. A similar study resulted in the primary recommendation that, "all soft wools and wool gabardines be cleaned in a separate load and add no water to the load." This same study showed that the shrinkage of wool gabardine cleaned in a system wherein the solvent relative humidity was 75% was progressive with successive cleanings and that within limits a portion of this shrinkage was recoverable by steaming and stretching. Another phase of this study indicated that for a soft sport coat type of fabric, three dry cleanings in a solvent having a relative humidity of 75% resulted in almost twice as much shrinkage (about 8%) as that obtained by the same number of dry cleanings using a solvent having a relative humidity of 65%. In the latter case, the shrinkage was about 4%. In both cases the length recovered by stretching was about 2%.

In considering any washing or dry cleaning test it would appear prima facie that in addition to dimensional changes the test material should also be evaluated for color loss, or color transfer. Even though the fabric shows acceptable dimensional stability, an alteration in shade may make the garment unsuitable for aesthetic reasons.

The manner of marking and measuring the test specimens for the various tests differs somewhat depending on the nature of the material undergoing test. All materials should be conditioned before both the initial and final measurements to minimize the effect of hygral instability.

The results of a shrinkage test on fabric may be expressed as a percentage of the original length in each direction or the shrinkage may be expressed as a per cent reduction in area. If the marked dimensions of the piece were 10 in. in both the warp and filling direction and the dimensions after the test were 9.5 in the warp direction and 9.8 in the filling direction, the shrinkage results would be expressed as 5% shrinkage in the warp and 2% shrinkage in the filling. The area shrinkage would be calculated as follows:

$$\frac{(10 \times 10) - (9.8 \times 9.5)}{10 \times 10} \times 100 = 6.9\%$$

When the shrinkage values for warp and filling are relatively low as in the above example, the sum of these values is very close to the area shrinkage.

In the evaluation of fabric for dimensional stability one must consider the magnitude of the change in dimension in relation to the projected end use of the fabric, bearing in mind that the ultimate consumer is concerned only with appearances and fit and not with the distinction between relaxation and felting shrinkage. For example, if a fabric intended for shirts has a combined relaxation and felting shrinkage of 2%, a size 15 collar would shrink 0.3 in. In the same manner a woman's skirt having a 25-in. waist would be reduced ¾ in. if the combined relaxation and felting shrinkage in this dimension of the fabric was 3%. The vast majority of woolen and worsted apparel fabrics made today are not treated to impart shrink resistance. These fabrics are intended to be cleaned by so called dry cleaning methods. This being the case, the primary shrinkage factor for these fabrics is the relaxation shrinkage.

However, the incorporation of small amounts of water in the dry cleaning mixture can result in the occurrence of some felting shrinkage.

Crease recovery and wrinkle resistance

The use of the terms wrinkle and crease frequently lead to misunderstanding. A logical distinction which is meaningful in the trade is the fact that in most cases a crease is intentionally inserted and expected to be reasonably durable, while a wrinkle is usually not inserted intentionally and is undesirable. ASTM D 1295-67 describes a method for measuring wrinkle recovery. The method consists of two parts: the specimen is folded and compressed under known time and load conditions, and then the amount of recovery from creasing is measured after a stated time. This method primarily measures the recovery.

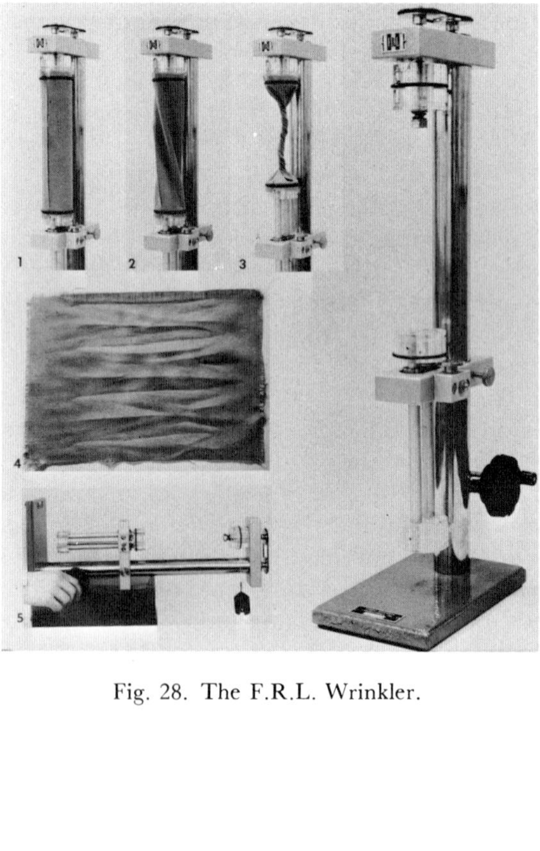

Fig. 28. The F.R.L. Wrinkler.

Another method is that developed by Celanese Corporation of America. In the latter method the relatively large samples are alternately squeezed and released in a manner simulating a hand squeeze. After some time interval the fabrics are removed and evaluated subjectively.

Abbott (1), working under the sponsorship of IWS, has developed a device known as the FRL Wrinkler. In use, a fabric sample is formed into a tube, and the tube twisted under a defined torque. The test specimen may be held under this force for a predetermined period of time following which the material is removed and evaluated subjectively. The apparatus and its operation are illustrated in Fig. 28. In the absence of chemical finishes, the extent to which fabrics wrinkle will depend on the fiber content of the fabric and the susceptibility of these to moisture. The latter is discussed by Farnsworth and Lindbert (40).

A device for the evaluation of creases in trousers has been developed by Landwehr (76).

A multiple-pleat test has been developed by Katz (62) for measuring the change in wrinkling properties of light to medium-weight wool fabrics following various finishing and chemical treatments. Accordion pleats are inserted in a fabric sample (6 in. $\times$ 3 in.) by two meshing sets of 30° wedges under uniform load. After holding under the required conditions, the sample is removed and the average pleat angle is measured optically at three intervals over a period of one day.

The complete wrinkling behavior is given by two terms, the percentage wrinkle recovery WR at a specified time (usually 15 min) and the recovery rate RR (WR change per unit log time).

A sequence of short conditioning and pleating times at 85% RH and 30°C, with recovery at 65% RH and 21°C, is used to develop a large wrinkling effect. The method requires fewer samples than conventional 180°-fold methods and is more versatile under the specified testing conditions. Results show that the change in wrinkling properties during storage or after steam-pressing must not be overlooked in interpreting the results.

Lako and Veer (72) developed a method and an instrument known as the WrinkleOmeter (Fig. 29), which created great interest because of its ease of operation and apparently good reproduction of actual wrinkling performance of textile fabrics. It has been adopted as a test method by the American Association of Textile Chemists and Colorists under Test Method AATCC 128-1968, Wrinkle Recovery of Fabrics: Appearance Method. The instrument, now known as the AATCC Wrinkle Tester, is obtainable through the headquarters of AATCC.

Thorsen, in searching for a method to record and evaluate fabric wrinkedness mathematically, developed a turntable wrinkle recorder and evaluator. The figure 30 shows the principle. The test fabric is placed on a precision turntable rotated at ½ rpm. A light counterbalanced arm with a glass tip rides on the fabric at 4 in. from the center and is attached to a microtorque, high-resolution potentiometer. As the tip follows the fabric contour, its motion produces in the potentiometer a voltage proportional to fabric height. This voltage, recorded against time, constitutes a record of

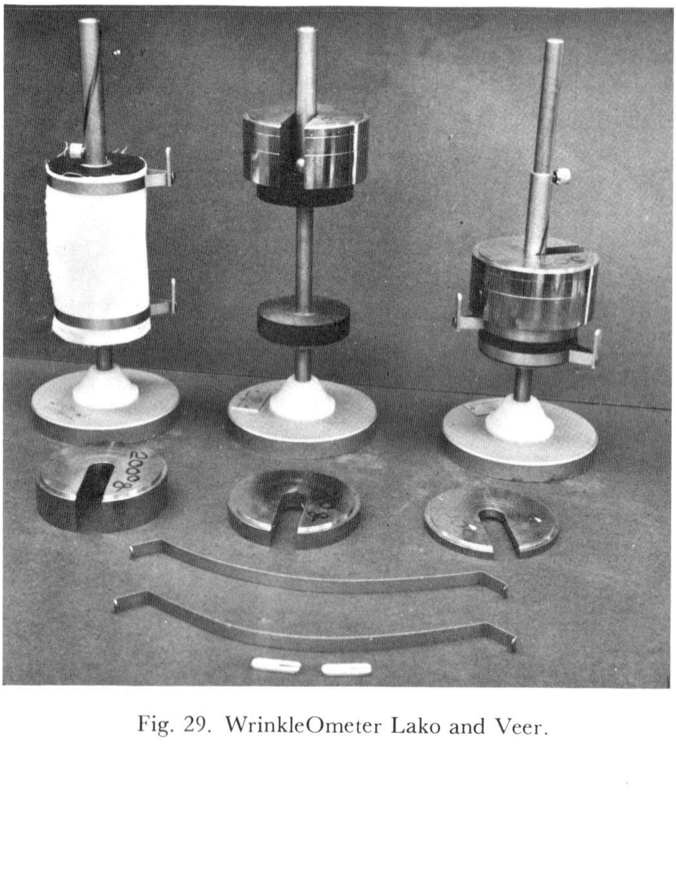

Fig. 29. WrinkleOmeter Lako and Veer.

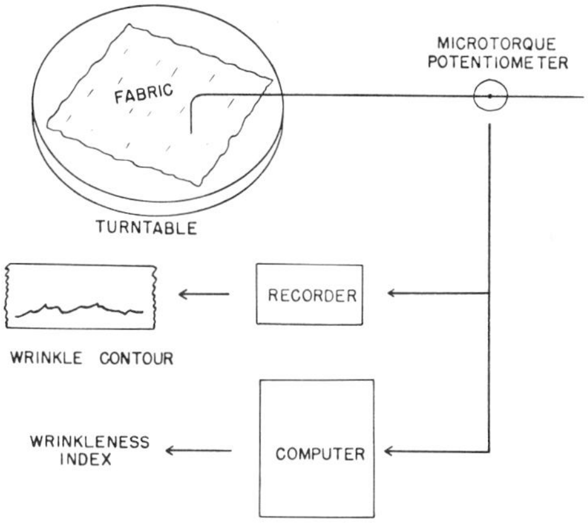

Fig. 30. Principle of fabric wrinkle dress evaluation. Fabric height is sensed mechanically, amplified electrically, recorded and evaluated by an electric analog computer.

the wrinkledness of the fabric. To obtain an index of wrinkledness, the voltage is fed into an analog computer.

The system used for wrinkle measurement is readily adapted for determining the sharpness of a crease. A platform moving at $1\frac{1}{2}$ in. min in a straight line is used instead of a turntable, with the crease laid at right angles to the direction of travel. A record is obtained of the crease contour. The computer can be programmed to give a value for the crease sharpness if a suitable formula can be provided. The three measurements used to define sharpness of crease are (1) the crease height, H; (2) radius of curvature, r, estimated by a circle closely approximating the contour at the apex; and (3) θ, the supplement of the angle formed by extending tangents to the crease contour at the points where it departs from the matching circle.

Twenty-four fabrics were experimentally creased with ethanolamine under conditions giving a wide range of sharpness. These were rated by a panel, and crease contours recorded instrumentally were evaluated in terms of the three parameters H, r, and θ. Crease sharpness indexes $H\theta/r$ and θ/r^2 tended to agree with the panel's judgment.

A similar method for measuring wrinkles has been developed at the Institute for Fibres and Forest Products Research, Jerusalem.

The method is based on the measurement of three characteristic parameters of the contours of the wrinkled surface: The height, the slope, and the frequency. These parameters are obtained as statistical means with the aid of the Sivim Wrinklemeter.

Modified Fiber

Various stages of treatment may modify the properties of wool. While some of these are designed to enhance certain characteristics, others are known to modify wool adversely. The various tests listed by Golub in Vol. 1 are in general applicable to wool fiber in all mechanical states of preparation from clean fiber to finished cloth. The test methods presented here serve to supplement the material given earlier.

The urea-bisulfite test developed by Lees and Elsworth has been standardized and adopted by the IWTO (G-12). The solubility of wool in this reagent is usually decreased by treatments such as exposure to neutral or alkaline conditions, steaming, dry heat, or treatment with cross linking agents. Oxidative or acid treatments usually increase the solubility of wool in this reagent.

A substantial portion of both past and current investigations related to the modification of wool has been concerned with "reduction type" treatment. Of substantial interest is the Siroset process which is described elsewhere in this volume. A method for the detection of prior treatment by this process has been developed by Gerthsent and Zahn (48). The method consists of spraying the fabric in question with a 0.5% aqueous solution of sodium nitroprusside and placing the test material so sprayed in an atmosphere of ammonia. The rapid development of a red-violet color on the material undergoing test is taken to be indicative of the presence of thiol. If it can be assumed that the material has not been exposed to other reagents containing or forming —SH, a positive reaction may be interpreted as indicative of material having been subjected to a Siroset type treatment. Fabrics treated by this process and subsequently extracted with water or methanol do not give this reaction. However, the extract will yield a positive reaction. Thus a negative reaction to the qualitative test made on the fabric does not necessarily indicated that the fabric has not been treated, since washing or dry cleaning would quite likely remove the —SH containing material. A paper chromatography of the total hydrolyzate of untreated wool will yield a spot characteristic of cysteine. A similar chromatograph for wool treated with thioglycollic acid will yield two additional spots, one for mixed disulfides and one for dithioglycollic acid.

A method for the analysis of wool for thio-groups has been developed by Zahn, Gerthsent, and Beck (128). The method uses a hydrolysis of relatively short duration (2–4) hr). Guanidinium chloride in an alkaline solution is added to an aliquot of the hydrolizate which is then titrated with phenylmercurihydioxide using sodium nitroprussite as an indicator. The authors of these papers give detailed description of the method of preparation of the reagents and have studied the errors in sampling and analysis. Zahn suggests that an interpretation of the combined results obtained by this method and that of Elsworth and Phillips may be used to prove the treatment of wool by sodium bisulfite or sodium sulfite. The latter method is for the determination of SO_2 content. Since sulfite treated wool contains more cystein groups than wool not so treated, the titrimetric method mentioned above will reveal this increase. Sulfite treated wool will also yield a higher sulfate content than wools not so treated when hydrolized by the method of Elsworth and Phillips. Zahn (128) further suggests that the interpretation of the results obtained by the latter method should be undertaken with caution since wools which have been carbonized will also yield higher values of sulfate content.

For the determination of total sulfur, the method developed by Mease (81) has been widely used. The principal steps in this method are: The combustion of the test material in an atmosphere of oxygen under pressure, the absorption of the oxides of sulfur by ammonium carbonate solution, the further oxidation to SO_4 by bromine and the precipitation of SO_4 as barium sulfate. This method requires the use of a Parr or similar bomb.

For laboratories not equipped with this type apparatus, the method of Meyers (82) provides a convenient means for the determination. This method is essentially a wet oxidation using a mixture of nitric acid and perchloric acid and a catalyst of potassium dichromate followed by the precipitation of barium sulfate.

Carpet Testing

The relatively unusual end use requirements of pile floor coverings as compared to apparel and other house furnishings requires a somewhat different selection of test methods. Anderson and Clegg have utilized a group of tests which show good correlation with end use.

Flammability

A fabric having a high rate of burning if used in certain apparel items may be a potential hazard to the wearer. For this reason the U.S. Government has established requirements for the flammability of clothing textiles. These are given in the Federal Trade Commission Rules and Regulations under the Flammable Fabrics Act, dated July 1954 and amended and expanded in December 1967. Certain states have established requirements which are more rigorous than the national standards. The details of the method of testing are given in ASTM D 1230-61.

Products of 100% wool do not burn at a rate which would approach the statutory limits. Sayers (97) discusses the interrelationship of fiber, fabric, and garment flammability.

Colorfastness

Colorfastness to light

Two basic methods are available for the assessment of colorfastness to light. These are exposure to daylight and exposure to some artificial light source.

The only method currently recommended by ISO is the daylight method. This is described in ISO Recommendation 105 entitled, Tests for Colorfastness of Textiles, Part II of which is concerned with Colorfastness to Light.

The AATCC Test Methods make a distinction between a sunlight test and a daylight test. These are given in AATCC 16B-1964T and 16C-1964T. In the sunlight test the samples are exposed only between the hours of 9 A.M. and 3 P.M. on sunny days between April 1 and October 1 in the Northern Hemisphere, or October 1 and April 1 in the Southern Hemisphere, while in the daylight test the test specimens are exposed continuously 24 hours a day.

For the latter there is no restriction on the time of the year the test specimens may be exposed. In either case the test specimens are protected from inclement weather by glass. A series of colorfastness standards are exposed simultaneously. This series consists of seven standards. They are identified as L2 (poor lightfastness) through L8 (outstanding lightfastness).

Artificial light sources

The daylight or sunlight test methods are limited in that inclement weather or seasonal occurrences will prolong testing. For this reason the use of artificial light has become common.

The instrument most widely used in the United States is the FadeOmeter. The details for use of this device are given in AATCC Method 16A-1960. In order to insure adequate calibration of this unit the National Bureau of Standards provides a set of calibration papers for use in calibrating this instrument.

In testing, the fabrics are exposed to the light source in the instrument usually for time periods which are multiples of 20 hr. The test specimens are examined for alteration in shade following each 20 hr exposure.

Relatively recently a new type illuminating source has been used. This source consists of Xenon gas in a quartz tube. At high current densities the spectral distribution of the radiation of this lamp is much closer to sunlight than any of the carbon arc lamps. Instruments incorporating this source are widely accepted in Europe.

Colorfastness to perspiration

This term is used to denote a color alteration of the material under test and/or a transfer of color from the material under test to another fabric when these are subjected to a set of test conditions which simulate exposure to natural perspiration.

The test method commonly used is described in AATCC Method 15-1962.

Colorfastness to crocking

The term crocking is used to denote the transfer of color by rubbing. The details of the method are given in AATCC Method

8-1961 which is the same as ISO recommendation 105-1959 No. 18. The test consists of two parts, a dry test and a wet test.

Extractable Matter

In addition to the determination of the naturally occurring fatty materials covered in Chapter 10, it is frequently desirable to measure the amount of "oil" present on a material in process or on a finished item. Many methods are available. All of these methods depend on the solubility of the "oil" in a selected solvent. Historically, the technique used was the extraction of a weighed sample with a solvent, usually diethyl ether, followed by the recovery and weighing of the material removed. For control purposes methylene chloride is preferred. Successive extraction first with ether and then with alcohol has been used as an indication of the amount of "fat" and "soap" respectively. Williams (121) has demonstrated that wools cleaned using soap and soda ash, solvent, or nonionic detergents either with or without an alkaline builder all contain about the same amount of alcohol extractables. This substantiates the work of Haller (51), Green et al. (50) and Veldsman and Palmer (111). The amount of alcohol and ether extractables found may provide a very reliable indication of the "cleanliness" of a sample but these values should not be referred to as oil and soap.

The classical method involves the use of a Soxhlet apparatus. A description of the apparatus and its method of use is given by Skinkle (135). Methods using this equipment are described in ASTM D 629-59T, D-2257-64T and D 1574-66 and Specifications of test methods approved by IWTO.

Other methods are available. These are the ACP method and the WIRA method. The WIRA method is a refinement of that proposed by Dickinson. Both of these are described by Golub in Chapter 10, Vol. 1. It must be recognized that these methods were developed for use on loose fibrous material and that their application to materials in other forms is likely to lead to low results. Wolf and Rankl (123) have developed a modification of the original Dickinson method. This modification includes the use of a compound lever system to increase the compression of the test specimen and thereby minimize channeling. Wolf (121) indicated the load used is sufficient to prevent channeling during the testing of some women's wear type fabrics.

Mease (80) has shown that alcohol is not completely removed from wool by drying at 105–110°C. In order to effect complete removal of alcohol, the wool must be rinsed in water prior to drying. It is quite likely that other strongly polar solvent will also show this phenomenon. Diethyl ether, stoddard solvent, and certain chlorinated solvents do not show this phenomenon. The computation used can either express the extract as a percentage of the weight of the dried test specimen adjusted to a commercial regain, or as a percentage of the weight of the dried test specimen adjusted to a commercial regain plus the weight of the extract.

Because a large number of proprietary products contain components which are either volatile under the conditions used for drying fibrous materials or whose solubility may be affected by heating, the test specimens should not be dried prior to extraction unless investigative work has demonstrated that the effect of heating prior to the test has a negligible effect on the test results.

Finishing Agents

The identification of various finishing materials applied to woolen and worsted fabrics is not a simple matter. In relatively recent years the number and type of finishing compounds on the market is enormous. Each of these is said to enhance certain properties of a fabric. Many of these proprietary materials are essentially identical in composition and the identification is confined to the type compound present. The identification of these materials is based on first, solubility and after classification by this means, type reactions. The most current scheme of identification is that given in AATCC 94-1965 T. A scheme also based on solubility but including additional tests has been developed by Krammes and Maresh (65).

Quantitative separation of fiber mixtures

The more commonly encountered blends of animal hair fibers with other fibers are typified by binary blends of wool with cotton, rayon, nylon, polyester, or acrylic fibers.

In some cases tertiary blends may be encountered. Elsworth (38) discusses some of the problems of assessing fiber composition. Vogler (112) presents a scheme for the systematic identification and separation of fibers. In general, the methods in most common use are "wet analysis" methods as opposed to the microscopic examination and counting of the fibers. However, in those cases where mixtures of specialty hair fiber with wool are encountered, microscopic examination and counting of the various fibers is at this time the only method available. Statlow et al. (94) has described special microscopic tests for the distinction between various animal hair fibers. Golub has developed techniques for assaying animal hair fiber blends.

The chemical methods consist of dissolving or dispersing one of the fibers in a reagent which has essentially no effect on the other component. The reagents and methods of use for the determination of the composition of selected binary mixtures is taken directly from AATCC Method 20A-1959. This method refers to the trademarked fibers by name. It is quite likely that the methods are equally applicable to mixtures of wool and other fibers of the same generic type as those indicated. However, before using these methods on other fibers it is advisable to verify the applicability thereof by the analysis of blends of known composition.

For the result of any quantitative separation to be meaningful the material must be freed from extraneous materials prior to the initial drying. After drying to constant weight at 105–110°C and weighing, the test specimen is treated by the appropriate procedure. These are:

Wool/Dacron

"Immerse a specimen weighing approximately 1 g in 100 ml of the NaOCl solution. Stir the specimen in this solution for 20 minutes and then filter through a tared fritted glass crucible. Wash thoroughly with water and remove excess water by suction. Dry in an oven at 105 to 110°C to constant weight." The NaOCl solution mentioned in this method must contain 5.0% of available chlorine. The residue is Dacron.

Wool/cotton or wool/rayon

"Immerse a specimen weighing approximately 0.5 g (Note 8) in 50 ml. of H_2SO_4 (59.5 per cent) in the 125-ml stoppered flask and shake vigorously for 1 minute. Let stand for 15 minutes at a room temperature of 15 to 25°C. (Note 9). Shake again and leave for another 15 minutes, shake for a third time and then filter the mixture through a weighed fritted glass crucible. Wash into the crucible any residue left in the flask, using three 10-ml aliquots of H_2SO_4 (59.5 per cent). Apply suction to drain excess liquor from the fiber residue after the addition of each aliquot. Wash the residue in the crucible with 50 ml of H_2SO_4 (1:19), then with water until the filtrate is neutral to litmus. Disconnect the suction and add to the crucible about 40 ml of NH_4OH, allowing the fiber residue to soak for 10 minutes before applying suction to drain it. Wash the residue with about 150 ml of water, allowing it to soak in the water for about 15 minutes. After the final washing, suck the fiber residue free from the rinse water and dry in an oven at 105 to 110°C. to constant weight." The residue is wool.

Wool/Acrilan

"Immerse a specimen weighing approximately 1 g in 100 ml of boiling NaOH solution and boil for 10 minutes. Stir the solution frequently during this period and then filter through a tared fritted glass filtering crucible using suction to remove excess liquid. Wash the fiber residue twice with 100-ml portions of NaOH solution, three times with 150-ml portions of water, twice with 100-ml portions of acetic acid solution, and three times with 150-ml portions of water. Drain with the aid of suction, and dry the residue to constant weight in an oven at 105 to 110°C." The residue is Acrilan.

Wool/nylon

"Immerse a specimen weighing approximately 1 g in 100 ml of HCl (sp gr 1.139) in a 500-ml stoppered flask, shake vigorously, and let stand for 5 minutes at a room temperature of 15 to 25°C. Shake again and leave for a further 15 minutes. Shake for a third time and filter the mixture through a weighed fritted glass crucible. Wash into the crucible any residue left in the flask using a little more HCl (sp gr 1.139). Apply suction to drain the excess liquor from the fiber residue. Wash the residue in the crucible with 40 ml of HCl (2:3) and then with water until the filtrate is neutral to litmus. Disconnect the suction and add to the crucible about 80 ml of NH_4OH allowing the fiber residue to soak for 10 minutes before applying suction to drain it. Wash the residue with about 250 ml of water, allowing it to soak in the water for about 15 minutes. After the final washing, suck the fiber residue reasonably free from the rinse water and dry in an oven at 105 to 110°C to constant weight." The residue is wool. The method of computation depends on whether the residue obtained is wool or some other fiber. The data have to be adjusted to the regains of the fibers present.

The commercial regains for some of the fibers are: rayon, 11%; cotton, 7%; nylon, 4.5%; acrilan, 1.5%; Dacron, 0.5%; wool, 13.6%.

Moth Resistance

It is well known that the animal hair fibers are susceptible to attack by the clothes moth larva and other "insects." A variety of materials have been developed which when applied to the fibers are said to minimize the susceptibility of these fibers to attack from these sources. In essence the method of testing the efficacy of these treatments consists of exposure of the treated material to specified larva, and after a stipulated time examining the material. The methods in common use in the United States are given in ASTM D 582. D1116-65 and AATCC 24-63 were as international; IWTO-9-62 and IWTO-14-64 are followed. Cole and Whitfield have compared five methods and concluded that the principal differences between the methods are primarily in the manner of interpretation of the findings.

Optical Methods

During the examination of textile materials either in process or in the finished fabric, the unaided eye is frequently inadequate. Magnification of even a relatively low order is frequently sufficient to reveal detail which is pertinent to the problem at hand. Microscopic examination of textile materials can provide a method for the detection of faults or the causal factors of some difficulties much better than can other methods. For the examination of large areas the relatively low power stereoscopic microscope may be used to advantage. The magnification range of this type instrument is from about 5 to about 30. For greater magnification, standard microscopes accepting a variety of lenses are necessary. The most useful magnifications for textile work are in the range of 100–500. The details of the techniques used for the examination of textile materials are given by Schwarz (101), Heyn (56), Preston (87) and Bigler (15). An ideal reference for the microscopist is the work of Chamot and Mason (30). Of interest are the techniques published by Berry (14) and Krammes and Marsh (67) for the preparation of replicas of fabric surfaces. These techniques facilitate the examination of fabric defects.

Millson and Turl (83) have developed a method for the observation and photographing of single fibers through the microscope while the dyeing process is in progress. This method is suitable for dyeing at temperatures up to about 210°F. Millson (84) has refined the system to permit observation and photographing during dyeing procedures up to 325°F. Rollins et al. (94) have described useful techniques for the investigation of changes during the finishing process. An excellent set of photomicrographs as well as a brief description of the methods is given in Ciba Bulletin No. 181 Sept.–Oct. 1962. Microscopic methods may be quantitative as well as qualitative. With appropriate accessories, yarn dimensions, as well as fabric dimensions, may be measured. Angular measurements are possible if the microscope is equipped with a rotatable stage or a special ocular holder.

In addition to the aid of magnification alone, other optical techniques are of use for both the microscopic and normal visual examination. These include the use of immersion liquids. Undyed textile materials immersed in a liquid having essentially the same refractive index as the fibers become practically invisible when viewed by transmitted light. This technique is extremely useful for the examination of some fabrics for the presence of contaminants. The technique may be used either on a macro or mciro scale. The transperizer developed by Wollner and discussed by Golub in Chapter 10, Volume 1, can be used for examining yarns or fabrics.

The use of illumination other than daylight or lamps used conventionally for room lighting can provide a means of detecting contaminants or the non-uniform distribution of some materials known to be present. The type of illuminant most useful is the ultraviolet lamp. Many materials fluoresce under ultraviolet light and an indication of the presence of these is easily determined. Those lamps having a broad range of wave length near the visible portion are preferred over those that produce either narrow bands or a spectral distribution relatively far from visible.

Identification and Estimation of Lambswool

The birth coat of a lamb consists of woolfibers with a pointed or tapering tip (Vol. 1, p. 166). Therefore the presence of such tips being characteristic of lambs wool can be used for its identification. The only known specification for lambs wool in garments especially in sweaters labelled "100% Lamswool" is an Australian custom regulation which stipulates that any item so labeled must show a minimum of 65% of fibers with tips. No study has been published to substantiate this rather high percentage.

Identification and Estimation of the Percentage of Pulled or Skinwool

The various methods by which pulled or skinwool are obtained from the sheep pelts are discussed in Vol. 1, p. 668. During the pulling process the root end cells are torn from their connection with the cells in the wool follicle bulb the papilla. When viewed under the microscope many of the root ends ending in a bulb, but most are strongly damaged and twisted.

Based on these differences it is possible to estimate the approximate percentage of pulled wool present in a wool blend, especially in top form by counting under the microscope the root ends of the pulled wool fibers and the cut ends of the shorn or regular fleece wool. Prestaining the top sample with Benzopurpurin (Vol. 1, p. 765) will be of considerable help as the damaged root ends will stain medium to dark red, whereas the cut ends show only a slight pink. The method of sampling the top is the same in principle as that used for measuring the fibers length making fiber draws. A properly squared end of a draw of about 150 fibers is convenient to be mounted in glycerin on a slide and covered with a square coverglass.

For examination and counting any projection microscope as used for measurement of the fiber fineness is suitable; but a magnification of 200–250 is sufficient to distinguish the different types of ends and have a large enough field for several fibers to be seen at once and enough length of each fiber to see the change over from the follicle part into the fully developed wool fiber. Data on studies made at Wira have unfortunately not been released, but they indicate that in a top blend of pulled and fleece wool which are close in diameter and length the counted numerical percentages of pulled wool come within a percent or two of the calculated numerical percentages for pulled wool content ranges from 5 to 50% by weight.

Photomicroscopy

For record or for demonstration at a later time, it is frequently desirable to have a picture. In many cases a simple sketch will suffice. However, a photographic record is most desirable. Photographic techniques have been described by Golub in Chapter 10, Vol. 1. For most black and white work a medium speed, fine grain, panchromatic film such as Eastman Kodak Plus X developed for normal contrast is suitable. No filters are usually necessary unless the contrast desired between colors is not possible with this type film. More critical conditions are required for color photography. The films available are balanced either for daylight or for artificial light of approximately 3200 K. If the illumination of the specimen is from a source other than the one for which the film is intended the color rendition may not be realistic. Special light balancing filters are available to correct the light source to the desired color temperature. Allen describes both the practical and theoretical aspects of photomicroscopy including photomicrography using phase contrast and interference techniques.

The newest instrument is the Scanning Electron Microscope known as the STEREOSCAN built by the Cambridge Instrument Co., in England described by Oatley and Smith, Wells and Reumuth.

The STEROSCAN technique represents a revolutionary step forward in the field of photomicrography. This is clearly seen in the three dimensional detail of Fig. 31*a* and *b*.

Electronic Fineness Testers

The instruments of the near future for the testing of the fineness of wool and other animal fibers are the electronic types, such as the Coulter Counter and the EFFI. The Coulter Counter has been proven by O'Connell and Martsch as an excellent instrument to determine fiber fineness distribution in Wool Top. This counter has already been described in Vol. 1, page 749. Fig. 32 shows a diagram of the instrument. The average fineness of wool tops, as determined

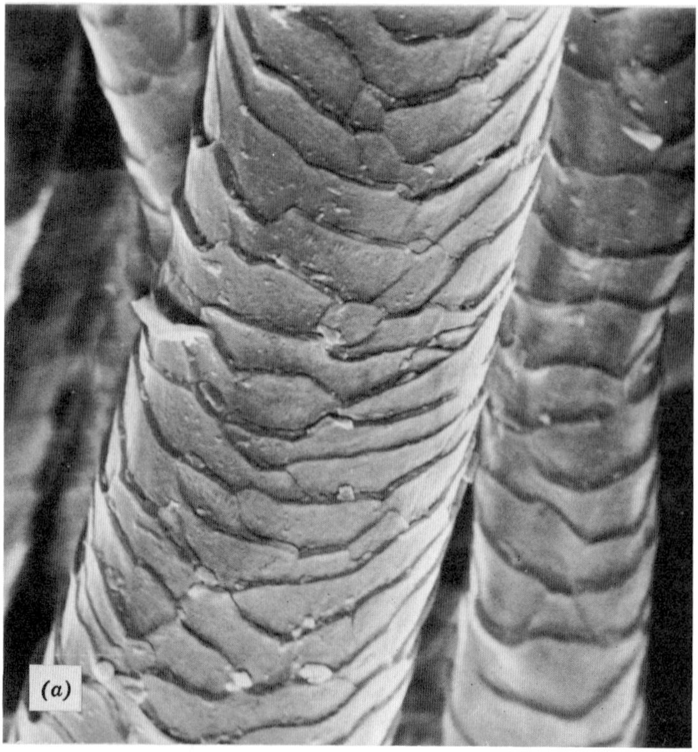

Fig. 31. Wool Fiber Steroscan. View (*a*) Courtesy of H. Reumuth.

by the Coulter Counter, can be estimated with a reproducibility that is equivalent to, or better than, that of estimates made by expert operators using the microcospe. Under a set of standard operating conditions, it is to be expected that the Coulter would produce a higher degree of reproducibility among operators or laboratories than the microscopic technique because operator judgment would not be a source of error.

The most critical points in the operation of the instrument lie in the control of the conductivity of the suspension medium and the stirring of the suspension of fiber particles. Cutting the fibers to uniform lengths was found to be of secondary importance. The removal of oils was found necessary for some tops to prevent clumping of fiber particles. The time required to analyze a wool

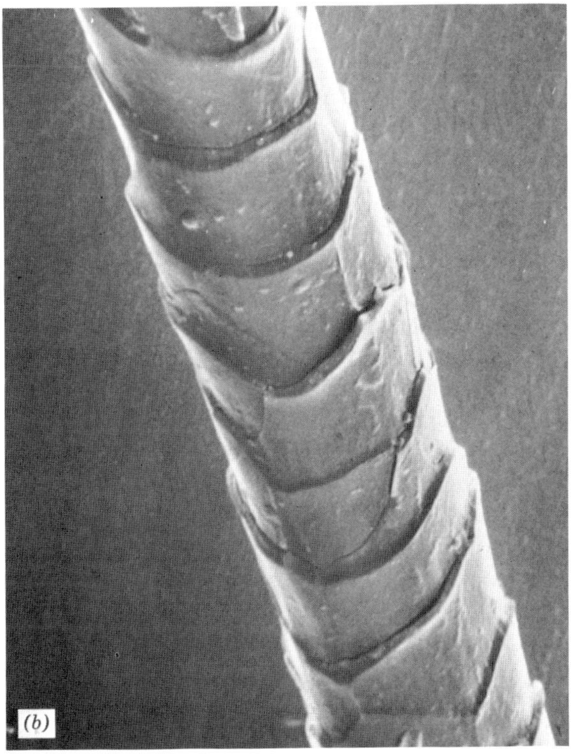

View (*b*) Courtesy of Deening Millikan Research Lab.

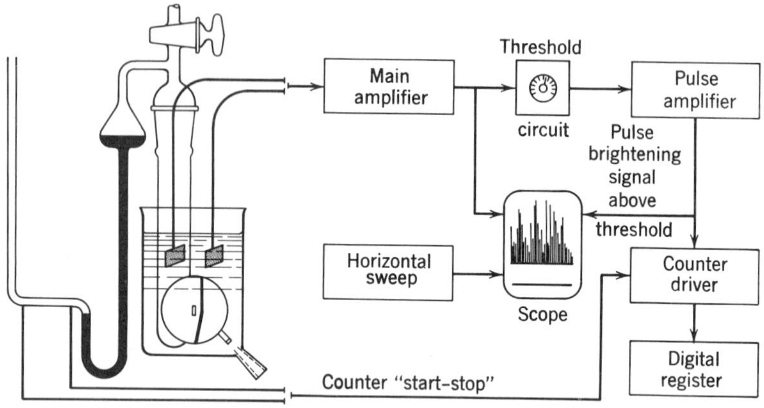

Fig. 32. Diagram of a Coulter Counter.

top using Model "A" including sample preparation, measuring and calculating is about 1 hr with about one-third of the time being used for each operation. Measuring and calculating time can be further reduced through use of more refined models of the instrument, which measure in numerous cells simultaneously and automatically analyze the resultant data.

A more refined instrument is Model "C" as seen in the set-up, Fig. 33*a* and *b*, used by Hourihan at U.S. Department of Agriculture Beltsville Laboratory. With this instrument the time for one Top analysis is approximately ½ hr. The cutting of the fibers is considered critical by Hourihan.

With a new multichannel Model "T" Coulter Counter, it is claimed possible to complete a 15 point size distribution analysis in 15 sec or less. This includes the tape-out of the data which should reduce the overall time to around 15 min. EFFI, the Electronic Fiber Fineness Indicator, is an optical, mechanical, and electronic system designed to automatically scan a microscope slide upon which a cut sample of wool has been floated in oil, such sample having been prepared in accordance with A.S.T.M. D2130-61 in preparation for the manual microprojection method of determining wool fiber fineness. Originally developed in 1956 by the Top Company in Boston it is at present under further development by Mary Hourihan and C. Terrill of the Sheep and Fur Animal

Fig. 33. Model C Coulter Counter. Courtesy of U.S.D.A.

Research Branch A.H.R.D., ARS, U.S. Department of Agriculture, Beltsville, Maryland. A total remodeling of the instrument was done by H. Neil of the Special Instrument Laboratory (Spinlab.) working under contract with the U.S. Department of Agriculture, Sheep and Fur Animal Research Branch. In the final report (in preparation for publication) the following physical description of EFFI are given:

> The system is housed in a metal desk console measuring 3 feet deep by 6 feet wide by 4 feet high with a total weight of approximately 600 pounds. It operates on 1150 AC, consuming approximately 600 watts of power. Data are displayed in electronic counters and fiber diameter distribution data are accumulated in a 100-channel multichannel analyzer. The prepared slide is dropped into a microscope stage which is equipped with high voltage electrodes which, by electrostatic forces, cause the cut fiber to align parallel to the electrostatic field. A synchronous motor programmed to scan back and forth across the slide while indexing toward the opposite side causes the fibers to move at a uniform rate perpendicular to their axes, so that their images may be picked up by photomultiplier tubes and converted

into electric pulses whose duration is proportional to the fiber diameter. The trick of simplification of this system is the use of polarized light. By arranging one polarizing filter above the stage and a second at right angles to the first below the stage, all light except that rotated by the birefringent wool fiber is filtered out, resulting in the wool fiber causing a "window" image to cross over a viewing slit.

Since total fiber alignment cannot be assured, two photomultiplier tubes are supplied and the slit image is split, the top one-half going to one phototube and the bottom half going to a second. A misalignment will cause the pulse from one phototube to lead the other so that a coincidence circuit can be used to discriminate against misaligned fiber.

A preset electronic counter is provided for totalizing fiber counted and the preset coincidence allows automatic termination of the program when a pre-determined number of fibers have been counted. A second electronic counter sums up the width of each fiber and is therefore called an integrating counter. By dividing the total fiber by the total diameter, the average fiber diameter can be obtained.

The multichannel analyzer is so programmed that the pulse width determines which channel a given bit of information is stored in. Adjustment of the gain of the analyzer allows a pulse of width representing a 31 micron fiber, for example, to be caused to be added to the memory of channel 31 so that at the end of the measurement this data can be recalled and caused to print out on paper tape. In addition, a visual display allows the data to be plotted on the face of an oscilloscope tube and thus produces a distribution diagram.

Quality Control

The term quality control as used here denotes an overall operation. The word quality is restricted to mean a characteristic of a material which can be evaluated and not a vague connotation of superiority. The word control in this expression indicates that some action will be taken to rectify a condition found to be undesirable. The combination of these words with the term statistical results in an expression which is the basis for all modern quality control techniques. The application of statistical methods to the results obtained from testing received great impetus during World War II. Experience gained at that time and since then has firmly established the superiority of these methods over a rule of thumb or subjective guesses when evaluating test data. The basic concept of statistical quality is given by Grant (49) and the ASTM Manual on Quality Control (6). Enrick (39) presents both theory and applications in the textile industry. Beardly and Cox (19) illustrate the technique with direct application to the woolen and worsted industry. Sufficient work has been done in the field so that the excuse heard frequently that, "It won't work in our industry," is without foundation. Statistical methods are not a substitute for test data. The proper use of these methods, however, will permit the reaching of valid conclusions with a minimum of testing.

Quality control is a managerial function and is related to such other managerial functions as time study, cost accounting, etc. To illustrate this point it might be well to consider the overall aspects of quality control and the intent of a quality control program. The primary purpose of a quality control program is to provide management with information, the use of which will increase profits. This information may be considered in two approaches. One approach is the insurance that the material manufactured will meet the requirements of the customer. This approach could consist of examination and testing of the final product, with the material not meeting the requirement being sold as seconds. Obviously if the incidence of seconds is high the profit position of the company is lessened. The second approach is the judicious sampling and testing of material in process with the intent of rectifying operating conditions which contribute to seconds. This can only be accomplished by a careful study of the incidence of defects or the occurrences of substandard material and relating these to an assignable cause. In an integrated mill the incidence of defects at the finish perch and the measured quality characteristics of the finished item can be analyzed to indicate the most likely places where an intensive investigation will be most fruitful. In a similar manner, the running records of the burling and mending department which are required for payroll purposes, and "loom stop" studies made by the time study group both provide information with respect to yarn quality as it affects production and therefore costs.

The methods used in an overall quality control program can be categorized as continuous instrumental methods, intermittent test methods and statistical analysis. The continuous instrumental methods include control of temperature, pH, liquid level, concentration, etc. These controls are available as blind, i.e. non-recording, non-indicating, indicating only, or indicating and recording. The cost of the latter is only slightly higher than the indicating type, but as may be inferred, these provide a record of the variations of the process variable. This record is of inestimable value when the problem of relating product variability to the process is encountered. The test methods used in a quality control program are in general those discussed earlier in this section. As indicated earlier, the test results are appraised by the appropriate statistical techniques.

In order to control the quality of an end item, the relationship between the quality characteristic of the intermediate products and the quality characteristics of the finished item must be established. The arbitrary setting of tolerances on a quality characteristic can lead to difficulties. Anyone can decide on a set of limits; however, if the limits established are not realistic with respect to the normal variability of the process, they cannot be met. The tolerance established for any quality characteristic must be realistic and be relatable to the variability of the end item. For example, in the manufacture of an all wool hard finished men's wear fabric made using 2/40s warp, a relationship was established between evenness of single yarn and both weaving efficiency and the incidence of defects in the finished goods. A maximum value of 19% was set as the upper limit for coefficient of variation of yarn unevenness of the single yarn. This number was not picked arbitrarily but was selected because yarns more uneven than this caused weaving efficiency to fall below a desired level and yarn unevenness below this level did not materially increase weaving efficiency or reduce the incidence of defects in the fabric. A broad quality control program consists of many studies. These may appear to be isolated independent investigations to the untrained. However, in the overall program each facet contributes to the whole picture. Many worsted manufacturers have established a relationship between the amount of short fiber in top and the unevenness of yarn produced and/or end down in spinning. The usual criteria is the amount of two inches. Some manufacturers have set values of not more than 20% while others require a lesser amount. An integration of the two prior examples illustrates that a quantitative relationship can be established between the staple length distribution of top and the weaving efficiency of yarns made therefrom.

Klock and Carter (63, 64, 65) have reported on the application of quality control methods in dyeing and woolen carding and have also described a method of presenting the concepts of statistical quality control to the statistically inexperienced. Lezinski (73) describes the application of quality control to visual characteristics of carpets. Richards (91, 92, 93) presents an application for the control of yarn number and twist of woolen spun yarn. Wilson (122) has compared the factors contributing to the unevenness of cotton card sliver. While this is not directly applicable to the woolen and worsted industry, it serves to illustrate the interrelationship between the various factors contributing to unevenness. This work revealed that the primary causal factors of unevenness were not those being rigidly controlled.

Two publications of value in determining the extent of sampling necessary are MIL-STD-105 and MIL-STD-414. These are U.S. Government publications. The first deals with sampling procedures for inspection by attributes and the latter with sampling procedures for inspection by variables and for percent defective.

Experience has indicated that the measurement of certain characteristics made on the raw material and intermediate products at certain points in the process will provide the most useful information.

The plans used at the Forstmann Woolen Company for quality control of production of the 18 oz olive drab U.S.Army serge and the 16 oz navy blue U.S. Navy melton are given as examples.

To show the points of control testing in the general manufacturing routines for the two cloths, the flow sheet in Table 10 is given. The test procedures for the various processes are found by reference to the key.

The control testing led to a study of the factors influencing the breaking strength of the two fabrics. It was found that the raw material is one of the primary factors. Yarns from fine domestic wools are generally weaker, due to their shorter fiber length, than Australian or Cape wool yarn of similar fineness. A change in the blend affected the strength level of the yarns and pieces as a whole.

TABLE 10

Flow Sheet of Manufacture and Testing Procedure

Worsted 18 oz. serge		Woolen 16 oz. melton	
Raw wool	1	Raw wool	1
Sorting	2	Sorting	2
Scouring		Scouring	
Combing	1,2,3	Stock dyeing	
Top dyeing	3	Carding	1
Spinning	4,5,6	Spinning	4,5,6
Weaving		Weaving	
Grey goods	4,7,8,9	Grey goods	4,7,8,9
Fulling	10	Fulling	10
Washing	4	Washing	4,12
Crabbing		Carbonizing	
Drying		Neutralizing	4
Specking	11	Decating	
Shearing		Drying	
Decating		Shearing	
Final examining		Pressing	
Finished cloth	4,7,8,9,13,14,15,16,17	Final examining	
		Finished cloth	4,7,8,9,12,13,14,15,16

	Key		A.S.T.M. Standards[a]
1	Fiber fineness	1	ASTM D 419-66, D 472-63, D 2130-61
2	Fiber length	2	ASTM D 519-66, D 1575-61
3	Tensile strength (bundle)	3	ASTM D 1294-63 T
4	Breaking strength	4	ASTM D 1578-67
5	Yarn number (count)	5	ASTM D 1907-67
6	Twist	6	ASTM D 1422-65T, D 1423-66 T
7	Ounce weight	7	ASTM D 1910-64
8	Piece length and weight	8	ASTM D 462-64
9	Picks per in.	9	ASTM D 1910-64
10	Fulling time	10	
11	Specking time	11	
12	Crocking	12	AATCC 8 1961
13	Width	13	ASTM D 1910
14	Ends per in.	14	ASTM D 1910
15	Weight and length loss	15	
16	Colorfastness to light	16	AATCC 16A-1964.
17	Extractable matter		

[a] 1967 Book of A.S.T.M. Textile Standards, Parts 24 and 25.

In the case of a 50/50 domestic–Australian wool blend, the strength was increased 5–8% over that of the 100% domestic blend.

The method chosen to dye the wool or top is another primary factor in the ultimate strength level. The length of time the wool is in contact with the dye bath and the temperature of the dye bath are the two most important factors. In this connection the metachrome dyeing process with its shorter time produced 6–8% stronger pieces than the top chrome dyeing process. Continuous indigo dyeing with its short immersion time and low temperature yielded pieces which averaged about 25% above the specified strength. In the case of the indigo-chrome combination the strength level was 10–15% above, whereas the chrome dyed wool produced pieces which were approximately 10% below specifications.

The variations introduced into the strength picture by the spinning process are of a varied nature. The yarn count directly affects not only the strength of the yarn itself but also that of the fabrics. A decrease of half a count metric in the two-ply serge yarn resulted in a 4% increase of the yarn strength.

Therefore, control over yarn count, that is, holding yarn fluctuations within small limits, will result in a reduction of the eventual variations in fabric strength. It was shown that the ply worsted yarn, by reason of the doubling twist, is about 25% stronger than twice the single yarn strength.

The yarn produced on ring spinning frames is 2% stronger than the yarn spun on mules. The amount of the difference, nevertheless, is not large enough to be of great importance. The following additional information is taken directly from the earlier edition.

"The yarn breaking strength varied directly with the amount of twist up to a certain maximum. Further insertion of twist then reduced the breaking strength. A definite end point for practical strength increase exists for each yarn size. Small differences in twist within the range that occurs in normal production have little or no influence on the strength of the yarn. It is only when large differences in twist occur that an effect on the strength will be noticed.

By the comparison of variations of worsted and wool yarn, indications of the inherent manufacturing uniformity levels and variations have been revealed. By reason of the complexity of the worsted yarn system small inequalities tend to be averaged out through the long preparatory processes before actual spinning takes place. This is not the case in the woolen system. Variations in roving off the cards is directly transmitted to the yarn as there are no intervening equalizing operations. Assignable causes of variations of this type, due to the manufacturing process, can be minimized by strict control but not completely eliminated without radical alteration of the process itself.

The foregoing comments on manufacturing operations previous to weaving indicate that the material with which the weaver has to work is by no means uniform. Since weaving is a combining operation, the yarn variations are carried over directly into pieces and form a part of the ultimate variations found in finished cloth.

To bring the factors influencing the breaking strength of Army and Navy cloth to a common ground, the coefficients of variation, that is, the relative variabilities, previously mentioned, are grouped in Table 11. The coefficients are completely different for the two spinning systems. Wool spinning has more than twice the variability of the worsted system. This generalization holds true for the piece goods. The data show that the strength variations in the yarns are directly responsible for the strength variations in the cloth. However, the worsted tends to increase in variability as manufacturing progresses, while, on the other hand, the trend in the woolen system is just the reverse. Evidently, the reaction of the product of wool spinning is toward a loss of variability, that is, a gain in uniformity.

TABLE 11

Coefficients of Variation (Percent) Throughout Manufacturing

	Worsted		Domestic blend		Woolen	
	Warp	Filling	Warp	Filling	Warp	Filling
Yarn Ring	5.5	—	3.4	—	—	—
Mule	—	4.5	—	4.0	8.6	8.7
Pieces Gray	4.4	4.0	4.4	4.0	5.8	7.0
Finished	4.3	5.2	4.5	4.4	6.0	8.1

The effect of fulling on worsted pieces is just opposite to that on woolen pieces. While the woolens gain in absolute strength, worsteds remain unchanged or become weaker, though both show a loss when breaking strength per thread is considered.

Carbonizing and neutralizing in piece form, when properly controlled, have very little effect. The effect of surface disturbing operations, such as sanding and napping, can be measured directly by the change in breaking strength. Operations designed to produce a certain touch such as decating and pressing have a complicated influence, the greatest part of which is due to moisture content of the cloth when testing is performed under mill conditions. There is a definite peak strength to be associated with moisture content of approximately 8 per cent. Higher or lower moisture content yields lower strength. The moisture level of control testing at 65 per cent relative humidity and 70°F is not that of maximum strength.

The results of control testing throughout manufacturing permitted the setting of minimum strength averages to be maintained. The figures in Table 12 give these levels. The finished piece

TABLE 12

Prediction of Strength[a]

	Worsted		Woolen	
	Warp	Filling	Warp	Filling
Yarn strength needed:				
Average	166[a]	180	87	72
Standard deviation	7.0	7.0	8.0	7.0
Gray piece strength needed:				
Average	143	130	55	52
Standard deviation	7.0	6.0	4.0	5.0
Finished piece strength needed to pass specification 100%				
Specification	120	110	58	46
Average	141	128	70	61
Standard deviation	7.0	6.0	4.0	5.0

[a] Actual production = 193 lb. According to the standard deviations given, these are 3 sigmas above the specification.

averages had to be set high enough so that the normal fluctuations would not cause any piece to fail specifications. The yarn and raw piece averages can then be adjusted to give the desired finished piece level. Greater standard deviations than those given necessitate the maintenance of higher average levels whereas smaller standard deviations permit maintenance of an average level closer to the specified minimum. From an economical point of view, this is the goal to achieve.

The use of statistical analysis in mill quality control work is the best weapon with which to overcome the rule-of-thumb methods still prevailing in many of the woolen and worsted plants. But the successful use of that weapon depends on a carefully trained staff of technologists, who must not only be able to interpret the test result obtained but also to sell the use of the result to the practical mill man.

One of the best ways of presenting data to the mill man is the use of control charts. Figure 34 shows the control charts used in the study of the production fluctuation of the U.S. Army W.A.C. cloth

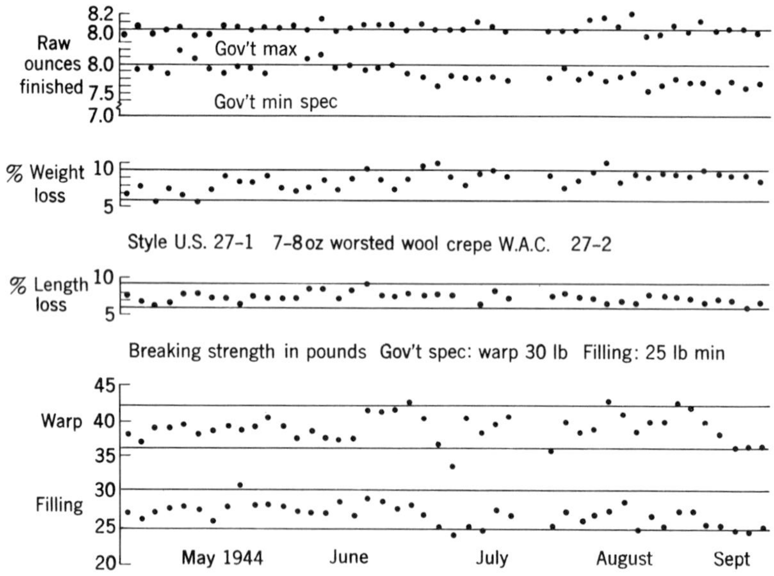

Fig. 34. Control Charts of Worsted-Wool Crepe.

in the Forstmann Woolen Company finishing plant. They illustrate much better than figures can ever do, whether a product is in control or not."

As indicated earlier, quality control is a managerial function. The integration of a program of this type with the overall operation of an organization requires not only technical knowledge and skill but also an appreciation of human values. Both supervisory personnel and production workers can either contribute to or hamper the success of a program. The recognition of people as human beings will assist to a considerable extent the rapid success of the program. In addition to the considerations given earlier, quality is also a state of mind. Each and every worker and supervisor must be constantly aware of his contribution to the quality effort. The time required to develop a successful program will depend on the type of operation. Two to four years is a reasonable estimate of the time to investigate the various aspects of the operation and institute the appropriate controls. The person in charge of the program should have freedom to investigate and be directly responsible to top management.

If adequate control is exercised over the manufacture and finishing of fabrics, and cooperation of the cutter and retailers can be obtained, promotional labeling can be used to advantage.

References

1. Abbott, Norman J., private communication.
2. Allen, Roy M., *Photo Micrography*, D. Von Nostrand, 1958.
3. Anderson, S. L., and D. G. Clegg, *Textile Inst. Ind.*, **1,** February 1963.
4. Annon. *Textile Ind.*, March 1965.
5. Annon. *Wool Sci. Rev.* No. 19, June 1961.
6. *ASTM Special Technical Publication* 15-c.
7. Baird, K., *Textile J. Australia*, February 20, 1962.
8. Baird, K., *Textile Res. J.*, **31,** July 1961.
9. Baird, K., *Textile Res. J.*, No. 11, 1961.
10. Barella, A., and A. Sust, *Textile Inst. Ind.*, **3,** July–December 1965.
11. Barella, A., P. Miro, and A. Sust, *Anale Scientifique Text. Belg.*, No. 3, 1965.
12. Basu, B. K., *J. Textile Inst.*, P 52, July 1961.
13. Beck, L., Belleli, J. Julifs, St. Jost, W. Tamschick, and Sieberichs, *Melliand Textile Ber.*, October 1965.
14. Berry, W. R., *Textile Res. J.*, **29,** November 1959.
15. Bigler, N., *Ciba Bull.*, No. 181, September–October 1962.

16. Bornet, G. M., *Survey of Short and Long Term Unevenness of Yarns*, Ontario Research Foundation, September 1961.
17. Bornet, G. M., *Textile Res. J.*, **34,** May 1964.
18. Breen, S., *Textile Inst. Ind.*, **2,** November 1964.
19. Brearley, A., and D. R. Cox, *An Outline of Statistical Methods for Use in the Textile Industry*, WIRA, Leeds, England.
20. Buckley, H. T., *Am. Dyestuff Rptr.*, **50,** 1961.
21. *Bull. Nat. Inst. Dry Cleaning*, T 198, 1948.
22. *Bull. Nat. Inst. Dry Cleaning*, T 319, 1953.
23. *Bull. Nat. Inst. Dry Cleaning*, T 323, 1954.
24. *Bull. Wool Ind. Res. Assn.*, **21,** No. 2, April 1959.
25. Burton, R. H., *Textile Inst. Ind.*, **3,** May 1965.
26. Burton, R. H., *Textile Inst. Ind.*, **3,** July 1965.
27. Burtt, J. F., and J. L. Woo, M.S. Thesis, Lowell Technical Institute, 1958.
28. Cassie, A. B. D., *J. Textile Inst.*, **37,** 556, 1946.
29. Cednas, M., *J. Textile Inst.*, T **51,** June 1961.
30. Chamot, E. M., and C. W. Mason, *Handbook of Chemical Microscopy*, John Wiley, 1958.
31. Cole, J. H., and F. G. S. Whitfield, *J. Textile Inst.*, **53,** November 1962.
32. Cooper, D. N. E., *Textile Res. J.*, **31,** 1961.
33. Coplan, M. J., and M. G. Bloch, *Textile Res. J.*, **25,** No. 11, 1955.
34. Crabtree, J. L., and A. Rae, *WIRA Wool Research*, Vol. 3.
35. Davidson, A. N., *J. Textile Inst.*, **45,** T 323, 1954.
36. Dillie, H., Mazingus, and Van Overbeke, *Bull. Trimstriel Centre Textile*, No. 64, 1964.
37. Douglas, K., *J. Textile Inst.*, **53,** March 1962.
38. Elsworth, F. F., *Textile Inst. Ind.*, March 1964.
39. Enrick, N. L., *Quality Control*, Industrial Press.
40. Farnsworth, A. J., and J. Lindbert, *J. Textile Inst.*, **31,** 687, 1961.
41. Feldman, H. D., and J. R. McPhee, *Am. Dyestuff Rptr.*, **53,** March 1964.
42. Ferguson, A. D., *Textile Inst. Ind.*
43. Foster, G. A. R., *J. Textile Inst.*, **48,** T 109, 1957.
44. Fourt and Harris, *Textile Res. J.*, **17,** 1947.
45. Fourt and Harris, *Textile Res. J.*, **23,** 763-9, 1953.
46. Fourt and Monego, *Textile Series Report*, 116 Q.M., R&EC.
47. Fuchino, K., and S. J. Veno, *Soc. Tex. Cellulose Ind.*, **12,** 1956.
48. Gerthsent, H. Zahn, *J. Soc. Dyers Colourists*, **75,** No. 12, 1959.
49. Grant, E. L., *Statistical Quality Control*, McGraw-Hill, 1964.
50. Green, T., R. P. Harker, and F. O. Howitt, *J. Textile Inst.*, **47,** T 110, 1956.
51. Haller, H. C., *Am. Dyestuff Rptr.*, **30,** 221, 1941.
52. Hamilton, J. B., and D. N. E., Cooper, *J. Textile Inst.*, **49,** T 411 and T 687, 1958.
53. Hearle, J. W. S., *Textile Res. J.*, **26,** 108, 1956.
54. Hearle, J. W. S., and P. H. Walker, *J. Textile Inst.*, **44,** P 811, 1935.
55. Henno, J., and R. Jouhet, *Bull. Inst. Textile France*, **92,** 106, 1961.
56. Heyn, A. N. J., *Fiber Microscopy*, Interscience, 1954.

57. Howell, H. G., K. W. Mieszkis, and D. Tabor, *Friction in Textiles*, Textile Book Publishers, New York, 1959.
58. Huberty, A., IWTO Technical Committee, *Proceedings*, **1,** 1947.
59. *ISO Recommendation* 139, January 1960.
60. *ISO Recommendation* 138, 1960 (E).
61. Jankowski, M., *Ciba Review*, 1963/6.
62. Klock, A. G., and C. W. Carter, *Ind. Q. C.*, **7,** 3, 1950.
63. Klock, A. G., and C. W. Carter, *Ind. Q. C.*, **7,** 2, 1950.
64. Klock, A. G., and C. W. Carter, *Ind. Q. C.*, **8,** 6, 1952.
65. Krammes and Maresh, *Am. Dyestuff Rptr.*, **42,** 317, 1953.
66. Krammes, R., and C. Marsh, *Am. Dyestuff Rptr.*, **43,** 1954.
67. Krasny, J., and A. E. Brown, *J. Textile Inst.*, **51,** No. 12.
68. Krasny, J., and H. E. Harris, *Textile Res. J.*, **30,** 1960.
69. Krasny, J. F., H. E. Harris, G. H. Lourigan, and E. D. White, *Am. Dyestuff Rptr.*, **48,** No. 18, 31, 1959.
70. Krasny, J. F., and J. Menkart, *Am. Dyestuff Rptr.*, **48,** No. 25, 114, 1959.
71. Lezinski, S. E., *Ind. Q. C.*, **10,** 1.
72. Lie, W., and B. Sumer, M. S. Thesis, Lowell Technical Institute, 1962.
73. Lighthipe, C. H., and R. G. Schaubhut, *Am. Dyestuff Rptr.*, **48,** No. 13, 31, 1959.
74. Landwehr, R. C., *Textile Res. J.*, **36,** No. 3, March 1966.
75. Linnert, A., *J. Textile Inst.*, **P52,** July 1961.
76. Mack, C., *J. Textile Inst.*, **46,** T 500, 1955.
77. Mackay, B. H., *Textile Inst. and Ind.*, **1,** June 1963.
78. Mease, R. T., *Ind. Enf. Chem., Anal. Ed.*, **5,** 317, 1933.
79. Mease, R. T., *J. Res. Nat. Bur. Standards*, **13,** 617, 1934.
80. Meyers, Ballenda, *J. Textile Inst.*, **50,** December 1959.
81. Millson, H. E., and Turl, *Textile Res. J.*, **21,** 685, 1951.
82. Millson, H. E., *Am. Cyanamid Dyes Tech. Bull.*, No. 856.
83. Newburgh, L. H., *Physiology of Heat Regulation*, W. B. Saunders Company, 1949.
84. Pointing, K. G., *World Wool Digest*, **17,** June 1966.
85. Preston, John Massey, *Modern Textile Microscopy*, Emmott & Co., 1933.
86. Preston, John Massey, *Fiber Science Textile Institute*, Manchester, 1949.
87. Ratcliffe and Wood, *Textile Inst. Ind.*, **3,** July 1965.
88. Rees, W. H., *J. Textile Inst.*, **36,** T 165, 1945.
89. Richards, R. T. D., *Textile Mercury*, **144,** 1961.
90. Richards, R. T. D., *Textile Rec.*, **79,** 1961.
91. Richards, R. T. D., *J. Textile Inst.*, **52,** November 1961.
92. Rollins, M. L., A. T. Moore, and I. V. deGray, *Am. Dyestuff Rptr.*, **52,** June 1963.
93. Ryberg, B. A., *Am. Dyestuff Rptr.*, September 1950.
94. Satlow, A., S. Cieplik, and G. Fichtner, *Faserforschung Textiltechnik*, **16,** H 3, 1965.
95. Sayers, L. W., *Textile Inst. Ind.*, **4,** July 1966.
96. Schiefer, H. F., *J. Res. Nat. Bur. Standards*, **10,** 105, 1933.

97. Schiefer, H. F., et al., *J. Res. Nat. Bur. Standards*, RP 1988.
98. Schiefer, H. F., *Symposium on the Functional Properties of Clothing Fabrics*, T.R.I., 1943.
99. Schwarz, E. R., *Textile and the Microscope*, McGraw-Hill, 1934.
100. Solmaz, G., M. S. Thesis, Lowell Technical Institute, 1960.
101. Sustmann, C. *Melliand Textileber.*, **41,** No. 3, 1960.
102. Sustmann, C., *Wool Science Rev.*, June 1961.
103. Sustmann, C., and J. Henning, *Melliand Textileber.*, **42,** No. 1, 1961.
104. Talpay, *Melliand Textileber.*, **10,** 1964.
105. Toner, R. K., C. F. Bowen, and J. C. Whitwell, *Textile Res. J.*, **17,** 7, 1947.
106. Toner, R. K., C. F. Bowen and J. C. Whitwell, *Textile Res. J.*, **18,** No. 9, 526, 1948.
107. Toner, R. K., C. F. Bowen, and J. C. Whitwell, *Textile Res. J.*, **19,** No. 1, 1, 1949.
108. Toner, R. K., and J. C. Whitwell, *Textile Res. J.*, **19,** No. 11, 755, 1949.
109. Veldsman, D. P., and R. C. Palmer, *Int. Wool Textile Conf.*, Aust. E 331, 1957.
110. Vogler, K., *Textilrundschau*, **17,** 1962.
111. Von Bergen, W., private communication.
112. Von Bergen, W., *Materials, Research & Standards*, **1,** No. 3, March 1961.
113. Von Bergen, W., and C. Clutz, *Textile Res. J.*, **20,** August 1950.
114. Vose, R. W., and C. H. Plummer, *J. Textile Inst.*, **36,** T 177, 1945.
115. Waters, W. T., *Textile Res. J.*, **25,** August 1955.
116. Wasastjerna, J. A., *Acc. Scientiarum Fennica. Comm. Phys. Math.*, **29,** 5, 1964.
117. Whelar, N. E., A. C. Goodings, and L. E. Turl, *Textile Res. J.*, **25,** 197–223, 1955.
118. Wiegerink, J. C., *Textile Res. J.*, **10,** 357, 1940.
119. Williams, V. A., *Textile Res. J.*, **31,** May 1961.
120. Wilson, C. C., *Ind. Q. C.*, **12,** December 1955.
121. Wolf, H. W., private communication.
122. Wolf, H., and Clutz, C.
123. Wolf, H. W., and R. S. Rankl, *Am. Dyestuff Rptr.*, **42,** 617, 1953.
124. Wool Industries Research Association, *Testing and Control in the Wool Industry*, Leeds, England, 1955.
125. Worner, R. K., *Textile Res. J.*, **26,** June 1956.
126. Zahn, H., T. Gerthen, M. Beck, and H. Meichel, *Melliand Textileber.*, **43,** 1179–1182, 1962.
127. Katz, H. T., *Textile Res.*, **36,** 874, 1966.
128. Lako, T., and L. S. Veer, *J. Textile Ind.*, **53,** 99, 1962.
129. Linnert, A., and B. K. Basu, *J. Textile Ind.*, **61,** 189–295, 1961.
130. Oatley, C. W., and K. C. A. Smith, *Brit. J. Appl. Phys.*, **6,** 391, 1955.
131. Reumuth, H., *Melliand Textileber.*, **48,** 489, 1967.
132. Wells, O. C., *Cambridge Diss.*, 1957.
133. *Wool Science Rev.*, **32,** 1937.
134. Terrell, G. C., *Textile Inst. Ind.*, **5,** 326, 1967.
135. Skinkle J. H., *Chemical Publishing Co., Inc.*, 1949.

CHAPTER 17

Marketing of Woolen and Worsted Fabrics

LELAND S. LIANG

United States of America

The end-use market for woolen and worsted fabrics and blends is confined chiefly to apparel. According to the latest available census figures, out of some 237,723,000 linear yards of woolen and worsted fabrics (including blends but excluding felts) produced in 1967, apparel fabrics accounted for 224,956,000 linear yards (95%), and non-apparel fabrics accounted for the remaining 12,767,000 linear yards (5%). See Table 1.

Furthermore, the women's and children's market (including misses', girls' and infants' market) alone accounted for 145,586,000 linear yards (61%) in 1967, and the men's and boys' market accounted for 69,504,000 linear yards (29%). Thus, these two markets together are the most significant, insofar as the marketing of woolen and worsted fabrics is concerned. Government orders accounted for 9,866,000 linear yards in 1967, amounting to 5%.

The non-apparel fabrics, accounting for 5% of the entire total production of woolen and worsted fabrics in 1967, consisted chiefly of blankets and blanketing, and transportation upholstery. Other end-use markets for cloth woven from wool are nil.

Thus, woolen and worsted fabrics and blends are used primarily for men's and women's apparel, and, therefore, by the garment manufacturers. Table 2 gives the production figures for the years 1960 to 1967.

TABLE 1

End-Use Markets for Woolen and Worsted Fabrics: 1967, United States

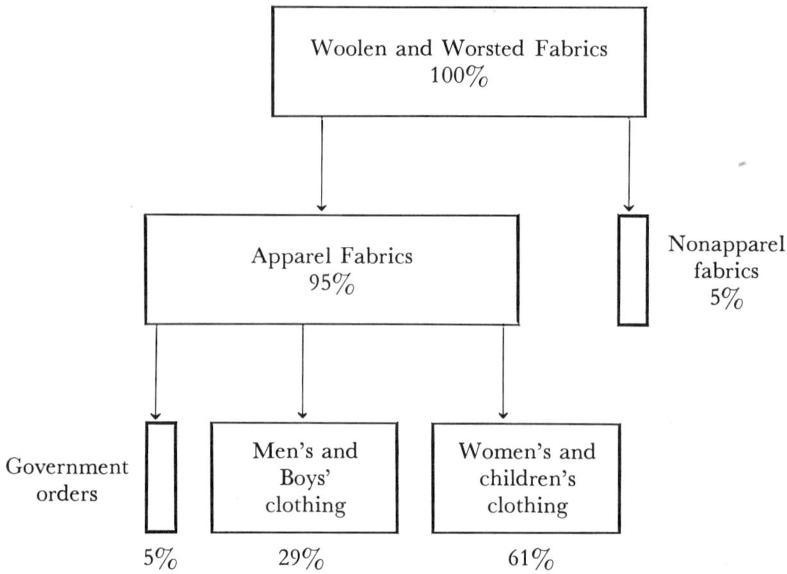

A further breakdown of the type of fabrics and ounce weights per linear yards is shown in Table 3.

Channels of Distribution

The structure of the woolen and worsted industry is made up of three large organizations, each with several mills, and a large number of small mills. According to the latest available data, the 1963 Census of Manufactures lists 69 scouring and combing, 144 spinning and twisting and 361 weaving, dyeing, finishing and sponging plants; a total of 574 woolen and worsted establishments and most of them are located in New England and the middle Atlantic states. It should be noted, however, that the three large organizations (Burlington Industries, J. P. Stevens & Co., Inc., Deering-Milliken, Inc.) have most of their woolen and worsted operations in the two Carolinas and Georgia.

TABLE 2

Production of Woolen and Worsted Goods, by Types of Fabrics: 1960–67

	1967	1966	1965	1964	1963	1962	1961	1960
				(1000 linear yards)				
WOOLEN & WORSTED FABRICS (EXCEPT FELTS), TOTAL:	237,723	264,874	267,343	259,691	286,550	309,861	286,883	286,456
Apparel Fabrics	224,956	250,348	257,933	251,918	275,487	301,238	280,141	280,757
Government Orders	9,866	11,198	5,354	2,677	3,503	8,563	4,254	3,298
Men's & boys' clothing	69,504	86,720	95,153	83,107	98,465	110,985	102,973	114,789
Women's & children's clothing	145,586	152,430	157,426	166,134	173,519	181,690	172,914	162,670
Non-Apparel Fabrics	12,767	14,526	9,410	7,773	11,063	8,623	6,742	5,699

Source: United States Bureau of the Census.

TABLE 3

Production in the United States of Woven Apparel Cloths Wholly or Chiefly by Weight of Wool, Reprocessed Wool, or Reused Wool[a]

(In thousands of linear yards 54–60 inches wide)

Type and weight per linear yard	Year[b] 1961	1962	1963	1964	1965	1966	1967[p]
Apparel cloths, total	280,141	301,238	273,939	247,322r	257,933	250,348	224,938
On government orders	4,254	8,563	3,779	2,465r	5,354	11,198	9,866
Other than government orders	275,887	292,675	270,160	244,857r	252,579	239,150	215,072
Men's and boys'							
Less than 9 oz.	9,876	11,373	8,952	6,693r	9,121	12,806	9,632
9–12.9 oz.	70,830	75,382	69,234	58,491r	68,244	59,642	48,317
13–15.9 oz.	8,853	9,285	8,178	6,044r	6,260	4,740	4,889
16–19.9 oz.	8,636	9,317	5,253	4,443r	5,016	4,668	3,675
20–25.9 oz.	4,165	5,111	4,951	5,311r	5,919	4,864	2,991
26 oz and over	613	517	607	519r	593	86,720	69,504
Total	102,973	110,985	97,175	81,501r	95,153	173,440	139,008
Women's and children's							
Less than 9 oz.	24,207	23,723	21,760	15,940r	10,924	11,239	11,846
9–12.9 oz.	88,243	94,521	92,369	89,863r	88,833	83,003	74,674
13–15.9 oz.	17,127	16,662	14,818	15,714r	14,053	14,119	19,408
16–19.9 oz.	26,665	26,230	21,786	21,651r	21,435	20,100	17,714
20 oz. and over	16,672	20,554	22,252	20,188r	22,181	23,969	21,926
Total	172,914	181,690	172,985	163,356r	157,426	152,430	145,568

[a] As defined by the Wool Products Labeling Act.

[b] Calendar years or 52-week or 53-week periods approximating calendar years.

[c] Source: Bureau of the Census, "Current Industrial Reports" Series. r, Revised.

[p] Preliminary.

This is reflected by the percentage production of woven cloth in the geographic areas and states given in Table 4.

Even though the woolen and worsted mills are located in New England, the middle Atlantic states, and the Southeastern states,

TABLE 4

Production in the United States of Woven Apparel and Nonapparel Cloths[a] Wholly or Chiefly by Weight of Wool, Reprocessed Wool or Reused Wool,[a] by Geographic Areas and States[b]

	Year		
States and areas	1955	1960	1965
United States	100%	100%	100%
New England	50	48	45
Maine	8	10	13
New Hampshire	9	11	12
Vermont	2	1	1
Massachusetts	22	18	11
Rhode Island	6	5	6
Connecticut	3	3	2
Middle Atlantic	10	6	4
New York	2	1 }	1 (New York and New Jersey)
New Jersey	2	1 }	
Pennsylvania	6	4	3
North Central[c]	6	2	2
South	32	42	48
Virginia	5	4	6
North Carolina	9	9	8
South Carolina	4	11	15
Georgia	9	13 }	19 (Georgia and All other)
All other[d]	5	5 }	
Western[e]	2	2	1

[a] Excludes felts.
[b] As defined by the Wool Products Labeling Act.
[c] Includes Ohio, Illinois, Michigan, Wisconsin, Minnesota, Iowa, and Missouri.
[d] Includes Maryland, West Virginia, Tennessee, Alabama, and Texas.
[e] Includes Utah, Washington, Oregon, and California.
Comparable data for 1963 were published in the 1964 *Bulletin*, Vol. XCIV, page 2–95.
Source: Derived from data published by the Bureau of the Census.

the marketing center, however, is now New York City. Most of their fabrics are sold through representatives in that city, or through traveling sales personnel managed from there.

The agencies or marketing intermediaries used by the woolen and worsted mills to dispose of their fabrics are the selling agents and the mill selling offices. The former is a separate entity, whose function is to sell the output of the mill it represents, while the latter is simply the sales department of the mill. It should be noted, however, that a fairly large proportion of the selling agents have financial interests in the woolen and worsted mills they represent. Thus, in some respects, the selling agents are really not so different from the mill selling offices. See Table 5.

By far the greater part of business done by the mill selling offices and selling agents is in direct sales to cutters or garment manufacturers. Most of these sales are made in New York City itself, since a large number of garment manufacturers have their offices and

TABLE 5

Marketing of Woolen and Worsted Fabrics: Channels of Distribution

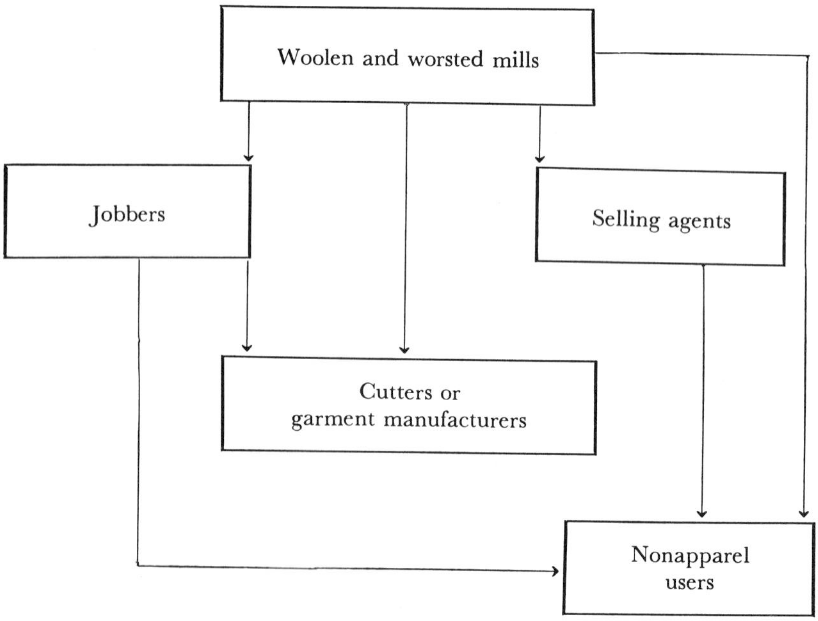

plants there. The others, with their plants elsewhere, send buyers into New York City from time to time to buy fabrics. Some of the large out-of-town men's clothing manufacturers maintain full-time buying offices in New York City to keep them advised of market developments and trends, in addition to handling the actual purchasing of the fabrics. In fact, the building at 1290 Avenue of the Americas in New York City has now become the center for the men's clothing manufacturers.

Both the mill selling offices and selling agents send salesmen out to the important cutting centers and maintain sales branches in those cities. These traveling salesmen and branch offices solicit trade especially from the smaller cutters or garment manufacturers who find it difficult to send buyers into the New York market.

There is very little use of brokers in the sale of woolen and worsted fabrics. On the other hand, wholesalers and jobbers are found in the industry. The former's activities are associated with the distribution of piece goods to retail stores for resale to household consumers, while the latter sell fabrics to the small cutters or garment manufacturers. It should be pointed out that the wholesalers are relatively unimportant in the trade. The jobbers who resell to cutters are somewhat more important, since they are able to handle the credit risks involved in dealing with a large number of small, weakly financed garment manufacturers.

Types of Cutters of Woolen and Worsted Fabrics: Men's Wear

Cutters or manufacturers of men's ready-to-wear clothing are by far the most important, although merchant tailors (sometimes known as master tailors) are still in existence. Most important among the former are the manufacturers of men's slacks, suits, topcoats, and overcoats. These manufacturers are scattered through a large number of cities. In terms of output, New York City is by far the most important. Other cities where such manufacturers are located include Philadelphia, Chicago, Baltimore, Cleveland, Boston, Cincinnati, Rochester, St. Louis, and Los Angeles.

There are many different types and sizes of these garment manufacturers. Many of these use the so-called "contract" method of operation, which means that several small, specialized shops make garments for a single contract manufacturer who, in turn, markets the finished product. Some of the garment manufacturers are very large firms operating huge, modern, and efficient plants; many are medium-sized; a great many are small. Most of these garment manufacturers are characterized by a high degree of specialization, both as to the kinds of clothing they make and as to quality.

Types of Cutters of Woolen and Worsted Fabrics: Women's Wear

Woolen and worsted fabrics and blends are used in the making of ready-to-wear women's clothing, particularly in unit-priced dresses, suits, and skirts. Some of these cutters or garment manufacturers make garments from their own fabrics, others make garments on contract from goods they do not own, while a third group buys fabrics and has them made up by the contract shops. Again, these garment manufacturers are quite heavily concentrated in New York City, ranging from 65 to 75%, depending on the type of garments made. The rest are scattered over the country, including Los Angeles, Chicago, Philadelphia, Boston, Baltimore, St. Louis, Cleveland, Dallas, among others.

Specialization by Mills

With the exception of the large mill organizations, almost all the woolen and worsted mills specialize in a fairly narrow range of fabrics. They will either produce 100% woolen fabrics, woolen blend fabrics, 100% worsted fabrics, or worsted blend fabrics. Within each of these fabric categories they will specialize further according to men's wear fabrics, women's wear fabrics, or some of the other products of the industry, such as blankets and transportation upholstery. They may specialize further by concentrating on staples or fancy goods, and by making low-, medium-, or high-price fabrics.

This kind of specialization in the woolen and worsted industry is part of the over-all pattern of the textile industry, and such specialization is both advantageous and necessary because the marketing problems of the various kinds of fabrics are quite different. In staple items, the problem is primarily one of price competition, and the mill weaving such fabrics must be efficient enough to meet both domestic and foreign competition. The marketing effort here is to get large-volume orders to keep the mill busy throughout the year.

On the other hand, a fancy goods mill must operate on a different basis, since the emphasis is not on price and volume but upon novelty and innovation. The woolen and worsted mills serving this market must be organized to make up designs which will meet the current demands of fashion promptly, deliver fabrics quickly to the garment manufacturers, get rid of goods before demand shifts to something else, and anticipate the demand of the new fashion trends.

The large organizations which make all types and kinds of woolen and worsted fabrics and sell to many different types of customers can do so successfully only if they achieve an effective specialization by departments equivalent to the specialization made by the mills themselves if they are small. Thus, a large organization dealing in woolen and worsted fabrics may have to be organized along the lines of men's wear department, women's wear department, transportation department, and so forth. This involves some difficulties of organization and administration and would require exceptional executive talents to carry out the program successfully.

Style and Fashion Problems

Styles and fashion problems in woolen and worsted fabrics are no different from other branches of the textile industry. It is particularly important in women's wear fabrics, and such fabrics account for some 60% of the entire woolen and worsted industry.

It should be noted especially that the woolen and worsted fabrics have no gray goods market. On account of technology, spinning, weaving and finishing of woolen and worsted fabrics are integrated. Thus, the weavers cannot make fabrics in the gray and leave to converters the problem of putting the style into them. The result is that the mills must make up the designs many months before sale to garment manufacturers. For this reason, the styling functions performed by the mill's stylists are extremely critical.

In a woolen and worsted mill, a stylist's functions should encompass many areas. He is a merchandiser, a salesman, a fabric development man, a designer, a colorist, and a hand and finish critic. Such are the demands of styling in today's markets where not only woolen and worsted fabrics are involved, but the blending of the various man-made fibers with wool has created complicated fiber-combination problems. A stylist in a woolen and worsted mill, working on blends, may concern himself with four polyester fibers, four acrylic fibers, several versions of nylon—each with its own fiber characteristics, not to mention the several spandex fibers. All of these man-made fibers are available in several lusters and a variety of staple lengths and deniers.

In fact, styling in the woolen and worsted industry, especially where man-made fiber blends are involved, is closely aligned with merchandising and sales.

Shifts in Weights and Constructions of Woolen and Worsted Fabrics*

The production of civilian men's and boys' wear woven wool fabrics in the United States was cut back sharply in 1951, when Korean War procurement of military fabrics engaged a large part of woolen and worsted production capacity. At this point, the introduction of Dacron to the men's summer suit market posed the first major challenge to all wool fabrics in the tailored clothing market. Since then, there have been peaks in production, namely in 1953, 1956, 1959, and 1962, reflecting the changing fortunes of woven wool fabrics in the civilian men's and boys' markets; a fifth peak is thought to have occurred in 1965. Detailed figures for 1954 onward are given in Table 6, which shows that, during the 1954–1965 period, there has been an overall downward movement in output. Over and above the basic impact of fluctuations in the United States economy, the major influences in this long-term decline were, first, the expanding volume of synthetic fabrics competing for traditional wool fabric markets; second, increasing price competition between raw wool and synthetic fibers; third, socio-economic changes resulting in shifts from tailored to casual wear, from heavier to lighter weight clothing and from woven to knitted constructions; fourth, increasing imports; and fifth, the distortion of the age composition of the male population, which is now recovering from the losses in the young adult market following World War II.

* Taken from Department of Economics and Statistics, The Wool Bureau, Inc. (1).

TABLE 6

Total Production of Civilian Men's and Boys' Woven Wool Fabrics: 1954–1967[a]
Millions of Linear Yards

Year	Less than 9 oz	9–13 oz	13–16 oz	16–20 oz	20 oz and over	Total
			Total Production			
1954	6.5	76.9	26.6	7.3	5.9	123.2
1955	8.3	90.3	25.3	10.2	7.0	141.1
1956	9.3	87.5	31.1	15.7	11.1	154.7
1957	7.3	80.9	21.1	10.4	9.5	129.1
1958	8.4	70.5	12.8	10.1	5.4	107.3
1959	7.8	90.3	15.5	12.5	6.6	132.8
1960	9.7	76.2	11.2	11.7	6.0	114.8
1961	9.9	70.8	8.9	8.6	4.8	103.0
1962	11.4	75.4	9.3	9.3	5.6	111.0
1963	9.0	69.2	8.2	5.3	5.6	97.3
1964	6.7	58.5	6.0	4.4	5.8	81.5
1965	9.1	68.2	6.3	5.0	6.5	95.2
1966	12.8	59.6	4.8	4.7	4.9	86.7
1967	9.6	48.3	4.9	3.7	3.0	69.5
			Worsteds			
1954	5.3	42.0	15.8	1.2	*	64.2
1955	6.0	45.0	12.5	2.4	*	65.9
1956	6.8	46.9	16.3	2.9	*	72.9
1957	4.7	41.6	6.0	1.2	*	53.5
1958	6.3	37.4	4.1	.9	*	48.6
1959	5.2	52.1	4.0	1.1	*	62.4
1960	5.7	43.7	4.3	1.1	*	54.8
1961	7.4	45.0	3.3	1.1	*	56.8
1962	9.0	46.7	4.1	1.4	*	61.2
1963	7.4	45.6	3.7	1.1	*	57.8
1964	5.9	36.7	2.8	1.2	*	46.6
1965	8.2	42.5	4.3	1.7	*	55.8
1966	11.7	37.4	2.4	1.1		52.8
1967	9.1	30.1	3.2	1.0		43.4
			Woolens			
1954	1.3	34.9	10.8	6.1	5.9	59.1
1955	2.3	45.3	12.7	7.9	7.0	75.2
1956	2.4	40.5	14.8	12.8	11.1	81.7
1957	2.6	39.2	15.1	9.3	9.5	75.6
1958	2.2	33.2	8.7	9.2	5.4	58.6

TABLE 6—(continued)

Year	Less than 9 oz	9–13 oz	13–16 oz	16–20 oz	20 oz and over	Total
1959	2.6	38.2	11.5	11.4	6.6	70.4
1960	4.0	32.5	6.9	10.6	6.0	60.0
1961	2.5	25.9	5.6	7.5	4.8	46.2
1962	2.4	28.7	5.2	7.9	5.6	49.8
1963	1.6	23.6	4.4	4.2	5.6	39.4
1964	.8	21.8	3.2	3.3	5.8	34.9
1965	.9	25.7	2.9	3.3	6.5	39.3
1966	1.1	22.2	2.3	3.5	4.9	33.9
1967	.6	18.2	1.7	2.7	3.0	26.1

* Included with woolens. Note: Totals do not always equal sum of items due to rounding.

[a] Source: Bureau of The Census, Current Industrial Reports, Woolen and Worsted Woven Goods.

Changes in the relative importance of worsted and woolen constructions in the production of men's and boys' wear wool fabrics reveal cyclical movements. From 1948 to 1957, worsteds gradually gave way to woolens, the latter reaching a peak of 59% of total production in 1957. Since that year, a reverse trend has been in progress, with worsteds reaching a peak of 62% of 1967 production. The pattern of distribution of total production between worsteds and woolens varies in the different fabric weight ranges. Owing to the increasing concentration of production in the 9–13 oz range, fabrics of these weights constitute the major influence in the cyclical pattern seen in total production.

In Table 7 are shown the striking differences which exist between the two classes of apparel fabrics in the distribution of worsted and woolen constructions.

Under the same table are given data showing the percentage of the 100% wool fabrics for both apparel fabric types in the most frequently used weight class.

It has been an increasing concentration of output in the 9–13 oz per linear yd range, which, in 1965, amounted to 72% of the total compared with 62% in 1954 and around 40% in the early postwar years. The corresponding decline in other weights was confined almost entirely to the 13–16 oz group, which fell from

TABLE 7

Percentage of Yarn Types in Woven Wool Fabrics

Year	Men's and Boys' Worsted	Men's and Boys' Woolens	Women's and children's Worsted	Women's and children's Woolens
1954	52	48	12	88
1960	48	52	6	94
1965	59	41	6	94
1967	62	38	6	94
	Percent of Woven Wool Fabrics Containing 100% Wool of Main Weight Class 9–13 Oz			
1960	83	45	NA	75
1965	94	61	NA	62
1967	91	48	NA	44

22% of the 1954 total to 6% of the 1965 total. Similar shifts occurred, in different degrees, in both woolen and worsted constructions, thus reflecting the trend toward lighter weight fabrics for men's cool weather clothing. There has also been a trend towards lighter weights in tropical worsteds, so that the less than 9 oz range made up 15% of worsted output in 1965, compared with 8% in 1954.

TABLE 8

Relative Importance of Construction in Different Weights of Woven Wool Fabrics Containing 50% or More Wool Percent of Total Construction by Weights

Year	Less than 9 oz	9–13 oz	13–16 oz	16–20 oz	20 oz and over	Total
			Men's and Boys'			
1954	5	62	22	6	5	100
1960	9	62	10	10	5	100
1965	10	72	6	5	7	100
1967	14	70	7	5	4	100
			Women's and Children's			
1954	13	42	20	16	9	100
1960	14	50	11	17	9	100
1965	7	56	9	14	14	100
1967	8	51	14	12	15	100

Table 8 shows the differences in the distribution of the various fabric weight ranges between the two classes of apparel fabrics.

Details for the years 1960–1967 of the relative importance of 100% wool content woven fabrics for men's and boys' wear, as opposed to wool blends, show that in the largest weight range, namely the 9–13 oz group, 100% wool fabrics were both absolutely and relatively more important than wool blends. In 1960, total production of worsted and woolen 9–13 oz cloth came to 76.2 million linear yd, of which 66% (50.6 million linear yd) were 100% wool and 34% (25.5 million linear yd) were wool blends. By 1965, out of the somewhat reduced total of 68.2 million linear yd of 9–13 oz cloth, 100% wool fabrics amounted to 81% of the total (55.6 million linear yd), with wool blends falling correspondingly to 19% (12.7 million linear yd). In 1960, worsted 9–13 oz fabrics amounted to 43.7 million linear yd, of which 83% (36.0 million yd) were 100% wool and 17% (7.6 million yd) wool blends; in 1965, after rising in the intervening years, the total was a little lower, at 42.5 million linear yd, with 94% (39.9 million yd) 100% wool and only 6% (2.6 million yd) wool blends. Woolen cloths of the 9–13 oz group amounted to 32.5 and 25.7 million linear yd, in 1960 and 1965, respectively. The share of 100% wool fabrics rose from 45% (14.6 million linear yd) in the former year to 61% (15.7 million linear yd) in the latter, while that of wool blends fell correspondingly from 55% (17.9 million linear yd) to 39% (10.0 million yd). Looking at the picture another way, that is, taking the percentage distribution of 100% wool cloth of the 9–13 oz range between worsteds and woolens one sees that of the 55.6 million linear yd of 100% wool cloth produced in 1965, some 72% were worsteds, compared with 71% of the 1960 total of 50.6 million yd, and 77% of the 1961 total. Woolens fell from 29 to 23% between 1960 and 1961, but thereafter rose to 28% of total 100% wool cloth yardage by 1965. In the case of wool blend fabrics, worsteds fell from 30% of the 1960 total of 25.5 million linear yd to 12% in 1964, but showed some recovery in 1965, rising to 21% of that year's total. Woolens rose from 70 to 87% between 1960 and 1964, but then fell slightly to 79% of the 1965 total of 12.7 million linear yd.

As far as the second largest category of men's and boys' civilian woven wool fabrics production, namely the less than 9 oz range, is concerned, full figures are only available for worsteds in the years 1963–1965; worsteds, however, make up the major part of less than 9 oz yardage. Worsted fabrics weighing less than 9 oz amounted to 7.4 million linear yd in 1963, 5.9. million yd in 1964, and 8.2 million yd in 1965. The share of 100% wool cloths fell from 39% (2.9 million linear yd) in 1963 to 20% (1.6 million yd) in 1965, but that of wool blends rose from 61% (4.5 million yd) to 80% (6.6 million yd). Full details for woolens are only available for 1964: in that year, 0.8 million linear yd of less than 9 oz woolens were produced, of which 79% (0.6 million yd) were 100% wool and 21% (0.2 million yd) wool blends.

The production of women's and children's woven wool apparel fabrics containing 50% or more wool, like that of men's and boys' fabrics, was sharply curtailed during the Korean War. Since then, however, production rose fairly steadily to reach a major peak in 1962; after that year a downward movement was recorded, reflecting the strong shift to knitted fabrics. Details for the years 1954 to 1965 are given in Table 9, which reflects fashion preferences in the large proportion of the total made up by woolens. The growth in output of women's and children's fabrics contrasts with the decline in output of men's and boys' fabrics over the same period, mentioned in our previous issue. Recessions in the production of the former are seen to be less severe than those in men's and boys' wear wool fabrics; this is due to fashion dictates in general, the increase in the number of working women, who have need, desire and ability to purchase more clothing than nonworking women, and the inclusion of the greatly expanded children's wear market in this category of production.

Changes in the relative importance of worsted and woolen constructions in United States output of women's and children's wear wool fabrics reveal a fairly definite trend in favor of woolens. Table 8 shows that the share of woolens in the total rose from 88% in 1954 to 94% in 1960; there was some fluctuation in the next few years, but by 1965 the share of woolens was up to 94% again. The dominance of woolen construction is seen to prevail in most weight ranges in varying degrees. In the less than 9 oz group, however, worsteds have recently recovered some of their earlier importance, thus replacing less than 9% woolens to some extent. This change reflects the development of lightweight wool fabrics in new weaves

TABLE 9

Total Production of Civilian Women's and Children's Woven Wool Fabrics: 1954–1967[a]

Millions of Linear Yards

Year	Less than 9 oz	9–13 oz	13–16 oz	16–20 oz	20 oz and over	Total
			Total Production			
1954	18.0	58.7	27.8	21.7	21.1	138.3
1955	16.2	68.9	30.7	22.5	13.1	151.4
1956	16.9	74.3	28.1	22.4	14.1	155.7
1957	16.2	71.5	24.6	22.0	14.6	148.9
1958	18.2	68.7	20.6	26.6	15.8	149.9
1959	20.4	79.9	21.1	31.8	15.8	169.1
1960	20.9	81.7	18.4	27.0	14.7	162.7
1961	24.2	88.2	17.1	26.7	16.7	172.9
1962	23.7	94.5	16.7	26.2	20.6	181.7
1963	21.8	92.4	14.8	21.8	22.3	173.0
1964	15.9	89.9	15.7	21.7	20.2	163.4
1965	10.9	88.8	14.1	21.4	22.2	157.4
1966	11.2	83.0	14.1	20.1	24.0	152.4
1967	11.8	74.7	19.4	17.7	21.9	145.6
			Worsteds			
1954	3.0	11.3	0.7	1.1	*	16.2
1955	4.2	9.3	0.8	1.0	*	15.4
1956	4.4	9.6	1.0	0.7	*	15.6
1957	3.6	7.6	1.0	1.1	*	13.2
1958	3.7	6.2	0.9	0.7	*	11.4
1959	4.1	9.3	0.4	*	*	13.7
1960	3.1	6.7	0.7	*	*	10.5
1961	5.2	7.5	0.7	*	*	13.3
1962	5.5	8.4	0.8	1.4	*	16.0
1963	5.7	5.8	0.7	0.7	*	12.9
1964	5.6	5.8	0.8	0.8	*	12.9
1965	3.1	4.1	0.8	1.6	*	9.6
1966	3.4	3.9	1.4	1.9	*	10.7
1967	3.0	3.0	2.1	1.2	*	9.2
			Woolens			
1954	15.0	47.4	27.0	20.5	12.1	122.1
1955	11.9	59.5	29.9	21.5	13.1	136.0
1956	12.5	64.7	27.1	21.7	14.1	140.0
1957	12.7	63.9	23.6	20.9	14.6	135.7

TABLE 9—(continued)

Year	Less than 9 oz	9–13 oz	13–16oz	16–20 oz	20 oz and over	Total
1958	14.6	62.5	19.7	25.9	15.8	138.5
1959	16.3	70.6	20.8	31.8	15.8	155.3
1960	17.7	75.0	17.7	27.0	14.7	152.1
1961	19.0	80.8	16.4	26.7	16.7	159.5
1962	18.3	86.1	15.9	24.8	20.6	165.6
1963	16.0	86.6	14.1	21.1	22.3	160.1
1964	10.4	84.1	15.0	20.8	20.2	150.5
1965	7.9	84.7	13.3	19.8	22.2	147.9
1966	7.8	79.0	12.7	18.2	24.0	141.8
1967	8.9	71.7	17.3	16.5	21.9	136.3

* Included with woolens. Note: Totals do not always equal sum of items due to rounding.

[a] Source: Bureau of The Census, Current Industrial Reports, Woolen and Worsted Woven Goods.

and patterns suitable for holiday wear and for evening and warm weather garments.

As with men's and boys' fabrics, there has been an increasing concentration of output in the 9–13 oz per linear yd range, which in 1965 accounted for 56% of the total as compared with 42% in 1954. The corresponding decline was confined largely to the 13–16 oz group, which dropped from 20% of the 1954 total to 9% of the 1965 total, although there were also small decreases (from 13 to 7%, and from 16 to 14%, respectively) in the less than 9 oz and the 16–20 oz ranges. Since the 20 oz and over fabrics rose from 9% of the total in 1954 to 14% in 1965, however, the heavy-weight fabrics as a whole (that is, fabrics of 16 oz and over) have slightly increased their overall importance. Changes in total output are, of course, largely influenced by the pattern of changes in the dominant woolen sector, so that the growth in the importance of 9–13 oz fabrics took place despite a decline in the share of these fabrics in the small worsted sector. In the latter, fabrics weighing 9–13 oz fell from 70% of total output of worsteds in 1954 to 43% in 1965. By contrast, less than 9 oz fabrics rose from 19% in 1954 to 45% in 1963, in which year this weight range and the 9–13 oz fabrics were of equal importance in worsted constructions. After

1963, however, less than 9 oz fabrics also declined in importance, falling to 32% in 1965. This reduction was offset by a sharp rise to 17% of total worsteds in 1965, recorded for the 16–20 oz group, which in previous years was considerably less important.

Details for the years 1960 to 1965 of the relative importance of 100% wool content fabrics for women's and children's wear, as opposed to wool blends, show that 100% wool fabrics made up approximately half to two-thirds of the large woolen sector. Total output of woolen fabrics came to 152.1 million linear yd in 1960, of which 68% (103.9 million linear yd) were 100% wool; by 1962, 100% wool fabrics made up 72% (119.2 million linear yd) of that year's total of 165.6 million linear yd, but by 1965 they were down to 56% (83.5 million linear yd) of the 1965 total woolen output of 147.9 million linear yd. In the largest single group of woolens, the 9–13 oz fabrics, 100% wool cloths made up 75% (56.5 million linear yd) of the 1960 total of 75.0 million linear yd, fell to 72% the next year, but then rose to 79% (68.2 million linear yd) of the 1962 total of 86.1 million linear yd. By 1964, the proportion was down to 59%, but in 1965 a partial recovery to 62% (52.5 million linear yd) of the total of 84.7 million linear yd occurred. In the 9–13 oz weight range it is seen, therefore, that wool blends are tending to gain in importance, as they rose from 25% in 1960 to 41% in 1964, fell slightly to 38% in 1965 but increased to 56% in 1967. In the 13–16 oz and the 16–20 oz groups, wool blends are also becoming more important, rising from 16% in 1960 to 37% in 1965 in the former group, and from 20 to 43% in the latter. In the less than 9 oz group, however, it is the 100% wool fabrics which are gaining in importance, rising from 34 to 53% of the total. A similar increase in the importance of 100% wool fabrics in the 20 oz and over group took place between 1960 and 1964, when their share of the total rose from 34 to 46%, respectively, but in 1965 s sharp reduction to 33% was recorded. As far as the worsted sector in concerned, a breakdown between 100% wool cloths and wool blends in available only for the less than 9 oz cloths in 1963: in that year, worsted fabrics of this weight totaled 5.7 million linear yd of which 89% (5.1 million linear yd) were 100% wool fabrics and only 11% (0.6 million linear yd) wool blends.

Looking at the picture another way, that is, taking the distribution of 100% wool cloths and of wool blends amongst the different weight ranges, it is seen that at least half of each in the woolen sector go into the 9–13 oz range. Some 63% of all 100% wool fabrics of woolen construction weighed 9–13 oz in 1965 compared with 54% in 1960, while 50% of wool blends of woolen construction went into this group in 1965 compared with 58% in 1964 and 39% in 1960. The second most important group for 100% wool cloth of the woolen type in 1965 was the 16–20 oz range, which took 13% (21% in 1960), followed by the 13–16 oz group, the 20 oz and over group, and the less than 9 oz group. In the case of wool blends of the woolen type, the second largest group in 1965 was the 20 oz and over range, with 23% of wool blend woolens (20% in 1960), followed by the 16–20 oz group, the 13–16 oz group and the less than 9 oz group.

International woolen and worsted industry

Table 10 gives a summary of the production in the United States of woven cloths and woven felts, wholly or chiefly made of wool.

Table 11 gives the production in certain countries of woven apparel and non-apparel cloth on the woolen and worsted systems from 1960 through 1966. Table 12 reflects the worldwide trade, import as well as export, in woven wool fabrics including blankets covering 35 countries. The EEC countries, the United Kingdom, and Japan are the leading exporters of wool fabrics. In 1966 some 94% of the world's export trade (excluding the Communist countries) was accounted for by these countries, little changed from 1965.

In the appendix charts I to VII (pages 1247 to 1253) illustrate in graphical form the main facts and activities in the wool textile industry of the world as prepared by the Commonwealth Secretariat commodities branch. They are the results of the 21st annual questionnaire of the International Wool Textile Organization and International Wool Study Group and were prepared for the 37th International Wool Conference in Montreux, Switzerland, May 1968.

TABLE 10

Production in the United States of Woven Cloths and Woven Felts Wholly or Chiefly by Weight of Wool, Reprocessed Wool or Reused Wool[a]

	Woven cloths						Woven felts		
	Nonapparel cloths								
Year[b]	Blanketing[c]	Transportation upholstery cloths	All other cloths	Total	Total apparel cloths	Total	Papermaker felts	All other felts	Total woven felts
	In millions of linear yards[d]						In millions of pounds		
1951	31.4	23.2	4.4	59.0	316.4	375.4	7.9	3.4	11.3
1952	24.4	16.5	2.5	43.4	308.0	351.4	6.3	2.2	8.5
1953	28.7	16.4	2.2	47.3	290.7	338.0	7.4	2.6	10.0
1954	12.6	5.1	4.1	21.8	262.4	284.2	7.2	2.6	9.8
1955	11.8	4.4	3.6	19.8	297.7	317.5	7.7	3.1	10.8
1956	8.8	2.0	.9	11.7	312.6	324.3	7.8	3.1	10.9
1957	8.7	2.0	.9	11.6	282.9	294.5	7.2	1.7	8.9
1958	7.2	1.1	.8	9.1	262.2	271.3	6.8	1.5	8.3
1959	5.5	.9	1.0	7.4	303.4	310.8	7.8	1.8	9.6
1960	4.1	.3	1.3	5.7	280.8	286.5	7.2	1.6	8.8
1961	4.5	.9	1.4	6.8	280.1	286.9	7.3	1.4	8.7
1962	5.9	.6	2.1	8.6	301.3	309.9	7.5	1.7	9.2
1963	7.6	.1	2.8	10.5	273.9	284.4	7.0	1.6	8.6
1964	5.3r	2.6		7.9r	247.3r	255.2r	6.3r	1.6	7.9r
1965	6.8	2.6		10.4	257.9	267.3	6.0	1.8	7.8
1966	10.4	4.2		12.8	250.3	264.9	8.7	2.2	10.9
1967	8.5	4.3		14.5	224.9	237.7	7.7	2.2	9.9

[a] As defined in the Wool Products Labeling Act.
[b] Calendar years or 52-week or 53-week periods approximating calendar years.
[c] Includes crib blanketing.
[d] Blanketing in 72-in. width or equivalent. Other fabrics in 54–60-in. widths or equivalent 54-in. width.

Source: Bureau of the Census, "Current Industrial Reports" Series. r, Revised.

TABLE 11

Production in Certain Countries of Woven Apparel and Non-Apparel Cloth on the Woolen and Worsted Systems Whether or Not Entirely of Wool*
(In Millions of Square Yards)

Country	1960	1961	1962	1963	1964	1965	1966
Australia[b]	39.6	32.2	36.6	36.0	35.2	34.5	31.1
Canada[a]	28.0	18.8	29.4	34.2r	37.4	37.4	
France	199.2	209.5	216.5	236.0	229.5	193.0	219.5
Germany (Federal Republic)[e,g]	209.5	202.7	198.7	177.7	174.2r	185.3	178.5
Hungary[g]	32.5	35.1	35.5	34.9	36.0r	34.5	
Irish Republic[a]	7.9	7.4	7.5	8.9r	8.6	9.1	8.6
Italy	241.9	239.6	257.7	253.2	234.1	206.5	260.6
Japan[g]	382.2	404.8	402.8	429.8	423.1	413.7	418.7
Netherlands	68.2	71.8	70.0	74.2	73.5	66.8	66.6
New Zealand[a,f]	4.6	4.4	4.9	4.7	4.4	NA	
Portugal	22.9	22.5	23.4	25.3	27.4r	28.8	28.0
Roumania	38.1	44.5	40.2	45.8r	48.9	48.9	52.9
Soviet Union	525.0	544.0	560.9	563.0	564.5r	557.3	608.8
Union of South Africa	8.3	4.0[c,h]	4.0[c,h]	4.0[c,h]	4.2[c,h]r	4.5[c,h]	
United Kingdom[d]	398.4	385.5	360.8	360.4	358.3	355.8	332.8
United States[e,g]	465.7	466.6	504.4	460.8	412.8r	434.2	427.8
Yugoslavia	39.2	35.7	34.3	39.2	44.9	46.2	45.9

* Source: Commonwealth Economic Committee.

r Revised. NA, Not available.

[a] Apparel only.

[b] Year ending June of year shown.

[c] Excludes blankets.

[d] Deliveries.

[e] Comprises throughout cloth containing by weight 50% or more of virgin or reprocessed wool. (U.S. figures are converted from linear yards.)

[f] Year ending March of year shown.

[g] Excludes woven felts.

[h] Linear yards.

TABLE 12

Trade in Woven Wool Fabrics (including Blankets)
(Million lb)*

Country	Imports 1964	Imports 1965	Imports 1966	Exports 1964	Exports 1965	Exports 1966
Argentina	0.02	0.03		0.33	0.10	0.06
Austria[c]	5.92	6.02	6.91	2.28	2.06	2.95
Belgium	10.65	11.49	13.19[c]	28.67	16.93	16.71[e]
Canada	10.78	10.41	9.05	0.03	0.12	0.03
Colombia						
Denmark	7.98	8.11	8.94	1.55	1.49	1.35
Egypt[b]	0.18	0.12	0.17	0.01	—	—
Finland[b]	3.39	3.68	4.80	0.47	0.36	0.70
France	15.62	13.47	14.61	14.40	13.51	17.10
German F.R.[f]	55.05	64.40	65.04	8.89	9.18	10.74
Greece[b]	1.51	1.54	1.62	0.19	0.04	
Iceland	0.32	0.28	0.36	0.09	0.10	—
Irish Republic[b]	2.07					
Italy	6.25	4.32	4.82	122.64	133.04	137.29
Japan[d]	3.71	2.68	2.78	21.71	27.02	24.47
Jordan	1.91	2.62	1.48	0.03	0.03	
Lebanon[b]	2.48			0.10		
Mexico[b]	0.04	0.01	0.06	0.02	0.01	0.10
Morocco	3.58	0.13		0.22		
Netherlands	21.88	19.90	23.35	13.12	13.63	13.90[e]
Norway[b]	3.01	3.19	3.30	0.69	0.60	0.77
Peru	0.15	0.22	0.65	0.08	0.06	0.04
Portugal	0.21	0.16	0.22	0.99	1.29	1.58
Spain	0.32	0.47	0.40	1.85	1.43	1.86[e]
Sweden	8.37	9.60	9.51	2.06	2.03	2.33
Switzerland	5.73	5.48	5.75	3.11	3.43	3.97
Syria	3.07	3.72		0.09	0.10	
Taiwan	0.53	0.26	0.25	0.10	0.33	0.22
Tunisia	0.86			0.08		
Turkey	—	—		0.08	0.06	0.04
United Kingdom	18.27	20.10	27.89	50.09	50.89	47.58
United States	20.34	26.45	25.58	0.99	1.02	0.77
Uruguay	0.02			0.43	0.75	0.75
Venezuela	0.26	0.25	0.11			
Yugoslavia	0.09	1.21	2.73	1.24	1.14	0.69
Total[a]	215.39	225.60	242.70	276.63	281.30	284.90

* *World Wool Digest*, **18**, No. 15 (1967).

[a] Including estimates for those countries listed for which statistics are not available. [b] Excluding blankets. [c] Excluding velvets and woven felts. [d] Includes wool/silk mixture fabrics. [e] Provisional. [f] Series revised to include pile fabrics and blankets.

The Woolmark of Quality

In August of 1964, the Wool Bureau announced that a new symbol would appear on selected items of men's and women's pure wool apparel. That symbol was the Woolmark. In the few years since its introduction it has gained recognition by consumers and the trade alike as the mark of the world's best, pure virgin wool.

What prompted the introduction of this international symbol? For what does it really stand? How successful has it been?

In the years after World War II, the wool industry saw wave after wave of new synthetic fibers penetrate product areas that once belonged almost exclusively to wool. Although the use of wool by the American public grew from 440 million pounds in 1954 to 491 million pounds in 1964, or 11%, the use of all fibers rose in the same period 28%.

More importantly, the synthetic fibers were being introduced at a prestige level that was casting doubts on wool's traditional superiority. Synthetic fibers were promoted as being "like wool," "as good as wool," or having such characteristics of wool as durability, warmth, wrinkle resistance, resilience or stain resistance. Consumers were becoming confused between wool and its imitators.

It was time to meet the synthetic fibers head on to protect the quality image of wool and to distinguish it from those attempting to trade on wool's established reputation.

The key to helping the public recognize the "real thing" amid the confusion of fiber names was the introduction of the Woolmark. But backing it up, the wool growers of Australia, New Zealand and South Africa taxed themselves to greatly increase the funds available for wool promotion.

It is the wool growers in these three countries which support the International Wool Secretariat with headquarters in London, England, and with branches in 19 countries around the world. The Wool Bureau, Inc., with offices in New York, Chicago and Los Angeles and laboratories in Woodbury, Long Island, New York, is the largest of the I.W.S. branches.

The United States was in the first group of countries to launch the Woolmark in 1964 on a limited and pilot basis. Apparel initially approved for the Woolmark was from only a handful of leading men's and women's wear manufacturers. There are now hundreds of licensees in the United States.

Besides the United States, only Britain, Holland, Belgium, West Germany and Japan were countries in which the "Woolmark" was first introduced. Further expansion has included all other countries where the I.W.S. has branch offices to administer the program.

In 1965, the United States expanded its Woolmark licensing program to include carpet manufacturers.

The selection of the Woolmark design was by a panel of six international design experts who conferrred in London for more than ten hours with the aid of interpreters. Eighty-six proposals were submitted to the panel and the design by Francesco Saroglia of Milan, Italy, was the winning choice. With its classically simple and swirling lines, it has a constant movement symbolizing both the timelessness of wool and its modernity in the space age. After its selection, thousands of man hours were spent in a worldwide search to ascertain that no other trade symbol was in use with which the Woolmark might be confused.

Most importantly, the Woolmark goes far beyond simply meaning that the product on which it appears is made of pure virgin wool. Products bearing the Woolmark must pass rigid tests to make certain they meet performance requirements in areas such as strength and color fastness. It is part of the plan that Woolmark requirements be gradually raised, after industry consultation and official notice, to include additional performance standards to meet consumer needs. By March 1968 certain items of knitwear were required to meet a washability standard to qualify for the Woolmark (2).

Under the Woolmark program, manufacturers are required to furnish certification that fabrics meet standards and to agree to testing by inspectors of random samples of their merchandise. Any failure to meet these tests jeopardizes the manufacturer's license for future use of the "Woolmark."

After the product preuse inspection has been made and the necessary qualification samples have been tested, the use of the Woolmark is currently granted to such consumer products as woven and knit apparel, handknitting yarns, blankets, carpets, upholstery fabrics, fabrics for home sewing, and both apparel and millinery felts. These products are evaluated against the designated combination of specifications such as Fiber Content, Tensile Strength, Colorfastness to Light, Colorfastness to Water, Washability, Solvent Extractable Matter, Pile Weight, Fabric Weight. The proper application of the Woolmark label is then policed during the manufacturing season and once again at the retail level with the frequency of sampling and/or purchasing being determined by the results of the inspection.

It has always been I.W.S. policy in the United States and elsewhere to promote wool as wool without reference to or regard for country of origin of the fiber. Wool Bureau programs are closely coordinated with those of the American Wool Council division of the American Sheep Producers Council, Inc., and the two organizations even employ the same advertising agency. Joint activities of several wool organizations are discussed at regular meetings of the Coordinating Council for Wool Promotion in the United States.

Slightly over two years after the Woolmark was introduced, the Wool Bureau commissioned a nationwide survey of merchandise managers in all department and specialty stores having assets over $125,000 to determine the degree of recognition and understanding of the symbol.

The Woolmark ranked first among six leading product symbols in men's and women's wear. Among more than 600 men's and women's wear retailers who replied, 83% correctly identified the Woolmark and 38% said it made customers more (or "considerably more") willing to buy. Questionnaires were distributed by and returned to an outside organization with nothing in the questionnaire to connect it with the Wool Bureau. Most of the other symbols in the questionnaire were in existence longer than the Woolmark and had been extensively promoted.

Another questionnaire in the carpet field showed 69% of retailers correctly identifying the "wool mark" compared with 90% for the leading product symbol. But considering that the "wool mark" was introduced to the carpet industry only a year before the survey, the degree of recognition was amazingly good.

Perhaps the purposes, aims and goals of the Woolmark are best summed up by Felix J. Colangelo, president of the Wool Bureau. "In a rapidly developing world of skyrocketing textile needs," he said in a speech, "the demand for wool inevitably must expand. Unlike some of our competitors, we have no delusions about capturing the whole market. Our strategy is to dramatize the difference between real wool and synthetic fibers by placing our stress on quality, thus ensuring a proper level of profit for wool growers, industry,a nd trade. That is what the Woolmark of quality is all about."

In 1968 the Woolmark program operated in 22 countries around the world. The numbers of manufacturers who have sought, and have been granted licenses to use the Woolmark now amounts to approximately 17,000. In 1966, there were 7,600 licensees and in the year before that there were only 1,700.

With relatively few exceptions, these licensees are the largest, highest quality, and most well known garment and textile manufacturers in the world.

Another measure of the program's widespread success is the number of Woolmark labels and hangtags being used. Since manufacturers pay the total cost of the labels they sew into garments, label sales are a gauge to the production of Woolmark products. In the first few months of 1967, labels were sold at the rate of 5 million per month. In addition, Woolmark hangtags were ordered at the rate of 4 million per month. Each month has seen a gain in Woolmark label and hangtag orders over the previous month.

It is generally acknowledged that the Woolmark has succeeded in gaining public acceptance faster, on a wider front, and in more countries than any other textile program in history.

References

1. Tarkendoff, R., U.S. Apparel Wool Review and Outlook. The Wool Bureau, Inc., 1968.

2. Holdsworth, R., The Influence of Woolmark on Textile Processing. T.S.D.C. 84, 301, 1968.

Appendix A

Tariff Schedules of the United States on Wool and Selected Wool Manufactures

The Tariff Reclassification Act of 1962 authorized the President to proclaim the reclassification of United States tariff schedules developed by the Tariff Commission under the Customs Simplification Act of 1954. The new schedules came into effect at the close of August 1963 under a proclamation issued August 21.

The Trade Expansion Act of 1962 authorized the President to enter into trade agreements until July 1, 1967 in which he could cut July 1, 1962, duties by 50% over a five-year period and could increase duties to 50% over July 1, 1934, rates. This was the latest extension of authority under the trade agreements program initiated in 1934. In this latest extension, beyond the basic tariff reduction authority, the President was authorized to make greater reductions in case of (1) articles for which the rate of duty or its ad valorem equivalent was not more than 5% (T.E.A. sec. 202); (2) tropical agricultural and forestry commodities (T.E.A. sec. 213); (3) nonagricultural articles which the United States and the European Economic Community together accounted for more than 80% of the aggregate world export value (T.E.A. sec. 211); (4) certain agricultural products in an agreement with the European Economic Community (T.E.A. sec. 212); and (5) any article, within narrow specified limits, in order to simplify the computation of duty (rounding authority of T.E.A. sec. 254).

Under this authority the United States participated in several years of tariff bargaining under the General Agreement on Tariffs and Trade in the so-called "Kennedy Round." An agreement was concluded on June 30, 1967. The tariff concessions the United States agreed to reach by 1972 on wool and the principal wool manufactures, except carpets and rugs, are listed on the pages that follow. Not reduced were items 306.31, 306.32, 306.33, and 306.34—wools finer than 44's (see Vol. 1, page 662).

Kennedy Round Tariff Reductions

On December 16, 1967, President Johnson issued a Proclamation making effective on January 1, 1968, the modifications in the Tariff Schedules of the United States as provided for in Schedule XX (Geneva—1967).

The following table shows rates effective for articles provided for therein which are entered in or withdrawn from warehouse, for consumption on and after the date at the head of the column in which such rate is set forth. Except for rates in the final column, such rate is superseded by the rate for the item in the next column, effective for articles which are entered in or withdrawn from warehouse, for consumption on and after the date at the head of such next column.

TSUS Item No.	Description		Rates of duty effective on and after January 1: 1968	1969	1970	1971	1972
	WOOL						
	Aleppo and other named wools; and other wools not finer than 40's						
306.10	On the skin	11¢	9.9¢	8.5¢	7.5¢	6.5¢	5.5¢ per clean lb
	In the grease or washed						
306.11	Not sorted	13¢	11.5¢	10.4¢	9¢	7.5¢	6.5¢ per clean lb
306.12	Sorted	14¢	12¢	11¢	9¢	8¢	7¢ per clean lb
306.13	Scoured	16¢	14.4¢	12.5¢	11¢	9.5¢	8¢ per clean lb
306.14	Carbonized	23¢	20.7¢	18.4¢	16.1¢	13.8¢	11.5¢ per lb
	Other wool						
	Finer than 40's but not finer than 44's						
306.20	On the skin	15¢	13.5¢	12¢	10.5¢	9¢	7.5¢ per clean lb
	In the grease or washed						
306.21	Not sorted	17¢	15¢	13.5¢	11.9¢	10¢	8.5¢ per clean lb
306.22	Sorted	18¢	16¢	14.4¢	12.5¢	10.5¢	9¢ per clean lb
306.23	Scoured	20¢	18¢	16¢	14¢	12¢	10¢ per clean lb

306.24	Carbonized	27¢	24.3¢	21.6¢	18.9¢	16.2¢	13.5¢ per lb
	Finer than 44's						
306.30	On the skin	24¢	21.5¢	19¢	16.8¢	14.4¢	12¢ per clean lb
	HAIR OF THE CAMEL						
306.40	On the skin	22¢	19.8¢	17.6¢	15.4¢	13.2¢	11¢ per clean lb
	In the grease or washed						
306.41	Not sorted	24¢	21¢	19¢	16.5¢	14¢	12¢ per clean lb
306.42	Sorted	25¢	22.5¢	20¢	17.5¢	15¢	12.5¢ per clean lb
306.43	Scoured	27¢	24¢	21¢	18.5¢	16¢	13¢ per clean lb
306.44	Carbonized	34¢	30.6¢	27.2¢	23.8¢	20.4¢	17¢ per lb
	HAIR OF THE ALPACA AND LIKE HAIR OF OTHER ANIMALS INCLUDING THE HAIR OF THE LLAMA AND VICUNA						
306.50	On the skin	5¢	4¢	3¢	2¢	1¢	Free per clean lb
	In the grease or washed						
306.51	Not sorted	6¢	4¢	3¢	2¢	1¢	Free per clean lb
306.52	Sorted	6.5¢	5¢	5¢	4¢	3¢	3¢ per clean lb
306.53	Scoured	8¢	7¢	6¢	5¢	4¢	4¢ per clean lb
306.54	Carbonized	11.5¢	10¢	9¢	8¢	6.5¢	5¢ per lb
	HAIR OF THE CASHMERE GOAT AND LIKE HAIR						
306.60	On the skin	16¢	14.4¢	12.8¢	11.2¢	9.6¢	8¢ per clean lb
	In the grease or washed						
306.61	Not sorted	18¢	16¢	14¢	12¢	10¢	9¢ per clean lb
306.62	Sorted	19¢	17¢	15¢	13¢	11¢	9¢ per clean lb
306.63	Scoured	21¢	18¢	16¢	14¢	12¢	10¢ per clean lb
306.64	Carbonized	28¢	25.2¢	22.4¢	19.6¢	16.8¢	14¢ per lb

TSUS Item No.	Description	Rates of duty effective on and after January 1				
		1968	1969	1970	1971	1972
	HAIR OF THE ANGORA GOAT, ETC.					
306.70	On the skin . . . . 20¢	18¢	16¢	14¢	12¢	10¢ per clean lb
	In the grease or washed					
306.71	Not sorted . . . . 22¢	19.5¢	17.5¢	15¢	15¢	11¢ per clean lb
306.72	Sorted . . . . 23¢	20.7¢	18.4¢	16.1¢	13.8¢	11¢ per clean lb
306.73	Scoured . . . . 25¢	22¢	20¢	17¢	15¢	12.5¢ per clean lb
306.74	Carbonized . . . . 32¢	28.8¢	25.6¢	22.4¢	19.2¢	16¢ per lb
	HAIR OF THE ANGORA RABBIT					
306.80	On the skin . . . . 32¢	28.8¢	25.6¢	22.4¢	19.2¢	16¢ per clean lb
	In the grease or washed					
306.81	Not sorted . . . . 34¢	30¢	27¢	23¢	20¢	17¢ per clean lb
306.82	Sorted . . . . 35¢	31¢	28¢	24¢	21¢	17¢ per clean lb
306.83	Scoured . . . . 37¢	33¢	29¢	25¢	22¢	18¢ per clean lb
306.84	Carbonized . . . . 44¢	39.6¢	35.2¢	30.8¢	26.4¢	22¢ per lb
	WASTES OF WOOL OR HAIR AND ADVANCED WASTE OF WOOL OR HAIR					
	Burr and card waste					
307.02	Not advanced . . . . 9¢	8¢	7¢	6.3¢	5.4¢	4.5¢ per lb
307.04	Advanced . . . . 14.5¢	13¢	11.5¢	10¢	8.5¢	7¢ per lb
	Noils					
307.06	Not advanced . . . . 12¢	10.5¢	9.5¢	8.4¢	7¢	6¢ per lb
307.08	Advanced . . . . 16¢	14.4¢	12.5¢	11¢	9.5¢	8¢ per lb
307.10	Top, slubbing, roving, and ring waste, whether or not advanced . . . . 28¢	25¢	22¢	19.5¢	16.5¢	14¢ per lb
307.12	Fibers, pickered, garnetted, or pickered					

	and garnetted, whether or not otherwise advanced	13¢	11.5¢	10.4¢	9¢	7.5¢	6.5¢ per lb
306.16	Yarn waste, whether or not advanced, but not including such wastes pickered, garnetted, or pickered and garnetted	10¢	9¢	8¢	7¢	6¢	5¢ per lb
307.18	Other	9¢	8¢	7¢	6.3¢	5.4¢	4.5¢ per lb
390.40	Rags: of wool	9¢	8¢	7¢	6.3¢	5.4¢	4.5¢ per lb
	YARNS OF WOOL OR HAIR						
307.62	Of angora rabbit hair	40¢	36¢	32¢	28¢	24¢	20¢ per lb
		plus 10%	9%	8%	7%	6%	5% ad valorem
	WOOL FABRICS						
	Woven fabrics of wool						
336.10	Handwoven less than 30″—loom width, not weighing over 4 oz. per sq yd, with warp wholly of vegetable fibers		30¢	30¢	30¢	30¢	30¢ per lb
			plus 22%	20%	17%	15%	12.5% ad valorem
336.15	Other		37.5¢	37.5¢	37.5¢	37.5¢	37.5¢ per lb
			plus 22%	20%	17%	15%	12.5% ad valorem
336.20	Serges for religious apparel		30¢	30¢	30¢	30¢	30¢ per lb
			plus 22%	20%	17%	15%	12.5% ad valorem
336.25	Other		37.5¢	37.5¢	37.5¢	37.5¢	37.5¢ per lb
			plus 22%	20%	17%	15%	12.5% ad valorem
336.30	Other: weighing not over 4 oz						
	Valued not over $1.26⅔ per lb		30¢	30¢	30¢	30¢	30¢ per lb
			plus 54%	48%	42%	36%	30% ad valorem
336.35	Val. over $1.26⅔ but not over $2.00 per lb		98¢	90.5¢	83¢	75¢	68¢ per lb
336.40	Valued over $2 per lb		30¢	30¢	30¢	30¢	30¢ per lb
			plus 34%	30%	26.5%	22.5%	19% ad valorem
	Wool fabrics of special construction						
346.52	Pile fabrics: of wool		33¢	33¢	33¢	33¢	33¢ per lb
			plus 22%	20%	17%	15%	12.5% ad valorem

TSUS Item No.	Description	Rates of duty effective on and after January 1: 1968	1969	1970	1971	1972
346.82	Tufted fabrics: of wool	33¢	33¢	33¢	33¢	33¢ per lb
		plus 22%	20%	17%	15%	12.5% ad valorem
347.40	Narrow fabrics: of wool	37.5¢	37.5¢	37.5¢	37.5¢	37.5¢ per lb
		plus 18%	16%	14%	12%	10% ad valorem
	Webs, wadding, batting and nonwoven fabrics: of wool					
355.15	Felts and articles of felt					
	Valued not over $1.50 per lb	22.5¢	22.5¢	22.5¢	22.5¢	22.5¢ per lb
		plus 18%	16%	14%	12%	10% ad valorem
355.16	Valued over $1.50 per lb	30¢	30¢	30¢	30¢	30¢ per lb
		plus 18%	16%	14%	12%	10% ad valorem
	Woven tapestry: of wool					
357.10	Valued not over $2 per lb	37.5¢	37.5¢	37.5¢	37.5¢	37.5¢ per lb
		plus 16%	14%	12.5%	10.5%	9% ad valorem
357.15	Valued over $2 per lb	37.5¢	37.5¢	37.5¢	37.5¢	37.5¢ per lb
		plus 12.5%	11%	9.5%	8%	7% ad valorem
	Clothing for papermaking: of wool					
358$30	Woven	37.5¢	37.5¢	37.5¢	37.5¢	37.5¢ per lb
		plus 13%	12%	10%	9%	7.5% ad valorem
358.35	Other	28.5%	25.5%	22%	19%	16% ad valorem
	Blankets: of wool					
363.10	Not over 3 yd in length	30¢	30¢	30¢	30¢	30¢ per lb
		plus 27%	24%	21%	18%	15% ad valorem
363.15	Over 3 yd in length	37.5¢	37.5¢	37.5¢	37.5¢	37.5¢ per lb
		plus 22%	20%	17%	15%	12.5% ad valorem
363.20	Other	38%	34%	29.5%	25%	21% ad valorem

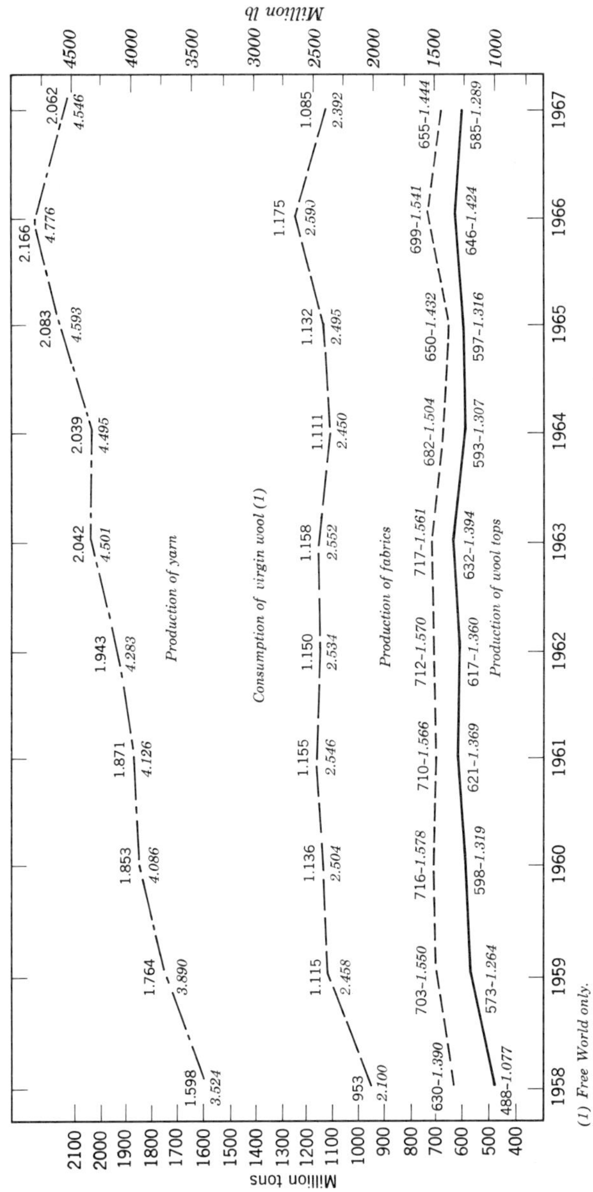

Chart 1. Activity in the wool textile industry of the world.

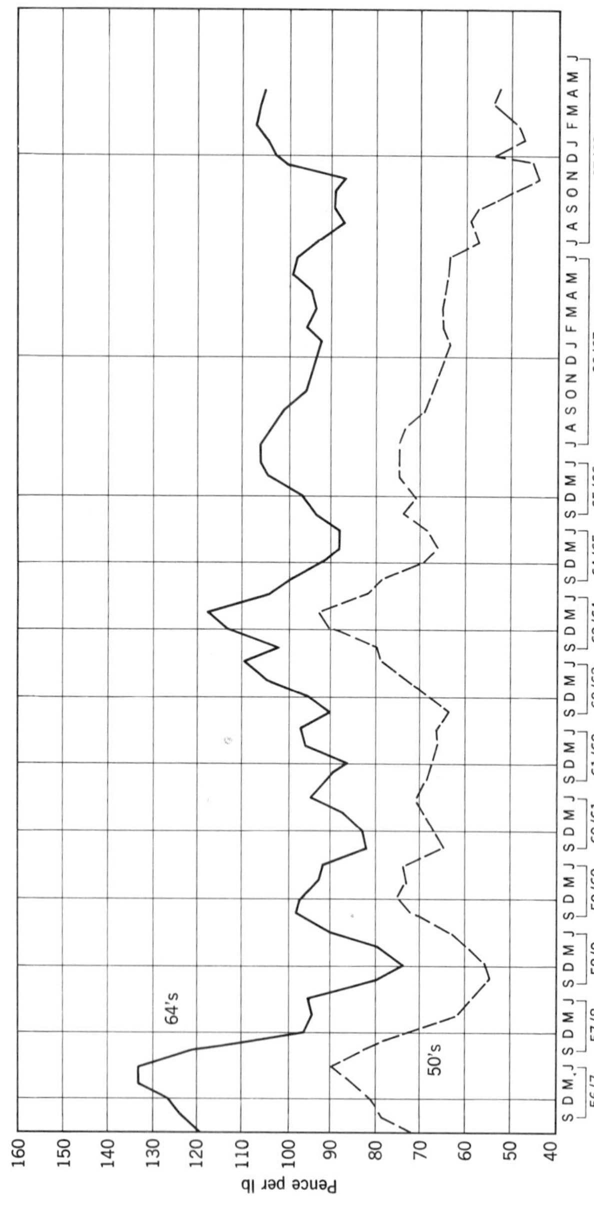

Chart 2. Raw wool prices.

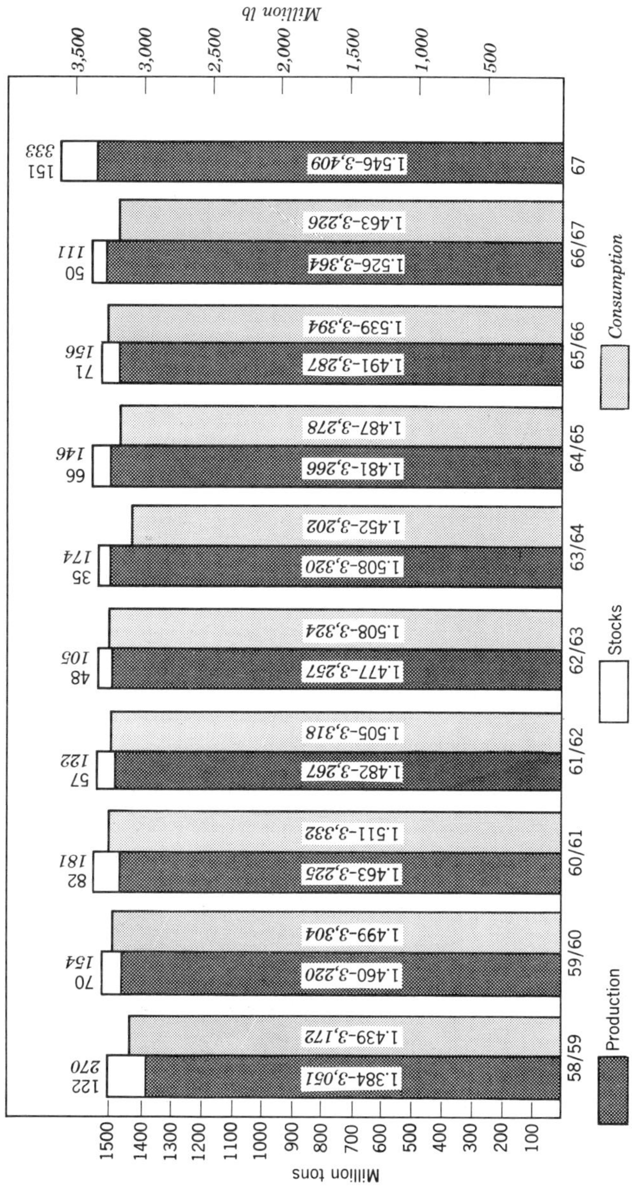

Chart 3. World wool production, consumption and supply stocks.

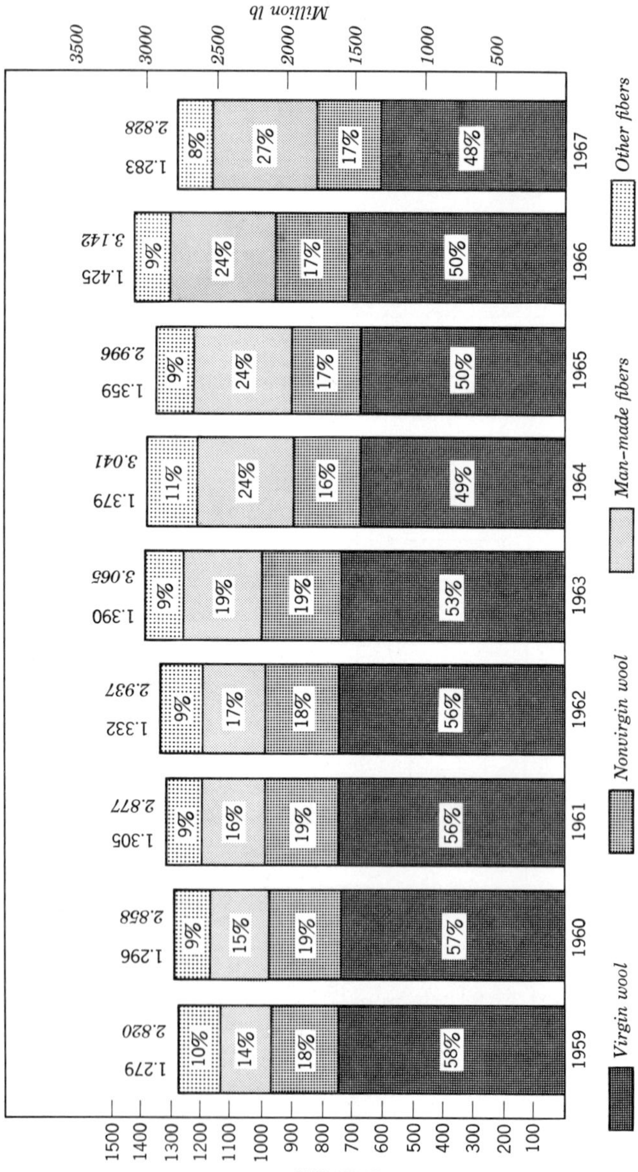

Chart 4. Consumption of materials (wool and others) by the wool textile industries.

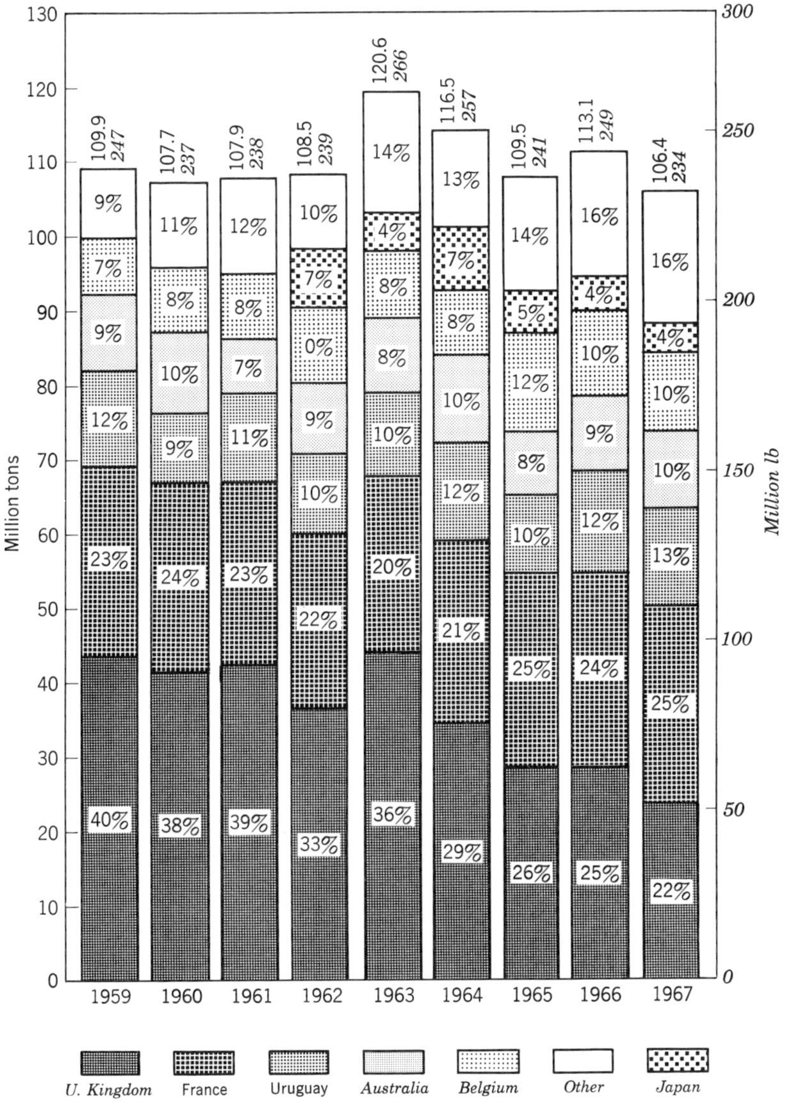

Chart 5. Export of wool tops.

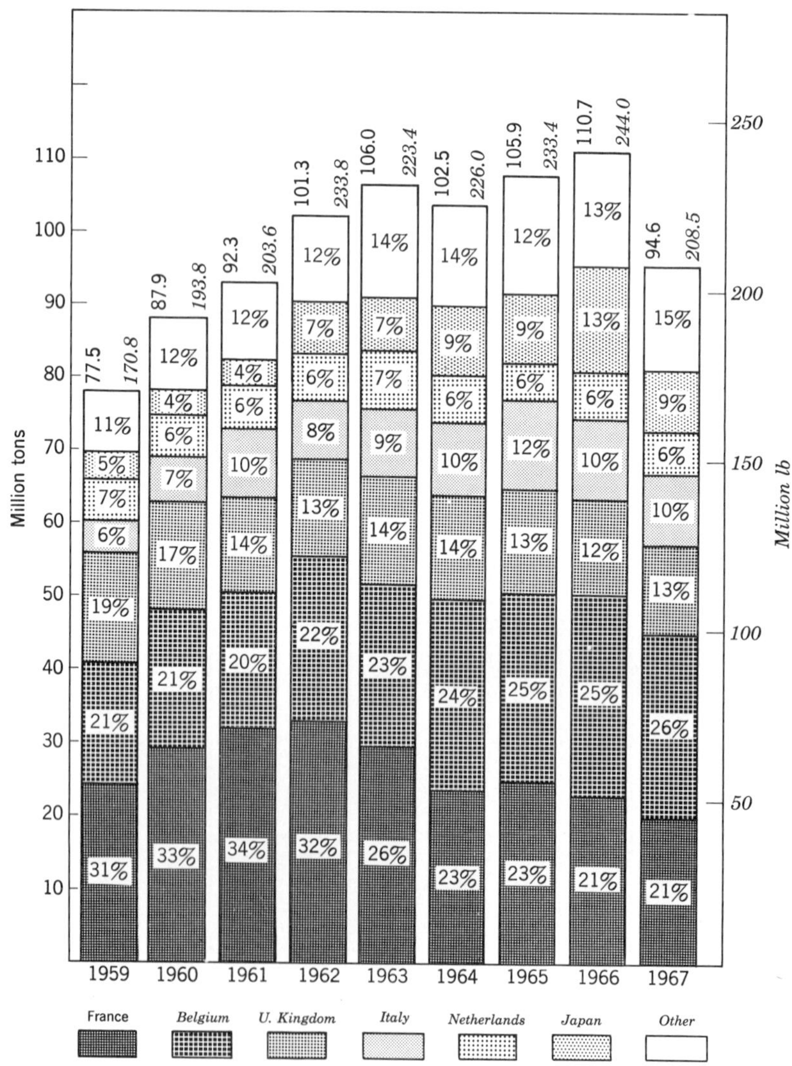

Chart 6. Export of wool yarn.

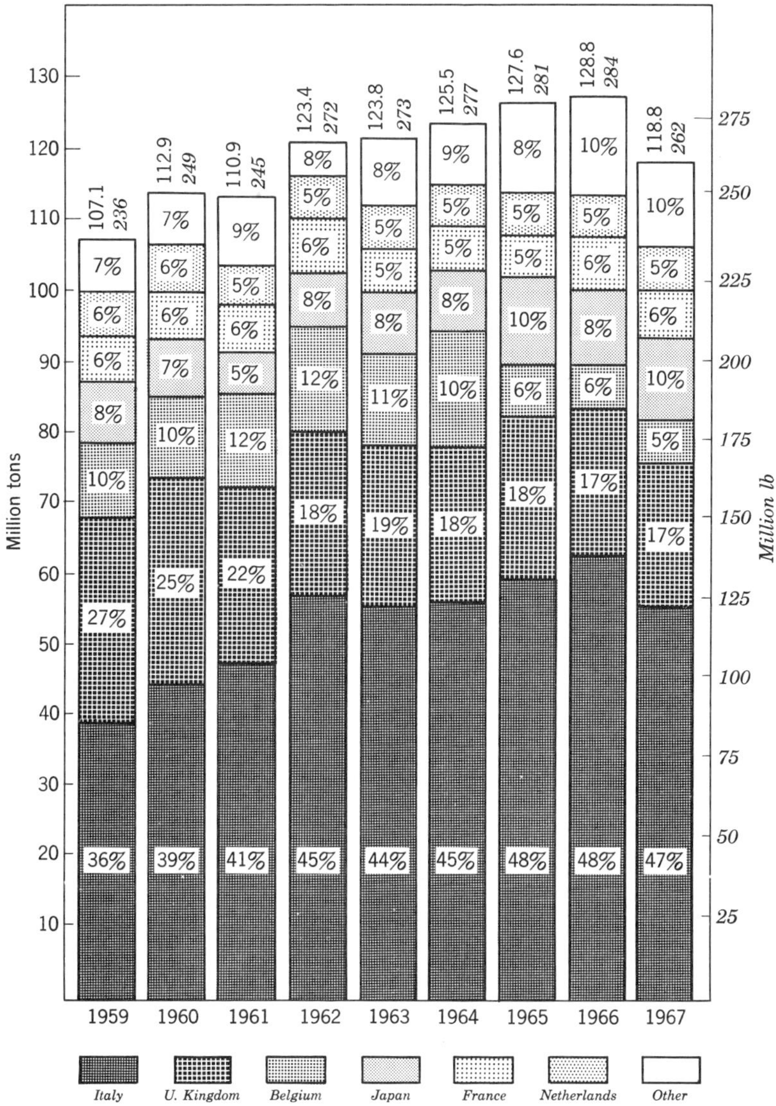

Chart 7. Export of wool fabrics.

Appendix B

International Wool Textile Organization

The International Wool Textile Organization (I.W.T.O.) has its headquarters in Bradford, England. The designation, seat and objects are set forth in its statutes as adopted at the International Wool Conferences, September 10 and 11, 1930. The following are excerpts from these statutes covering Articles 1 to 4.

International Wool Textile Organisation

Registered Office: Commerce House, Bradford, England

STATUTES

Adopted at the International Wool Conference, 10th and 11th September, 1930

DESIGNATION, SEAT, OBJECTS

ART. 1.

The International Wool Textile Organisation is an Association of Organisations interested in the production, commerce and industry of wool and allied fibres and their finished products.

ART. 2.

The duration of the Organisation is unlimited.

ART. 3.

The registered office of the Organisation is at Bradford and may be transferred to any other town or country by the decision of the Annual Congress.

Art. 4.

The Organisation has for its objects:—

(*a*) To maintain a permanent connection between the Organisations in the wool textile industry of each country.

(*b*) To represent the production, commerce and industry of wool textiles in all branches of international economic activity.

(*c*) To promote, support or oppose measures affecting the production, commerce and industry of wool textiles and allied fibres.

(*d*) To promote the study and solution of economic and commercial questions affecting the aforesaid interests.

(*e*) To ensure the functioning of the International Arbitration Agreement in the wool textile industry.

(*f*) To collect and disseminate statistical and other information relating to the production, commerce and industry of wool allied fibres and of their finished products.

(*g*) Generally to initiate, promote and support measures recommended by the competent Associations, and Organisations and to do all such other lawful things as may be conducive to the extension of production, commerce and industry incidental to the attainment of any of the above objects.

The Organizational Chart of the I.W.T.O. as of January 1968, and a list of the member countries as of January 1968 are given below.

International Wool Textile Organization—Member Countries

ARGENTINA

Federacio Lanera Argentina, Avenida Paseo Colon 823–5, Buenos Aires.

AUSTRALIA

Australian National Committee, International Wool Textile Organisation, 578, Bourke Street, Melbourne, Victoria.

AUSTRIA

Fachverband der Textilindustrie, Bauernmarkt 13, Vienna 1.

BELGIUM

Comité Central Belge de la Laine, 24, Rue Montoyer, Brussels.

CANADA

The Wool Textile Association of Canada, 1080, Beaver Hall Hill, Suite 1002, Montreal 1, Quebec.

Organisational Chart of the International Wool Textile Organisation

January 1968

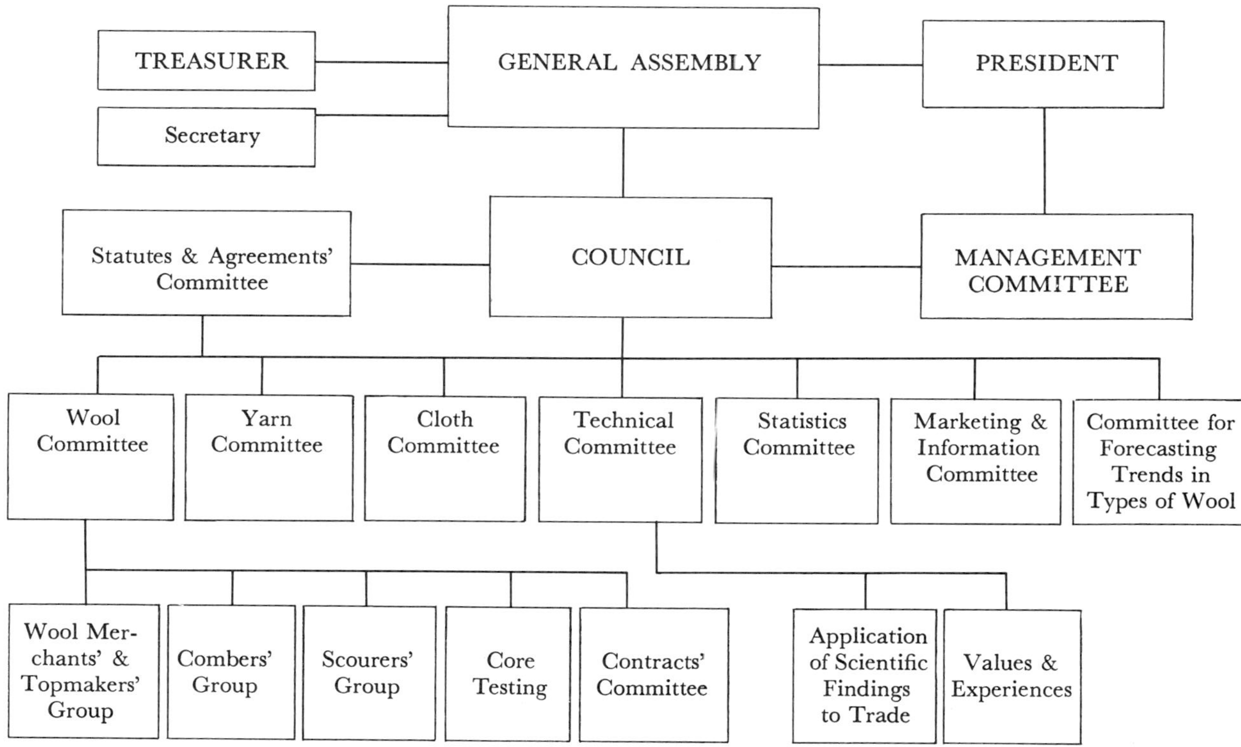

DENMARK

Danske Textilfabrikanforeningen, Vestre Farimagsgade 20, Copenhagen V.

FINLAND

Villetehtailijain Yhdistys Ry., Etelaranta 10, Helsinki.

FRANCE

Comité Central de la Laine, 12, Rue d'Anjou, Paris 8.

GERMANY

Deutsche Wollvereinigung, 863, Hohenstein, Post Coburg.

INDIA

The Indian Woollen Mills, Federation, G–4 Advent Building, Foreshore Road, Bombay 1.

IRELAND

The Irish National Committee, I.W.T.O., Glasnevin House, Ballymun Road, Dublin 9.

ISRAEL

The Textile Department, The Manufacturers' Association of Israel, P.O. Box 297, Tel-Aviv.

ITALY

Comité National Italien de la F.L.I., Via Borgonuovo 11, Milan.

JAPAN

Japan Wool Industry Conference, Senikaikan Building, 3–Chome, Honcho, Nihonbashi, Chuo-Ku, Tokyo.

MEXICO

Camara Nacional de la Industria, Seccion Lanera, Rio Rhin 27, 2 y 3, Mexico 2.

NETHERLANDS

Algemene Vereniging voor Wolhandel en Wolindustrie in Nederland, Reitse-plein, I, Tilburg.

NORWAY

De Norske Tekstilfabrikkers Hovedforening, Skoyenveien 68, Postboks 188, Skoyen, Oslo.

PORTUGAL

Federacio Nacional dos Industriais de Lanificios, Rua. D. Estefania 17.3, Lisbon.

SOUTH AFRICA

The South African Wool Textile Organisation, Adderley Street, P.O. Box 2221, Port Elizabeth.

SPAIN

Delegacion Espanola en la F.L.I., Gremio de Fabricantes de Sabadell, Sabadell.

SWEDEN

Textilradet, Biblioteks Gatan 6–8, Box 7007.103.81, Stockholm.

SWITZERLAND

Verein Schweizerischer Textilindustrieller, (Wolle-Synthetics), Parkring 12, 8002 Zürich.

UNITED KINGDOM

The British National Committee, Commerce House, Bradford 1.

URUGUAY

Camara Mercantil de Productos del Pais, Avenida General Rondeau 1908, Montevideo.

U.S.A.

National Association of Wool Manufacturers, 386, Park Avenue South, New York, N.Y. 10016.

The information relating to the production, commerce, and industry of wool are published in three documents. The White Paper, the Blue Book, and the Red Book.

Subject Index